The Author

W Robert Griffiths KC, SC is a practising barrister. Born in the Mumbles, Swansea, and brought up in Pembrokeshire, he attended Haverfordwest Grammar School from which he was awarded an Open Scholarship to St Edmund Hall, Oxford. He was called to the Bar in 1974 and appointed a Junior Counsel to the Crown (Common Law) in 1989 and appointed a QC in 1993. He was appointed a Senior Counsel (SC) in New South Wales, Australia in 1999. He was The Times Lawyer of the Week (2007) for his representation of the Australian umpire Darrell Hair in his claim against the International Cricket Council in respect of the ball tampering affair at the Oval. He is a Master of the

Bench of Middle Temple and a Fellow of the Erasmus Forum in recognition of his contribution to law, philosophy and sport. He was Chairman of the MCC Development Committee for four years Chairman of the Laws Sub-Committee for six years and a member of the MCC Committee for twenty years. He was a director of The London Chamber of Commerce and President of The London Chamber of Arbitration. He was formerly a trustee of The Lord Taverner's and a special adviser to The Prince's Regeneration Trust. He is a Vice President of Crawshays Rugby Club, President of Johnston (Pembrokeshire) Cricket Club and he is a Freeman of the City of London.

To my late parents, John and Megan, for their love and encouragement always and to my much-loved children, Anna, Helena and Charles, and to my younger brother, Allen.

Acknowledgements

Plato, Socrates, Muhammad Ali, José Ortega y Gasset, Nietzsche, Wittgenstein El Cordobés and my former Pupil Master, Gary Flather QC, for their inspiration.

I would also like to give my heartfelt thanks and gratitude to Simon Barnes, Sir John Major, Gerald Davies, the late Barry John, Rebecca Ramsey Owens, Nicola Strachan, Michael Atherton, Matthew Syed, Ivo Tennant, my typist Ann Kavanagh and my colleagues and clerks at Six Pump Court Chambers, especially the Junior clerk, Danny Lamb, the Publishers and various contributors to the Journal of the Philosophy of Sport and Times Newspapers Limited.

Table of Contents

The Man in the Arena

It is not the critic who counts, not the man who points out how the strong man stumbles, or where the doer of deeds could have done them better. The credit belongs to the man who is actually in the arena, whose face is marred by dust and sweat and blood who strives valiantly; who comes short again and again, because there is no effort without error and shortcoming but who does actually strive to do the deeds; who knows great enthusiasm; the great devotions; who spends himself in a worthy cause; who at the best knows in the end the triumph of high achievement, and who at the worst, if he fails; at least fails while daring greatly, so that his place shall never be with those cold and timid souls who neither know victory or defeat.[1]

Theodore Roosevelt, 23 April 1910

A man can be destroyed but not defeated.
Ernest Hemingway, The Old Man and the Sea

You base football player.
William Shakespeare, *King Lear*, Act 1, Scene 4: Kent taunts Gonerill's servant, Oswald.

Foreword

It's an accepted fact of English life: only stupid people like sport. If you attempt to discuss sport intelligently, you're faking it, or at least the intelligence part. By extension, if you like art, you can't possibly like sport. If you talk about sport and art in the same breath, you're a phoney.

Albert Camus was a goalkeeper. So was Vladimir Nabokov, Yevgeny Yevtushenko, Che Guevara, and Pope John-Paul II. So for that matter was Julio Iglesias. Samuel Beckett is the only Nobel Prize winner to have made the pages of *Wisden Cricketers Almanac*. Note that none of the above is English.

Many stupid people like sport. But then many stupid people also like sex, and you seldom hear intelligent people rejecting sex for that reason. But the equation of stupidity with sport is all-pervasive: one more snobbery in a country already overladen with such stuff. It follows that to despise sport is a prima facie claim of lofty intellectual status.

The gulf between the arties and the hearties is both deep and wide. In *Brideshead Revisited*, Anthony Blanche reads *The Waste Land* 'in languishing tones' to the 'sweatered and mufflered throng' of students heading to the river to train for rowing races. That showed them who's boss.

And that's why this book brings a special pleasure, for it's a discussion of sport by a writer who makes no bones—quite the reverse—about his intelligence and learning. We have Ruskin, Wittgenstein, Picasso, Shaw and Dr Johnson before he has even cleared his throat.

The fact that you appreciate James Joyce, Bach and van Gogh doesn't mean that you can't or shouldn't appreciate Usain Bolt, Simone Biles and Ben Stokes. And vice versa. A person who fails to appreciate great art is a philistine; but so too is a person who fails to appreciate great sport.

You can watch sport without giving your mind the day off, without reverting to primeval barbarity, without forfeiting your credentials for intelligence. Sport is part of a whole person's pleasure in life: it needs neither apology nor dressing up.

So here is an unapologetically intelligent person's book about sport. Lace up your intellectual running shoes and see how long you can keep up.

Simon Barnes

Governments invariably underestimate the value and the power of sport. They should not: Sport—in all its many forms—is the primary social interest of millions of people in nearly every corner of the world. Robert Griffiths loves sport—perhaps, especially cricket—but more importantly, he understands how it resonates in the human spirit. How. And why. He is a stranger to clichés and commonly accepted views, and brings an original opinion to every issue he touches. He can be controversial—it is, for him, the stuff of life. But his controversy has purpose, insight and wisdom. What he lacks is any capacity to be boring. The book Robert Griffiths has written is one that cries out to be read.

The Rt Hon Sir John Major KG CH

Sport as Philosophy and Philosophy as Sport

I made art a philosophy and philosophy an art: I altered the minds of men and the colours of things: there was nothing I said or did that did not make people wonder... Oscar Wilde – De Profundis.

This book is about sport as philosophy and philosophy as sport. It is about the wonder of sport.

"All are lunatics but he who can analyse his delusions is called a philosopher." So said the controversial American writer Ambrose Bierce (1842–1914) in his Epigrams.

Philosophy is about thoughts, ideas and concepts. Throughout history there have been a plethora of them. I set out below a small sample of these in epigrammatic quotations so as to cajole the prospective reader into reading a book about thinking about sport as philosophy and philosophy as sport. The relationship between the two concepts is touched upon in this book as a recurring theme. However, I should stress that I have no delusions about the task I have embarked upon especially as on any definition I am not a professional philosopher or sportsman. But I hope what I have written about, a subject I love contains sufficient wisdom to be properly characterised as philosophy bearing in mind the etymology and the meaning of philosophy is the love of wisdom.

I leave the reader to decide whether I have achieved that, whatever your view of what I have written I hope it gets you thinking. Not thinking is an anathema to me. Descartes got it right: "I think therefore I am;" if I don't think, I don't exist.

Martin Heidegger (1889–1976), a somewhat unfashionable philosopher these days, said that he was troubled by the sense that modern technology was divorcing people from what he called the 'nearness and shelter of being', making life 'inauthentic' in a way that had not been true for their more primitive

15

ancestors who had lived in 'happy' communion with 'nature'. In conclusion, he observed, "the most thought-provoking thing in our thought-provoking time is that we are still not thinking."

That observation is in my view even more pertinent in our modern world of computers and most recently quantum computing. As I have said philosophy is about thinking and thinking about thinking. I therefore hope that this book whether you agree with its ideas or not, if nothing else makes you think about sport and philosophy in a way that you have not thought about it before.

I always have a quotation for everything—it saves original thinking.
Dorothy L. Sayers.

O to be a dragon a symbol of the power of Heaven.
Marianne Moore

The human body is the best picture of the human soul.
Ludwig Wittgenstein

Though man a thinking being is defined,
Few use the grand prerogative of mind,
How few think justly of the thinking few
How many never think, who think they do.
Jane Taylor, 1783–1824

Truth, like a torch, the more it is shook it shines.
Sir William Hamilton

History is philosophy teaching by example.
Thucydides

On earth, there is nothing great but man in man there is nothing great but mind.
ibid Sir William Hamilton (1788–1856)

If I were charged with introducing an alien form of life to the epitome of human form potential creativity perseverance and spirit, I would introduce that alien lifeform to Michael Jordan.
Harry Edwards.

Of all those arts in which the wise excel, nature's chief masterpiece is writing well.
Sheffield

A man may write at any time if he will set himself doggedly to it.
Samuel Johnson

Conception, my boy, fundamental brainwork, is what makes the difference in all art.
Dante Gabriel Rossetti

Metaphysics is the finding of bad reasons for what we believe upon instinct.
F.H. Bradley

No man's knowledge can go beyond his experience.
John Locke

What we think, or what we know or believe is, in the end, of little consequence. The only consequence is what we do.
John Ruskin

Concept is a vague concept.
Ludwig Wittgenstein

What I understand by philosopher; a terrible explosive in the presence of which everything is in danger.
Nietzsche

Look to the essence of a thing, whether it be a point of doctrine of practice or of interpretation.
Marcus Aurelius

Death is not an event in life: we do not live to experience death.
Ludwig Wittgenstein

The thought of suicide is a great consolation by means of it one gets successfully through many a bad night.
Nietzsche

Men have mistaken words for concepts, and concepts for realities.
Hamann

To have a great idea, have a lot of them.
Thomas Edison

There is an inverse relationship between the clarity and the truth of a concept.
Niels Bohr

The sound must seem an echo to the sense.
Pope

I am now convinced that theoretical physics is actual philosophy.
Max Born

Play is the highest form of research.
Einstein

Work is a necessity for man. A horse does not go between the shafts of its own accord. Man invented the alarm clock.
Picasso

There are only three sports; bull fighting, car racing and mountain climbing. All the others are mere games.
Ernest Hemingway

An artist cannot talk about his art more than a plant can discuss horticulture.
Jean Cocteau

The people never give up their liberties but under some delusion.
Edmund Burke

Games are for people who can neither read nor think.
George Bernard Shaw

It was remarked to me…that to play billiards was a sign of an ill-spent youth.
Herbert Spencer

Sport is something that unbends the mind by turning it off from care.
Samuel Johnson

Think like a man of action, act like a man of thought.
Henri Bergson

Thus so wretched is man...and so frivolous is he that though full of a thousand reasons for weariness, the least thing, such as playing billiards or hitting a ball, is sufficient enough to amuse him.
Blaise Pascal

His only fault is that he has none.
Pliny The Younger

We should treat all trivial things of life very seriously, and all the serious things of life with sincere and structured triviality.
Oscar Wilde

I, too dislike there are things that are important beyond all this fiddle.
Marianne Moore

Toys and games are the prelude to serious ideas.
Charles Eames

You may not know cricket who only cricket knows.
C.L.R. James
Isn't football the toy department of life? Mike Wallace

Doubt not but angling will prove to be so pleasant that it will prove to be like virtue, a reward in itself.
Izaak Walton

Are we human or are we dancers?
The Killers

The athlete is a monster, he is the man who laughs, the geisha with the compressed and atrophied foot, dedicated to the instrumentalization.
Umberto Eco

Philosophy, as I have hitherto understood and lived a voluntary living in ice and high mountains.
Nietzsche

All philosophy is a meditation on death.
Socrates

A book that furnishes no quotations is me judice, no book but a plaything.
Thomas Peacock

The frivolous world of polished idleness.
James Mackintosh

A little nonsense now and then is treasured by the wisest men.
Roald Dahl

Prologue

Some men see things as they are, I dream things that never were and say "why not?"
George Bernard Shaw

We will either find a way, or make one.
Hannibal

The only way to discover the limits of the possible is to go beyond them into the impossible.
Arthur C. Clarke

There is nothing worse than a sharp image of a fuzzy concept.
Ansell Adams

Man's mind stretched to a new idea never goes back to its original dimensions.
Oliver Wendell Holmes

Plato, probably the most cerebral human being to walk the earth (who had been a wrestler), in Theaetus majestically proclaimed "philosophy begins in wonder."

This book in many ways is a celebration of the wonder of sport.

It could also be viewed as a digression from sport. Philosophy is a digression. It takes you where it will. It has no bounds. I hope and believe that Laurence Sterne was right in his epigram: "progress as I digress." Aristotle in Metaphysics said that a man is free whose life is lived for his own sake not for others and that the only free pursuit is philosophy. Sport exists for its own sake. The crown of life is not being mortal, to think mortal thoughts, but rather to immortalise and to live to the highest element in oneself. Both the philosopher and sports person strive to achieve that goal.

In Greek philosophy, the term 'metaphysics' originally meant "that which comes after physics." Meta in ancient Greek means 'after'.

Aristotle's Metaphysics was found, untitled, placed after his treatise on physics. But metaphysics soon came to mean those topics that lie 'beyond physics'. Today, it might be said 'beyond science'. So metaphysics was the study of topics about physics (or science generally) as opposed to the scientific subject itself. Paul Davies more recently[2], in The Mind of God has characterised metaphysics as dealing with how the world of appearances presented to our senses relates to its underlying reality. That is what I have attempted to do in relation to the concept of sport. All categories, comparisons and judgments are constructs of the human mind. They only exist as abstract concept and labels for complex gestalts of experience.

As to the title of this book, '*Beyond the Concept of Sport*', my thesis is that reality always stands on the horizon. Meister Eckhart said, "If you seek the kernel you must break the shell." The truth and significance of a concept can only be properly assessed by looking beyond the concept of what is sought to be clarified. Niels Bohr said "There is an inverse relationship between the clarity and the truth of a concept." It is, I think, what CLR James had in mind when he entitled his seminal work on cricket 'Beyond a Boundary' and what Pope meant in the Iliad of Homer in opining "The best of things beyond their measures cloy." Having written most of this work and having chosen its title I noticed in one publication[3] that the writer's view was that "To be a genius you need a strong impulse to see beyond horizons envisioned by your fellow humans." I therefore want to make it clear at the outset that notwithstanding my impulse to see beyond the horizons of sport I do not purport to be a genius and pace Wittgenstein I am not being self-effacing! Readers, I am sure, will share that opinion! In philosophy the horizon recedes as we advance and research is always incomplete.

And with apologies to Lewis Carroll, the book represents a "beyond the looking glass reflection of the concept of sport." Marcel Proust put it in this way "The real voyage of discovery consists not in seeking new landscapes but in having new eyes."

The truth of this 'beyond axiom' was brought home to me some years ago when being driven by car in the Northern Territories in Australia. I saw a road sign which simply stated "You are now entering Beyond the Outback."

While metaphors are often thought of as verbal expressions, they also use non-verbal material and imaginative metaphors. These are embodied in movements, postures, sounds, objects, structures in the environment, mental images and feelings—metaphors with words, sport is a metaphor without words.

Welcome to the metaphorical outback of sport!

The fundamental question I ask is: what is the thematic horizon of sport? I think a short answer to that question was encapsulated when Picasso defined art as "not the application of a canon of beauty but what the instinct and brain can conceive beyond any canon." In C.S. Lewis' terms, sport lies in the shadowlands of human existence.

Sport is no stranger to the odyssey of human experience. It has been a significant feature of most every culture and most every epoch in human development. The sportsman is the Man in the Arena and this book is about his world and its significance philosophically.

Bertrand Russell observed that philosophy is a stage in intellectual development, and is not compatible with mental maturity. If that is the case, a book about philosophy and sport may be indicative of arrested or retarded intellectual development. I hope not! At one level there is an absurdity about taking seriously something as superficially trivial and non-serious as sport. But as Camus noted there is a metaphysical honour in enduring the world's absurdity.

I agree with Camus that there is in the human condition a basic absurdity as well as an implacable nobility. The two coincide as is natural. Both of them are represented in the ridiculous divorce separating our spiritual excesses and ephemeral joys of the body. Camus also thought "there is but one truly serious philosophical problem and that is suicide." Judging whether life is or is not worth living amounts to answering the fundamental question of philosophy. All the rest comes afterwards.[4]

Why do we have sport?

What existential purpose does it serve? Nietzsche said "We have art in order not to die of the truth." It may be that sport is a form of madness and genius. He also said genius and heroism are madness.

I have also referred to Socrates' opinion that all philosophy is a meditation on death. Is sport itself a symbolic meditation on death? Death being 'beyond life' and 'beyond the horizon' takes us to a different world as does sport. On that basis sport is a metaphor for life, death and another world and exists in order for us not to die of the truth and to make life worth living. To experience pain and pleasure and treat them both the same. Sport is a form of sacrifice and tragedy. Friedrich Schlegel said "The hidden meaning of sacrifice is the annihilation of the finite because it is finite. In order to demonstrate that this is its only justification, one must choose to sacrifice whatever is most noble and most beautiful; but particularly man, the flower of the earth...hence man can only sacrifice himself. In the enthusiasm of annihilation, the meaning of the divine creation is realised for the first time. Only in the midst of death does the lightning bolt of eternal life explode."[5]

Terry Eagleton in his stimulating and erudite work Radical Sacrifice analyses this fascinating concept but does not directly draw an analogy with the nature of sport.[6] However, in my view, there is an obvious and very real one. Eagleton refers to Eugene O'Neil's observation that "What I am after is to get an audience leaving the theatre with an exultant feeling from seeing somebody on the stage facing life, fighting against the eternal odds, not conquering, but perhaps inevitably being conquered. The individual's life is made significant just by the struggle."[7] The same could be said of sport.

In sport as in theatre, "the spectators can know the self-lacerating delights of being torn to pieces, transferring their own mortality onto his shoulders; but at the same time they can exult in the sadistic pleasure of surviving him, thus indulging an agreeable, fantasy of immortality."[8] Eagleton continues "...in framing and stylising death, tragic art intimates a passage beyond it. It's very artistic form is a species of theodicy, constituting a symbolic victory over the terrors of its content."[9]

Carl Jung said "What we call a symbol is a term, a name or even a picture that may be familiar in daily life, yet that has specific connotations in addition to its conventional and obvious meaning. It implies something vague, unknown or hidden."[10]

Sport is a symbolic and existential metaphor for what is and what is to come afterwards the beyond. Nietzsche was right when he said "what is done out of love takes place beyond good and evil." The sportsman does what he does out of his love of the sporting contest which is therefore beyond good and evil. The true sportsman occupies the metaphysical high ground of human endeavour and existence.

Sport operates at the interface of what has been called mind and body. It is played by man and not animals. Its territory is not limited to the world of consciousness and purpose. It extends to what has been labelled the unconscious, the beyond consciousness, the no purpose, the without thought, the intuitive, counter-intuitive and instinctive. It provides some fundamental data on the essence of man. It also is a symbol as a physical reality with a meaning or a significance beyond itself.

Jeremy Bentham, in his Introduction to the Principles of Morals and Legislation[11], said "Nature has placed mankind under the governance of two sovereign matters pain and pleasure." Maybe the appeal of sport is that it constitutes both.

Sport is intimately related to and frequently mistaken for such as game playing, bodily movement, recreational activity, physical education and dance.

A philosophical study of sport is about the fundamental nature and significance of sport. The book does not provide a sporting narrative but is a conceptual philosophical analysis of philosophy and sport. I have reminded myself constantly of what Epictetus, one of my philosopher kings, said when he posed the question "What is the first business of one who studies philosophy?" His answer was "to part with self-conceit." "For it is impossible for anyone to begin to learn, what he thinks that he already knows." Overconfidence in one's conclusions should not be the hallmark of philosophy.

A 'complete' account of sport[12] depends necessarily on a philosophical examination of it. It is and only on this basis, that the other essential features of sport (i.e. the biological, psychological, sociological, religious, artistic and historical features) and sports itself can be made wholly intelligible[13] [14] In the Structure of Magic 1, Richard Bandler and John Grinder discuss the idea of "a model of our model of our world," or, simply a meta-model and pay homage to Alfred Korzybski and his famous dictum, "the map is not the territory."

The book's title should therefore not be interpreted to mean that it is solely a philosophical work. Conceptual thinking is not the sole prerogative of philosophy. I am reminded of Keat's line that philosophy will 'clip an Angel's wings'. In that sense, the title could have been "beyond the concept of sport and philosophy."[15] But I decided to keep it simple!

The idea that philosophy is somehow the counterpart of sport might be surprising to many, but it is nothing new to Plato. In several dialogues, he uses athletic imagery and metaphor to depict Socrates as an athletic hero of the soul. In the Republic, the dialectic is compared with a wrestling match of three falls, the third "dedicated in Olympic fashion to Olympian Zeus the Saviour" (583b). Indeed when the discussion started about the kind of education needed to turn the soul from night to day, to draw the soul beyond the world of appearance and towards "what is," Socrates recalls that the rulers needed to be "athletes in war when they're young" (521d).

Philosophy, according to Plato, required harmony of the soul—and this is achieved in the earliest phases of education by simple games and dances combined with music and poetry. Eventually, however, the soul's harmony must become robust—galvanised by the endurance, focus, and resoluteness characteristic of athletes. It is for this reason that young citizens dedicated 2 to 3 years of their lives exclusively to physical training and to the cultivation of athletic virtues that might serve their eventual pursuit of wisdom. As the citizens grow older, Socrates explains they should increase their mental exercise, eventually retiring from military and public life to "graze freely in the pastures of philosophy and do nothing else" (498b).

In the meantime, gymnastics serves the cause of virtue. Socrates says of the just person:

> "He won't entrust the condition and nurture of his body to the irrational pleasure of the beast within or turn his life in that direction, but neither will he make health his aim or assign first place to being strong, healthy and beautiful, unless he happens to acquire moderation as a result. Rather it is clear that he will always cultivate the harmony of his body for the sake of the consonance of his soul (591c)."

Most recently, Nassim Taleb has considered the concept of 'skin in the game' as being "mostly about justice, honour and sacrifice, things [that] really are existential for humans[16]." This was not stated in the context of sport but it could have been and may represent its intrinsic and symbolic appeal and importance to civilisations of all ages. We recognise that technically all words represent concepts. However, we maintain that in practice there is an experiential distinction between words which are processed sensorially, conceptually or symbolically. For example, whether the word 'mother' refers to the person who is my mother or the concept of a mother, or to the symbolic 'Mother Earth' makes all the difference. This book looks at sport sensorially, conceptually and symbolically.

In his seminal work, Homo Ludens, Johan Huizinga argues that sport is a form of play. This view is widely accepted as evidenced by the use of terms such as 'non-serious', 'autotelic', and 'gratuitous'. Its intrinsic character seems inconsistent with the modern world, which takes sport very seriously, puts it in the service of deliberate ends, and views it (or competition at least) as essential for human thriving. Huizinga saw this and noted:

> "Now, with the increasing systemisation and regimentation of sport, something of the pure play quality is inevitably lost. The sport of the professional is no longer the true play spirit; it is lacking in spontaneity and carelessness…they push sport further and further away from the play sphere proper until it becomes a thing sui generis: neither play nor earnest. In modern social life, sport occupies a place alongside and apart from the cultural process. The great competitions in archaic cultures had always formed part of the sacred festivals and were indispensable as health and happiness bringing activities. The ritual then has now been completely severed, sport has become profane, 'unholy' in every way and has no organic connection whatever with the structure of society…"[17]

However, our modern concept of sport seems to better resemble ancient Greece, where athletic contest (agōn) served specific political and educational goals. Huizinga claimed that the ancient Hellenes simply became unaware of their contests' autotelic character; the concern is that we are now becoming unaware of—or indifferent to—sport's contemporary *ends*. Insofar as we value the social and educational potential of sport in the modern world, we can still benefit from a study of its corresponding function in the ancient world. What has been said is the study of these phenomena reveals that sport's social and educational benefits derive not from it playful characters, but from its philosophical origins as a knowledge-seeking activity.[18]

Like philosophy, democracy, and other forms of competitive truth seeking that emerged in ancient Greece, athletic contests display the characteristics of authentic questioning, impartial testing, and public demonstration of results; features that endure in such modern practices as courtroom trials and scientific experiments. Indeed we may better put sport in the service of humanity today, by viewing it not merely as playful, but also as philosophical; as an expression of what Aristotle called the natural and universal human desire to learn and know[19].

"Olympia, déspoin' alatheias" (Olympia, mistress of truth). So begins Pindar's eighth Olympian Ode[20]. The ancient association between Olympia and knowledge seeking derives partly from the existence of an oracle at the site, but also from the less tangible sentiment that athletic results from Olympia were reliable indicators of truth about the gods' wishes and the relative merits of athletes and their tribes. There was nothing new or revolutionary in the association of athletics and truth. Our earliest accounts of sport-like activities (up to a millennium before the Olympic Games) among Mesopotamians, Egyptians, Assyrians, Minoans, and Hittites, show royals using athletic display as public evidence for social standing and worthiness to lead. Rarely, if ever, though was the worthiness of the ruler actually challenged. What was distinctive about Hellenic athletics and Olympic style contests was that they were knowledge-seeking, rather than presumption-affirming. Their outcomes were generally uncertain. They were governed by impartial rules, and they were subject to public scrutiny.

As a result, Hellenic athletics were from the start subversive. But what they subverted specifically were dogmatic and relativistic standards for truth (i.e. those controlled by worldly rank and power) and what they promoted were more impartial and universal standards, capable of settling disagreements among diverse and even warring tribes.

The Greek term 'philosophia', which literally means 'love of wisdom'[21], seems to have been coined in the 6th century BCE by Pythagoras, who used it to describe those rare thinkers, like himself, who acknowledged not their wisdom but rather their ignorance[22]. It was of course Socrates who made this conception of philosophy famous by declaring more than a century later than his renowned 'wisdom' derived precisely from the awareness that he lacked knowledge. We cannot truly love and desire what we think we already possess; so we are philosophers only as long as we pursue authentic questions with uncertain answers. Sport, likewise, is philosophical only as long as it is actually open to finding answers which may conflict with what people already believe. The contest must not be designed simply to affirm the status-quo, or any other preferred outcome: it has to reflect the spirit of really wanting to know. When challengers boxed the pharaoh in Ancient Egypt, the question of who would win was neither authentic nor was its answer uncertain. Although such contests were intended to reassure subjects of the pharaoh's divine invincibility, they begged their own question. Sport begins with authentic questions derived from real uncertainty about outcomes.

But where did such 'authentic questions' come from? What prompted Presocratic philosophy and contemporary athletics as described in Homer and practised at Olympia to embrace the uncertain, impartial, and public pursuit of truth? The answer quite simply is: closely competing claims among divergent stakeholders. Mycenaean funeral games, perhaps the earliest form of philosophical sport, settled competing claims to the deceased's property. Patroclus' funeral games as depicted in Homer's *Iliad* take this concept further by negotiating Achilles' and Agamemnon's competing claims to honour and authority. Later, at Olympia, the religious puzzle of who should have the honour of lighting the sacrificial flame came to be solved by a simple footrace from the edge of the sanctuary to the altar.

Although modern sport no longer addresses questions about religious favour or worthiness to lead, it still negotiates competing claims to excellence and often decides the distribution of cash, prizes, and educational opportunities.

The basic features of Olympic-style sports, such as common starting lines and level playing fields, exhibit the philosophical drive for rational impartiality. Already in Homer's Bronze Age, the fair construction of contests is emphasised. In the chariot race, for example, there was no permanent track, so a common starting line was literally drawn in the sand and the reliable elder Phoenix is sent off to referee the turnaround point. The starting positions were determined by drawing lots, and when young Antilochus recklessly cuts off Menelaos at a narrow stream crossing, a dispute erupts over the validity of the results. A serious discussion and redistribution of prizes ensues until the community is satisfied with the end result.

Modern sports rules generally respect the principles of impartial testing; competitors even switch sides in field and court games just in case some advantage has slipped through the cracks. Just as with scientific experiments, the value of the results depends on the integrity of the test. Not only must competitors obey the contest rules, officials must meticulously enforce them.

A characteristic of sport is public observation of the contest and the effect this has on their acceptance of the results. Rooting for one's favourite athlete or team is as much a part of sports as arguing for one's thesis as part of philosophical inquiry. In neither practice, however, should the winner be determined by who supports it or even how many support. Rather, each candidate should be subjected to the rational and impartial test before everyone's eyes. The public interest in accurate results requires that popular opinion defer to demonstrable evidence. Sport must be open to authentic questions, secure the impartiality of its tests, and strive for public transparency in its results.

In ancient Greece, the social function of athletics was already well-developed by the time that gymnastics became an integral part of education for excellence (arête). Confusion abounded even in Plato's time about how exercises apparently focussed on the body could build the moral strength we like to call 'character'. No doubt it was the youth's obsession with sports that led Socrates into Athena's gymnasia where he learnt and adapted tricks from the athletes' trade to turn young men's souls away from victory and towards wisdom. But this philosophical journey, at least for Plato, does not leave athletics behind. Rather the force of character revealed and developed through sport seems essential for those who would become philosopher-kings in the *Republic*. This is because the body (soma) was to the ancient mind inanimate. Intentional physical movement was a product and expression of the mind/soul (psyche). The fit athletic body, as a product of voluntary and intentional movement, is merely testament of the arête of the soul.

Socrates turned the natural investigations of the Presocratic towards the explicitly educational ends of moral philosophy. There was also the connection between the Socratic method known as elenchus and athletic contest. Plato's persistent use of the athletic settings and metaphors (see later for the importance of metaphor in sport) is more than literary window-dressing. The Socratic dialogues exhibit the same characteristics of truth-seeking as athletics described above. Like competitive sport, they expose imperfections, test for improvement, and provide public evidence of their findings. Socrates adapts this athletic framework along with its attendant lust for victory (*philonikia),* away from the relativistic goal of defeat and towards the idealistic goals of truth and virtue, that is *philosōphia.*

The arête sought in Plato's *Republic* is described as the healthful and harmonious organisation of the intellectual, spirited, and appetitive parts of the soul. Plato seems to think athletics can achieve this because they require the intellect to apprehend the rules of the game and then to recruit the spirit and appetite to its cause. In another dialogue, *Phaedrus*, this virtuous harmony is illustrated by the athletic metaphor of a two-horse chariot in which the intellect drives a noble and spirited horse alongside the strong but less obedient appetitive horse. Since athletic success depends on the taming of selfish appetites and the directing of honour or spirit towards the noble ends apprehended by the intellect, sport could train the soul for higher education and ultimately public service.

Athletics in the *Republic* are neither playful nor autotelic. Plato uses them explicitly to train souls and select a social elite who will go on to distinguish themselves in academia and, ultimately, public service.

We can also look to Rome as well as Greece to see our own athletic values reflected in antiquity. There sports were primarily entertainment enjoyed by masses of inactive spectators and exploited by politicians who sought public favour. But even the bloody spectacle of gladiatorial fights preserved the truth-seeking and educational functions that connect sport and philosophy. While the Emperor saluted the Roman spectators, who were seated in tiers according to social class, the contest itself challenged that hierarchy. It gave the lowly 'socially-dead' gladiator the opportunity to prove his social worth by prevailing in a publicly observed and strictly regulated test of relevant virtues. The condemned gladiator who received the wooden sword of freedom from the emperor as the community shouted its approval from the stands is an enduring symbol of sport's ancestral ties to philosophy.

There is an ancient resemblance between sport and philosophical inquiry. This connection recalls the important social and educational functions of ancient Greek and Roman athletics and it challenges us to preserve the integrity of sport as a knowledge-seeking practice capable of serving noble human ends.

Deep in our western tradition is the feeling that athletes can be superhuman. "To the Greeks, the gods were athletic, the athletes must try to be godlike."

Rhythm, harmony, flow, gracefulness, and such terms are usually subsumed under the concept of 'unity'. The unity of any artistic expression, be it in the plastic arts, the literary arts, or the movement arts, is an alternative statement of its wholeness or totality, and therefore of its aesthetic presence as an entity.[23]

Several levels of interpretation operate in any reasoned explanation of unity. For instance, unity belies the mind-body duality; further unity is homeostasis or psychological balance. Unity can be symmetrical or asymmetrical, so long as the balance of forces acting in the aesthetic expression are not overwhelming discordant or dissonant.

Unity in one shape or another is the ultimate goal of all human thought or endeavour.

With this conclusion in mind, C.E. Robinson[24] says:

"With (the Greeks) Unity of Design was the first condition of Beauty. Exact symmetry, in short, was essential. Contrast, of course, there will be…in the use of the spoken word, no less than in the visual arts, unity of form was the invariable aim. Speeches were planned from prelude to peroration with an eye to that total effect. The Greek orator was a master of balance and symmetry. In Drama, the Law of Unity was carried still further. The Unities of Time and Place served to enhance the unity of the plot[25]."

The mind-body principle pervaded Greek life sufficiently for us to believe that athletics was not independently regarded for its emphasis on physical performance. Sport for the Greeks was a total mind-body experience, just as today we can accept the message of karate and mushin. Mushin is 'the moon on water', an entity that is neither the one nor the other. There is harmony between the two component parts, found only in the image created by the reflection—a reflection caught only by one in a position to see and upon seeing, perceive. Ripples on the water make the 'moon' dance gracefully; the rhythm is one with cadence that we do not find disturbing, either visually or psychologically. If the rhythm were to break, be splintered by a splash, the unity would be destroyed, all gracefulness, vanish, and the harmony lost.

Harmony, in Greek and Roman mythology, was the daughter of Venus and Mars, representing the social balance between love and war.

Robert Thuma (1897) in his The Grace of Man, identified harmony with the flexibility of the body, the physical component in his 'concept of unity'[26]. He spoke of the 'harmony of poise', which he based on the Delsartian Theory that there can be traced a system of infinite harmony in all things; man, animals and nature and which he claimed takes one towards a theological harmony.

This is but an extension of the thoughts of the eighteenth-century German philosopher, Leibnitz.

The theory of the nature of harmony that Leibnitz developed found its inspiration in the belief held by the Greeks that musical harmony was reflective of a more inclusive universal or cosmological harmony.

To substantiate his new theory, Leibnitz had to develop a new dimension of mathematics accounting for the selective teleological choice of the most pleasingly perfect tones and combinations. He believed that the highest possible degree of perfection of harmony could be achieved by dependence on a selection of principles more limited in scope than those associated with ordinal sequence. In other words, harmony as perfection reflected the beauty of simplicity. There is a view that mathematics is the guarantor of precision and objectivity and that it represents the ultimate reality of the natural order.

In the twentieth century, the Leibnitz-Thuma development of reasoning on harmony found statement in the thinking of Huizinga (1950) who applied the underlying theoretical construct to his analysis of the role of play in culture. Huizinga stated:

"In play, the beauty of the human body in motion reaches its zenith. In its more developed forms [viz, sport] it is saturated with rhythm and harmony, the noblest gifts of aesthetic perception known to man[27]."

Using Plato and Aristotle in his thesis exploring the 'golden age of the body' Fairs states:

"The high ethical and aesthetic value assigned to the body, particularly to the body of the youthful athlete, was an extraordinary happening in the cultural history of the body[28]."

For the Greeks, the pursuit of physical excellence was the avenue to a balanced and harmonious personality.

The Greeks were sufficiently inquisitive of nature to want to estimate and explain the universe according to rational laws and principles. For this purpose, they developed an intense interest in mathematics. The Platonic philosophy of education insisted that mathematics was necessary as a basis for many other branches of study. One of the most significant discoveries in mathematics was the Pythagorean theory of proportion of commensurable qualities. Applying the aesthetic standard of measurable proportion to the development of the human body that each part of the body had a measurable relationship with other parts of the body.

"The purpose of the aesthetic standard of measurable proportion was the development of bodily grace and symmetry."[29]

Qualities which are demanded of all athletes are sense of balance and timing and control of mind and body in rapid movement. Sports demand of their exponents a sense of positional play, or one might say of pattern and design in movement, flowing and continuous though often interrupted and changed, but still basically creative and alive.

It is clear that in all sports, especially where real skill is displayed, these elements of balance, controlled movement, and interrelated and interdependent patterns of action exist as some of the qualities of design which are searched for by the artist.[30]

Sport takes us to different realms. It has been said of rugby that the game can either be painted in golden epic colours of international drama, in pastel shades of casual rural scenes, or sometimes viewing the Pickwickian shape of the trundling tight head prop, in bold and splashing tints of humour and even of farce.[31]

One way of looking at this book is to consider life as an attempt to overcome the serious constraints acting upon us in the way we live. The concept of Rousseau that 'man is born free and everywhere he is in chains.'[32]

This book particular the chapter on Constraint Theory and Boxing should be looked at in that context. The constraint theory of sport reflects the nature of Man and his existence and the concept of overcoming constraints.

Sport and Game Play Theory

In Ortega's "Sportive Origin of the State" (1924), 'Play' is developed into a 'game theory'. He anticipated postmodern 'play theory'.[33] Philosophy is, then, no more than an activity of theoretic knowledge, a theory of the universe. "We are attempting a theory, or, what is the same, a system of concepts about the universe," but the attitude appropriate for theory, "not only for philosophy but for all the sciences," is 'sportive', or only half-serious (though 'rigorous'), like an intellectual 'game' with rules[34]. That applied also to 'scientific convictions', which (as rational) are different from the firm, deeply vital, even 'blind' or irrational convictions of traditional faith, or belief. Theory also differs from belief much as do 'idea' or 'concept'[35] and such is so for 'thinking and believing' too. But the game *is* serious, not jest.

Theory (or concept) is not reality but an 'instrument' for understanding reality. "Theory…begins by denying reality…it is an ideal reduction of the world to nothing…and a redoing backwards the path of its genesis[36]." Philosophy's theoretical turning of reality into nothing, or into absolute problems, however, was the opposite of the sciences' *practical* task of exploiting reality so as to give existence to what did not exist. 'Radically theoretical' is a *homo theoreticus* of 'pure contemplation'. The philosopher did not have to pose his theoretical problems from any practical necessity or 'utilitarian principle'. In contrast, sciences that treated only partial worlds and partial problems manifested a varying residue of active utilitarianism[37]. It was like the difference between the biblical 'Mary and Martha' between the apparently superfluous and the useful, but as sisters of the same vitalist family, they compare with the metaphor on 'the Gemini', or 'twins' of 'knowing and being'. The basic reality is human life[38], yet said Fichte, living is not philosophising or theorising, which is nevertheless a '*form* of living'—"theoretic life, contemplative life"[39]—but the sciences are oriented mainly to 'action' and praxis.

In *Leibniz*, Ortega said that "theory is also life," is a "way of living," but it is only a small "part of our [self-conscious] life[40]." As one plays in the "Pancratium and at hurling the discus"—purposefully—so one "plays at philosophising[41]."

He observed that both Descartes and Leibniz were devoted to chess and card games and inspired their mathematician followers to work *very seriously at games.*

Poincaré thesis was that mathematics is 'playful'. Theory has in it *something* of a 'game'. As a concept, 'game' has "an extraordinary wealth of styles, ingredients, dimensions." "It breaks into the plural: games" ranging from 'baby play' (both animal and human) to climbing Mount Everest, or to playing torero in death-defying games of courage[42].

Perhaps the best example of play theory is the discovery of quantum mechanics, Schrodinger's Cat.

The Mind and Intelligence

The true, strong, and sound mind: is the mind that can embrace equally great things and small.
Samuel Johnson

On earth, there is nothing great but man; in man there is nothing great but mind.
Sir William Hamilton

Daniel Dennett has described Gilbert Ryle's '*The Concept of Mind*', as one of the most original and influential works of philosophy of the century. The goal of the book was to quell just such sorts of confusions about mental events and entities, the confusions that had generated the centuries-old pendulum swing between Descartes's dualism and Hobbes's materialism, both sides correctly discerning the main flaws in the other, but doomed to reproducing them in mirror image.[43]

Ryle said that underlying this partly metaphorical representation of the bifurcation of a person's two lives there is a seemingly more profound and philosophical assumption. It is assumed that there are two different kinds of existence or status. What exists or happens may have the status of physical existence, or it may have the status of mental existence. It is a necessary feature of what has physical existence that it is in space and time; it is a necessary feature of what has mental existence that is in time but not in space. What has physical existence is composed of matter, or else is a function of matter; what has mental existence consists of consciousness, or else is a function of consciousness.

Ryle spoke of it as 'the dogma of the Ghost in the Machine' and hoped to prove that it was entirely false, and false not in detail but in principle. It was not merely an assemblage of particular mistakes. It was one big mistake and a mistake of a special kind. It was, namely, a category-mistake.

He first indicated what he meant by the phrase 'Category-mistake'.

He used cricket to illustrate this type of mistake.

A foreigner watching his first game of cricket learns what are the functions of the bowlers, the batsmen, the fielders, the umpires and the scorers. He then says "But there is no one left on the field to contribute the famous element of team-spirit and sees who does the bowling, the batting and the wicket-keeping; but I do not see whose role it is to exercise *esprit de corps.*" Once more, it would have to be explained that he was looking for the wrong type of thing. Team-spirit is not another cricketing-operation supplementary to all of the other special tasks. It is the keenness with which each of the special tasks is performed, and performing a task keenly is not performing two tasks. Certainly exhibiting team-spirit is not the same thing as bowling or catching, but nor is it a third thing such that we can say that the bowler first bowls *and* then exhibits team-spirit or that a fielder is at a given moment *either* catching *or* displaying *spirit de corps.*

Ryle tells us that when Galileo showed that his methods of scientific discovery were competent to provide a mechanical theory which should cover every occupant of space, Descartes found in himself two conflicting motives. As a man of scientific genius he could not but endorse the claims of mechanics, yet as a religious and moral man he could not accept, as Hobbes accepted, the discouraging rider to those claims, namely that human nature differs only in degree of complexity from clockwork. The mental could not be just a variety of the mechanical.

According to Ryle, Descartes and subsequent philosophers availed themselves of the following escape-route. Since mental-conduct words are not to be construed as signifying the occurrence of mechanical processes, they must be construed as signifying the occurrence of non-mechanical processes; since mechanical laws explain movements in space as the effects of other movements in space, other laws must explain some of the non-spatial workings of minds as the effects of other non-spatial workings of minds. The difference between the human behaviours which we describe as intelligent and those which we describe as unintelligent must be a difference in their causation; so, while some movements of human tongues and limbs are the effects of mechanical causes, others must be the effects of non-mechanical cause, i.e. some issue from movements of particles of matter, other from workings of the mind.

The differences between the physical and the mental were thus represented as differences inside the common framework of the categories of 'thing', 'stuff', 'attribute', 'state', 'process', 'change', 'cause', and 'effect'. Minds are things, but different sorts of things from bodies; mental processes are causes and effects, but different sorts of causes and effects from bodily movements.

As thus presented, minds are not merely ghosts harnessed to machines, they are themselves just special machines. Though the human body is an engine it is not quite an ordinary engine, since some of its workings are governed by another engine inside it—this interior governor-engine being one of a very special sort. It is invisible, inaudible and it has no size or weight. It cannot be taken to bits and the laws it obeys are not those known to ordinary engineers. Nothing is known of how it governs the bodily engine.

Ryle agreed that it is perfectly proper to say, that there exist minds, and to say that there exist bodies. But these expressions do not indicate two different species of existence.

Ryle sought to show that there are many activities which directly display qualities of mind, yet are neither themselves intellectual operations nor yet effects of intellectual operations. Theorising is one practice amongst others and is itself intelligently or stupidly conducted.

He argued that theorists have been so preoccupied with the task of investigating the nature, the source, and the credentials of the theories that we adopt that they have for the most part ignored the question what it is for someone to know how to perform tasks. In ordinary life, on the contrary, we are much more concerned with people's competences than with their cognitive repertoires, with the operations than with the truths that they learn.

He said there are certain parallelisms between knowing *how* and knowing *that*, as well as certain divergences. One of those divergences is knowing when one of the great Welsh rugby legends Cliff Morgan told another legend Gerald Davies that "you can teach someone how but not when." Cliff must have had Ryle's distinction in mind!

This point is commonly expressed by saying that an action exhibits intelligence, if, and only if, the agent is thinking what he is doing while he is doing it, and thinking what he is doing in such a manner that he would not do the action so well if he were not thinking what he is doing. This popular idiom is sometimes appealed to as evidence in favour of the intellectualist legend. Champions of this legend are apt to try to reassimilate knowing *how* to knowing *that* by arguing that intelligent performance involves the observance of rules, or the application of criteria.

To do something thinking what one is doing is, according to this legend, always to do two things; namely, to consider certain appropriate propositions, or prescriptions, and to put into practice what these propositions or prescriptions enjoin. It is to do a bit of theory and then to do a bit of practice.

Certainly, we often do not only reflect before we act but reflect in order to act properly. The chess-player may require some time in which to plan his moves before he makes them. Yet the general assertion that all intelligent performance requires to be prefaced by the consideration of appropriate propositions rings implausibly, even when it is conceded that the required consideration is often very swift and may go quite unmarked by the agent. Ryle argued that the intellectualist legend is false and that when we describe a performance as intelligent, this does not entail the double operation of considering and executing.

Sport is an excellent example of this paradigm. It is a multi-faceted combination of domains. It is multiple intelligences in practice. Each sport reflects a bespoke form of variegated intelligence in action perceptively qualified from a sporting perspective. Different sports illustrate different intelligences in action.

It is therefore possible for people intelligently to perform some sorts of operations when they are not yet able to consider any propositions enjoining how they should be performed. Some intelligent performances are not controlled by any interior acknowledgements of the principles applied in them.

The exercise of intelligence in practice cannot be analysed into a tandem operation of first considering prescriptions and then executing them. For there need be no visible or audible differences between an action done with skill and one done from sheer habit, blind impulse, or in a fit of absence of mind.

Learning of all but the most unsophisticated knacks requires some intellectual capacity. The ability to do things in accordance with instructions necessitates understanding those instructions. So some propositional competence is a condition of acquiring any of these competences.

A man knowing little or nothing of medical science could not be a good surgeon, but excellence at surgery is not the same thing as knowledge of medical science; nor is it a simple product of it. The surgeon must indeed have learnt from instruction, or by his own inductions and observations, a great number of truths; but he must also have learnt by practice a great number of aptitudes. Even where efficient practice is the deliberate application of considered prescriptions, the intelligence involved in putting the prescriptions into practice is not identical to that involved in intellectually grasping the prescriptions.

When a person plays sport, the actions are themselves the things which he is intelligently doing, though the concepts in terms of which the physicist or physiologist would describe his actions do not exhaust those which would be used by his pupils or his teachers in appraising their logic, style or technique. He is bodily active and he is mentally active, but he is not being synchronously active in two different 'places', or with two different 'engines'. There is the one activity, but it is one susceptible of and requiring more than one kind of explanatory description. There are three types of description—sensory, conceptual and symbolic. They each represent distinct yet interrelated ways of perceiving the world. Each domain has its own vocabulary, its own logic and brings forth its own perceptions.

The statement "the mind is its own place," as theorists might construe, is not true, for the mind is not even a metaphorical 'place'. On the contrary, the chessboard, the platform, the scholar's desk, the judge's bench, the lorry-driver's seat, the studio and the football field are among its places. These are where people work and play stupidly or intelligently. 'Mind' is not the name of another person, working or frolicking behind an impenetrable screen; it is not the name of another place where work is done or games are played; and it is not the name of another tool with which work is done, or another appliance with which games are played.

Beyond the Horizon Complementarity

The test of a first-rate intelligence is the ability to hold two opposed ideas in the mind at the same time, and still retain the ability to function.
F. Scott Fitzgerald

It is clear that this complementarity overthrows the scholastic ontology of what is truth? We pose, Pilate's question not in a sceptical, antiscientific sense, but rather in the confidence that further work on this new situation will lead to a deeper understanding of the physical and mental world.
Arnold Sommerfeld

Complementarity is the concept that one single thing when considered from different perspectives, can seem to have very different or even contradictory properties. It is an attitude towards experiences and problems. There are mind-expanding insights of complementarity. It highlights and deepens our understanding of reality and concepts such as sport. It takes you beyond the metaphysical horizon. Its precepts are and the basic messages are:

(1) The questions you want answered mould the concepts you should use.
(2) Different, even incompatible, ways of analysing the same thing can ease after making valid insights into its nature and significance.

Picasso and the cubists created visual complementarity pictorially. By taking up different perspectives in the same picture, they were liberated to bring out with great freedom aspects of the feel in portrait.

So complementarity in an invitation to consider different perspectives. Unfamiliar questions, unfamiliar facts or unfamiliar attitudes, in the spirit of complementarity, gives us opportunity to try out new points of view and to learn from what they reveal. They foster mind expansion beyond the horizon. This is what I have tried to do in this book.

Phenomenology and Sport

The great slogan of phenomenology is "Life is not a problem to be solved but a reality to be experienced."

Phenomenology explains not some abstract notion of what a given phenomenon is 'in itself', without reference to the human experience, but the given phenomenon as it is experienced, or "as it appears." Edmund Husserl, a 20th century German philosopher is often regarded as the founder of phenomenology.

The phenomenologist does not seek an explanation of the 'unconscious' psychoanalytic roots of the experience, much less of the relevant neuron firings, but rather an accurate and adequate description of the experience itself, of its qualities, characteristics and structures. The conclusion reached is the way we actually experience ourselves is neither a 'dualism' of mind and body somehow mysteriously interacting, nor a 'mere body', but as a unity of mental and physical activity. It has been put in this way:

> "The phenomenon in question is my body as concretely lived. The body as immediately apprehended is not a corporeal substance which is in some way attached to, or united with, another substance, variously called in the tradition a 'soul', 'mind' or 'self'. The body thus conceptualised by a later abstraction and objectivization, which is phenomenologically eviscerated and epistemologically problematic. I experience my body first as a complex of life-movements which are indistinguishable from my experience of selfness…The distinctions between soul and body, or mind and body, as they have been formulated in the tradition (particularly by Descartes) are reified and objectivised distinctions, foreign to man's experience as it is immediately lived."[44]

Human activity can only be adequately understood by thinking of human experience as a unified whole and it is not the sum of its parts.

According to his view, "an analysis of man's incarnation reveals that man is an opaque and partially concealed body subject without clear and precise points of demarcation for the various aspects of his being; he is a unity of physical, biological, and psychological relationships necessarily interrelated and only meaningfully investigated when analysed as a whole."[45]

Our bodies, as the French philosopher, Gabriel Marcel, has said, are not something we have; I am my body.[46] He was the first to insist on the insufficiency of abstract intelligence to grasp the richness of experience. And in the words of Sartre "Existence comes before essence. Man is nothing else but that which he makes of himself." Kierkegaard, likewise, thought 'Truth is subjectivity'.

There could be no better evidence for the claims of phenomenologists of the unity of the lived body than in our experience of embodiment in sport. This is as has been explained as follows:

> "…precisely what characterises most athletic activity is the fully unified co-presence of mental and physical activity, a unity so deep that the two components of the activity can no longer be separated."[47]

The lived body unity seems especially appropriate for the successful experience of the skilled athlete, who, in the achieved unity of mental and physical activity, 'no longer has to think', in the explicit analytical sense, about what he or she does. There the unity is manifest. But what about the novice, skiing for the first time, who desperately tries to remember all the instructions about keeping one's skis parallel, unweighting etc and transform them into action, usually with comical failure? Is the novice not one who has failed to achieve the phenomenological unity, for whom there is all too much of, a 'mind-body distinction'?

Paul Weiss in his "Sport: a Philosophic Inquiry" characterises this desire to overcome the mind-body dualism as part of the very appeal of sport.

On that reasoning, then it is not so much that the lived body is a unity but a desire.

> "It is rather a special achievement out of the experience of dualism which, on this understanding, becomes a defective mode of human experience, but a mode nevertheless."[48]

If this is accepted, we would seem to be both dualist and a unified lived body, but at different times, and in a hierarchical order such that the lived body unity is a superior achievement, something for which we should strive as the explicit overcoming of the mind-body dualism which we often experience as alienating.

It was for that reason that Hyland concludes that:

"Sport, combining in such intimacy our mental and physical capability offers a marvellous testing ground for some of the major theories that have been put forward to explain this intimate connection between the mental and physical in human beings. Does it resolve the problem in favour of one or another theory? Hardly. But it does offer us that arena for asking those questions which can move us towards at least a plausible position on this important issue."[49]

This book looks philosophically at the man in that arena as described by Theodore Roosevelt in his magnificent speech in 1910 and extracted in the Frontispiece. But it also looks at how sport can take us 'beyond the horizon' both conceptually, symbolically and by way of sensory experience. Sport can transport us into different realms and domains both serious and sometimes farcical. Gerald Davies, the great Welsh rugby three-quarter said in the context of rugby and its players. "These are people with a hinterland, not exclusively one dimensional cherishing something beyond their narrow, though expert, field of endeavour."[50]

Chapter 1
The Greeks and Sport

To Greece we give our shining blades.
Thomas Moore (1779–1852) in 'Evenings in Greece'.

Athens, the eye of Greece, mother of Arts.
Milton

The beauty which old Greece and Rome, surely pointed wrought, lies close at home.
John Whittier (1807–1892)

It has been said that every man is born either a Platonist or Aristotelian.
William Barrett

By taking a philosophical approach to sport athletes of all ages, shapes and sizes can reclaim the educational value of athletics as it was championed in ancient Greece.
Heather L. Reid

It is a curious thing that God learnt Greek when he wished to turn author—and that he did not learn it better.
Nietzsche

In this chapter, the philosophical nature of sport is examined in the light of several key concepts from ancient Greek philosophy: the pursuit of excellence (*arête*), the idea that the virtuous life lies in a type of moderation (*sophrosyne*), the importance of the power (*dynamis*) to both passively accept one's bodily limitations and actively try to improve one's body through sport's discipline (*askesis*), the concept of play (*paidia*) and the concept of *kalokagathia*[51]. But the chapter, I believe, tells us as much about the ancient Greeks as it does about sport. Ancient Greek thought always took refuge behind the conception of limits. It never carried anything to extremes, neither the sacred nor reason. It took everything into consideration balancing shadow and light. I should say at the

42

outset of this chapter that I totally disagree with Hegel who believed that "what experience and history teach is this—that people and governments never have learnt anything from history or acted on principles deducted from it."[52]

At the dawn of Greek thought, Heraclitus was already imagining that justice sets limits for the physical universe itself: "The sun will not overstep his measures; if he does the Erin yes, the handmaids of justice will find him out."

Socrates, facing the threat of being condemned to death, acknowledged only this one superiority in himself; what he did not know he did not claim to know. The most exemplary life and thought of those centuries closes on a proud confession of ignorance.

Kalokagathia is derived from the Greek words for "beautiful and good," kalos kai agathos.

The terms in this composite signify an admiration for physical and moral excellence, respectively.[53]

The ancient Greeks tended to be hylomorphists who gloried in both physical and mental achievement. Hence they were predisposed to value highly the ideal of kalokagathia.[54]

Johan Huizinga in *Homo Ludens* characterises as much of human culture as possible by way of the ludic: athletics as well as language and philosophy and even war.

This characterisation of human affairs as *sub specie ludi* is meant to highlight the thesis that *Homo ludens* (the human player) deserves equal footing with more well-known characterisations of humans as *Homo sapiens* (the human knower) and *Homo faber* (the human maker). I like Gerald Davies's observation that "I enjoy very much the crossing over of cultural boundaries, of Sals, a musician with whom we are familiar and recognise as the figure on the podium all dressed in black with a white dress shirt and bow tie conducting the evening's music only for us to notice him 'dressed down' with scarf and bobble cap among the maddening crowd at Murrayfield."

Huizinga draws attention to the fact that the *Homo ludens* hypothesis is firmly rooted in Plato's philosophy, specifically in the *Laws* (803–804): God alone is ultimately worthy of seriousness and we human beings are players.

But this does not mean that we are 'merely' players in that the ludic element in us is our essential feature. The *Homo ludens* hypothesis is perfectly compatible with the agonic tendency of the ancient Greeks in that an *agon* (contest or struggle) itself is a life-affirming, albeit competitive, type of play.[55]

In *Sport, Play, and Ethical Reflection* by Randolph Feezell the key insight is that Aristotelian moderation provides the best clue we are likely to get as to how to assess philosophically contemporary sport and athletics.

'Play' is derived from the Anglo-Saxon *plega* and involves the free movement of bodily exercise and the joy or delight in such movement.

'Athletics' by contrast, is derived from the Greek infinitive *athleuein* and involves the effort to contend for a prize and to endure in such an effort.

Play is a general category that can have (at least) three specific instances:

(1) aimless play or *frolic*, from the Old High German frolic, which signifies a pure outburst of fun with no rules;

(2) competitive, rule-governed play found in games and which in everyday discourse is called *sport* or *athletics* if the games in question test physical skill or prowess; and

(3) a sort of violent play that borders on *war*.

The original meaning of 'sport' from the Anglo-French *disporter*, is "to divert or to amuse oneself." The *Oxford English Dictionary* reveals many different historical meanings of 'sport' and its cognates, the most recent of which are compatible with the view of sport as *competitive* play.

Reflective equilibrium is a philosophical technique invented by Rawls and now widely used in practical philosophy.[56]

The fundamental idea is that no one consideration is foundational or fixed in advance and that several relevant considerations must be brought into some sort of compatibility.

Athletics was integral to the life of the ancient Greeks, just as it is integral to our life together today.

In neither case, would one do justice to the society in question in question if athletics were ignored.

The Olympic Games were reportedly started in 776 B.C.E., in the very century when Homer is presumed to have written his epics concerning the Trojan War.

Eventually games at Delphi. Isthmia, and Nemea were added to those at Olympia. The development of these stephanitic or crown games (from the Greek word for crown: *stephanos)* mirrors the development of Greek culture itself.

It might not have been easy for Plato to give up an athletic career as a wrestler at the Isthmian Games in order to become a philosopher,[57] (but thank God he did!) and we can easily understand why Aristotle condescended to include a list of Olympic victors among his works.[58]

The ancient Greek infinitive *athleuein* meant "to compete for a prize;" the noun *athlon* referred to the prize itself; and an *athlete* was the one who did the competing.

The ancient Greek athletes performed in the nude.

The word *gymnos* meant 'nude', and the verbal form, gymnazein, meant "to perform in the nude."[59]

A gymnasium was a place for nudity, specifically a place to train the body *and* the mind while nude.

Originally, an agon was a place to watch athletic or other competition, but eventually it came to refer to the competition itself. This word provides the basis for our word 'agony'.

Legend has it that one of the races in the ancient Olympic Games was the marathon, supposedly first run by Pheidippides from Marathon to Athens to declare the Greeks' victory over the Persians.

"But it is clear that the original 'marathon' never happened"[60] and that the marathon was not an ancient athletic event.

A twenty-six mile race would have violated the ancient Greek virtue of sophrosyne, or 'moderation': Nothing in excess![61]

The ancient Greek athletes who were lionised were males, but they were not necessarily heroes.

'Hero' was a technical term in ancient Greece that referred either to someone who had at least one divine parent (e.g. Herakles) or to someone who achieved a semi divine status.

Further, an ancient hero had to be dead before a hero cult could arise.

Not even the most famous ancient Greek athlete, Milo of Kroton, was a hero.

The Greeks had no admiration for athletes who failed in other aspects of life or in later life.

Pausanias claimed that those who gloried in their strength alone were doomed to perish.[62]

The goal was kalokagathia: bodily *and* moral or intellectual excellence.

The importance of athletic competition in ancient Greek culture is evidenced in the fact that every city-state had a gymnasion, and some had several.[63]

In addition, it was common to have a separate wrestling building called a palaestra, from the ancient Greek word for wrestling, *pale*.

The *palaestra* was also used for boxing and the *pankration.*

Typically, there was a central pit in the *palaestra* surrounded by colonnades, behind which there were bays (exedrai) that held classes in philosophy, rhetoric, and other disciplines.

"The fundamental feature of the palaestra-gymnasion [is that it] is a place where the mind as well as the body is exercised and trained."[64]

This should be an embarrassment to those contemporary scholars who want to separate entirely the academic world from the athletic one. They are at odds with the original Academy of Plato and the Lyceum of Aristotle, the first two institutions of truly higher learning in world history. The gymnasion was more specifically a place for physical exercise than the *palaestra,* but it is clear that the *palaestra-gymnasion* complex was a place for training in general: physical *and* mental.[65]

This legacy from the ancient Greeks, however, seems to be that the best way to deal with this problem would not rely on a Cartesian view of human being wherein body and mind (or soul) are seen as two radically different substances and hence need radically different training programmes.[66]

The Greeks were hylomorphists who saw the material part of a human being (byle) as integrally connected to, and informed by, the structure (*morphe)* given to it by mind (or soul).

This hylopmorphism was crucial in the effort to achieve the ideal of *kaloklagathia.* Not only Aristotle, with his obvious hylomorphism, but also Plato would have been committed to this ideal.

It would also be wrong to view Plato as a Cartesian in that the Academy itself was a *gymnasion-palaistra* complex that was the site of both bodily and intellectual training.

Weiss's theory is that athletics is best understood as competitive activity wherein young men seek bodily excellence.

Weiss's views are compatible with the athletics-as-play hypothesis.

> "There is…a seriousness exhibited even in the simplest and most innocent of games and play, and a freshness and spontaneity exhibited in most competitions. Both are subject to limiting conditions."[67]

His view is Aristotelian, in that it is centred around the concept of excellence, or *arête.*

He thought that we perfect ourselves as human beings by enhancing our character.

Sport and athletics are unique in the possibilities it provides for character development as being making possible a "character-controlled body."[68]

The character of athletes is tested in the plot of an athletic contest itself, which, like Greek tragedy, involves properties like recognition and reversal.

An athletic contest is even more 'tragic' than Greek tragedy in that it is not known beforehand who will lose[69].

Weiss seldom used the Greek word *arête.* But it is clear that the English word 'excellence', which roughly translates the Greek, is at the centre of his approach to athletics.

Weiss is an aretic pluralist.[70]

That is, he believed human beings can be excellent in many different ways. That can be compared with Gilbert Ryle's concept of multiple intelligence—see The Concept of Mind to which I have earlier referred.

His thesis was that young men, in particular, find it easier to "master their bodies and to achieve excellence in athletics than to achieve excellence in other areas."

He adopts Plato's famous definition of being in the *Sophist (247E)* as dynamic power (dynamis), specifically the power to affect others and to be affected by others, in order to explain how athletes occupy a middle position with respect to the possibilities regarding human activity and passivity.

One extreme is exhibited by the 'naturalistic mystic' or Stoic who passively gives up all discrimination and seeks only to be in harmony with whatever is. (See Chapter 2 for the expansion of this idea.)

The other extreme is the aggressive attempt to subjugate all realities that offer us resistance.

In reality, every human action has something of both activity and passivity in it (as Kant realised), with excellence in sport and athletics consisting in knowing when and how to subjugate one's own body to those of others, and when and how to accept one's own bodily limitations or those of others as they are.[71]

Huizinga highlights a view that is somewhat different from the athletics-as-pursuit-of-bodily-excellence hypothesis; there is, in addition, the athletics-as-play hypothesis. A view that can be traced back to Plato.

He relies on a key passage from the *Laws* (803) where 'the Athenian' (presumably Plato's spokesperson in this dialogue) compares what we today would call a plan of life with the work of a shipwright.

The shipwright begins the work by laying down the keel and then builds the outline of the ship.

The 'keel' and 'outline' (chemata) of the best life (bion arista) are provided by the advice that life should not be taken too seriously, from the ancient Greek word for seriousness, *spoudaios*.[72]

Or again, in the Sophist (222D) reference is made to the playful aspects of legal activity, an activity that Plato deals with at length in the later dialogues.

One can turn a bad cause into a good one by playing with words. No lawyer would accept that!

This jocular element is evidenced frequently in Plato's dialogues (e.g. Sophist 268D); Symposium 223D; Philebus 50B).

Like Shakespeare, Plato seems to have seen the whole of human life as a blend of tragedy and comedy.[73]

The hallmark of Huizinga's view is a certain fusion of play and seriousness that leads to an intellectually rich view of sport and athletics.

He is aware of the fact that the Greek word for play, *paidia*, is etymologically associated with the word for what is childish.

But there were other ancient Greek words that helped to solve this problem: *agon* (contest), scholazein (to take one's leisure), *diagoge* (to pass the time), and so on.

All of these retain in some sense the spirit of play, even if the ancient Greeks did not have one generic word that captured this spirit.

Spivey had the idea that for the ancient Greek Freeman the task in life was to figure out how to spend one's leisure time (schole), how to enjoy autotelic activity without the assumption that it had to produce something else.[74]

Huizinga's approach is meant to provide an alternative to two other views of human nature that have been dominant since the Enlightenment: the human being as Homo sapiens (the human knower) and Homo faber (the human maker). In his opinion, *Homo Ludens* (the human player) is a more powerful explanatory device than either of these.

A common misunderstanding of Huizinga's view is fostered by a mistake in the sub-title of his book that was forced on him by others: he is concerned not with the play element *of* culture. The former compartmentalises and trivialises play in ways that are inaccurate, he thinks.[75]

To him play is *really* basic in the sense that it colours most of life and can be traced back in evolutionary history to nonhuman animals.

This has led some to see play as something else, say, as preparation for the serious business of predation in the cases of dogs and cats, or as the contemporary opiate of the people, as in Marxism.

But the intensity and absorption of play activity, according to his argument, resists natural or social scientific efforts at reductionism.[76]

It was quite common in the seventeenth century to see all the world as a stage, all of us as players, and everything else *sub specie ludi*. As Shakespeare said in Macbeth "Life's but a walking shadow, a poor player that struts and frets his hour upon the stage and then is heard no more." "It is a tale told by an idiot full of sound and fury, signifying nothing." A sportsman's life can be couched in that motif, but I think not.

This is a variation on the biblical theme of the vanity of all things.

Hence, there are clear historical roots that lie beneath Huizinga's quite original contribution. This contribution lies primarily in an opposition to the current tendency to see play as the opposite of seriousness.

Football, to take an obvious example, is played and watched with "profound seriousness." That is, folly and comedy are not to be identified with play.[77]

There are three crucial characteristics of play that enable Huizinga to offer something that at least approaches an essential, rather than merely stipulative, definition of play.

First, contrary to Weiss's, opinion, play is *free* activity, not in the sense of free will versus determinism. Huizinga sees play as free in the sense that it is a liberating, enjoyable activity done at leisure.

Second, related to the freedom of play is the fact that it is *separated* from 'ordinary' or 'real' life. It involves a 'pretending quality'. It is, in a sense, disinterested, an interlude of sorts or an intermezzo, even if the play in question becomes a habitual part of life in general.

Third, and related to the separateness of play, it is *limited* in time and place.

It begins and ends, on the one hand, and it occurs in some specific place or other. Sometimes this specific place is special, even sacred, as has often been noted regarding the baseball diamond. It is such separateness and limitedness when considered together makes play a fecund ground for memory and for the dating of the passage of time, as in the ancient Olympic Games.[78]

As Bernard Suits has emphasised,[79] play governed by rules occurs in games.

These games are worlds within the world with their own limited perfections.

There is noticeably a proliferation of aesthetic terms used in every day discourse to describe these playful games: tension, poise, balance, harmony, contrast, monotony, variation resolution, and, of course, beauty itself.

But the tense, dramatic character of play is readily apparent, especially in the sort of competitive play that occurs in athletic contests[80].

Huizinga talks of play being *outside* of ordinary reality. This is the 'beyond' concept again. By this, he means *above* ordinary reality in a higher realm, a 'mystical' or ecstatic (literally *ek statis,* "outside of one's normal place").

This approach can be compared with Plato's view in the *Laws* (803–804).

The sacred space of play is not a place where the sacred is merely imitated; it is concretely enacted or performed.

To use the appropriate ancient Greek terms, play is not mimetic (from mimesis, or 'imitation'), but methexic (from methexis, or 'participation').

Huizinga's ideas can be compared with those of Romano Guardini, who was an influential theologian in the early decades of the twentieth century.

Guardini noted that earnest people often have problems with both athletic play *and* religious liturgy.

The allegation is that athletic play and liturgy are alike in being childish and aimless and full of superfluous pageantry, and they are trifling and theatrical for no reason. Guardini's response was to say that there are many worthwhile areas in life that are not purposeful.

49

But to say that athletic play or liturgy or a theatrical performance do not have a purpose is not to say that these lack meaning.

Although these things, lack purpose in that they are ends in themselves they still may have meaning.

That is, by escaping from the hegemony of purposiveness one may train the *psyche* to develop more playful, artistic, and religious sensibility.[81]

"Frivolity and ecstasy are the twin poles between which play moves."[82]

The ludic function is evident in both play and religious belief in that in both there is activity outside of (or beyond) the necessities of everyday life that must be taken seriously; and in both there is always an element of ritualised make-believe.[83]

Huizinga gets support for his *Homo Ludens* hypothesis by drawing attention to words from languages around the world and from different historical eras.

Ancient Greek has no less than three words for play, in general.

(1) *Paidia,* which is etymologically related to childishness, nonetheless denotes all kinds of play, including the highest and most sacred, as in Plato's *Laws. Paizein* (to play) and *paigma* or *paignion* (a toy) bring out the obvious light-heartedness and joy associated with all of the cognates of *paidia.*

(2) Another word for play, *aduro* or *adurma* stands for the strictly trifling or the nugatory.

(3) At the other end of the continuum of play would be *agon.*[84]

Huizinga's stance is rescued from the charge of Eurocentrism when the Sanskrit *kritati* is compared with the ancient Greek *paidia* (also the Japanese asobi, the Arabic *la'iha* the Hebrew *sahaq* and the Germanic *Spiel*), when the Chinese *cheng* is compared with *agon,* and when the Blackfoot *koani/kachtsi* is compared with the *paidial/agon* tension.

The Latin *ludus* (and its cognates) is a generic that seems to cover all of the Greek types of play that lack an umbrella term to unify them; hence the title of his book.

Further, in the Romance languages *ludus* is replaced with *jocus* (and its cognates). Kurt Riezler captures Huizinga's intent when he argues that in the evolutionary transition from worms (which presumably play at ultra-minimal levels) to cats to human beings, there is a progressive escape from dependence and an opening up to a world of voluntary rhythms, sounds, words, movements, games, works of art, and religious beliefs.

Art and athletics, he thinks in a Huizinga-like way, are not *merely* play, but they are types of liberating play nonetheless.

The *Merchant of Venice* highlights the importance of play in its interpenetration with the serious:

In the *Merchant of Venice,* the relation itself between play and seriousness is the core of the work.

In most performances, the tragedy of Shylock is put to the fore as the centre of the work framed by a playworld of live, fun, music, and sweetness.

Such performances can hardly be convincing.

"If the relation is reversed, the performance convinces—a world of play and love put to the fore against the background of a world in which Shylocks hate and suffer."[85]

Drew Hyland, alerts us to the fact that competitive play can serve the serious function of enabling us to acquire Socratic self-knowledge.

Very few activities in life make us aware of our limitations as quickly and as decisively as athletic competition. Likewise in the situation regarding the intimate hylomorphic relationship between mind and body.[86]

Kant was fond of speaking of "the play of ideas," the play of imagination, and "the whole dialectical play of cosmological ideas."

Given the wide range of 'play' or its equivalents in other languages, it makes sense in English and other languages to speak of athletic contests, theatrical presentations, and musical performances as examples of play.

The contrast is between the Greek *paidia* and *spoude* (play and seriousness, respectively).

A better way to relate these terms is to see the serious not so much as contrasting with play but as *heavy* play.

The Latin words *serius* and especially *gravitas* are helpful in bringing out this metaphor.

A game of checkers with a child is 'light' but not necessarily because it is play.

That is, "play can very well include seriousness[87]."

The Greek athlete merges different phenomena: exercise, struggle, contest, suffering, endurance, and indeed joyful play.

Serious play is thus a complicated matter that is typically trivialised in popular explanations.[88]

The play *of* culture (in contrast to the play *in* it) is thus, in a way, compatible with the aretic view.

Indeed, the ancient Greek *arête* is etymologically related to *aristos,* 'the best', "the most excellent."

Aristotle speaks of virtue as a prize (*Nicomachean Ethics* 1123D), and in Latin the word for virtue, virtus, is derived from the idea of (athletic) virility.

To this very day a courtroom is, as it was in ancient Greece, a hieros cyclos, a sacred circle for the re-enactment of an agon.

The judge's gown, like the distinctive jersey of a referee at an athletic contest, indicates the partially make-believe character of the event.

Justice (*dike*), whatever else it involves, includes a weighing of evidence and the process of deliberation, which are not unrelated to the athletic labours of the ancient heroes.

Urteil in German captures this, in that the word refers to both judgment and ordeal.

A trial is a test of sorts, an attempt to see whose rhetorical (rather than bodily) dexterity will win the day. As with the origins of life insurance, in several parts of the world trials are accompanied by wagering on the winner.

A *litigium* in Latin (the root for our word 'litigation') is continuous with other (including athletic) sorts of *agon*.[89]

The *Homo Ludens* hypothesis illuminates the relationship between athletics and violence.

There is a continuum of play from mere frolic, at one end, to violent play that borders on war, at the other (see later Chapter 29) 'Sport and War' for the fuller development of this analogy.

The competitive play that is found in athletic contests is between these two extremes.

Assertiveness involves moving freely in an athletic competition and taking advantage of one's game-specific rights, *aggression* involves the use of physical force as one asserts oneself in an athletic contest, and *violence* involves the intent to harm others through one's aggression.[90]

Indeed, it was Huizinga's work on the waning of the Middle Ages that initially led him to the *Homo ludens* hypothesis: chivalric orders and tournaments, and the banners and crests that accompany any noble game, have a residual meaning even today.[91]

Roland Barthes's book *What is Sport* shows how athletics and war are closely connected—as I have said for a full analysis see Chapter 29.

Although superior to the unreflective violence of nonhuman animals, human violence in athletic competition is, Barthes thinks, sometimes murderous.

Nonetheless, even Barthes admits (along with Spivey and other defenders of a bellicose view of athletics) that athletic competition is often paradoxically engaged in *with* others in a spirit of generosity and with a shared sense of play, say, if the competitors are members of the same nation, city, or even neighbourhood.

It is perhaps this paradox that makes athletic competition so appealing, to the point where the spectacle of athletic events, where athletes try to determine who is best, now provides the dramatic backdrop for culture once provided in ancient Greece by theatre.[92]

Poetry had its roots in bragging matches (themselves agonic) that very often had as their subject matter some mythic *agon*.

It was not easy for ancient Athenians to distinguish among three practitioners of *agon:* the competitive poet, the sophistical lawyer, and the philosophical dialectician.

The ancient Sophist was explicit in his effort to defeat his rivals in a public contest, indeed in an exhibition (*epideixis*). It is not surprising that Protagoras was compared to an athletic victor (*Euthydemus* 303A) and that Gorgias was aware of the fact that his activity was a game (*paignion*). We have also seen that the word 'school' grew out of the ancient word for leisure (schole). The ancient Greeks often spent their leisure dealing with a particular problem (problema), literally something that one put oneself as a defence or that one put before others as a challenge.[93]

The claim that philosophy is a noble game is compatible with the idea that it grew out of less noble (not exactly ignoble) games like riddle solving and sophistic play.[94]

Athletics-as-play view is very much compatible with the presence of 'competition'.

This word comes from the Latin *competitionem,* which points to two parties striving for the same object in a match meant to determine the relative excellence of the two parties.

The word is a compound of *petere* and *com:* to strive with, rather than against. In order to understand the athletics-as-play view, some sort of appeal to Aristotelian moderation is required in that athletics, in general, and the athletic virtue of sportsmanship, in particular, are means between two extremes (which are vices).

Aristotle himself saw play (*paidia*), the childlike (*paidia*), and education (*paideia)* as not only etymologically, but also conceptually related.

They involve a sort of light heartedness that includes wit or a sense of irony.

Aristotelian moderation is apparent in his treatment of the virtue of sportsmanship, which may very well be *the* most important virtue of an athlete and which deserves a place along with courage, justice as fairness, and the other major virtues.

The development of the virtue of sportsmanship often goes hand in glove with the development of virtue in general. Or the lack thereof.[95]

The phenomenon of bad sportsmanship is dealt with in Keating's influential 1964 article (later revised in 1978). Keating's fundamental distinction between sport and athletics (which we have seen to be largely based on etymological grounds) supports the idea that the attitudes and behaviours appropriate for playful, sporting activities are *"quite different from the norms and responses*

53

appropriate for participation in the deadly serious world of competitive athletics."[96]

That may explain why we speak in everyday English of a 'bad sport' but not of a 'bad athlete'.

There is something myopic about the widely held dichotomy between the seriousness of life and philosophy, on the one hand, and the non-seriousness of athletic competition, on the other.[97]

Athletic competition is itself, in one sense and within conceptual bounds, serious.

Although there is considerable evidence on both sides to indicate that the athletics-as-pursuit-of-bodily-excellence hypothesis is at odds with the athletics-as-play hypothesis, these two views are ultimately compatible; they mutually reinforce each other and can be brought into equilibrium.

The concepts of *arête* and *telos* emphasised by Weiss need not be seen (and should not be seen) as being at odds with Feezellian *sophrosyne*.

Weiss is primarily interested in what can be seen as a Platonic search for the nature of athletics, where Feezell is primarily interested in the phenomenology of athletic competition.

Feezell admits that "I have little doubt that the pursuit of excellence is an important element in sports, but I have doubts about whether this is the element that *defines* the nature of sport."[98]

Why does athletic competition "offer to many people the context of their hopes, the locus of their momentary reprieve from a burdensome reality, or the repository for the only kind of heroism that they can appreciate at this moment in history?"[99]

Presumably, we cannot answer this question on the basis of the pursuit-of-bodily excellence hypothesis.

This is because Weiss's view does not help us understand much outside of, say, some competitions at the Olympic Games.

What about village cricket, the city softball leagues, golf for hackers, after-work bowling, and pickup soccer games in South American slums?

Feezell, like Michael Novak, has a faith in athletics that seeks understanding.

There is no need to divorce interest in athletics from serious thought.

But how *are* we to understand athletics? Thus far I have emphasised the Greek ideal of moderation *(sophrosyne)* in Feezell's account, but this is only part of Feezell's view.

Despite the freedom experienced by athletes, Feezell considers whether athletics is 'absurd'.

As a first approximation of an adequate response to the issue of the absurdity of athletics, Feezell compares the athlete to Sisyphus in a Camus-like way.

In that, nothing comes of the competitive play in athletics (one innings after another, one spring training after another, etc.), just as nothing comes of Sisyphus's efforts with the rock, one is tempted to say that the absurdity of athletics might lie in "the incongruity between human purposiveness and necessary frustration."[100]

The absurdity of athletic play is better understood, Feezell thinks, in terms of Thomas Nagel's version of the absurd. Camus's Sisyphean absurdity arises not in a human person or in an irrational world, but in the relationship between the two, specifically in the incongruity between human aspiration and the recalcitrant reality that we confront.[101]

Nagel's absurd, by way of partial contrast, consists in a collision *within ourselves* between the seriousness with which we take our projects in life and the ever-present possibility of taking our projects as open to doubt.

That is, we act as if our activities are all that matter, but there is always a point of view from which this seriousness seems either gratuitous or absurd.

It is when these two viewpoints collide that sport seems absurd.

The absurd that arises in us is a function of our ability to disengage ourselves in reflection.

This reflective detachment, it should be noted, has a long history in the classics of spirituality (variously called *adiaphoria* by the Stoics, *nada* by Saint John of the Cross, indifference by Saint Ignatius of Loyola, or the no self-doctrine in certain Buddhists, etc.).

The two best known clichés in sport are: "Winning is not the most important thing: it's the only thing;" and "It's not whether you win or lose, but how you play the game." The latter is known as the Corinthian spirit.

The former pushes us towards the Aristotelian *agrios* or *skleros*.

And the latter pushed us towards Aristotelian *homolochos* by utterly trivialising the pursuit of victory; such pursuit is a legitimate *aspect* of sporting activity.

The task for the philosopher is to avoid all such partial truths.

If "life is to be played," in the absurd sense Feezell has in mind; that does not necessarily mean that I have capitulated to "the despair of *homo gravis.*"[102]

Just as Aristophanes in his play *The Clouds* lampooned Socrates by having the actor 'playing' Socrates put in a basket that was elevated over the stage so as to stimulate the heavens, so also it is easy to put athletes in baskets:

Putting a little ball in a small hole in the ground some distance away [golf].

Carrying a leather ball to a point many yards away—American football and rugby.

Hitting a thrown ball with the intention of allowing one to run a short distance—baseball and cricket.

From the standpoint of everyday life, games are by their very nature rather absurd.[103]

Life is not absurd[104] but some sort of Stoic or Neoplatonist or religious detachment is required even regarding those concerns that are most serious.[105]

In all Nagel's analogy.

"Watching the human drama is a bit like watching a Little League baseball game: the excitement of the participants is perfectly understandable but one can't really enter into it."[106]

Some people, at least, *do* enter into the excitement of the human drama with a Bergsonian élan vital or a Zorba-like zest.

It is very easy to slip into dis-equilibrium when we take sport too seriously or not seriously enough, as the ancient Greeks realised long ago.

The problem with sport is still not much different from the one they confronted.

What is new and exciting in Feezell's approach is the clarity with which he formulates two key components that must be brought into reflective equilibrium: one needs to be simultaneously (or at least sequentially) athletic participant and observer.

"It is like sitting in the stands watching the game and judging it to be trivial while at the same time playing or coaching with utmost seriousness, playing or coaching *as if* it really mattered."[107]

The problem one has is of preserving subjective immediacy, while recognising the paradoxical conclusion that participants and fans are unaware of the objective assessment of sport as being both serious and trivial.[108]

The goal is to live with a sort of grace and self-command while affirming both relevant perspectives. Both 'goalkeeper' Albert Camus and football manager Bill Shankly in their much cited quotations highlighted from their different perspectives this intrinsic paradox. Camus felt all he knew about morality and obligations he owed from football and Shankly believed that football was more important than life and death. In this respect, the writer of The Myth of Sisyphus and the great football guru shared the same view.

The irony involved here enables one to engage in competitive play *as if* it really mattered; hence, there is not as wide a gap as might initially seem to exist between athletic play and the theatrical play of an actor. Method actors, in particular, can, like sportsmen, easily get caught up in, and hence lost in, their play.

Competitive play is either 'serious non-seriousness' or 'non-serious seriousness' and one sometimes treats these two labels as equivalent.[109]

The important thing is to have one's sporting engagement held in check by objective detachment and one's objective detachment modified by that engagement and humility.[110]

It is to be noted that the virtue of humility is different from Aristotle's moderation in that, on Nietzschean grounds, humility is a Christian virtue and has no ancient equivalent.

It meant a moderation between what Nagel calls "nihilistic detachment and blind self-importance."

It is a type of Aristotelian moderation that requires a sort of self-knowledge very much in the tradition of the ancient imperative *Gnothi Seauton* (know yourself!)

There is no doubt, in Wittgenstein's terms there is a family resemblance between modern (or post-modern) irony, humility, and certain ancient Greek ideals.[111]

In Greek thought all of the cardinal virtues from Plato's *Republic* are required for an athlete to be virtuous.

Moderation is crucial. But courage (*andreia*) is also needed when athletes confront isolation and failure, as inevitably they must.[112] Questions of justice (Dike) aksi abound, especially because athletics is rule-bound activity.

The only point to discussing the absurd in sport is to elicit in sportsmen, coaches, and fans their latent potential for *Sophia,* for wisdom.[113]

By coming to understand sport better we do not thereby dry up our competitive juices in that as (Aristotelian) rational animals we are naturally— with large brains relative to body size and large cerebral cortices relative to brain size—to *think* while, or at last immediately after, we compete. But I note Yogi Berra's view that he could not hit and think at the same time.

The maniacal (Kierkegaardian) competitor and the crazed fan, who know nothing of moderation and humility, are not worthy models for sportsmen to follow.[114]

Chapter 2
Epicurean and Stoic Philosophy, Goethe and Sport

Man is born to live and not prepared to live.
Boris Pasternak

Live while you live, the epicure would say,
And seize the pleasures of the present day.
Philip Doddridge (1702–1751)

Men live by intervals of reasons under the sovereignty of humans and passion.
Sir Thomas Browne

The aim of the wise man is not to secure pleasure but to avoid pain.
Aristotle

All human race, from China to Peru
Pleasure, howe'er disguised by art, pursue.
Thomas Warton (1728–1790)

If you are distressed by anything external, the pain is not due to the thing itself
but to your own estimate of it; and thus you have the power to revoke at any
moment.
Marcus Aurelius

It is a clear gain to sacrifice pleasure in order to avoid pain.
Schopenhauer

Then the spirit looks neither ahead nor behind. Only the present is our happiness.
Goethe[115]

I want to begin this chapter with an illustration of how much philosophical power there is in humour, anecdote and metaphor. When I first became a barrister, I was a member of a Chambers in London, where the head of Chambers was an eminent QC and incidentally a Welshman, who became one of my heroes even though not a sportsman. His name was Sir Elwyn Jones (later Lord Elwyn Jones) who subsequently became Attorney General and Lord Chancellor. As a young man being a member of his Chambers, I was invited to and attended a dinner where he gave an after-dinner speech. He began that speech with the following observation by Chinese philosopher Confucius, very relevant to the subject matter of this chapter. It went as follows:

"My Lords Ladies and Gentleman, there are three great pleasures in life; to eat with nice people, to drink with nice people and to sleep with… (pause for comic effect) a contented mind!!!"

In a similar vein, he once advised me "never shoot until you see the whites of their eyes".

Montaigne conceived that "pleasures are to be avoided if greater pains be the consequence, and pains to be coveted that will terminate in greater pleasure." That sentiment may explain the appeal of sport.

Philosophical literature is littered with references to the relationship between pain and pleasure in our lives and indeed in sport but this chapter looks at a different but related concept—the living in the present moment. Adopting that philosophy, Wittgenstein in *The Tractatus* said, epigrammatically, "Eternal Life belongs to those who live in the present."

And as L.P. Hartley said, "The past is a different country: they do things differently there."

Putting it more philosophically it is not eternal life that matters but eternal vivacity. Both Faust and Don Quixote are eminent creations of that nobility. I find it metaphysically and metaphorically fascinating that Goethe on his death bed called for light reflecting possibly the biblical plea 'You are the light of the world' (Matthew 5:14-16).

The essence of Epicurean and Stoic philosophy was its focus on the living in the present instant. Forget the past and do not worry about the future. Sport is the practical embodiment of that philosophy. Live for the now and enjoy the experience of the moment. The value of the present instant is what sport teaches us about life. One can immerse oneself in the now and feel a heightened sense of existential self and happiness which transcends the norm of our day to day worries as to the future and the debilitating memories and regrets of the past. It provides us with a zest and enthusiasm which infuses us with the will to live and to handle adversity. It represents the joie de vivre of life and reflects the passion that can be found in our existence as may be experienced beyond our existence on this earth. This is in essence what sports represents metaphorically and spiritually. As Heather Reid has rightly observed the best way to experience the life of a philosophical athlete is to live it.[116] Somebody should tell us, right at the start of our lives that we are dying. Then we might live life to the limit every minute of every day. Whatever you want to do, do it now. There are only so many tomorrows.

Live life as a transcendental experience beyond the metaphorical horizon.

In Goethe's *Second Faust*, we find the following quotation:

"Then the spirit looks neither ahead nor behind. Only the present is our happiness."

In that line, we find an expression of the art of concentrating on and recognising the value of the present instant which in essence sport embodies. It is about the man in the present and in the arena, love, beauty, pain and pleasure experiencing in the now. It corresponds to an experience of time which was lived with particular intensity in such ancient philosophies as Epicureanism and Stoicism. In what follows we shall be especially concerned with this type of experience. We ought not, however, to forget the literary context in which these lines are spoken, the meaning they take on within the context of the *Second Faust*, and, more generally, within the work of Goethe. In the process, we will find that Goethe himself is a remarkable witness for the type of experience I have mentioned of living in the moment.

Although Epicurus is said to have taken the view that mental pleasures are superior to physical pleasures it is notable that many maverick but outstanding sportsmen have lived an Epicurean life ostensibly devoted to pleasure. The wonderful George Best is an outstanding example of this phenomenon. How can we ever forget his marvellous comment "I spent a lot of money on booze, birds and fast cars. I just squandered the rest". Sport is a pleasurable pursuit for both sportsmen and spectator. Pleasure and leisure are correlatives. It is a manifestation of Epicurean philosophy writ large. Sportsmen pursue pleasurable activities both at work and play. Bacchanalian activities of revelry such as drinking, singing and dancing, are often associated with the playing, watching and total enjoyment of the sporting activity.

This chapter owes much to the perceptive writings of the eminent French philosopher Pierre Hadot, especially his article 'the Present Alone is our joy: The Meaning of the Present Instant in Goethe and in Ancient Philosophy (published in French 1986). His detailed research is a consummate reminder of the relationship between poetry and philosophy. As Keats said "cold philosophy can clip an angel's wings". For me poetry can make it fly. I do not necessarily agree with everything Hadot has written about Epicurean or Stoic philosophy or indeed Goethe. In my view the pleasure of the present instant is often made more pleasing by its association with past memories. Sport illustrates that. Past sporting feats form the backcloth to many contemporary displays of sporting brilliance and virtuosity.

The verses quoted below mark one of the climaxes of the *Second Faust*; a moment when Faust seems to reach the culminating point of his "quest for the highest existence."[117] Beside him, on the throne which he has had built for her, sits Helen, whom he had evoked in the first act, after a terrifying journey to the realm of the Mothers, in order to amuse the emperor; but had since fallen hopelessly in love with her:

Has the Source of Beauty, overflowing its banks,
Flowed into the deepest recesses of my being?
To you I dedicate the stirring of all strength,
The essence too of passion;
To you, affection, love, worship, and madness.[118]

It is Helen for whom he has searched throughout the second act, throughout all the mythical forms of classical Greece.

It is then that the extraordinary encounter takes place between Faust and Helen; Faust, who although he appears in the guise of a medieval knight, is really the personification of modern man, and Helen, who, although she is evoked in the form of the heroine of the Trojan War, is, in fact the figure of beauty itself, and in the last analysis of the beauty of nature. With consummate mastery, Goethe has succeeded in bringing these figures and symbols to life, in such a way that the encounter between Faust and Helen is as highly-charged with emotion as the meeting between two lovers, as laden with historical significance as the meeting between two epochs, and as full of meaning as the encounter of a human being with his destiny.

The choice of poetic form is used very skilfully to represent both the dialogue of the two lovers and the encounter between two historical epochs. Since the beginning of the third act, Helen had been speaking in the manner of ancient tragedy and her words were set to the rhythm of iambic trimeter, while the chorus of captive Trojan women responded to her in strophe and antistrophe. Now, however, at the moment when Helen meets Faust and hears the watchman Lynceus speak in rhymed distichs, she is astonished and charmed by this unknown poetic form:

No sooner has one word struck the ear
Than another comes to caress its predecessor.[119]

The birth of Helen's love for Faust will, moreover, express itself in the same rhymed distichs, which Faust begins and Helen finishes, inventing the rhyme each time. As she learns this new poetic form, Helen learns, as Phorkyas says, to spell out the alphabet of love.[120] Helen begins:

Tell me, then, how can I, too, speak so prettily?
"That's easy enough," replies Faust;
It must come from the heart,
And when one's breast with longing overflows,
One looks around, and asks:

Helen: who shall enjoy it with us.
Faust begins again:
Now the spirit looks not forward, nor behind
Only the present
Helen: it is our happiness.
Faust: It is our treasure, our highest prize, our possession and our pledge.
But who confirms it?
Helen: My hand.[121]
The dialogue continues:
Helen: I feel myself so far away and yet so close;
And I say only too gladly: Here Am I! Here!
Faust: I can scarcely breathe; my words tremble and falter;
This is a dream and time and place have disappeared.
Helen: It seems to me that I am broken down with age,
and yet I am so new;
Mingled with you, I am faithful to the Unknown.
Faust: Don't wrack your brains about your destiny, so unique!
Existence is a duty, be it only for an instant.[122]

The dialogue may be understood at several levels. First and foremost, it is the dialogue between two lovers, who, as such, resemble all lovers everywhere. Faust and Helen are two lovers absorbed by the living presence of the beloved: they forget everything—both past and future—which is other than this presence. Their excess of happiness gives them an impression of dreamlike unreality: time and space disappear. We are entering the unknown, and it is the moment of love fulfilled.

On a second level of interpretation, however, the dialogue takes place between Faust and Helen as symbolic figures, representing, on the one hand, modern man in his ceaseless striving, and on the other, ancient beauty in its soothing presence; both are miraculously reunited by the magic of poetry, which abolishes the centuries. In this dialogue, Faust as modern man tries to make Helen forget her past, so that she may be wholly in the present instant, which she is incapable of understanding. She feels herself to be so distant and yet so close, abandoned by life and yet in the process of rebirth, living in Faust, mingled with him, and trusting in the unknown. Faust asks her not to reflect upon her strange destiny, but to accept the new existence which is being offered to her. In this dialogue between two symbolic figures, Helen becomes 'modernised', if one may say so; as she adopts rhyme, the symbol of modern interiority, she has doubts, and reflects upon her destiny. At the same time, Faust becomes 'antiquated': he speaks as a man of antiquity, when he urges Helen to concentrate on the present moment, and not to lose in it hesitant reflection on the past and the future. As Goethe said in a letter to Zelter, this was the characteristic feature of ancient life and art: to know how to live in the present, and to know what he called "the healthiness of the moment." In antiquity, says Goethe, the instant was 'pregnant'; in other words, filled with meaning, but it was also lived in all its reality and the fullness of its richness, sufficient unto itself. We no longer know how to live in the present, continues Goethe. For us, the ideal is in the future, and can only be the object of a sort of nostalgic desire, while the present is considered trivial and banal. We no longer know how to profit from the present; we no longer know—as the Greeks did—how to act in the present, and upon the present.[123] Indeed, if Faust speaks to Helen as a man of the ancient world, it is precisely because the presence of Helen—that is, the presence of ancient beauty—reveals to him what presence itself is: the presence of the world, "That splendid feeling of the present" Herrliches Gefühl der Gegenwart) as Goethe wrote in the East-West Divan.[124]

This, finally, is the reason why the dialogue can be understood at a third level. Here, it is no longer the dialogue of two lovers, nor of two historical figures, but rather the dialogue of man with himself. The encounter with Helen is not only the encounter with ancient beauty emanating from nature; it is also the encounter with a living wisdom and art of living: that "healthiness of the moment." The nihilist Faust had wagered with Mephistopheles that he would never say to an instant: "Stay, you are so beautiful!" But now, following after humble Gretchen, it is ancient, noble Helen who reveals to him the splendour of being—that is, of the present instant—and teaches him to say yes to the world and to himself.

I must now define the ancient experience of time which was expressed in the above-quoted verses from Faust. Basing ourselves on the letter from Goethe to Zelter I cited above, it may be thought we are dealing with a generalised, common experience of ancient man, and that it was natural for ancient man to be familiar with what Goethe calls the "healthiness of the moment." Moreover, following Goethe, many historians and philosophers, from Oswald Spenger[125] to the logician Hintikka,[126] have alluded to the fact that the Greeks "lived in the present moment" more than did the representatives of other cultures. In his book Die Zauberflöte,[127] Siegfried Morenz gives a good summary of this conception:

"No one has better characterised the particular nature of Greece than Goethe…in the dialogue between Faust and Helen: '…then the spirit looks neither backwards nor ahead. Only the present is our happiness.'"

The Greeks in general gave particular attention to the present moment, and this attention could take on several different ethical and artistic meanings. Popular wisdom advised people both to be content with the present, and to know how to utilise it. Being content with the present meant, on the one hand, being content with earthly existence. Aristotle stated "Moral virtue manifests itself in pleasure and pain." This is what Goethe admired in ancient art, particularly in funerary art, where the deceased was represented not with his eyes raised towards the heavens, but in the act of living his daily life. On the other hand, knowing how to utilise the present meant knowing how to recognise and seize the favourable and decisive instant (*Kairos*). *Kairos* designated all the possibilities contained within a given moment: a good general, for example, knows how to strike at the opportune *Kairos*, and sculptors fix in marble the most significant *Kairos* of the scene which they wish to bring to life.

The Greeks paid particular attention to the present moment. This, however, does not justify us in imagining—as did Winckelmann, Goethe, and Hölderlin—the existence of an idealised Greece, the citizens of which, because they lived in the present moment, were perpetually bathed in beauty and serenity. As a matter of fact, people in antiquity were just as filled with anguish as we are today, and ancient poetry often preserves the echo of this anguish, which sometimes goes as far as despair. Like us, the ancients bore the burden of the past, the uncertainty of the future, and the fear of death. Indeed, it was for this human anguish that ancient philosophies—particularly Epicureanism and Stoicism—sought to provide a remedy. These philosophies were therapies, intended to provide a cure for anguish, and to bring freedom and self-mastery, and their goal was to allow people to free themselves from the past and the future, so that they could live within the present. An experience of time wholly different from the common, general experience of man. This experience is expressed in the verses from *Faust*: "Only the present is our happiness…don't think about your destiny. Existence is a duty." It is a philosophical conversion, implying a voluntary, radical transformation of one's way of living and looking at the world. This is the true "healthiness of the moment," which leads to serenity.

Despite the profound differences between Epicurean and Stoic doctrine, there is an extraordinary structural analogy between the experiences of time as it was lived in both schools. This analogy will perhaps allow us to glimpse a certain common experience of the present underlying their doctrinal divergences. Both Epicureanism and Stoicism privilege the present, to the detriment of the past and above all the future. They posit as an axiom that happiness can only be found in the present, that one instant of happiness is equivalent to an eternity of happiness, and that happiness can and must be found immediately, here and now. Both Epicureanism and Stoicism invite us to resituate the present instant within the perspective of the cosmos, and to accord infinite value to the slightest moment of existence.

Epicureanism is a therapy of anguish and pain and suffering, and a philosophy which seeks, above all, to procure peace of mind. Its goal is consequently to liberate mankind from everything that is a cause of anguish for the soul: the belief that the gods are concerned with mankind; the fear of post-mortem punishment; the worries and pain brought about by unsatisfied desires; and the moral uneasiness caused by the concern to act out of perfect purity of intention. As William Hazlitt said: "The art of pleasing consists in being pleased."

Epicureanism is the philosophy of pleasure. The gods, it affirms, live in perfect tranquillity. They are not troubled by the worry of producing or governing the universe, since the latter is the result of a fortuitous coming together of eternally existence atoms. Epicureanism asserts that the soul does not survive the body, and that death is not an event within life. Desires only trouble us to the extent that they are artificial and useless. We must reject all those desires which are neither natural nor necessary and satisfy—with prudence—those of our desires which are natural but necessary. Above all, we are to satisfy those desires which are indispensable for the continuation of our existence. As for moral worries, they will be completely appeased once we realise that man, like all other living beings, is always motivated by pleasure. If we seek for wisdom, this is simply because it brings peace of mind: a pleasurable state. Nietzsche's view was that "the secret of reaping the greater fruitfulness and the greatest enjoyment of life is to live dangerously."[128]

Daniel Klein in his charming book Travels with Epicurus (2012) points out, which is a particular comfort to me in my dotage, that it is not the young man that should be considered fortunate but the old man that has lived well, because the young man in his prime wanders much by chance, vacillating in his beliefs, while the old man has docked in the harbour, having safeguarded his true happiness.

Epicureanism proposes a form of wisdom, which teaches us how to relax and to suppress our worries. We must renounce a great deal, in order that we may desire only that which we are certain of obtaining, and submit our desires to the judgment of reason. What is required, in fact, is a total transformation of our lives, and one of the principal aspects of this transformation is the change of our attitude towards time. According to Epicureanism, senseless people—that is, the majority of mankind—are tormented by vast, hollow desires which have to do with wealth, glory, power, and the unbridled pleasures of the flesh.[129] What is characteristic of all these desires is that they cannot be satisfied in the present. This is why, for the Epicureans:

> senseless people live in hope for the future, and since this cannot be certain, they are consumed by fear and anxiety. Their torment is the most intense when they realise too late that they have striven in vain after money or power or glory, for they do not derive any pleasure from the things which, inflamed with hope, they had undertaken such great labours to procure.[130]

According to an Epicurean saying, "The life of a foolish man is fearful and unpleasant; it is swept totally away into the future."[131] Thus, Epicurean wisdom proposes a radical transformation, which must be active at each instant of life, of mankind's attitude towards time. We must, it teaches, learn how to enjoy the pleasure of the present, without letting ourselves be distracted from it. If the past is unpleasant to us, we are to avoid thinking about it, and we must not think about the future, insofar as the idea of it provokes in us fears or unbridled expectations. Only thoughts about what is pleasant—of pleasure, whether past or future—are to be allowed into the present moment, especially when we are trying to compensate for current suffering. This transformation presupposes a specific conception of pleasure, peculiar to Epicureanism, according to which the quality of pleasure depends neither on the quantity of desires it satisfies, nor on the length of time it lasts.

The quality of pleasure does not depend on the quantity of desires it satisfies. The best and most intense pleasure is that which is mixed to the least extent with worry, and which is the most certain to ensure peace of mind. It can therefore be procured by the satisfaction of natural and necessary desires; that is, those desires which are essentially and necessary for the preservation of existence. Now, such desires can easily be satisfied, without our having to rely on the future for them, and without our being exposed to the worry and uncertainty of lengthy pursuit. "Thanks be to blessed nature, who made necessary things easy to obtain, and things which are hard to obtain unnecessary."[132]

Not only does pleasure not depend upon the quantity of satisfied desires, but—above all—it does not depend upon duration. It has no need to be long-lasting in order to be perfect: "An infinite period of time could not cause us more pleasure than can be derived from this one, which we can see is finite."[133] "Finite time and infinite time bring us the same pleasure, if we measure its limits by reason."[134] This may seem paradoxical, but it is founded on a theoretical conception. As the Stoics were to repeat, a tiny circle is no more of a circle than a large one.[135] The Epicureans thought of pleasure as a reality in and for itself, not situated within the category of time. Aristotle had said that pleasure is total and complete at each moment of its duration, and that its prolongation does not change its essence.[136] For the Epicureans, a practical attitude is joined to this theoretical representation: if pleasure limits itself to that which procures perfect peace of mind, it attains a summit which cannot be surpassed and it is impossible for it to be increased by duration. In the words of Guyau: "In enjoyment, there is a kind of inner plenitude and over-abundance which makes it independent of time, as well as of everything else. True pleasure bears its infinity within itself."[137]

Thus, pleasure is wholly within the present moment, and we need not wait for anything from the future to increase it. Everything we have been saying so far could be summed up in the following verses from Horace: "Let the soul which is happy with the present learn to hate to worry about what lies ahead."[138] The happy mind does not look towards the future. If we limit our desires in a reasonable way, we can be happy right now. Not only *can* we be happy, but we *must*: happiness must be found immediately, here and now, and in the present. Instead of reflecting about our lives as a whole, calculating our hopes and worries, we must seize happiness within the present moment. The matter is urgent; in the words of an Epicurean saying:

> We are only born once—twice is not allowed—and it is necessary that we shall be no more, for all eternity; and yet you, who are not master of tomorrow, you keep on putting off your joy? Yet life is vainly consumed in these delays, and each of us dies without ever having known peace.[139]

Mae West obviously had this saying in mind when in contemporary mode and in her inimitable style she notoriously said "You only live once but if you do it right once is enough".

Once again, we find the echo of this idea in Horace: "While we are talking, jealous time has fled. So seize the day [*carpe diem*], and put no trust in tomorrow."[140] Horace's *carpe diem* is by no means, as is often believed, the advice of a sensualist playboy; on the contrary, it is an invitation to conversion. We are invited to become aware of the vanity of our immensely vain desires, at the same time as of the imminence of death, the uniqueness of life, and the uniqueness of the present instant. From this perspective, each instant appears as a marvellous gift which fills its recipient with gratitude:

Believe that each new day that dawns will be the last for you:
Then each unexpected hour shall come to you as a delightful gift.[141]

There is perhaps an echo here of the Epicurean Philodemus: "Receive each additional moment of time in a manner appropriate to its value; as if one were having an incredible stroke of luck."[142]

The secret of Epicurean joy and serenity is to live each instant as if it were the last, but also as if it were the first. "If the whole world were appear to mortals now, for the first time; if it was suddenly and unexpectedly exposed to their view; what could one think of more marvellous than these things, and which mankind would less have dared to believe?"[143] In the last analysis, the secret of Epicurean joy and serenity is the experience of infinite pleasure provided by the consciousness of existence, even if it be only for a moment. In the words of an Epicurean saying: "The cry of the flesh is: Not to be hungry, not to be thirsty, not to be cold. Whoever has these things, and hopes to keep on having them, can rival in happiness with Zeus himself."[144] The lack of hunger and thirst is thus the condition for being able to continue to exist, being conscious of existing, and enjoying this consciousness of existing. God has nothing more than this. It could be objected that God's pleasure consists in his knowledge that he has the happiness of existing forever. Not so, replies Epicurus; for the pleasure of one instant of existence is just as total and complete as a pleasure of infinite duration, and man is just as immortal as God, because death is not a part of life.[145]

In order to show that one single instant of happiness is enough to give such infinite pleasure, the Epicureans practised telling themselves each day: "I have had all the pleasure I could have expected." In the words of Horace: "He will be master of himself and live joyfully who can say, every day: 'I have lived.'"[146] Seneca also takes up this Epicurean theme:

> When we are about to go to sleep, let us say in joyous cheerfulness: "I have lived; I have travelled the route that fortune had assigned to me." If God should grant us tomorrow as well, let us accept it joyfully. That person is most happy and in tranquil possession of himself who awaits tomorrow without worries.[147] Whoever says: "I have lived," gets up every day to receive unexpected riches.[148]

We can also see here the role played in Epicureanism by the thought of death. To say, every evening: "I have lived," is to say "my life is over." It is to practise the same exercise as that which consists in saying: "Today will be the last day of my life." Yet it is precisely this exercise of becoming aware of life's finitude which reveals the infinite value of the pleasure of existing within the present instant.

The Epicureans had their own particular vision of the universe. As Lucretius put it: thanks to the doctrine of Epicurus, which explained the origin of the universe by the fall of atoms in a void, the walls of the world bust open for the Epicurean: he saw all things come into being within the immense void,[149] and traversed the immensity of the all. Alternatively, he exclaims, in the words of Metrodorus: "Remember that, born a mortal, with a limited life-span, you have risen up in soul to eternity and the infinity of things, and that you have seen all that has been and all that shall be."[150] Here again, we encounter the contrast between finite and infinite time. Within finite time, the sage grasps all that takes place within infinite time, or as Léon Robin puts it in his commentary on Lucretius: "The sage places himself within the immutability of eternal nature, which is independent of time."[151]

Thus, the sage perceives the totality of the cosmos within his consciousness of the fact of existing. Nature gives him everything within an instant, and since she has already given him everything, she has nothing left to give him, as she says in Lucretius' poem: "You must always expect the same things, even though the span of your life should triumph over all the ages; nay, even were you never to die."[152]

The fundamental attitude that the Stoic must maintain at each instant of his life is one of attention, vigilance, and continuous tension, concentrated upon each and every moment, in order not to miss anything which is contrary to reason. We find an excellent description of this attitude in Marcus Aurelius:

Here is what is enough for you:

1. the judgement you are bringing to bear at this moment upon reality, as long as it is objective;
2. the action you are carrying out at this moment, as long as it is accomplished in the service of the human community; and
3. the inner disposition in which you find yourself at this moment, as long as it is a disposition of joy in the face of the conjunction of events caused by extraneous causality.[153]

This is the spiritual exercise Marcus himself calls "delimiting the present."[154] Delimiting the present means turning one's attention away from the past and the future, in order to concentrate it upon what one is in the process of doing.

The present of which Marcus speaks is a present delimited by human consciousness. The Stoics distinguished two ways of defining the present.[155] The first consisted in understanding the present as the limit between the past and the future: from this point of view, no present time ever actually exists, since time is infinitely divisible. This, however, is an abstract, quasi-mathematical division, with the present being reduced to an infinitesimal instant.

The second way consisted in defining the present with reference to human consciousness. In this case, the present represented a certain 'thickness' of time; corresponding to the attention-span of lived consciousness. When Marcus advises us to "delimit the present," he is talking about this lived present, relative to consciousness. This is an important point: the present is defined by its reference to man's thoughts and actions.

The present suffices for our happiness, because it is the only thing which belongs to us, and depends upon us. For the Stoics, it was essential to distinguish between what does and does not depend upon us. The past does not belong to us, since it is definitely fixated, and the future does not depend on us, because it does not yet exist. Only the present depends on us, and it is therefore the only thing which can be either good or bad, since it is the only thing which depends on our will. Since the past and the future do not depend on us, they do not come under the category of moral good or evil, and must therefore be indifferent to us.[156] It is a waste of time to worry about what is long gone, or what will perhaps never occur; we must therefore "delimit the present." "All the happiness you are trying to achieve by long, roundabout ways: you can have it all right now…that is, if you leave everything past behind you, entrust the future to providence, and if you arrange the present in accordance with piety and justice."[157]

Elsewhere, Marcus describes the exercise of delimiting the present in the following terms:

> if you separate from yourself, that is, from your thought…everything you have said or done in the past, everything that disturbs you about the future; all that…attaches itself to you against your will…if you separate from yourself the future and the past, and apply yourself exclusively to living the life that you are living—that is to say, the present—you can live all the time that remains to you until your death, in calm, benevolence, and serenity.[158]

Seneca describes the same exercise as follows:

> Two things must be cut short:[159] the feat of the future and the memory of past discomfort; the one does not concern me anymore, and the other does not concern me yet.[160]

The sage enjoys the present without depending on the future…Liberated from the burden of worries which torture the mind, he does not hope for or desire anything. He does not plunge forward into the unknown, for he is happy with what he has [i.e. the present, which is all that belongs to us]. And don't believe that he is content with not very much, for what he has is everything.[161]

Like pleasure for the Epicurean sage, the happiness of the Stoic sage is perfect. It lacks nothing, just as a circle, whether it is large or small, still remains a circle. The same is true of a propitious or opportune moment or favourable opportunity: it is an instant, the perfection of which depends not on its duration, but rather on its quality, and the harmony which exists between one's exterior situation, and the possibilities that one has.[162] Happiness is nothing more or less than that instant in which man is wholly in accord with nature and sees what is beyond and on the horizon Sport gives us that Epicurean and Stoical experience.

Just as was the case for the Epicureans, one instant of happiness is, according to the Stoics, equivalent to an eternity. In the words of Chrysippus: "If a person has wisdom for one instant, he is no less happy than he who possesses it for an eternity."[163] Similarly, as for the Epicureans, so for the Stoics: we will never be happy if we are not so right now. It's now or never. The matter is urgent: we must hurry, for death is imminent, and all we require in order to be happy is to *want* to be so. The past and the future are of no use. What is needed is the immediate transformation of our way of thinking, of acting, and of accepting events. We must think in accordance with truth, act in accordance with justice, and lovingly accept what comes to pass. In the words of Marcus Aurelius: "How easy is it to find oneself, right away, in a state of perfect peace of mind."[164] In other words, it is enough just to want it. It is, to achieve it.

For the Stoics, as for the Epicureans, it is the imminence of death which gives the present instant its value. "We must carry out each action of our lives as if it were the last."[165] This is the secret of concentration on the present moment: we are to give it all its seriousness, value, and splendour, in order to show up the vanity of all that we pursue with so much worry: all of which, in the end, will be taken away from us by death. We must live each day with a consciousness so acute, and an attention so intense, that we can say to ourselves each evening: "I have lived; I have actualised my life, and have had all that I could expect from life." In the words of Seneca: "He has peace of mind who has lived his entire life every day."[166]

We have just seen the first reason why the present alone is sufficient for our happiness: namely, that one instant of happiness is equivalent to a whole eternity of happiness which may lie beyond. The second reason is that, within one instant, we possess the totality of the universe. The present instant is fleeting—Marcus insists strongly on this point[167]—but within this flash, as Seneca says, "we can proclaim, along with God: 'all this belongs to me'."[168] The instant is our only point of contact with reality, yet it offers us the whole of reality; precisely because it is a passage and a metamorphosis, it allows us to participate in the overall movement of the event of the world, and the reality of the world's coming-to-be and transcendence into eternity and infinity.

In order to understand the preceding, we must bear in mind what moral action, virtue and wisdom meant for the Stoics. Moral good—for the Stoics, the only kind of good there is—has a cosmic dimension: it is the harmonisation of the reason within us with the reason which guides the cosmos, and produces the chain of causes and effect which makes up fate. At each moment, we must harmonise our judgment, action, and desires with universal reason. In particular, we must joyfully accept the conjunction of events which results from the course of nature. At each instant, we must therefore resituate ourselves within the perspective of universal reason, so that, at each instant, our consciousness may become a cosmic consciousness. Thus, if one lives in accord with universal reason, at each instant his consciousness expands into the infinity of the cosmos, and the entire universe is present to him. "For the Stoics, this is possible because there is a total mixture and mutual implication of everything with everything else: Chrysippus, for example, spoke of a drop of wine being mixed with the whole of the sea, and spreading to the entire world.[169] He who sees the present moment sees all that has happened from all eternity, and all that will happen throughout infinite time."[170] This explains the attention given to each current event, and to what is happening to us at each instant. Each event implies the entire world: "Whatever happens to you has been prepared for you from all eternity, and the mutual linkage of cause and effect has, from all eternity, woven together your existence and the occurrence of this event."[171]

One could speak here of a mystical dimension of Stoicism. At each moment and every instant, we must say 'yes' to the universe; that is, to the will of universal reason. We must want that which universal reason wants: that is, the present instant, exactly as it is. Some Christian mystics have also described their state as a continuous consent to the will of God. Marcus, for his part, cries out: "I say to the universe: 'I love along with you'."[172] We have here a profound feeling of participation and identification; of belonging to a whole which transcends our individual limits, and gives us a feeling of intimacy with the universe. For Seneca, the sage plunges himself into the whole of the universe *toti se inserens mundo).*[173]

It is quite remarkable that the two schools of Stoicism and Epicureanism, in other respects so opposed, should both place the concentration of consciousness upon the present moment at the very centre of their way of life. The difference between the two attitudes consists only in the fact that the Epicurean enjoys the present moment, whereas the Stoic wills it intensely; for the one, it is a pleasure; for the other, a duty.

In his conversations with Falk,[174] Goethe had spoken of certain beings who, by virtue of their innate tendencies, are half Stoic and half Epicurean. He found nothing surprising, he said, in the fact that they accepted the fundamental principles of the two systems at the same time, and even that they tried to unite them as far as possible. One might say that Goethe himself, in his way of living the present moment, was also "half Stoic and half Epicurean." He enjoyed the present moment like an Epicurean, and willed it intensely like a Stoic.

In Goethe, we re-encounter most of the themes we have enumerated above; in particular, the delimitation of the present followed by expansion into the totality of the cosmos, which we observed in Epicureanism and in Stoicism. In this regard, Goethe could have mentioned an opposition that was dear to him: that between 'systole' and 'diastole'.

First of all, let us consider concentration on and delimitation of the present. In moments of happiness, these processes take place spontaneously: "Then the spirit looks neither forward nor behind." This verse from *Faust* is echoed by a poem dedicated to Count Paar:[175]

Happiness looks neither forward nor backwards;
And thus the instant becomes eternal.

The present instant is perceived as a grace which is accorded us, or an opportunity we are offered.

The mind may also, however, turn voluntarily away from the past and the future, in order to more fully enjoy the present state of reality. Such is the attitude of Goethe's Egmont:[176]

> Do I live only in order to think about life? Must I prevent myself from enjoying the present moment, that I may be sure of the one that follows, and then waste that one, too, in cares and useless worries?…Does the sun illuminate me today, that I may ponder what happened yesterday? That I may guess at and arrange that which cannot be guessed nor arranged: the fate of the oncoming day?

This is the same secret of happiness which Goethe formulated in the 'Rule of Life':[177]

Would you model for yourself a pleasant life?
Worry not about the past
Let not anger get the upper hand
Rejoice in the present without ceasing
Hate no man…

And the future? Abandon it to God.

The 'rule of life'—that 'high wisdom'—consists in looking neither forward nor behind, but in becoming aware of the uniqueness and incomparable value of the present. In Goethe, then, we find the same exercise of delimitation of the present that we had encountered in ancient philosophy. This exercise is, however, inseparable from another exercise, which consists in becoming aware of the inner richness of the present, and of the totality contained within the instant. By delimiting the present, consciousness, far from shrinking, swells to fill the dimensions of the world; for that vision which "looks the instant in the eye" is the disinterested vision of the artist, the poet, and the sage, which is interested in reality for its own sake.

Enjoying the present, without thinking about the past or the future, does not mean living in total instantaneousness. Thoughts about the past and the future are to be avoided only insofar as rehashing past defeats, and cowering in fear of future difficulties, can cause distractions, worries, hopes or despair, which turn our attention away from the present, where it ought to be concentrated. When we do concentrate our attention on the future, however, we discover that the present itself contains both the past and the future, insofar as it is the genuine passage within which the action and movement of reality are carried out. It is this past and this future which are seized by the artist's vision, in the instant which he chooses to describe or to reproduce. It is beyond. The artists of antiquity, says Goethe, knew how to choose the 'pregnant' instant, heavy with meaning, "which marks a decisive turning-point between time and eternity."[178] To use one of Goethe's favourite terms, such instants 'symbolise' an entire past, and an entire future.

Likewise, when an artist seizes an instant of the movement of a dancer, it allows us to glimpse both the 'before' and the 'after': "The marvellous suppleness with which a dancer moves from one figure to another, and provokes our admiration in the face of such artistry: it is fixated for a moment, so that we can see, simultaneously, the past, the present and the future, and we are thus transported into a supraterrestrial state."[179] It is the same with the sportsman and athlete.

Whoever practices the art of living must also recognise that each instant is pregnant: heavy with meaning, it contains both the past and the future; not only of the individual, but also of the cosmos in which he is plunged. This is what Goethe gives us to understand in his poem 'The Testament':[180]

Let reason be present everywhere
Where life rejoices in life.

This point at which life rejoices in life is nothing other than the present instant.

Goethe is even more explicit on this point in one of his conversations with Eckermann:[181] "Hold fast to the present. Every circumstance, every instant is of infinite value, for it is the representative of an entire eternity." And it is a profound insight into the world beyond.

In order to explain the Goethean notion of the instant as representative of eternity, we must rather thank of the Epicurean and Stoic tradition of which I have spoken above. This tradition affirmed, in the first place, that one instant of happiness is equivalent to an eternity; and secondly, that one instant of existence contains the whole eternity of the cosmos. In Goethean terms, this second idea could be expressed by saying that the instant is the symbol of eternity. Goethe defined the symbol as "the living, instantaneous revelation of the unexplorable,"[182] but we could just as well define the instant as "the living symbol of the unexplorable." The idea of the 'unexplorable' corresponds to what, for Goethe, is the inexpressible mystery at the basis of nature and of all reality. It is its very fleetingness and perishable nature that make the instant the symbol of eternity, because its ephemeral nature reveals the eternal movement and metamorphosis which is, simultaneously, the eternal presence of being: "All that is perishable is only a symbol."[183]

It is here that the thought of death comes into play, for life itself and sport in particular is perpetual metamorphosis, and, inseparably, the death of every instant. Sometimes, for Goethe, this theme takes on a mystical tone:

In order to find himself in the Infinite
The individual willingly accepts to disappear.
It is a pleasure to abandon oneself.[184]
I would praise the living creature who aspires to death in the flame.[185]

In the last analysis, then, it is eternity—that is, the totality of being—which gives the present moment its value, meaning, and pregnancy. "If the eternal remains present to us at each instant, we do not suffer from the fleetingness of time."[186] The ultimate meaning of Goethe's attitude towards the present is thus, as it was for ancient philosophy, the happiness and the duty of existing in the cosmos. It is a profound feeling of participation in and identification with a reality which transcends the limits of the individual. That is the true nature of beyond the concept of sport. "Great is the joy of existence, and greater yet the joy we feel in the presence of the world."[187] "Throughout all things, the Eternal pursues its course. Hold on to Being with delight!"[188]

However, the reader may rightly ask why have I spent so long considering Epicurean Stoic philosophy and Goethe. Maybe I got carried away with the poetic significance of Faust which seemed to form an important backcloth to sport and poetry in motion. But it was more than that. The concepts of pain and pleasure pervade sport. The relationship between the two is so close and integral to the sporting existential experience. They cannot be separated. They form a metaphysical whole in a dualistic relationship. Epicurean and Stoic philosophy are the bedrock of sporting endeavour. They cannot be separated or distinguished as visceral forms. The experience of pain and pleasure transcends the pursuit of excellence in sport. Exposure to both manifests a desire in man to live in the moment and to treat victory and defeat just the same in the existential reality of love, beauty, pain and pleasure. To adapt the philosophy of Goethe, I believe that in all acts of initiative and creation which sport represents there is one elementary truth, namely that the moment one definitely commits oneself then, providence moves too and we live beyond the horizon.

Finally, by way of conclusion to this chapter I would like to refer to the most recent work on Stoic philosophy,

The Stoic Mindset: Ten Ancient Lessons for Modern Life- by Mark Tuitert (published by Penguin Books 2024) The work was perceptively reviewed by Anna Maxted in The Times on the 20th April 2024. It has become an international best seller, not surprisingly bearing in mind the writer and the subject matter. Tuitert from the Netherlands won an Olympic Gold medal at the winter Olympics in Vancouver in 2010. In the book he gives practical advice on playing sport and living life. He draws extensively on Greek and Roman philosophy including the writings of Epictetus, Seneca and especially to the Roman philosopher Marcus Aurelius to establish certain tenets of Stoic wisdom. These include "What stands in the way becomes the way"- Marcus Aurelius.

"We are all stones in the bridge"- Seneca.

"It's not the things that troubles us it is our judgement of them"- Epictetus.

"Just that you do the right thing; the rest doesn't matter"- Marcus Aurelius.

Tuitert rightly identifies key Stoic values that we can all aspire to- courage, temperance, justice and wisdom, and which are to be found displayed in the concept of sport and beyond. I highly recommend the reading of this book written by a sportsman with an insight into philosophy and sport and sport and philosophy, a central theme of this book.

Chapter 3
The Concept of Play

Life ain't all beer and skittles, and more's the pity but what's the odds, so long as you are happy.
George du Maurier

Catch only what you've thrown yourself, all is
mere skill and little gain;
but when you're suddenly the catcher of a ball
thrown by an eternal partner
with accurate and measured swing
towards you, to your centre, in an arch
from the great bridgebuilding of God:
why catching them becomes a power
not yours, a world's
Rainer Maria Rilke

The play's the thing wherein lies the conscience of the king.
Shakespeare (*Hamlet*)

The supreme accomplishment is to blur the distinction between work and play.
Arnold Toynbee

One man in his time plays many parts.
Shakespeare

In every real man, a child is hidden who wants to play.
Nietzsche

We don't stop playing because we grow old; we grow old because we stop playing.
George Bernard Shaw

The juxtaposition of the serious and the non-serious lies at the heart of humanity. The concept of play including sport reflects that dichotomy. Sid Waddell the pre-eminent darts commentator explained "when Alexander of Macedonia was 33 he cried salt tears because there were no more worlds to conquer. Eric Bristow is only 27". Here we find Alexander the Great and Eric Bristow the World Champion Darts player being put amusingly on the same heroic pedestal, as if like were being compared with like in a serious and non-serious parallel.

William Temple was Archbishop of Canterbury between 1942-44. He was the author of a number of influential, philosophical and theological works, including an enlightening essay Mens Creatrix (1917). He was a proponent of what was called situation ethics, a pragmatic and relativist mode of philosophy based on man's core duty towards his fellow man. He opined "personally I have always looked upon cricket as organised loafing".

The Oxford English dictionary defines Loafing as "to spend time idly".

It is interesting that a theologian should make such a comment of cricket bearing in mind the words of J.B Priestly the great novelist, that "It is hard to see where the MCC ends and the Church of England begins."

It may be that William Temple was influenced by the biblical mantra put in verse by the poet Isaac Watts in his Divine Songs (Song XX) "For Satan finds some mischief still for idle hands to do".

Sir Pelham Warner saw that cricket as with all sport is much more than a mischief for idle hands to do. He said in 1911 on the way to Australia with the MCC via Columbo "cricket has become more than a game. It is an institution, one may almost say it is a religion. It has got into the blood of a nation and wherever British men and women are gathered together in the world there will the stumps be pulled".

Cricket and sport generally are much more than spending time idly. And although I accept that idle hands can make mischief idleness can also be a liberating, creative and productive force in the world.

Perhaps that is what the distinguished Archbishop Temple had in mind but I somehow think not.

The Roman poet Lucan wrote "Leisure creates varied thought". Historically philosophers have frequently eulogised the concepts of idleness and leisure. Montaigne in his essay on Of Idleness refers to it as "Like a horse that has broken from its rider". Hamann commented that true idleness takes courage and fortitude. I prefer to characterise it as 'passivity'. Many early thinkers thought idleness was responsible for the production of profound ideas. In that regard I am reminded of the stanzas from W.H. Davies' poem Leisure (1911) which reads as follows:

What is this life If, full of care,
We have no time to stand and
stare?

No time to stand beneath the
boughs
And stare as long as sheep and
cows:

No time to see, when woods
we pass,
Where squirrels hid their nuts
in grass:

No time to see, in broad
daylight,
Streams full of stars, like skies
at night:

No time to turn at Beauty's
glance:
And watch her feet, how they
can dance:

No time to wait till her mouth
can
Enrich that smile her eyes
Began:

A poor life this if, full of care,
We have no time to stand and stare.

I set these stanzas out in full as they are so germane to what I have to say not only in this chapter but generally in the book. Furthermore, it was my dear father's favourite poem as it reflected his personality and philosophy and brings back memories of how much sport, especially cricket meant to him, the love of which he passed on to me.

Bertrand Russell in his perspicacious essay 'In Praise of idleness' (1935), scolds us for not using our free time for fun. He said "this is a condemnation of our civilisation". In this computer world that sentiment has become even more relevant to the way we live our lives. The concept of play is intrinsic to the nature of man and his existence.

Montaigne said that children at play are not playing about; their games should be seen as their most serious activity.

Einstein's description of thinking as playing with concepts is supported by the French mathematician Henri Poincaré and other creative thinkers who appear to agree that finding solutions to mathematical and physical problems requires that concentrated work needs to be followed by a period of passivity. Writers and composers describe how new ideas 'come to' them as solutions to artistic problems after long periods of playing around with different possibilities. Nietzsche said "I know of no other way of dealing with great tasks than that of play: this is a sign of greatness, an essential precondition."[189]

Gadamer looks at an idea that has played a major role in aesthetics: the concept of *play*. He sought to free this concept of the subjective meaning that it has in Kant and Schiller and that dominates the whole of modern aesthetics and philosophy of man. When we speak of play he says in reference to the experience of art, this means neither the orientation nor even the state of mind of the creator or of those enjoying the work of art, nor the freedom of a subjectivity engaged in play, but the mode of being of the work of art itself.[190]

He believed it is possible to distinguish between play and the behaviour of the player, which, as such, belongs with the other kinds of subjective behaviour. Thus it can be said that for the player play is not serious: that is why he plays. We can try to define the concept of play from this point of view. What is merely play is not serious.

Play has a special relation to what is serious. It is not only that the latter gives it its 'purpose': we play "for the sake of recreation," as Aristotle says.

More importantly, play itself contains its own, even sacred, seriousness. Yet, in playing, all those purposive relations that determine active and caring existence have not simply disappeared but are curiously suspended.

The player himself knows that play is only play and that it exists in a world determined by the seriousness of purposes.

Play fulfils its purpose only if the player loses himself in play.[191]

Seriousness is not merely something that calls us away from play; rather, seriousness in playing is necessary to make the play wholly play.

Someone who doesn't take the game seriously is a spoilsport.

The mode of being of play does not allow the player to behave towards play as if towards an object.

The player knows very well what play is, and that what he is doing is "only a game;" but he does not know what exactly he 'knows' in knowing that.

Our question concerning the nature of play itself cannot, therefore, find an answer if we look for it in the player's subjective reflection.

It is not aesthetic consciousness but the experience of art and thus the question of the mode of being of the work of art that must be the object of our examination.

But this was precisely the experience of the work of art that is maintained in opposition to the levelling process of aesthetic consciousness: namely that the work of art is not an object that stands over against a subject for itself. Instead, the work of art has its true being in the fact that it becomes an experience that changes the person who experiences it.

The 'subject' of the experience of art, that which remains and endures is not the subjectivity of the person who experiences it but the work itself.

For play has its own essence, independent of the consciousness of those who play.

Play—indeed, play proper—also exists when the thematic horizon is not limited by any being-for-itself of subjectivity, and where there are no subjects who are behaving 'playfully'.

The players are not the subjects of play; instead play merely reaches presentation through the players.[192]

We can already see this from the use of the word, especially from its many metaphorical usages.

If we examine how the word 'play' is used and concentrate on its so-called metaphorical senses, we find talk of the play of light, the play of the waves, the play of gears or parts of machinery, the interplay of limbs, the play of forces, the play of gnats, even a play on words. In each case what is intended is to-and-for movement that is tied to any goal that would bring it to an end.

The movement backwards and forwards is obviously so central to the definition of play that it makes no difference who or what performs this movement.

The movement of play as such has, as it were, no substrate.

It is the game that is played—it is irrelevant whether or not there is a subject who plays it. The play is the occurrence of the movement as such.

Thus we speak of the play of colours and do not mean only that one colour plays against another, but that there is one process or sight displaying a changing variety of colours.

Hence the mode of being of play is not such that, for the game to be played, there must be a subject who is behaving playfully.

Rather, the primordial sense of playing is the medial one.

Thus we say that something is 'playing' somewhere or at some time, that something is going on or that something is happening.

This linguistic observation is an indirect indication that play is not to be understood as something a person does.

As far as language is concerned, the actual subject of play is obviously not the subjectivity of an individual who, among other activities, also plays but is instead the play itself.

However, modern anthropological research has conceived the nature of play so broadly that it has almost gone beyond viewing play as subjectivity.

Huizinga has investigated the element of play in all cultures and most important has worked out the connection of children's and animal's play to 'holy play'.

That led him to recognise the curious indecisiveness of the playing consciousness, which makes it absolutely impossible to decide between belief and non-belief.

Here the *primacy of play over the consciousness of the player* is fundamentally acknowledged and, in fact, even the experiences of play that psychologists and anthropologists describe are illuminated afresh if one starts from the medial sense of the word 'playing'.[193]

Play clearly represents an order in which the to-and-fro motion of play follows of itself.

It is part of play that the movement is not only without goal or purpose but also without effort.

It happens by itself. The ease of play—which naturally does not mean that there is any real absence of effort but refers phenomenologically only to the absence of strain—is experienced subjectively as relaxation.

The structure of play absorbs the player into itself, and thus frees him from the burden of taking the initiative, which constitutes the actual strain of existence.

This is also seen in the spontaneous tendency to repetition that emerges in the player and in the constant self-renewal of play, which affects its form (e.g. the refrain).

The fact that the mode of being of play is so close to the mobile form of nature permits us to draw an important methodological conclusion. It is obviously not correct to say that animals *too* play, nor is it correct to say that metaphorically speaking, water and light play *as well*.[194]

Rather, on the contrary, we can say that *man* too plays. His playing too is a natural process. The meaning of his play too, precisely because—and insofar as—he is part of nature, is a pure self-presentation. Thus in this sphere it becomes finally meaningless to distinguish between literal and metaphorical usage.

But most important the being of the work of art is connected with the medial sense of play (also, game and drama). Inasmuch as nature is without purpose and intention, just as it is without exertion, it is a constantly self-renewing play, and can therefore appear as a model for art.

As Friedrich Schlegel wrote, "All the sacred games of art are only remote imitations of the infinite play of the world, the eternally self-creating work of art."[195]

Another question that Huizinga discusses is also clarified through the fundamental role of the to-and-fro movement of play: namely the playful character of the contest.

It is true that the contestant does not consider himself to be playing. But though the contest arises the tense to-and-fro movement from which the victor emerges, and thus the whole becomes a game.

The movement to-and-fro obviously belongs so essentially to the game that there is an ultimate sense in which you cannot have a game by yourself.

In order for there to be a game, there always has to be, not necessarily literally another player, but something else with which the player plays and which automatically responds to his move with a countermove. Thus the cat at play chooses the ball of wool because it responds to play, and ball games will be with us forever because the ball is freely mobile in every direction, appearing to do surprising things of its own accord.

In cases where human subjectivity is what is playing, the primacy of the game over the players engaged in it is experienced by the players themselves in a special way.

Once more it is the improper, metaphorical uses of the word that offer most information about its proper essence. Thus we say of someone that he plays with possibilities or with plans.

What we mean is clear. He still has not committed himself to the possibilities as to serious aims. He still has the freedom to decide one way or the other, for one or the other possibility. On the other hand, this freedom is not without danger.

Rather, the game itself is a risk for the player. One can play only with serious possibilities. Obviously this means that one may become so engrossed in them that they outplay one, as it were, and prevail over one.

The attraction that the game exercises on the player lies in this risk. One enjoys a freedom of decision which at the same time is endangered and irrevocably limited. One has only to think of jig-saw puzzles, games of patience, etc. But the same is true in serious matters. If, for the sake of enjoying his own freedom of decision, someone avoids making pressing decisions or plays with possibilities that he is not seriously envisaging and which, therefore, offer no risk that he will choose them and thereby limit himself, we say he is only "playing with life."

This suggests a general characteristic of the nature of play that is reflected in playing: all playing is a being-played. The attraction of a game, the fascination it exerts, consists precisely in the fact that the game masters the players.

Even in the case of games in which one tries to perform tasks that one has set oneself, there is a risk that they will not 'work', 'succeed', or 'succeed again', which is the attraction of the game. Whoever 'tries' is in fact the one who is tried.

The real subject of the game (this is shown in precisely those experiences in which[196] there is only a single player) is not the player but instead the game itself.

What holds the player in its spell, draws him into play, and keeps him there is the game itself.

This is shown also by the fact that every game has its own proper spirit. But even this does not refer to the mood or the mental state of those who play the game.

Rather, the variety of mental attitudes exhibited in playing various games and in the desire to play them, is the result of not the cause of the differences among the games themselves. Games differ from one another in their spirit.

The reason for this is that the to-and-fro movement that constitutes the game is patterned in various ways. The particular nature of a game lies in the rules and regulations that prescribe the way the field of the game is filled.

This is true universally, whenever there is a game.

It is true, for example, of the play of fountains and of playing animals.

The playing field on which the game is played is, as it were, set by the nature of the game itself and is defined far more by the structure that determines the movement of the game from within than by what it comes up against—i.e., the boundaries of the open space—limiting movement from without.

Apart from these general determining factors, it seems to me characteristic of human play that it plays *something*.

That means that the structure of movement to which it submits has a definite quality which the player 'chooses'.

First, he expressly separates his playing behaviour from his other behaviour by *wanting* to play.

But even within his readiness to play he makes a choice. He chooses this game rather than that.

Correlatively, the space in which the game's movement takes place is not simply the open space in which one "plays oneself out," but one that is specially marked out and reserved for the movement of the game.

Human play requires a playing field. Setting off the playing field—just like setting off sacred precincts, as Huizinga rightly points out—sets off the sphere of play as a closed world, one without transition and mediation to the world of aims.[197]

First and foremost play is self-presentation.

As we have seen, the self-presentation of human play depends on the player's conduct being tied to the make-believe goals of the game, but the 'meaning' of these goals does not in fact depend on their being achieved.

Thus there are games which must be called representation games, either because, in their use of meaningful allusion, they have something about them of representation (say "Tinker, Tailor, Soldier, Sailor") or because the game itself consists in representing something (e.g., when children play cars).[198]

All presentation is potentially a representation for someone.

That this possibility is intended is the characteristic feature of art as play.

The closed world of play lets down one of its walls, as it were.

A religious rite and a play in a theatre obviously do not represent in the same sense as a child playing.

Their being is not exhausted by the fact that they present themselves, for at the same time they point beyond themselves to the audience which participates by watching.

Play here is no longer the mere self-presentation of an ordered movement, nor mere representation in which the child playing is totally absorbed, but it is "representing for someone."

The directedness proper to all representation comes to the fore here and is constitutive of the being of art.

In general, however much games are in essence representations and however much the players represent themselves in them, games are not presented for anyone—i.e., they are not aimed at an audience.

Children play for themselves, even when they represent.

And not even those games (e.g., sports) that are played before spectators are aimed at them.

Indeed, contests are in danger of losing their real play character precisely by becoming shows.

A procession as part of a religious rite is more than a spectacle, since its real meaning is to embrace the whole community.

And yet a religious act is a genuine representation for the community; and likewise, a drama is a kind of playing that, by its nature, calls for an audience.

The presentation of a god in a religious rite, the presentation of a myth in a play, are play not only in the sense that the participating players are wholly absorbed in the presentational play and find in it their heightened self-representation, but also in that the players represent a meaningful whole for an audience.

Thus it is not really the absence of a fourth wall that turns the play into a show.

Rather, openness towards the spectator is part of the closedness of the play.

The audience only completes what the play as such is.

This point shows the importance of defining play as a process that takes place "in between."[199]

We have seen that play does not have its being in the player's consciousness or attitude, but on the contrary play draws him into its dominion and fills him with its spirit.

The player experiences the game as a reality that surpasses him.

This is all the more the case where the game is itself 'intended' as such a reality—for instance, the play which appears as *presentation for an audience*.

According to all that we have observed concerning the nature of play, this subjective distinction between oneself and the play implicit in putting up a show is not the true nature of play.

Rather, play itself is a transformation of such a kind that the identity of the player does not continue to exist for anybody. Everybody asks instead what is supposed to be represented, what is "meant."[200]

The players (or playwright) no longer exist, only what they are playing.

Even Plato, the most radical critic of the high estimation of art in the history of philosophy, speaks of the comedy and tragedy of life on the one hand and of the stage on the other without differentiating between them.

For this difference is superseded if one knows how to see the meaning of the play that unfolds before one.

The pleasure of drama is the same in both cases: it is the joy of knowledge.

"Reality" always stands on a horizon of the desired or feared or, at any rate, still undecided future possibilities. Hence it is always the case that mutually exclusive expectations are aroused, not all of which can be fulfilled.[201]

The undecidedness of the future permits such a superfluity of expectations that reality necessarily lags behind them.

Now if, in a particular case, a context of meaning closes and completes itself in reality, such that no lines of meaning scatter in the void, then this reality is itself like a drama.

Likewise, someone who can see the whole of reality as a closed circle of meaning in which everything is fulfilled will speak of the comedy and tragedy of life.

In these cases, where reality is understood as a play, emerges the reality of play, which we call the part of art.

The being of all play is always self-realisation, sheer fulfilment, energeia which has its telos within itself.

"The world of the work of art, in which play expresses itself fully in the unity of its course, is in fact a wholly transformed world. In and through it everyone recognises that this is how things are."[202] Sport can achieve that status. Carwyn James according to Frank Keating described the great Barbarians try against the All Blacks on 27th January 1973 which resulted from the great fly half Phi Bennett picking up the ball under his posts and Gareth Edward scoring, the most memorable rugby movement in the following way. It was "rare and unforgettable, when you can play at a level outside the conscious, when everything is instinct, but as clear as a bell because you have practised it so often and especially dreamed it that unique moment when sport, lovely sport, not only achieves but assumes an art form."[203] Oscar Wilde said "A work of art is the unique result of a unique temperament." In my experience and understanding great sportsmen have a distinctive and unique temperament which is creative and a form of art. As in the creativity game of art you are not a player unless you have something to say. Great sportsmen have something to say.

Chapter 4
Fair Play, The Hand of God and Sex

There ain't no sin and there ain't no virtue, there's just what people does.
The Preacher in Steinbeck's *The Grapes of Wrath*

Rules are like vestal virgins—unless they are violated there will be no issue.
Hamann

A Parliament can do anything but make a man a woman, and a woman a man.
2nd Earl of Pembroke c.1534–1601

Fair is foul, and foul is fair.
Shakespeare, *Macbeth*

One should always play fairly when one has the winning cards.
Oscar Wilde

*If everything about England was destroyed, except the laws of cricket, English
society could still be created from them.*
Neville Cardus

Justice is truth in action.
Disraeli

*The Common Law of England has been laboriously built about a mythical
figure—the figure of the Reasonable Man.*
A.P. Herbert

George Orwell said "serious sport has nothing to do with fair play. It is bound
up with hatred, jealously, boastfulness, disregard of all rules and sadistic pleasure
in witnessing violence[204]." It is 'war minus the shooting'. Literary giant as he
was, I am not so sure how much guidance we get from him on sport and/or fair
play. Grantland Grace in his poem Alumnus Football (1941) said "For when the

Great scorer comes to mark against your name. He writes not that you won or lost but how you played the game."

The International Council for Sport and Physical Education defines fair play, first and foremost, "as respect for the rules of the game and requires, as a minimum that the competitor shows strict, unfailing observance of the written rule and that respect is due to the spirit rather than the letter of the rules." As Rawls said "It's not the rules but what's between the rules."

This understanding of fair play is important in that it draws attention to the rule governed and defined nature of sport and law. Games exist and are defined by their rules. This position is also important because it acts as the foundation of the logical incompatibility thesis, the thesis which demonstrates that one cannot win if one cheats. But there is a different view the great rock guitarist Jeff Beck said "I don't care about the rules. In fact, if I don't break the rules at least ten times a day I'm not doing my job properly."[205]

Fair play as respect for rules is, however, an inadequate formulation for capturing some of our intuitions about the idea. Fair play, as respect for the rules cannot account for actions we take to be required by fair play, but which are not covered by any rule. As Jonathan Duke-Evans in An English Tradition has recently written "the cult of fair play is an English exceptionalism." In a detailed study of the history of the phrase 'fair play' he traces broader ideas of justice, equity and honour philosophically and historically, taking us back to Homer, Virgil and Livy and to Beowulf and Malory. My focus is on its contemporary significance in sport, law and philosophy.[206]

"The concept of unfairness is a romantic notion and was part of the philosophy of the Enlightenment Movement. However, in our modern world it has taken on a harder form which belies its philosophical significance as a hall mark of morality. Sport especially cricket for the most part has been the emblem of fair play not just according to the law but to the implicit moral code which is beyond the law and is a higher ethical force which should guide our treatment of fellow man. Lord Harris a doyen of MCC cricket, in his book A Few Short Runs (1921), put it beautifully when he said in the context of cricket 'it's not cricket' the brightest gem ever worn by any pursuit; in constant use on the platform, in the pulpit, parliament and the press, is to dub something as not fair, not honourable, not noble. What a tribute for a game to have won, but what responsibility for those who play and manage it."

The penalty kick was first introduced in soccer in 1891. A contemporary source is quoted as saying,

"It is a standing insult to sportsmen to have to play under a rule which assumes that players intend to trip, hack and push their opponents and to behave like cads of the most unscrupulous kind. I say that the lines marking the penalty area are a disgrace to the playing field of a public school."

The reasoning behind this statement was as follows. A player might trip, hack and push his opponent by accident or by design. If it occurred by accident, no penalty was required. No gentleman sportsman would ever intentionally consider such behaviour. Sport was played by gentlemen, so either way, the penalty areas were clearly unnecessary and insulting.

Writing in The Times, Michael Atherton, a friend and outstanding cricket correspondent—described the Spirit of the Game in cricket as a "lot of well-meaning guff."[207]

He says:

"...other than encouraging basic human qualities such as showing respect for your opponents and discouraging violence on the field, nobody is left any the wiser what the Spirit of Cricket is. It is something to be nurtured and cherished for sure, but precisely what that something is, is not clear?"

Later he put forward his attempt at cutting the waffle, a definition.

"The Spirit of Cricket is best expressed when there exists due respect for the umpires, the Laws of the Game and those who play it."[208]

Atherton's concern is that it is a grey area and asks captains to make moral judgments that are sometimes contrary to the laws of the game. To him 'the spirit of the game' and the phrase 'it's not cricket' are no more than a remarkably successful marketing strategy started by the Victorians to cleanse the game from its match fixing crisis in the early 19th century and latterly by the ICC keen to exploit the opportunities offered by something it reckons differentiates cricket from other sports.

Conceptually, his view is that the "Spirit of cricket demands interpretation which in turn renders it worthless." As all law demands interpretation, I assume Michael's anarchic view is that it should also be characterised in the same way!

John Woodcock offered his interpretation as "What is thought of as being honourable and less than honourable."[209]

In Christopher Martin-Jenkins' view, "cricketers from village green to the test arena have a shrewd idea, in 95 cases out of 100, what is and is not within the Spirit of the Game. That spirit, distilled, is simply fair play."[210]

In the Republic, Plato declares that "there are two arts which I would say some god gave to mankind music and gymnastics not for the soul and body incidentally, but for their harmonious adjustment."

As I have mentioned earlier, Albert Camus, centuries later, said that he learnt ethics from sport. Harold Pinter thought cricket was god's greatest gift to mankind.

Does or should sport reflect values and if so, what are or should they be? Some say that sport simply reflects values already expressed in our society.

In a capitalised society, it is not surprising that it will embody the notions of intense competition and rivalry. Others argue that sport has a set of values peculiar to itself. In my opinion, it has its own 'logic' and intrinsic moral content.

The phrase 'it's not cricket' was originally used by the Rev. James Pycroft in 1851 to express his displeasure and disapproval of overarm bowling. It was not until 1864 that it was legal to bowl overarm.

130 years later in February 1981 in a one-day match with New Zealand, with the visitors needing six from the last ball for a tie, Australian Captain Greg Chappell ordered the bowler, his brother Trevor, to deliver a 'mullygrubber', that is an underarm delivery along the ground. Trevor complied and "A stunned New Zealand batsman blocked it."

Australia won. On television, former Australian Captain Richie Benaud called Chappell's action "the most gutless thing I have ever seen on a cricket field." New Zealand's Prime Minister Sir Robert Muldoon called the underarm delivery an 'act of cowardice' and stated that he now understood the reason why Australia's One Day uniform is 'yellow'. Yet what Chappell did was lawful. As a result the laws of One-Day Cricket were changed so that it could not be repeated.

It has been said that "no other case in recent cricket memory has so epitomised the conflict between law and morality." Rex Mossop in the Sydney Morning Herald said "one ball dents Australia's image as a sports nation."

Don Oslear, the English test umpire condemned Greg Chappel for violating the spirit of the game. Asked how he could have reacted had he been standing in the match, he replied "I might have said, come on Greg. It's been a tremendous game. Let's bowl it up." Chappel himself subsequently said that he was disappointed that it was against the spirit of the game.

Oslear took the position that umpires must play an active role in upholding the spirit of the game. But, as has been pointed out, had he been present had Chappel rejected his imprecations and bowled underarm, whatever his own moral view, he could not as an umpire, have acted any differently from the two umpires who were actually present. This has led David Fraser in his book 'Cricket and the Law' to conclude that:

"This idea, that the spirit of the game is superior to strict adherence to the formal content of the Laws of Cricket, whatever its moral, deontological status, has never in such circumstances, had an actual legal consequence or effect." (ibid p.1511).

This, if I may say so, misses the point as to what the Spirit of the Game is about.

The point of the Spirit of Cricket is not to override the Laws of Cricket. It is there to supplement the Laws and assist in their interpretation, application, and development by players, officials and administrators. In some instances, it will not be in the Spirit of the Game to rely upon the strict rigour of the law—a formalistic adherence to the letter of the law may result in an injustice being done to an opponent. Players should have regard to this is the way they conduct themselves on the field of play.

A recent example of this phenomenon of the incident which took place at Lords cricket ground in an Ashes Test on Sunday 2nd July 2023. It would be added to the rich folklore of cricket and sporting history. It caused massive controversy and criticism from the proponents of the Spirit of the Game. I cite the events that took place as an example of how significant philosophically and psychologically sport is to many people's lives, The incident resulted in the Australian players being castigated in a form of moral opprobrium which belies the view of cynics that sport is a trivial pursuit not worthy of serious characterisation. I wholly disagree.

To many what happens on the sports field is of the utmost seriousness. Shame on them I can hear the dissidents exclaim but that is the reality.

I have referred elsewhere in this book to the great football manager Bill Shankly's response, when it was put to him that football is "a matter of life and death." He succinctly replied "no it is much more important than that."

I now briefly turn to the facts and indeed the law and my conclusions of the true significance of what happened on that day.

The Australian wicket keeper Alex Carey stumped the English batsman Johnny Bairstow allegedly unfairly and it was said not within the spirit of the game, notwithstanding the fact that if one applied the laws of cricket strictly he would have been lawfully dismissed.

Matthew Syed writing in The Times on the 4th July 2023 commented:

"I hoped that this Ashes series would bring out the best in the game, I hoped that the Australians might be taken at their word when they said after various controversies, that they would not merely focus on the rules of the game but the spirit. I hoped they were telling the truth when their former head coach Justin

Langer said 'cricket is not just about being good cricketers but good people that play the right way.'"

In the same newspaper on the same day, the Australian journalist Gideon Haigh, not surprisingly took a different view and wrote

"Carey's stumping was full of dexterity and involved originality. The unusual is sometimes mistaken for the unethical. BJT Bosanquet who invented the googly, some decried the notion of pretending to spin the ball one way but actually spinning the ball the other way. Bosanquet himself joked that the googly was 'not illegal but merely immoral.'"

In my opinion this incident again reflects the fact that sport raises issues of metaphysical import. The Hart and Devlin debate about law and morality and the Hart and Dworkin jurisprudential dialogue about the nature of law and rules are apposite and cover the same issues as the two cricketing incidents to which I have earlier referred. They are paradigm examples of sport as philosophy and philosophy as sport which is the subject matter of not only this chapter but the book generally. Sport is fertile territory for philosophical discourse. Hence its appeal for both its cerebral and physical content and manifestation. That Cartesian dualism of mind and body is at its essence.

The categories of potential injustice are not closed. No code of laws can anticipate every possible eventuality. Furthermore, all laws require interpretation and good judgement is needed in their application. In some instances, there is scope for the exercise of discretion. Umpires should be encouraged to do that having regard to the Spirit of the Game.

Law makers and administrators likewise should ensure that the game develops in such a way that the Spirit of the Game and its traditional values are enshrined in the laws and playing conditions. That principle was set out in law 42 which stated "the responsibility lies with the captains for ensuring that play is conducted within the spirit and traditions of the game and that the umpire will be the sole judges of fair and unfair play."

The game of cricket is changing. No-one should be averse to change provided the intrinsic nature and value of the game is not lost.

Writing in The Times (November 23rd 2009), Simon Barnes, a writer of one of the forewords to this book, and the author of an outstanding book 'The Meaning of Sport',[211] its Chief Sports writer says:

"Cricket is becoming a boundary-hitting contest, the bowlers human tees. If cricket doesn't try to redress the balance, the game will lose its point. Every format of the game needs to be a contest between bat and ball. The fall of a wicket—not a six—is the single most exciting thing in cricket. We're losing the point of the whole damn thing."

Some say that 20–20 and The Hundred is clear evidence that the point and purpose of the game is in danger of being lost.

Respect for the essence of the game, its traditional values should guard against what would be a tragic loss to sport. Recourse to the Spirit of the Game is invaluable. It provides essential guidelines as to the way the game should be played, administered and developed.

The rules of sports, famously characterised by Bernard Suits in his seminal works *The Grasshopper and Elements of Sport,* create obstacles for the purpose of challenging the competitors. Sport on this view is no more than a test of a competitor's ability to meet the challenge created by the rules. It is an attempt to overcome (excel) the sport specific obstacles created by the rules. Is this pursuit of excellence in itself a value? Dedication, self-examination and honesty about one's strength and weaknesses that success in competitive sport calls for, are virtues. Is this the moral content?

Success at being a burglar may involve some of the same factors but lack moral standing. There is no doubt that excellence can be used for improper purposes. But that no more shows that it is not a value than the fact that there can be justice among thieves shows that justice lacks moral standing. What gives moral force to the virtues of excellence in sport is their connection within the practice to have respect for certain qualities of human beings.

Jay Cookley[212] argues that "athletes are systematically encouraged to over-conform to a unique set of norms embodied in what might be called a 'sport ethic'." He identifies these norms as

(a)   Being an athlete involves making sacrifices for 'the game'.
(b)   Being an athlete involves striving for distinction.
(c)   Being an athlete involves accepting risks and playing through pain.
(d)   Being an athlete involves refusing to accept limits in the pursuit of possibilities.

He says:

"It is seldom recognised that many ethical problems and forms of deviance in sports are not due to athletes denying or rejecting social values or norms. Instead, they are due to athletes accepting and committing themselves without question or reservation to the normative guidelines that constitute what might be called the 'ethic' of many sports cultures today..."

His view of performance-enhancing drugs is instructive:

"The use of these substances is not the result of defective characters among athletes, or the existence of too many material rewards in sports, or television coverage, or exploitation by coaches and managers, or moral weakness among athletes. After all, users are often the most dedicated and committed people in sports! Instead...most substance and abuse is clearly tied to an over-commitment to the sport ethic itself; it is grounded in over-conformity that leads injured distance runners to continue training even when training may cause serious physical problems, and American football players to risk their bodies through excessively violent physical contact week after painful week in the NFL, and figure skaters to risk leg injuries by doing triple after triple after quadruple jumps in their quests for 'perfection'."

Fair play is often the phrase used to capture the view that sport teaches values but the content of those values depends upon the way in which sport is played, taught and practised. It has been said that because sport is conducted by human beings, it falls within the realm of morality.

Sport is thus amenable to moral discussion. However, what would count as a violation of the moral order is determined both by the nature of sport itself and the agreement of the competitors to modify or suspend the scope of general moral rules. For example in rugby one accepts that one will be subjected to and must inflict actions that would be both illegal and immoral in other contexts but it is no less a 'moral code' for that.

The central feature of sport is its nature of being set apart insert or being 'beyond' as used in the title of this book. It is freely chosen and entered into for its own sake. The attitude is therefore one of playfulness. This means that the purpose of sport is the creation of enjoyment of the activity and that the appropriate attitudes are generosity and magnanimity. It is the balanced recognition and acceptance of the essential non-seriousness of sport combined with the utmost dedication and commitment in its pursuit.

Kant distinguished between constitutive and regulative rules. The former constitute an activity the existence of which is logically dependent on the rules as opposed to those rules which regulate a pre-existing activity, an activity whose existence is logically independent of the rules. This is why in the case of cricket the Spirit of the Game is antecedent to and underlies the Laws.

Fairness prescribes how to act in play. Rule conformity is a necessary condition of game playing but does not necessarily evidence a moral reason for abstaining from rule violation. However, game-playing requires co-operation.

It follows that it is mutually advantageous to obey the rules. This on one view is the intuitive idea of fairness. It is an implicit agreement amongst participants that there is a moral obligation to adhere to the playing rules of the game. All sports have the hallmark of this tradition. However, sports in their rules offer a minimal morality. They attempt to frustrate the excesses of egoistic desires in the form of sanctions.

One theory is that a game is only realised as such if it is played in every detail as defined by its rules. This is called formalism. However, the problem with this is that the game becomes an ideal type which is never realised in practice. What constitutes fairness in sport must relate to more than the formal playing rules.

Wittgenstein pointed out a rule cannot determine its own application.

One challenge to formalism is the idea of ethos.

It has been defined as a shared group of norms on the interpretation of the rules in a game. The ethos of an activity distinguishes between permissible acts that are in accordance with the rules, acceptable acts of rule violations that are 'part of the game', and rule violations that are considered unacceptable.

Fairness therefore not only requires participants to keep the formal playing rules of the game but to keep the shared ethos of the sporting games in which they take part. This would differ from sport to sport. All sport is defined by a firm framework of rules. But between the lines of the rules, certain fundamental normative principles and values can be found. Any meaningful and workable ethos of a sporting game needs to respect these fundamentals. All sporting games have as their institutional goals to measure, compare and rank participants according to their performance. The predominant norm is of playing to win. But one seeks to conserve and recreate the best traditions of the practice and the supportive character traits that entrench those traditions anew in each generation. This will involve the blending of instrumental virtues such as courage, resolve and tenacity as well as the other virtues of compassion, empathy and fairness.

According to this argument fair play has to include a sense of internal goods, a sense for the good game permeated by uncertainty of outcome and in which the participants perform and act pursuant to the standards of excellence of the game. It is only this fuller appreciation of virtues suffusing both practices and their institutions that can generate the ethos where fair play can flourish.

The logic of sports binds us to formal criteria of inefficiency with respect to the means we are allowed to use to attain its ends, and normative criteria of virtuous action with respect to the just, honest and temperate ways we are to conduct ourselves in practice. Bernard Suits saw sport as an expression of a lusory (playful) attitude. When we are engaged in sport we should not just play fair, but play to win with a sense for and respect for the internal goods of the game.

It is important to underline the diversity of the game's ethos. They make sporting games exciting and open arenas for creativity and exploration. These are critical criteria for evaluating ethos—they are sometimes called 'discourse ethics' which attempts to take the pluralism of modern life seriously. The most prominent theorist in this tradition, Jurgen Habermas, describes discourse ethics as a search for consensus on moral grounds and underlines the importance of real discourse. He formulates 'the distinctive idea' of the ethics of discourse as follows.

"Only these norms can claim to be valid that meet with the approval of all affected in their capacity as participants in a practical discourse."

In a sporting context players enter a situation of real, embodied discourse in which they search for consensus on how the game should be played. Understood in this way, a game discourse builds on arguments or conduct based on experience, skill and established standards of excellence, and the traditions and social norms of the game to rise above other arguments and exert a cultivating effect on all participants.

How then should sporting games be played?

Sporting games are practices that embody a range of ethical norms wherein moral agents explore their own limits, physical and moral, that link them to the past and future generations. In considering the meanings of a game, of winning and losing there is a need to distinguish between two categories of goals and goods.

Internal goods are realised within the very practice of game playing. Internal goods take the character of experiential value such as the excitement of the tight game, the joy of a well conducted stroke, the kinaesthetic pleasure of rhythm in bowling. The realisation of internal goods depends upon the realisation of the game according to the relatively shared ethos that informs it. It is not to be understood as an attempt to pinpoint some kind of 'essence' or 'core value' of sport. It is an articulation of the phenomenal structure of the good sport experience.

The cheat may gain the prize, the wealth and adoration, but never the internal goods. One cannot experience internal goods without realising the practices according to the rules and normative values and by striving towards the standards of excellence to which they belong.

External goods, by contrast, are those benefits that accrue from engagement in sport, but which can be achieved in a number of fields of endeavour. The most obvious of these are prestige, status, medals and money. They can be achieved through adhering to the ethos of the game, though not exclusively so. The point for the externally motivated participant is not to play the game, but simply to end up on top of the final ranking of competitors. The cheater can get away with cheating and still win the prize. The realisation of the cheater's intentional goals does not presuppose fairness. The predominant norm is simply "win—even at the cost of unfair play" and this is precisely to be distinguished from the norm of playing to win. The debate that is about the Spirit of the game has some similarities to what is known to lawyers as the Dworkin-Hart debate. In his writings Professor Dworkin, an American from Harvard, criticised legal positivists particularly the Oxford Professor H.L.A. Hart for holding too narrow a view of the nature of law. Legal positivists, like Hart identified law with a formal structure of rules. Such an approach is similar to the formalistic approach to the analysis of games and sport.

Formalism characterises games, such central elements of games as winning and losing and allowable moves within the activity, primarily in their formal structure and particularly their constitutive rules. This approach has helped to understand the nature of games but issues of fair play often go beyond conformity to the formal rules of a sport. It also has problems with rule changes and rule formation. Rules of many sports include what Professor Hart calls 'rules of change'. A rule of change for cricket might say that a proposal becomes a law of cricket when it is accepted through established procedures by the MCC and ICC. But although such rules of change will establish when a change becomes official, they do not necessarily establish what is good for the game. In emphasising the formal constitutive rules of given sports, formalists have ignored the implicit conventions that apply to the sports in question. Conventionalists make a contribution to our understanding of sporting practice by exploring the role of the ethos and cultural context of games.

The question raised is whether the ethos/the spirit of the game has normative tone. In other words, should the conventions express what ought to take place as well as describe what does take place in sporting practice. One problem with conventionalism is the ethical status of the conventions themselves. This is true even where conventionalism is plausible; for example, strategic fouls in sports such as football. Critics of strategic fouling acknowledge that they are opposing a widely accepted practice, as well as the conventional understanding upon which practice is based, but they argue that appeal to central values implicit in the logic of sports requires the reform of existing conventional behaviour.

However, it is doubtful whether conventionalism can respond any better than strict formalism to an evaluation of proposed changes in the rules or conventions of a sport. Just as appeal to the existing rules cannot settle the issue of whether a proposed rule change is or is not an improvement so appeal to existing conventions cannot be the sole basis for evaluating proposals for reform.

One of Hart's major contributions is to show the diverse rules that make up the law; these include criminal sanctions, rules of change and adjudication and the Rule of Recognition, which identifies the rules of the legal system and distinguishes them from non-legal rules.

Professor Dworkin argued that in addition to rules, there are legal principles, which have normative force within the legal system. What makes these principles legal ones, rather than simply moral principles imported from beyond the law and applied according to the particular and ethical commitments of individual judges, is that they are either presupposed by the legal system or are required to make sense of its key elements.

An analogous position in the philosophy of sport has been developing for some time.[213]

In "Fair Play as Respect for the Game,"[214] Robert Butcher and Angela Schneider maintain that "if one honours or esteems one's sport…one will have a coherent conceptual framework for arbitrating competing claims regarding the fairness…of actions." They suggest that sports themselves have interests; athletes show respect for the game when they make its internal interests their own. They claim that "the idea of the interests of the game provide a hearing for judging one's own action in relation to the sport…Taking the interests of the game seriously means that we ask ourselves whether or not some action we are contemplating would be good for the game, if everyone did that."[215] Appeal is made to the norms or principles internal to the idea of sport. These are not conventions nor are they the formal rules of the game.

In "Are Rules All an Umpire has to Work With" J.S. Russell goes beyond explicit formal rules. He says "we might try to understand and interpret the rules of a game…to generate a coherent and principled account of the point and purposes that underlie the game, attempting to show the game in its best light." He cites as an example of a principle that might underlie good competitive sports, the injunction that "rules should be interpreted in such a manner that the excellences embodied in achieving the illusory goal of the game are not undermined but are maintained and fostered."

The view is that in addition to the constitutive rules of sport, there are other resources connected closely—perhaps conceptually—to sport. These resources can be used to adjudicate moral issues in sports and athletics. On this view,[216] sport has an independent 'gratuitous logic' of its own that makes it more than a mirror reflecting the values of society. A fundamental component of that logic is respect for the traditional values of the game.

What is respect? One meaning of respect is merely observing or following. For example respecting the rules of the road by adhering to the speed limit. Another meaning is stronger and carries connotations of honouring, holding in regard, esteeming or valuing. In the context of sport, it is easy to run the two senses together. Because sports are games made up by their rules, there is the requirement that we respect the rules of the game. This could mean that we treat the rules of the game in the same way as we treat the rules of the road. We observe or follow them, perhaps for the sake of expediency or as a courtesy. It is not obvious how one could honour or esteem traffic regulations. But it is precisely in the latter sense that the idea of fair play can be understood in terms of respect for the game.

In 'After Virtue' by A. MacIntyre a practice is defined as:

> "Any coherent and complete form of socially established co-operative human activity through which goods internal to that form of activity are realised in the course of trying to achieve those standards of excellence which are appropriate to and partially determinative of, that form of activity, with the result that human powers to achieve excellence, and the human conceptions of the ends and goods involved, are systematically extended."

MacIntyre stresses that practices have histories and traditions and form living, vibrant and changing entities. Practices change—they must—but the change comes from within and operates inside the context formed by tradition and practice. Respect for one's game and its traditions requires that one takes on and assumes the interests of that game.

The sportsperson takes on the interests of his or her sport. Those interests become the interests of the sportsperson. If you respect the game, you honour and take seriously the standards of excellence created and defined by that game. The idea of the interests of the game provides a means of judging one's own actions in relation to the sport. People approach any activity with mixed motivations and interests. Taking the interests of the game seriously means that the player asks himself whether or not some action he is contemplating would be good for the game concerned, if everyone did it. For example, in cricket, the 'Spirit of the Game' provides guidelines as to not only what is done but what ought to be done.

From a player's perspective, it will influence actions on the field of play, attitudes towards one's own level of commitment to the game. It also has implications for actions and decisions at the level of policy.

It has been argued that the rules of sport are the only legitimate source of an official's authority and action. Again the argument is based on the view that games are rule-governed institutions that have been designed to be especially orderly and predictable.

They are laid down to settle authoritatively the terms for co-operation and competition. They settle fully what conduct is permissible and what is not.

This is erroneous. It obscures that rules in sport face the same indeterminacies that rules do in other contexts (see Dworkin Law's Empire).[217]

The exercise of discretion by judicial officials has long been recognised "as a moral and practical necessity."[218]

Rules are devices that are intended to guide human conduct but that they can fail to be definitive guides for a variety of other related reasons. There is obviously some core of agreed meaning if they have any use as guides to action.

But because language is an imprecise instrument, the core of agreed meaning may break down. This has been called the 'open texture' of rules by H.L.A. Hart.[219]

Some rules are simple and precise enough to be fully determinative but other rules may be open textured.

There are uncertainties of meaning and sometimes purposes that rules are designed to serve. These indeterminacies determine the uncertainties regarding the scope of a rule's application.

Furthermore, all the circumstances in which a rule might be applied cannot be foreseen in advance. This means that the application of the rules may be different from the purposes that the rules are intended to serve.

It might be possible to minimise, if not eliminate, the area of discretion by adopting very narrow interpretations of rules and applying them inflexibly to all cases where their application is in question.

But that would be as Hart pointed out:

> "at the cost of blindly prejudging what is to be done in a range of future cases, about whose composition we are ignorant. We shall thus indeed succeed in settling in advance, but also in the dark, issues which can only reasonably be settled when they arise and are identified."[220]

It is easy to see how such formalism can invite observations like Mr Bumble that "the law is an ass."

Indeterminacies in the laws means that adjudicators are sometimes faced with practical and moral necessities to exercise discretion where rules fail to provide sufficient guidance by themselves.

It has been said by Dworkin that the supreme virtue of law is integrity.[221] This means that the law should speak in a coherent and principled way. It means that law must be interpreted to aim ultimately towards "some single comprehensive vision of justice." Translated to sport J.S. Russell put forward a list of principles that are relevant to the tasks of adjudication in sport. It is not exhaustive but a very good start.

Principles of Adjudication in Sport

1. Rules should be interpreted in such a manner that the excellences embodied in achieving the lusory goal of the game are not undermined but are maintained and fostered.

By the 'lusory goal' of the game is meant winning by overcoming certain obstacles or inefficiencies that the game sets in its participants' path. Competition in a game is designed to test how well the participants are able to surmount these obstacles. If so, then rules that establish games should be interpreted to create a context that allows for the realisation of those obstacles and the related excellences (the 'lusory means' that are available to overcome them).

2. Rules should be interpreted to achieve an appropriate competitive balance.

This does not mean setting out to ensure a basic quality of skill among participants in a contest. It is the more formal idea of competitive balance to games. Like the balance between bat and ball in a cricket match. For if we think of sports as games involving the mastery of certain physical excellences that aim to be rewarding and challenging to be judged worthwhile and an ongoing source of enjoyment in the pursuit of a lusory goal, the competition should be designed to provide fair and meaningful opportunities for participants to exercise such skills. The opportunities should be fair in the sense that the rules should not unduly prejudice the outcome from the beginning in favour of some of the participants. And the opportunities should be meaningful in the sense that they genuinely allow for the exercise of sport-related physical skills by participants. To the extent that either of these conditions fail, competition will be undermined along with the opportunity that it affords for the development, mastery and display of a game's distinctive physical excellences.

These ideas follow from the previous two principles and should be interpreted in the light of them. It does not follow that if the rules of a game fail to specify exhaustively the limits, say of the use of deception, any deception is permissible where the rules are silent. The principle of "what is not specifically ruled out must be permitted" is called the 'libertarian principle'. But it must be weighed against other principles. Giving it absolute weight would permit all manner of unfairness to enter into games. The principles of sportsmanship and fair play take precedence over the libertarian principle. This principle is strongest when it supports, or at least does not detract from the first principle of game adjudication.

3. Rules should be interpreted to preserve the good conduct of games.

The idea here is that rules should be interpreted to ensure that games are conducted in a timely manner and with an appropriate degree of civility on the part of participants. This principle is also related to the previous ones.

4. Participants should not be permitted to benefit directly from breaking rules.

All these principles need to be interpreted and afforded weight. This is where the notion of integrity comes in.

The Integrity of Games

It has been suggested by Russell that rules should be interpreted by some notion of integrity[222] akin to Dworkin's theory of law. It has been said of this theory that because the competing conceptions of morality that are evident in the law do not admit any satisfactory unitary and comprehensive interpretation that while moral principles are part of the law, it is in effect a patchwork quilt of rules and principles that cannot be sewn into a seamless whole by employing the 'pious fiction' of integrity[223]. This may be so of a holistic legal system where so many factors operate. However, the basic idea of a game and its particular goals tend to be relatively clear and well-defined. In these circumstances, the virtue of integrity is not simply a fiction but a useful, and indeed required guide to resolving issues.

The arguments above are instructive for not only adjudicators but those that write the rules of games. It should demonstrate that the rules they write or oversee cannot pretend to be definitive and that a formalistic approach is at odds with their own aims as legislators or administrators.

The exercise of discretion must be recognised as a practical and moral necessity in these circumstances. Sacrificing the integrity of games to blind rule worship cannot be justified. Acknowledging the need for the exercise of discretion and not insisting on a formalistic interpretation of the rules will encourage intelligent discussion of the rules and of the aims of sport, while formalism does not.

Rules should be adopted to indicate that adjudicators have authority to rule on matters affecting the integrity of the game, not just on matters that "are not specifically covered by the rules." The latter position while it affords adjudicators some discretion, also sends the wrong message, since it permits a formalistic reading that implies that the rules are authoritative in every case apparently covered by them.

The 'Hand of God'

Claudio M. Tamburrini's the *Hand of God?* takes its title from Diego Maradona's characterisation of his controversial handball goal in Argentina's World Cup match against England in Mexico, 1986. Indeed, a revealing photograph of that incident appears on the cover of the book, appropriately reflecting the provocative nature of its contents.[224] Tamburrini takes unconventional positions in controversial debates about cheating, doping, nationalism, elitism, and sex-integration in sport. He offers unexpected perspectives on familiar issues, viewing professionalism and commercialisation with optimism while criticising international and sex-segregated competition.

His general approach is to review then rebut conventional criticisms of commercialisation and professionalisation in modern sports. He argues boldly and provocatively, defending widely disdained practices such as cheating and doping, while criticising such popular phenomena as national teams and sex-segregated competition.

He begins with a reference to the 'hand of god' incident. He makes the unexpected claim that "Maradona as a matter of fact acted *rightly* when he scored his goal, from a universalistic moral point of view." He explains that much of the 'unsporting behaviour' excoriated by critics is in fact morally acceptable from the point of view of a 'practitioners' ethos'. The author then launches his examination of cheating in sport, taking into account such issues as intention, deception, and obedience to the formal rules. Ultimately Tamburrini provides three conditions under which cheating or bad sportsmanship deserves condemnation:[225]

a. it introduces unfairness in competition;
b. it spoils the game;
c. it exposes sport practitioners to (an increased risk of) unnecessary physical injury.

He examines such controversial incidents as Maradona's handball goal according to these conditions, noting first of all that we should adopt the ethos of game practitioners rather than the ethos of game authorities.[226] Since handling the ball is relatively frequent in football, Tamburrini reasons, it is acceptable within the practitioners' ethos, and therefore does not constitute an unfair advantage. Nor does Maradona's goal violate the last two conditions for blameworthy cheating because it enriches rather than spoils the game and does not increase the risk of injury. On the contrary, he argues that Maradona's goal actually heightened the competitive spirit of subsequent England-Argentina matches, while diffusing some of the tension caused by the Falklands/Malvinas war. 'Thus', Tamburrini concludes, "his goal had positive effects not only for the game of football, but even for the international community at large."[227]

He further argues against the current doping ban on the grounds that the substances included, the methods of implementation, and the restraints put on sports development are all arbitrary. He emphasises the loss of transparency caused by the current ban, explaining that "in elite sports, we may have arrived at a situation in which we often celebrate not the most excellent but the most cunning athlete, the one who uses dope and gets away with it."[228] Next come critiques of many conventional arguments for the ban, including the claims that doping is harmful to athletes, to society, and to the fairness of competition itself. He addresses the frequent complaint that doping is artificial and goes over concerns about coercion, paternalism, and informed consent and challenges the arguments of Robert Simon's classic article, "Good Competition and Drug-Enhanced Performance," as well as Angela Schneider's and Robert Butcher's paper, "Why Olympic Athletes Should Avoid the Use and Seek the Elimination of Performance-Enhancing Substances and Practices From the Olympic Games."

He says "A professional athlete aims to become excellent in her discipline and to achieve the external goals—mainly prestige and money—that usually follow such victory. Given the hard competition that characterises professional sports, doping is not only rational, but even necessary, for securing these goals."[229]

The chapter entitled 'The Return of the Amazons' is an extended discussion of sex-segregation in sports competition that culminates in a 'gender equity' proposal for the gradual sex-integration of sports. After introducing the topic and outlining his argument, Tamburrini explains the 'conservative positions' he will be arguing against. "Right-wing sport sex conservatives" he says, claim that sex integration in sport would harm women by defeminising them and risking their health. He rightly dismisses this position as implicitly sexist, since it understands women in terms of what is attractive to men and denies them the opportunity to make their own decisions about personal risk, including the risks surrounding performance-enhancing technologies.

Jane English's article, 'Sex Equality in Sports', is then discussed as an example of 'feminist sport sex conservatism'. Although he generally agrees with

English's goals of equal access to the benefits of sport, including self-respect, he disagrees that such goals can be achieved as long as sports remain sex-segregated. He assumes that this difference of opinion stems from a disagreement about whether the current performance gaps between male and female athletes can be overcome. He suspects that they can be overcome (at least in many sports) but only if women have the opportunity and incentive to compete with men. He levels a similar criticism at Betsy Postow's article, "Women and Masculine Sports," which argues that we conserve sex-segregation in existing sports, while introducing new sports more naturally adapted to women. He points out that many existing sports, such as ice-skating and rhythmic gymnastics, already seem to favour female athletes. Furthermore, he says, Postow's stance discriminates against more powerful women by barring them from measuring their skills against male athletes. Finally he notes that if separate classes are made for sex characteristics, then separate classes should be made for all relevant physical characteristics, an eventuality which would make sports too "fragmented and specialised."[230]

Tamburrini's over-arching argument against the conservatives with which I agree is that sex-segregation in sport perpetuates gender stereotypes and forfeits an excellent opportunity for women to break "the male monopoly of physical strength."

> Rather than exhorting women not to participate in sports that favour male physique, the appropriate course of action is to increase female participation in strength-based sport disciplines. Such a development might allow us finally to come to terms with reactionary aesthetic and normative stereotypes that relegate women to a passive role, as objects to be admired on grounds of their grace and beauty, rather than as subjects to be respected for their active role in life.[231]

But the author stops short, again in my view correctly, of calling for immediate sex-integration in all sports. Noting that sex-segregation in sport aims to promote wider social justice, he adopts a more moderate "gender equity position," which aims for the eventual integration of all sports disciplines but allows for sex-segregated competition in disciplines where women perform at a lower level than men. By working gradually towards integration in all sports while emphasising positive role models in co-ed child and juvenile sports, he hopes that women will eventually reach the level where they can compete with men in all sporting disciplines. Offering a new twist in his comparisons between professional sports and conventional professions, he defends limited sex segregation in sports by noting that in other professions, women already perform as well as men. That is something with which I wholeheartedly endorse.

In conclusion, he reviews his earlier arguments, offering them as support for the wider conclusion that there is no need for 'moral panic' about the professionalisation and commercialisation of modern sports. But I disagree with the bold conclusion that he makes that based on his observations about sports and nationalism, we should "rid the world of sports of international competitions."[232] That would be an absurd idea. Nationalism is at the heart of sport and its competitive motif and ethos. In that respect sport reflects man's political and societal instincts and belonginess. It is a form of communal self and the herd mentality of man. He is hardly pessimistic in his vision of the future of sports, however. Having 'exonerated' sport from the most frequent charges of corruption through cheating and doping, he envisions a world where technology will mitigate the performance differences generated by the genetic lottery and lead to a re-emphasis on athletes' moral virtues. Finally, he says that we might choose to view the future of the techno-athlete as "a universe in which humans, and not the whims of nature, have the power at last to decide over their own destinies."[233] I share that perceptive vision and optimism.

Chapter 5
Sport and Reality

If God didn't exist, it would be necessary to invent him.
Voltaire

The important thing is not to stop questioning. Curiosity has its own reason for existing. One cannot help but be in awe when he contemplates the mysteries of eternity, of life, of the marvellous structure of reality. It is enough if one tries merely to comprehend a little of this mystery every day. Never lose a holy curiosity.
Albert Einstein

A common man marvels at uncommon things; a wise man at the common place.
Confucius

It is often stated that of all the theories proposed this century, the silliest is quantum theory. In fact, the only thing that quantum theory has going for it is that it is unquestionably correct.
Michio Kaku

The propositions of natural science have nothing to do with philosophy.
Wittgenstein

He that knows not what the world is, knows not where he is himself. He that knows not for what he was made, knows not what he is nor what the world is.
Marcus Aurelius

We are all captives of the picture in our head—our belief that the world we have experienced is the world that really exists.
Walter Lippmann

Nothing is one thing just by itself.
Socrates

What everyone believes is true.
Aristotle

The use of travelling is to regulate imagination by reality, and instead of thinking how things may be, to see them as they are.
Samuel Johnson

There is no reality there is only perception.
Gustave Flaubert

Nothing is real.
John Lennon

Nothing is more real than nothing.
Samuel Beckett

Look at that old photograph, is it really you?
Krist Kristofferson, *This Old Road*

The real is rational and the rational is real.
Hegel

Aristotle is Plato diluted by common sense.
Bertrand Russell

There is no such thing as philosophy-free science.
Daniel Dennett, *Darwin's Dangerous Ideas*

One thing I have learnt in a long life: that all science measured against reality is primitive and childlike.
Einstein

At a certain subliminal level, sport is an escape from reality but metaphysically it is at its essence. It is no more an escape from reality than art, music, religion or even science itself. They are simply interacting domains of the same reality. We play and watch sport not to escape reality but for reality not to escape us.

The problem with 'reality' is from whose perspective? I have referred earlier to Gustave Flaubert's aphorism that there is no reality only perception. It depends on whose perception is being considered. Maybe that is the problem with the way we look at things in life generally. Atticus another literary favourite of mine, in that wonderful novel 'To Kill a Mockingbird' by Harper Lee, said you never really understand a person until you get inside their skin and walk around in it. The same could be said of reality. Carlo Rovelli in his book Anaximander and the Nature of Science, first published in French in 2009, referred to "the pernicious modern separation between science and the humanities."

As a theoretical physicist, albeit often described as the poet of physics, he took the view that Anaximander's contribution as a scientist had been underestimated because of the difficulty specialists in history or philosophy have "in evaluating the importance of insight whose nature and legacy are intimately scientific."

As a result, they rarely try to grasp the interaction between sport, science and other disciplines.

Whether that is true of the contribution Anaximander made to science, I know not but I agree entirely with the implicit proposition that the concept of reality is a holistic one which can only be evaluated by adopting a multidisciplinary approach not limited to the scientific, philosophical, psychological or biological. That is what I have attempted to do in this book. Rovelli acknowledges in his book that the concept of reality is 'beyond appearances'. Poetically he expresses it in the following way:

"beyond the plain blue sky there is an immense space full of galaxies, black holes, and neutron stars. But the uncertainty of our knowledge and the variability of the scientific pictures of the world tell us that we are not getting to an ultimate picture of reality."

In other words, science does not provide the complete answer to the meaning of reality.

Stendall said "God's only excuse is that he does not exist." Reality is the study of our existence and who we are and the nature of things and the universe. What is real? John Lennon said "Nothing is real." I disagree, he was real, as was Elvis Presley, Bob Dylan, Buddy Holly, Roy Orbison, Tom Petty, Pavarotti, Joan Sutherland, Maria Callas, Mozart, Tchaikovsky, Beethoven, Wagner, Debussy, Chopin. They live inside our heads and are as real as real can be. So, too, do George Best, Pelé, Bobby Moore, Bobby Charlton, Jack Charlton, Roger Federer, Mark Spitz, Puskas, Maradona, Di Stefano, Rocky Marciano, Joe Louis, Tommy Farr, Sugar Ray Robinson, Muhammad Ali, George Foreman, Mike Tyson, Anthea Gibson, Fred Perry, Arthur Ashe, Lew Hoad, Rod Laver, John

McEnroe, Bjorn Borg, Boris Becker, Serena Williams, Venus Williams, Babe Ruth, Joe DiMaggio, O.J. Simpson, LeBron James, Stirling Moss, Emerson Fittipaldi, James Hunt, Michael Schumacher, Cliff Morgan, Stanley Matthews, Billie-Jean King, Martina Navratilova, Nadia Comaneci, Olga Korbut, Fangio, Emil Zatopek, Roger Bannister, Sebastian Coe, Steve Ovett, Daley Thompson, Lester Piggott, Frankie Dettori, Johan Cruyff, Tom Daley, Steve Redgrave, Barry John, Gerald Davies, David Duckham, Gareth Edwards, Mike Gibson, Tom Kiernan, Jonah Lomu, Brian O'Driscoll, WG Grace, Don Bradman, Walter Hammond, Harold Larwood, Frank Tyson, Freddie Trueman, Gary Sobers, Wes Hall, Charlie Griffith, Dennis Lillee, Jeff Thompson, Rod Marsh, Ian and Greg Chappell, Mark and Steve Waugh, Shane Warne, Mike Gatting, Len Hutton, Dennis Compton, Ken Barrington, Colin Cowdrey, Peter May, Ian Botham, Michael Atherton, Geoff Boycott, David Gower, Kevin Pietersen, Joe Root, Ben Stokes, Steve Davis, Alex Higgins, Cristiano Ronaldo, Lionel Messi, Harry Kane, Gareth Bale, Ally McCoist, Lee Trevino, Seve Ballesteros, Tony Jacklin, Jack Nicklaus, Gary Player, Greg Norman, Ben Hogan, Sam Snead, Arnold Palmer, Tiger Woods.

There is a lot in a name and sport is no exception. These names conjure upon remembrances of things past of which Proust would have been proud, They are embedded memories and although subjective are a form of reality and with each new name a new is reality is created.

Epictetus (circa 60 A.D) observed that appearances to the mind are of four kinds. Things either are what they appear to be or they neither are, nor appear to be; or they are, and do not appear to be; or they are not, and yet appear to be. He concluded that "Rightly to aim in all these cases is the wise man's task." Have wise men made much progress since that time? Notwithstanding the advance of science we are still struggling to get a grip on reality. Nobel prizewinning physicist Richard Feynman once described the quest to understand reality as a bit like watching a game of chess without knowing the rules. By observing the game, we slowly get to grips with what the pieces are and how they are allowed to move and interact. With the developments of the past century, we've identified many of the pieces and new moves—but we're still far from completing the rulebook.

Dr Johnson said famously that we know something exists because you can kick it.

Quantum physics and other attempts by physicists zoologists and biologists have only shifted the true nature of reality yet further away from our experience of it, is quite a feat given the extent to which quantum theory and relativity have already done so. You can't kick it.

As Einstein commented "Quantum mechanics is very impressive. But an inner voice tells me it is not the real thing."

Roger Penrose, the physicist, has recently expressed that his view is that there are three different kinds of reality; the physical, the mental and the Platonic-mathematical, with something (as yet) profoundly mysterious in the relations between the three. He says of quantum theory "…the 'reality' that quantum theory seems to be telling us to believe is so far removed from what we are used to that many quantum theorists would tell us to abandon the very notion of reality when considering phenomenon at the scale of particles, atoms or even molecules."[234]

His view is similar to Karl Popper in his Objective Knowledge (1972). He distinguished the world 1 containing physical objects and world 2 mental objects and world 3, objective contents of thought, especially of scientific and poetic thought and artworks or art.

Richard Henry Stoddard (1825–1903) put it more poetically in The Castle in the Air: "We have two lives about us, two worlds in which we dwell, within us and without us, Alternative Heaven and Hell, without, the sombre Real, Within our Hearts, the beautiful ideal." More recently, as we will see later, Daniel Kahneman has spoken of the two selves—System 1 and system 2—thinking fast and slow. Sport reflects that reality of the two selves as do games such as chess and bridge. John Fowles in The Magus saw the dangers of science in limiting the concept of reality. "There had always been a discord in me between mystery and meaning. I had pursued the latter, worshipped the latter but then I saw an attempt to scientise reality, to name it to vivisect it out of existence was like trying to remove all the air from the atmosphere. In the creating of the vacuum, it was the experimenter that died."

Science has not yet provided anything like a complete answer to the question what is reality. It may be the reason for this is that the question is not simply a scientific question but a philosophical one. As Penrose points out "where does quantum non-reality leave off and the physical reality we actually seem to experience begin to take over?" Understandably Penrose, a physicist, says "Present day quantum theory has no satisfactory answer to this question." "He adds present day (2020) quantum theory is not quite right and that as the objects under consideration get more massive, then the principle of Einstein's general relativity begins to clash with those of quantum mechanics and the notion of reality that is more in accordance with our experiences will begin to emerge."

Quantum physics may assist with cosmological issues as to the nature of the universe and such issues as to whether there is a multiverse.

It may be that the problem is as encapsulated by Raphael Bosso of the University of California, Berkeley, who says "You never know anything for sure when you are working on the edge of knowledge."

Ontological issues relate to the study of man as we know him. It is in that context that I look at what sport teaches us about man and his place in mankind. As Pope said the proper study of mankind is man.

In The Trouble with Physics (2006) Leo Smolin, a renowned theoretical physicist, argues that physics (the basis of all other science) has lost its way. We know nothing more than we knew in the 1970s. Quantum theory has led to a number of scientific conjectures as to the nature of the universe and physical reality based not on scientific observation but on intuition or depending on your views counter-intuition. But, arguably, quantum mechanics makes things worse. It is not possible, for example, to attribute a definite motion and position to an atom at the same time. Atoms and subatomic particles inhabit a shadowy world of half-existence.

In Unweaving the Rainbow (1998), Richard Dawkins, the zoologist says "…the parts of our brains responsible for doing intuitive statistics are still back in the stone age." He says the same may be true of intuition generally and refers to the views of Lewis Wolpert, the distinguished embryologist who in The Unnatural Natures of Science (1992) argued that science is difficult because it is more or less systematically counter-intuitive.

This view is contrary to the view of T.H. Huxley in Darwin's Bulldog who saw science as "nothing but trained common sense, different from the latter only as a veteran may differ from a raw recruit." For Huxley, the methods of science "differ from those of common sense only as far as the guardsman's cut and thrust differ from the manner in which a savage wields his club."

Wolpert maintains that science is deeply paradoxical and surprising, an affront to common sense rather than an extension of it. He describes Newton's law that objects stay in motion unless positively stopped as counter-intuitive, as is Galileo's discovery that when there is no air resistance light objects fall at the same rate as heavy objects, so, he says is the fact that solid matter, even a hard diamond consists almost entirely of empty space.

Dawkins acknowledges the counter-intuitive nature of quantum theory. He concludes on this issue as follows:

> "More profoundly difficult are the conclusions of quantum theory, overwhelmingly supported by experimental evidence to a stupefyingly convincing number of decimal places, yet so alien to the evolved human mind that even professional physicists don't understand them in their intuitive thoughts."

There are many studies of sporting science to which I refer later in the book. I do not look solely to science to identify the contribution sport makes to understanding the fundamental nature of man or any insight into the intrinsic essence of sports. That essence is in my view principally to be found by putting sport in the domain of philosophy, the arts and aesthetics.

Sport is a product of human skill and ingenuity which in some way is emotionally charged. It shares that characteristic with a work of art such as poetry, or musical drama. The effects of this on the spectator may be consoling, pulsating, entertaining, disturbing.

What is aesthetically relevant is the connection between the sport and the feelings of those who are exposed to its effects.

Sport has a unique aesthetic feeling, a thrill which indicates that one is receiving a genuine aesthetic stimulation, due to a kind of built-in device in the emotional equipment of some people who are sufficiently sensitive to pick up these signals in the proper order and intensity.

Sport has been treated by many philosophers as absurd, superficial and trivial as they have with some forms of art such as music and poetry. George Boas in Philosophy and Poetry (1945) said "the ideas in poetry are usually stale and often false and no one older than sixteen would find it worth his while to read poetry merely for what it says."

But this compels us to ask whether it is not so much the truth that we are valuing as something else, namely the impact of the total experience which may contain affirmation of the truth but is much more, and which is enjoyed only in so far as we are actually entering into the life of the poem as uttered or the drama as acted. The same can be said of our enjoyment of sport.

The truth of a statement is its quite impersonal relationship to the independent real; what we enjoy in the poem or a drama and similarly sport is not the recording of an abstract relationship, but a total participation, felt and understood through a single experience, in a fresh insight. A truth, as such, once recorded, can be docketed and assumed; it does not need to be rediscovered unless it has been lost; once is enough. The insight of a poem or drama and likewise sport, on the other hand, are in a sense discovered in a fresh enjoyment each time; once is not enough.

Music does set a special problem for any theory which claims that art is 'true' or 'revealing' 'reality' since the world of pure music seems to be so different from our world outside it. On the other hand, the 'reality' revealed by the non-representative arts is a fresh construction of reality, not a repetition of the ordinary patterns of the world of everyday practical experience; and it is 'reality' which is apprehended inseparably from the medium in which it is embodied.

That is precisely the same with sport. It is a fresh construction of reality not dissimilar from poetry and drama, abstract painting and sculpture.

The naïve or commonsense model of truth is a statement or picture conforming to an existing reality and 'reality' is supposed to be of the commonsense, or it might be of the scientific kind. This 'correspondence' view of truth and reality is wholly inadequate, but it is a commonsense view and it is on that assumption that I am proceeding. On this approach there is an assumption that somehow or other 'we can get at' 'behind' our knowledge of it, and so check our knowledge, whereas in fact in so far as we can 'get at' reality at all it must be through our knowing of one kind or another. This, as I said in the Prologue, in part explains the title of this book with the use of the word 'beyond'.

A commonsense judgement, for example 'The book is on the table' is checked by another judgement of perception, not by some bare fact or reality behind it.

We have to assume that there is a world which in part at least, is given to us, and has an independent structure. But this given world, we do not know directly, or as in a mirror. We come to know it, in some sense and some degree, we come to terms with it, through our own active symbolic constructions, the constructions of commonsense, science, philosophy, as well as through a number of other ways of knowing.

Sense perception is one kind of construction.

The senses themselves are highly selective; in a manner they manufacture the qualities which enable us to adjust practically to the external world, and perception is, in ways familiar to philosophers and psychologists, a highly active synthetic constructive affair. Science is in its turn, in its attempts to come to terms with certain aspects of the world of nature, a system of intellectual construction; tested against observation and experiment, and enabling prediction and control.

Philosophy takes a wider sweep, trying to understand, the nature and claims of values as well as facts, and the relations of values to facts, as well as surveying the different types of knowledge and experience.

Philosophy attempts some sort of synoptic perspective, making its own constructions and having its own tests of truth.

Each form of knowledge has its own symbols and uses them in its own ways, and through them hopes to cope a little better with the mysterious reality which is infinitely greater than any, of our necessarily limited and abstract constructions.

These are all attempts to know and we call them, in a general and not very accurate way 'knowledge'. The question I want to explore is "Is sport knowledge?" Does it increase insight into the world? Does sport, like other attempts at knowledge, achieve some revelation of reality?

I believe the answer is certainly "yes," but "yes" only in a sense which does justice to the uniqueness of sport and to its differences from other knowledge. In that sense knowledge is ultimately a richer concept than truth.

Through sport we apprehend dimensions of life not apprehended before we are initiated into experiences of the nature of man or the form of the external world (including forms and movements of the human body). With sports, human growth in stature is unlimited; without them it would be greatly impoverished; education in them is far more important than in contemporary times is commonly recognised. Perhaps the lack of recognition of the importance of sport may be in part a confused belief lingering from the eighteenth century that they are not, knowledge, but legitimately pleasant diversions for leisure time. In an age of science and technology, they are invariably relegated to the category of 'extras' in education.

But in claiming that sports are a major form of knowledge, we must not weaken the claim by trying to assimilate them into other models of knowledge. They are not true in the sense of conforming to a 'reality' which we know perfectly well in other ways. Sport is not a picture of life but something which may result in us seeing and feeling life as we never saw it before. It would be absurd in assessing sport as knowledge to measure it against science with its empirical tests of validity.

Sport is a discovery of its own kind and in its own ways and is a basic and incredible form of knowledge. It is a world providing an insight into the statics and dynamics of bodily poise, of rhythm, of the relations of shapes, spaces and time, of the symbolism of movement, consciousness, and beauty. It is a delight in the incarnate life of man. This delight is extended to the inner life of the unconscious and imaginative, which drawing in part from the outer life, outdistances it beyond measure. Sport can open up new perspectives of conception, a new understanding of life, a new vista of the life of the senses and experience and all that can symbolise—the Greek hero against fate, the Shropshire land against the world and his inevitable end—metaphysical ideas, beauty, adoration, joy, pain and suffering, grief and mortality. These and other human themes come into being with the flesh and blood of living human beings. Sports can submit their insights to the test of reality outside of themselves.

Why should we limit reality to what the plain matter of fact man assumes it to be—ready made and complete, what commonsense and science can show?

This assumption underlies the mistaken conception of truth—of a world of solid substantial reality, to which insubstantial fabrications of mind have to conform, to which being must submit. This is far too superficial knowledge. The physical aspects of the 'real world' of an Einstein, Newton, Schrodinger, Niels Bohr, Stephen Hawking are a marvellous construction of intellect but our knowledge of reality should not be limited to those insights wonderful as they are. Mark Johnson, The Body in the Mind wrote "Our reality is shared by the patterns of our bodily movement, the contours of our spatial and temporal orientation, and the forms of our interactions with objects."[235]

Sport and the sportsman can create a new reality and a new vision of reality. The real world for me consists of the whole ongoing endless discovery which is its cultural history. It is ongoing and endless and is why it is not absurd to speak of a 'new reality'.

Plato, William James and Bergson spoke of the unfinished universe. But whatever is true of the physicists' world, it is beyond doubt that the reality both made and revealed by sport is an expanding reality. Its universe is always expanding, on the horizon, and beyond.

Describing his life shortly before his death, Sir Isaac Newton put his contributions this way "I don't know what I was seen to the world, but, as to myself, I seem to have been only like a boy playing on the sea shore, and diverting myself and now and then finding another pebble or a prettier shell than ordinary, while the great ocean of truth lay undiscovered before me."

Memory is the strongest and highest form of reality. William Faulkner captured it beautifully when he said "The past is never dead. It's not even past." Sporting memories, in particular, seem to last forever an eternity. Memory is charged with symbolism. It is an essential component of our perceptual experience of reality. The perceptual present is outside time altogether which may coincide with the actual present, with a memory or with an expectation of a future event. Flaubert said "There is no reality only perception." That is another way of saying our memories are reality. The represent our deepest existential experience of the world we live in. Although the memory may not reflect the reality or the truth, it becomes the reality of man's existential self and is charged with symbolism. The entire metaphorical landscape comes alive, filled with relationships and relationships within relationships. In that sense memories become the embedded gestalt of my perceptual experience. Like dreams they are vivid and imaginative. Day dreaming is part of philosophy. Sometimes called passivity about which I have written at length. It takes you beyond the metaphysical horizon. There is no limit to where, when and in what form a perceiver can exist. A symbolic perceiver may be floating beyond, above and behind intensely involved in reliving an incident from childhood or in interacting with an abstract representation of a symbol planted in our brains—a memory of a person, an event, or an abstract idea, concept or phenomenon. I agree with Aeschylus that "memory is the mother of all wisdom."

Chapter 6
The Aesthetics of Sport

The excellence of every art is its intensity, capable of making all disagreeables evaporate from their being in close relationships with beauty and truth.
Keats

Always the beautiful answer who asks a more beautiful question.
E.E. Cummings

Give me beauty in the inward soul; may the outward and the inward man be at one.
Socrates

The aesthetic is a kind of fictive or heuristic realm where we can suspend the force of our usual powers, imaginatively transferring qualities from one drive to another in a kind of free-wheeling experiment of the mind.
Terry Eagleton

Beauty in things exists in the mind which contemplates them.
David Hume

Beauty is truth, truth beauty that is all we know on earth, and all ye know need to know.
Keats, *Ode on a Grecian Urn*

A thing of beauty is a joy forever. Its loveliness increases, it will never pass into nothingness.
Keats, *Endymion*

All the beauty of the world is skin deep.
Ralph Verney

Beauty is that to which the human mind responds at its deepest and most profound.
Subrahmanyan Chandrasekhar

Remember the most beautiful things in the world are the most useless; peacocks and lilies for instance.
Ruskin

Beauty seen is never lost.
John Whittier (1807–1892)

Art never expresses anything but itself.
Oscar Wilde

Art is not the application of a canon of beauty but what the instinct and the brain can conceive beyond any canon.
Picasso

Art must be parochial in the beginning to become cosmopolitan in the end.
George Moore

Angling is somewhat like poetry—men are to be born so.
Izaak Walton (1592–1683)

Cricket is a classic ideal art form which reconciles mind and body at least on the playing field as the finest of other art does in the theatre.
Paul Buhile

What happens when a man or for that matter an animal, has no need to work for a living? On the genetic approach the simplest case is that of a domesticated cat—a paradigm of affluent living. Its needs provided for almost before they are expressed. It is protected against danger and inclement weather. Its food is there before it is hungry or thirsty. What then does it do? How does it pass its time?

We might expect that having taken its food in a perfunctory way it would curl upon its cushion and sleep. But no, it does not just sleep. It prowls the garden and the woods killing young birds and mice. It enjoys life in its own way. The fact that life can be enjoyed and is most enjoyed by many living beings in the state of affluence draws attention to the dramatic change that occurs in the working of the organic machinery at a certain stage of the evolutionary process. This is the reversal of the means-end relation in behaviour in the state of nature the cat must kill to live. This happens with men. When men have no need to work for a living there are only two things left to them. They can 'play' and they can cultivate the arts. These are their two ways of enjoying life. It is true that many men work because they enjoy it, but in this 'work' has changed its meaning. It has become a form of 'play'. Play is characteristically enjoyed in for its own sake—without concern for utility or any further end. 'Work' is characteristically directed to the production of some utility in the simplest and easiest way. Hence the importance of ergonomics and work study—the objective of which is to reduce and save time. But as Picasso said about his work "worst of all that he never finishes. There's never a moment when you can say "I've worked well and tomorrow is Sunday." As soon as you stop, it's because you started again. You can put a picture aside and say you won't touch it again. But you can never write THE END."[236]

In play the activity is often directed to attaining a pointless objective in a difficult way, as when a golfer, using curious instruments guides a small ball into a not much larger hole from remote distances and in the face of obstructions deliberately designed to make the operation as difficult as may be. This involves the reversal of the means-end relation. The 'end'—getting the ball into the hole—is set up as a means to the new end, the real end, the enjoyment of difficult activity for its own sake. Art has this in common with play.

The enjoyment of the arts take two forms. There is the enjoyment of the creative artist who enjoys for its own sake the activities of creation. There is the enjoyment of those who do not themselves create the objects the perception and contemplation of which they enjoy for their own sake. In both instances it enriches our lives. Oscar Wilde put it dramatically when he pronounced that "It is through Art and through Art only that we realise our perfection: through Art and Art only that we can shield ourselves from the sordid perils of actual existence."

The aesthetic element in play and sport and the association between art and play are indicated by the fact that many forms of play are described as arts—boxing (the noble art), archery, fencing, fishing, gymnastics, skating, cricket for example.

There is another way of describing the transition from barbarism to civilisation. When men no longer seek to work to satisfy their needs, when there is leisure to be enjoyed, the 'organic machine' begins to work in a different way. Configurations which appear in the field of perception cease to be treated as signals, cease to be merely releasers of instinctive or acquired disposition, they become objects of interest in themselves like the scent of a rose or the song of a bird. The instruments of response now work to produce new precepts which can be enjoyed for their own sake. They work in the practice of an art. They also work for the enjoyment of the activity itself.

There is a long history of debate as to whether sports in general are truly of their very nature aesthetic. A classic position is that held by David Best.[237] Best argues that although any phenomenon like sport could be viewed from an aesthetic point of view, that does not make those things intrinsically possessed of aesthetic properties.

The word 'aesthetics' (from the Greek *aisthanesthac*, things perceptible) was introduced into philosophical terminology about the middle of the eighteenth century by Alexander Gottlieb Baumgarten (1714–62), a pupil of the Leibnitzian codifier Christian Wolff (1679–1754) and it is because of the odd use to which he put the word that the branch of philosophy which is concerned with investigation into the nature and principles of beauty now bears this incongruous name. Baumgarten put forward the idea that the current system of philosophical discipline was incomplete and needed to be rounded off by the addition of a science of the 'inferior cognition' which is mediated by the senses on the analogy of logic which, as the science of the "clear and distinct cognition" mediated by intellect, stood as a general introduction at the beginning of the four Departments of Ontology, Cosmology, Ethics and Psychology into which Wolffian metaphysics was divided.

He wrote a book *Aesthetica* (1750–58). Etymologically, he ought to have given the name 'aesthetics' to the study of perception. Instead he gave it to the theory of beauty, even defining 'aesthetics' in the opening paragraph as "the theory of the liberal arts, the science of sensory cognition."

Kant criticised Baumgarten for restricting the word to the field of taste and proposed to apply it to its true etymological meaning to sense perception generally.

It has now ceased to be merely a technical term in philosophy and it is now one of those vague but useful semantic signs whose meaning everyone seems to know though no one is able to define. Even in more or less specialised writing the word is now used as if its meaning were more precise than that of the older terms of criticism and appreciation. John Dewey wrote in Art as Experience 1934, p.48

"To be truly artistic, a work must also be aesthetic."

In A Glossary of Art Terms (1950) by J. O'Dwyer and Raymond de Mage it is stated "that art which is principally concerned with the production of works of aesthetic significance is distinct from useful or applied art which is utilitarian in intention."

Plato and Aristotle discussed the nature of beauty and art, and the place of artists in the community. They assumed as having no need of argument that art is to be judged by its usefulness in the community.

Collingwood in The Principles of Art said "that there is no connexion at all between beauty and art."

Sport like art is symbolic. There is more to it than meets the eye.

It is a symbol—a physical reality which is endowed with a meaning or a significance beyond itself.

It was Kant's Critique of Judgement which first distinguished aesthetic qualities from the categories of the useful, the pleasant and the good.

From the plausible position that the aesthetic quality of a work of art is not dependent on its practical usefulness or even its congruence with conventional morality some came to the extreme position somewhat paradoxically that a work of art must not serve any purpose.

Il n'y de vraiment beau said Gautier in the Preface to Mademoiselle de Maupin "que ce qui ne pent servier à rien."

The ancient Greeks, from whom many of our aesthetic and critical principles descend, were interested primarily in the educational and social uses of works of art. European tradition has emphasised the religious, cultural, aesthetic, educational, humanistic and entertainment functions. The most characteristic feature of the modern school of painting is its rejection of the representational function of art. Whether a picture represents something else or not the degree of accuracy with which its represents are matters of indifference to aesthetic judgement. No Greek believed that music or architecture must represent something not themselves.

Charles Pierce defined aesthetics as the basic narrative science which institutes a comparison among ultimate values in the "science of ideas or that which is objectively admirable without any ulterior reason."

Few have followed him in giving this very wide meaning to the word. It is not usually considered to be part of the function of aesthetics to consider why we value the appreciation of beauty or what rank we give to it among other valuable human activities. There is also the view that beauty is a superficial concept as reflected in the adage "All the beauty of the world is but skin deep."[238]

Among modern theories that which perhaps comes closest to explain why we value art is the theory that art is the language of symbolism. The modern theory has little affinity with the medieval theory of symbolism since it is maintained without the religious postulate that what is symbolised is the nature of the Deity.

The appreciation of beauty is a cognitive act and it is instinctive, sometimes counterintuitively, intuitive or non-conceptual awareness of that which is symbolised. For example, it has become the beautiful game. The term 'beauty' here signifies a whole series of ascribed qualities of the sport, yet because the adjective 'beautiful' is used in a rather loose metaphorical sense, it is difficult to pinpoint what the term might mean as regards the aesthetic components of the sport.

Iris Murdoch in The Sovereignty of Good published in 1970 looked philosophically at the concept of beauty and characterised it as "to step outside the self" beyond its particular conceptions of beauty. I deal at length in chapter 45 with the relationship between the self, beauty and virtue.

Athletic Beauty[239]

Beauty, as David Hume said, "is in things that exist in the mind which contemplates them." It is a subjective concept and more colloquially is in the eye of the beholder.

Beauty in movement in time and space is at the heart of sporting endeavour and its aesthetic appeal to so many.

No one has yet invented a 'beauty meter' that can measure the aesthetic things without referring to human criteria. No doubt there are all sorts of biological and psychological factors at work in framing our impressions of what is beautiful. The structure and operation of the brain may also dictate what is pleasing to the eye or ear. Music may reflect cerebral rhythms. The same may be true of sport but beauty is more than mere biology at work. Our aesthetic appreciation stems from contact with something firmer and more pervasive.

In an article in The Times (August 19th 2020) entitled "O'Sullivan proves the power of beauty" Matthew Syed distinguished between the beauty and quality of sport in the following passage:

> "Beauty is difficult to define in sport but it is perhaps fair to say that it isn't quite the same thing as quality."

He cited the following examples of this distinction:

> "Novak Djokovic being an admirable a tennis player as Federer, but nothing like as beautiful."
>
> Steve Smith, the cricket batsman arguably a superior batsman to Brian Lara—more dogged, more difficult to get out a higher test average but his style in the eyes of many rather ugly.
>
> Sebastian Coe more beautiful than Emil Zatopek but not the finest distance runner of all time which probably goes to the latter "who was nothing like as fluent."

In Syed's opinion beauty in sport has a number of ingredients including economy of movement.

> "Is there any more beautiful in sport than Ronnie O'Sullivan gliding around the snooker table potting balls from all angles, his cue moving like a piston in a Rolls Royce engine."

Beauty also he says involves something more difficult to define—"the sense that they have all the time in the world even when operating under severe constraints."

He gave an example of this phenomenon the Wimbledon final in 2008 between Federer and Rafael Nadal. He reminds us of Federer at one match point down, the Spaniard serving wide receiving a short reply and then bullying his way to the net with a heavy forehand. He says that even now when you watch on replay, there is a moment when the world seems to stop. Federer sizing up the situation, his body coiled before playing a backhand into a corridor roughly two inches wide, Nadal unable to make contact at the net.

O'Sullivan too creates this illusion. Syed thinks that his unusually agile brain is able to think multiple moves ahead enabling him to avoid the more staccato rhythm of his peers in his maximum 147 in 5 mins 20 seconds remains one of the most astonishing feats in sport.

He describes O'Sullivan as the most aesthetic player of all, a man who has six world titles and conquered his demons over a career of almost three decades. He concludes:

> "As with Federer, when you watch him play, you are transported into a higher state of consciousness, not dissimilar to looking at the Renaissance works at the National Gallery."

I agree with Syed that "O'Sullivan shows us that beauty—for all its inherent mystery which is central to sport, a core part of which we watch, adding indefinable frisson to our lives."

In Praise of Athletic Beauty by Gumbrecht is a scholarly treatise that extols the beauty of sport. The author asserts that sport has been a transfiguring power in his life. Despite a very rich career in academia, he has found little as intense, stressful, or invigorating as watching an athletic event. I would add as long as we appreciate our heroes having feet of clay, our sports heroes inhabit another world beyond ours and they can and should be objects of our admiration. Like Barry John (see later) they often operate in a different dimension of time and space. There is, Gumbrecht notes, "nothing wrong with praising our heroes now and then."

Gumbrecht was the Albert Guérard Professor in Literature, professor of French and Italian and comparative literature at Stanford University, and co-editor of *A New History of German Literature*. He combined his love and joy of sport with a rich history of cross-cultural, multidisciplinary scholarly work.

His book is as seminal as Novak's *The Joy of Sports*. *In Praise of Athletic Beauty* considers the beauty of sport, our admiration for athletes, and the fun of being a spectator.

He acknowledges that sport appreciation is not part of high culture in society. But he argues that this is a mistake caused by a metaphysics that divides our world into the physical and the spiritual. The spiritual aspects of our lives are deemed intellectually acceptable, whereas the physical elements are denigrated. He suggests that intellectuals are interested in sport only as it enters the sacred and subordinates the physical.

Athletic movement is fully encompassed within Kant's three-part definition of 'beautiful', since sport provides "pure, disinterested satisfaction." The connection between sport and beauty depends exclusively on an individual's inner feeling of pleasure or displeasure, and there is a community of shared feelings.

Athletes and spectators get "lost in focussed intensity." Athletics can be defined in terms of three components: a distraction from the everyday world, bodily movement, and the Greek values of agon (competition) or *arête* (striving for excellence).

Gumbrecht focuses on sport in Western civilisation and points out that the changes in the structure and nature of sport have been caused by differences in the civilisation. He characterises sportsmen as 'Demigods' (the Greek Olympics), 'Gladiators' (Roman sport as uneven warfare), 'Knights' (knights as a Middle Ages counterculture), 'Ruffians' (early boxing), 'Sportsmen' (emergence of team sports), 'Olympians' (the early 20th-century Olympic movement), and, finally, 'Customers' (globalisation, television, and the fan as co-actor).

In his attempt to delineate just what fascinates us about sports, he refers to the seven aesthetic and philosophical fascinations of sport. These include 'Bodies' (the sculpting of bodies and development of harmony), 'Suffering' (the freely willed pain of sport), 'Grace' (the beauty of bodily flow), and 'Tools' (the various tools that enhance movement during competition, such as track and field implements). Each sport and, indeed, each sport contest relates independently to the seven fascinations. In each event one fascination tends to dominate the moment, but all sports carry more than one fascination for the spectator.

Spectators, like athletes, become "lost in focussed intensity." There are two types of spectator. There are those who analyse strategies, techniques, and so on and those who commune with the team, engaging in the joy and pain of the athlete and fellow spectators. These two types of spectatorship relate to Nietzsche's Apollonian and Dionysian attitudes. The Apollonian spectator appreciates the beauty of unusual movement and forms and the creation and use of space and time. The Dionysian spectator, in contrast, abandons his individuality and becomes 'enraptured' or even 'intoxicated' with the athletic experience. I leave the reader to decide into which category they fall. I have no doubt that I am a schizophrenic Apollonian and Dionysian—a Jekyll and Hyde and not just in sport!

But athletic beauty is not confined to the sportsman. Athletic beauty is to be found in the world of dance especially in ballet. To watch Rudolph Nureyev or Michael Baryshnikov was to witness the highest level of athletic movement combined with artistic creation and aestheticism. The sportsman too can reach those heights in an equally focussed but more reactive way. Male ballet dancers have immense physical strength, great timing and grace. These characteristics are invariably found in the most outstanding sportsman in various athletic pursuits including but not limited to boxing, football, rugby, ice hockey and gymnastics. Flow and being in The Zone, which I consider later along with the concepts of movement, cognition and perception combines in various proportions to produce athletic and aesthetic experiences of beauty and pleasure. In Chapter 39 Consciousness, Perception and the Senses, I provide numerous examples in practice of this phenomenon.

Beauty as a Guide to Truth[240]

Mathematics, rightly viewed, possesses not only truth, but supreme beauty—a beauty cold and austere, like that of sculpture.
Bertrand Russell (The Study of Mathematics)

I believe my fascination with cricket from an early age was its geometrical character and its mathematical precision. Cricket is beauty in form and action. It is about lines, angles and shapes.

Paul Davies has written that it is widely believed among scientists that beauty is a reliable guide to truth and many advances in theoretical physics have been made by the theorist demanding mathematical elegance of a new theory. Sometimes, he says, where laboratory tests are difficult, those aesthetic criteria are considered even more important than experiment.

Einstein, when discussing an experimental test of his general theory of relativity was once asked what he would do if the experiment didn't agree with the theory. He said "so much the worse for the experiment. The theory is right." Paul Dirac, the theoretical physicist whose aesthetic deliberations led him to construct a mathematically more elegant equation for the electron said "it is more important to have beauty in one's equations than to have them fit the experiment." Another great philosopher, Bertrand Russell said "Mathematics, rightly viewed, possesses not only truth but supreme beauty—a beauty cold and austere, like that of sculpture."

Applying that to beauty in sport, elegance in performance is often the hallmark of outstanding athletic prowess. It's not just about winning but doing it in style and in an aesthetically pleasing way. As in so many practices it is not what you do but the way that you do it. Sport is no exception.

One thing is certain, beauty is in the eye of the beholder. It is a subjective concept. It is often, but not necessarily, a matter of taste and predilection.

Symmetry, flow, harmony and unity can be pleasing to the eye. But so can discordance, dissonance and asymmetry.

Harmony is a close ally of unity and beauty and can be defined as "disparate elements organised convincingly and pleasingly." St Augustine wrote in *De Vera Religonie* (391 AD), "In all the arts, that which pleases is harmony, which… invests the whole [of a work] with unity and beauty, either through the resemblance of symmetrical parts, or through the gradual arrangements of unequal parts."

Sport can represent harmony of body and soul and be poetry in motion. But in the same way that pain and pleasure are correlatives, so are beauty and ugliness.

Dylan Thomas, in many ways a philosopher poet, recognised this when he described his and my home town Swansea, as "ugly, lovely, or so it was and is to me" as quoted from his 1943 radio broadcast "Reminiscences of Childhood."

People are incongruously paradoxical as was so magnificently dramatized in Sergio Leone's film entitled, "The Good, the Bad and the Ugly."

Sport is a symbolic and representational manifestation of that idiosyncratic yet beautiful human trait. Therein, in my view, lies an additional form of aestheticism not necessarily found in conventional creative artistic domains but found in abundance in sport. It can take you beyond the aesthetic horizon. It was said about Carwyn James the great Welsh and British Lions coach that "he had his own standards of excellence and a coherent picture of what might be the artist's glimpse of total possibilities—this as an inspiration vision as if like Shelley he was aware of the fading coals of creation of the first moment of seeing which ever after is never quite so bright and the task somehow always paler than the brilliance of the first concept."[241]

Sport as Culture

Terry Eagleton, a writer to whom I have already referred in relation to his illuminating book Radical Sacrifice, put it succinctly in his lecture 'Where does culture come from?' (2024) he said "Our art or culture can issue a powerful rebuke to society not so much by virtue of what it says but because of the strange pointless, intensely libidinal thing that it is. It is one of the few remaining activities in an increasingly instrumentalised world that exists purely for its own sake". Sport is an emblematic and symbolic manifestation of that concept.

As the word cultivate suggests it is something that grows out of the pursuit of an activity whether it is work or play or a state of mind or body. It is a beyond concept. Culture is related to not only civilisation but to a plethora of activities such as customs, lifestyle, mores, society, the arts and way of life, accomplishment, breeding, education, elevation, erudition. It is a thesaurus of concepts. It is probably the broadest concept of them all. Gilbert Ryle as I earlier pointed out talked about multiple intelligences. We live in a world of multiple cultures. Much as I admire Eagleton's writings I do not agree that culture is limited metaphysically to the aesthetic and non-instrumental. Culture can extend to work place activities for example the culture of the police, politics or indeed the law.

Chapter 7
Sport and Identity

In oneself lies the whole world and if you know how to look and learn, then the door is there and the key is in your hand. Nobody on earth can give you either that key or the door to open, except yourself.
Krishnamurti

The master in the art of living makes little distinction between his work and play, his labour and leisure, his mind and body, his education and his recreation, his love and his religion. He hardly knows which is which, he simply pursues his vision of excellence in whatever he does, leaving others to decide whether he is working or playing. To him he is always doing both.
Buddha

The sweet freedom which is freedom's self.
Shelley

The boundaries of your life are merely creations of the self.
Robin Sharma, *The Monk who Sold his Ferrari*

Whenever I climb, I am followed by a dog called 'Ego'.
Nietzsche

There are three mirrors of life: the first reflects how I see myself; the second mirror reflects how others see me; the third mirror reflects the truth.
Chinese proverb

Listen to me for I am thus and thus. Do not above all, confound me with what I am not.
Nietzsche

Man has been conditioned through the years to think of the body as an inferior part of the individual.[242]

However, one writer has said that:

"...kinaesthetic perception of movement is one of the sources from which man derives the meanings of his life as he carries on the uniquely human mental process of transforming sensory perception into human thought."[243]

The same writer believed that the thing that is really at stake in sport is the self, our self-concept, or the mastery of the self.[244]

Sartre's view of his own identity was that "In fashioning myself I fashion man."

Man seeks meaning, and the only way to attain it is to test himself to see what is there.

"Play is risk challenge, creativeness, fantasy, self-discipline, caring, involvement and a thousand other things all tied up in total being."[245]

As Heidegger said "Every man is born as many men and dies as a single one."

There is probably no other area in our society where "the totality of man" is tested so severely or so often as in sport. On the horizon, we die as one and alone.

"It is clear that everything which the individual experiences may be said to have a twofold nature. It is both a single thing and also a sum of parts—a unity and a plurality."[246]

Death unites plurality.

It is here, in the full commitment and complete involvement, that man begins to sense the wholeness of his being. He no longer is in the water. He swims in the water.

As Nietzsche said: "He who has a why to live can bear with almost any how."

One does not just play mentally, or emotionally, or physically. One plays period. There is a oneness in athletics, and because of the participation of the total person, there is the chance for integration of the person. There is the chase for wholeness[247].

In order to exist as an individual at all, one must conceive of some thread of unity on which the various parts, aspects or experiences of one's self may be said to be strung. What this means is that one must have some unified concept of oneself as an entity.[248]

"The performer, as he moves, transforms non-being into being in the world of nothingness."

To propose that man is in sport because he should be or that he participates as a result of knowledge of and commitment to a value, vastly oversimplifies the situation. Man is in sport on grounds independent of the practical or rational.[249]

The man of sport is not simply in space and propelled by force but simultaneously aware and concerned with time and with goal orientation; he uses the matrix of time, force and space to order his world[250].

Paradoxically, man becomes more human as he seeks to go beyond his rational thought which has set him apart from other living things. He goes beyond that which he knows and which he can explain in relational terms, and as a result humanises himself. His achievement is not in accumulating facts but in doing.

There is a fine line that is hard to describe in sport. This line separates the performance itself from the symbolic meaning it represents. What we actually do may hold meanings of great importance to us that we are not consciously aware of.[251]

The athlete fulfils many animal needs just as non-athletes do. But the athlete steps beyond his animal instincts. He must actually learn to limit his biological and animal needs in order to perform as well as possible. He does this through the ability to symbolise. His innermost thoughts become expressed through the overt activity going on during play.[252]

In Hindu thought, the body is also called the physical self.

Many Hindu yoga texts praise the body as the perfect instrument for the realisation of the divine self. They speak of the spirituality of the body.

Today many sportsmen practise yoga gymnastics to achieve a higher physical resistance, and also to improve their physical constitution. Yoga positions together with proper breathing influence glands, nerves and spinal column and contribute to a better body balance.

According to these teachings, there exists between body and soul an indissoluble reciprocity and a two-way responsibility. An acceptance of the body is at the same time an acceptance of the unity of body, soul and spirit. There is no existence for the body outside this unity.

It follows that a sport which helps the body become an element for the achievement of this unity has therefore a potential for the realisation of the divine self.

There are four elements involved:

-   the body, whose limits the sportsman learns only by experiment;
-   the will—or better, the enthusiasm—to throw himself into the game without thought of winning or losing; this enthusiasm can lead to opening oneself up;

- the feeling of responsibility to the team whoever acts from this ideal grows out of himself and improves his individual performance for the team performance;
- the possibility of carrying the spirit of sport over into life, so that the sportsman can think of life as sport, where one plays well, without fear, cunning, or ill-will, where one always tries to advance, but does not worry about success or failure.

In these points, we have the elements of a religion of transcendence, which calls for the spiritualisation of life.

The pre-Buddhist philosophy of India posited an antagonistic relationship between body and spirit. There its religious practice sought the obliteration of the body through asceticism in order to bring about the union of the indwelling spirit, that is, the immortal individual self, the atman, with the essentially identical eternal spirit of the world, the Brahman.

In contrast, Buddha Sakyamuni rejected this body-soul dualism and Buddhist contemplative practice had as its object, not the obliteration, but rather the regulation of bodily activity.

This tendency was strengthened even more in Mahayana, which arose in India at the turn of the era and became the definitive new direction in Buddhism.

With the planting of this thought in the cultural soil of old China there followed a certain decline of the speculative spirit of India in favour of a more practical concrete stance, which again directed interest towards the corporeal side of man.

The agent of this Chinese transformation of Buddhism was the Ch'an, the meditation school, which borrowed much from the old Chinese, and especially from the Taoist tradition and is best known in its Japanese form of Zen-Buddhism.

The role of the body in Zen religious practice and the types of sport based on this doctrine are best explained by reference to a book by the German philosopher and Japanese expert, Eugen Herrigel, in which he considers the sport of archery.[253]

In this book he describes his training in archery by a Japanese master of the art.

There are three steps.

First, one concentrates only on the drawing of the bow; then comes the release of the arrow; and only thirdly, does one concern oneself with the target.

The first step of drawing the bow is one's physical constitution. Most importantly, here for Zen is the problem of looseness. When we engage in Zazen, that is Zen meditation, in a contemplative position with folded legs, our bodily condition is actually the same as in drawing the bow. The important thing is the principle of looseness filled with a lively tension. Strength is thereby concentrated only in the 'koshi', that is, in the small of the back and in the loins. The upper body, supported by this solid foundation is held free and upright, but not drawn upwards; for it is from the shoulders and all other parts of the body that the strength must be supplied. This was already being taught by the Chinese Zen master Ch'ang-lu Tsung-I in his 10th century instructions for Zazen[254].

In archery, where one tries to retain this stance when drawing the bow, the problem of breathing soon arises. Zen meditation lays great importance on breathing. It is the breath which joins body and spirit.[255]

Even as early at the 6th century, a Chinese text[256] mentions the counting of breaths as a help for meditation for beginners, and most especially in the following manner: one inhales quietly and counts one as he exhales. This he does again, only this time with the number two. He repeats this to the count of ten, after which he starts over with one. By means of analogous rhythmical breathing, co-ordinated with the phrases of the necessary motion, Herrigel was able to draw even the strongest bow with full looseness of the entire body.

The second step is the training for the release.

Herrigel says what is involved in these words:

> "The first step…has resulted in bodily looseness, without which the correct drawing of the bow would not be achieved. In order now to bring about the proper release, this bodily looseness must be transformed into mental and spiritual looseness, whose purpose is not only to make the mind flexible but also to make it free: flexible for the sake of freedom, free for the sake of an ever more primary flexibility. This primary flexibility is essentially different from everything one usually envisions under mental flexibility. There is a great difference in level between these two conditions, bodily looseness on the one hand and mental freedom on the other, which cannot be bridged simply by breathing alone, but rather only by cutting oneself off from all bonds, of whatever sort they may be; by becoming selfless to the very core in which a way that the soul, turned in upon itself, anticipates in the fullness of power of its nameless origin."[257]

This is the great theme of 'becoming selfless', and Herrigel's words actually capture the psychological dimension of this fundamental Buddhistic idea; namely, the 'selflessness', the 'muga' as it is called in Japanese.

The above passage makes clear that this 'selflessness' is not simply a denial of the ego. Rather it concerns not only bodily, but much more mental and spiritual looseness to accomplish the final goal of rendering the mind not only flexible but free from all bonds.

It is exactly this that Buddhism calls 'muga'—'selflessness' (or with another word, 'mushin', best translated in this context as 'purposelessness').

Activity derived from it is a purposeless activity, an "activity without acting."

In the art of archery, the marksman must call forth this purposeless activity, which itself is the highest goal of Zen or of Buddhism, and which forms the nucleus of its mystical thinking. This ultimate goal of striving towards selflessness serves on the second level of archery training as an exercise in concentration of consciousness, which is the integral part of Zen[258].

This condition is "disinterested self-recollection" and is threatened from the outset by distractions, which Herrigel describes as follows:

"Uninvited and out of nowhere appear moods, feelings, wishes, worries, and even thoughts in nonsensical mixtures, and the more remote or strange they are, and the less they have to do with the task at hand, the more obstinately they remain. It is as though they were taking revenge by disturbing depths of concentration which they otherwise could never reach."[259]

Herrigel learnt to disarm these distractions in a manner suggested by a Chinese meditation manual:

"When thoughts pop up, acknowledge them, and when you have acknowledged them, they will disappear of themselves."

This acknowledgement means that, rather than try consciously to supress them, one should realise their actual lack of substance, and let them peacefully come and go, without paying the least attention. In this way they extinguish themselves. This is called "white-cloud-samâdhi" ('samâdhi' being the Sanskrit word for concentration).

By means of this 'white-cloud-samâdhi', it is possible to master these distractions. Herrigel writes that one arrives at a twilight condition so like the state of falling asleep, one slips into this condition with such finality that one runs the risks of being waylaid by it.

These, or perhaps only the twilight situation which Herrigel describes, are the dangers of exercises for concentration of consciousness.

"One comes in contact with this by a peculiar leap of concentration, perhaps comparable to the jerk a tired person gives himself, who knows that his life depends on the alertness of all his senses. Once this leap has been successfully executed, it can be safely repeated. By it the soul is led almost of itself to a carefree state of self-oscillation, which is capable of being intensified to a feeling of matchless lightness only seldom experienced in dreams, and which triggers an exhilarating certainty of being able to direct energy in any direction, and to raise or lower tensions in gradual adjustments."[260]

This leap of concentration is actually a very sudden inner experience. It is the perception of the 'primary flexibility of the spirit' the 'right presence of mind'.

For the sake of this right presence of mind, the Zen-Buddhist subjects himself to strict exercises. For its sake, the archery marksman takes on a long period of training. In order to be able to shoot correctly, namely, in truly purposeless action, he shoots 'without shooting' or as Herrigel puts it, 'it shoots' so that this 'it' is perceived not as an objective something, as something other in contrast to me. It is much more the 'true I' existing outside myself, and in this sense this purposeless action is a 'selfless' action.

Should the marksman achieve this complete selflessness, he has already mastered the third step, in which in purposeless sighting of the target, he releases *the shot*. For in the *completely* purposeless shot the completely sure aim is already assured. As Herrigel's teacher said, "there are levels of mastery," and therefore what is essentially one must be divided into two steps during training.

"You can become a master archer, even if not every shot is on target. The reaching of the target is only an outward test and certification of highly refined purposelessness, selflessness, concentration, or whatever else you wish to all this condition."

The hitting of the mark corresponds to the inner spiritual condition of purposelessness or selflessness, and when the spirit has reached the highest level of completion, then every shot must be a bull's eye.

This is the fulfilment of the third step. I, bow, arrow, and target become one, that is *I am* bow, arrow and target. *Outside* of me there is no bow, no arrow, no target. They are in me. The outside is the inside.
This is the 'great liberation' in Zen, and therefore it says in its scriptures:
"Do not seek Buddha on the outside, nor on the inside, nor in between!"

And so in complete selfless action, the spirit is the body and the body is the spirit. As the spirit completes corporeal action, so also the sureness of masterful action confirms spiritual completion. From the viewpoint of Zen, this expresses the highest form of relation of corporal body and spiritual soul.

Zen means 'action with awareness', being completely in the present moment. The qualities that accompany the Zen experience include expansive vision, effortless focus, a feeling of equanimity and timelessness, abundant confidence, and complete freedom from anxiety or doubt. Interestingly, this is exactly the way champion athletes describe "being in the Zone." It is also strikingly similar to the way golfers describe the feeling of a perfectly struck golf shot, a feeling every golfer wants to have again and again.[261]

A young man had read all the books he could find about Zen. He heard about a great Zen master and requested an appointment with him to ask for teachings. When they were seated, the young man proceeded to tell the master everything he had understood from his reading, saying that Zen is about this and Zen is about that, on and on. After some time, the master suggested that they have tea. He performed the traditional tea ceremony while the student sat at attention, bowing when served, saying nothing. The master began to pour tea into the student's cup. He poured until it was full, and kept pouring. The tea ran over the edge of the cup and onto the table. The master kept pouring as the tea ran off the table and onto the floor. Finally, the student couldn't contain himself any longer. He shouted, "Stop! Stop pouring! The cup is full—no more will go in."

The master stopped pouring and said "Just like this cup, your mind is full of your own opinions and preconceptions. How can you learn anything unless you first empty your cup?"
How big is your mind? Is it the same size as your brain? Does it have a shape? Is it located in a particular spot?
A Zen master asked a student, "Where is your mind?" The student said, "When I perceive my thoughts, it is as if someone were speaking inside my head. So my mind must be in my head."

The master motioned for the student to approach him. When the student stood right in front of him, the master banged his fist down on the student's big toe and said, "Now where is your mind?"

In Buddhism, mind and awareness are synonymous. Awareness is open and spacious, the container of whatever is experienced. Like a mirror, it does not have any particular colour or content of its own but reflects whatever appears. Thoughts, sense perceptions, emotional feelings, and dreams all appear in your mirror-like mind. None of these *are* the mind; they all arise as *contents* of the mind. You have thoughts, but you are not your thoughts.

Chapter 8
The Grasshopper

The Joyes of Earth and Aure are thine entire,
That with thy feet and wing,
doest hop and flye:
And when thy Poppy works thou dost retire.
To they Carv'd Acron bed to lye
Richard Lovelace, *The Grasshopper*

Chiefs who no more in bloody fights engage
But wise through time and narrative with age,
In summer days like grasshoppers rejoice,
A blood less race, that send a feeble voice
Alexander Pope

One day it will have to be officially admitted that what we have christened reality
is an even greater illusion than the world of dreams.
Salvador Dali

I am rich beyond the dreams of avarice.
Edward Moore

Bernard Suits' *The Grasshopper* is a cult classic in sporting philosophy.

The book begins with a puzzle in the form of a recurring dream and ends with its solution. Sandwiched between the dream and its meaning, Suits defines 'game playing' as a voluntary attempt to overcome unneeded obstacles.

Suits solves the dream by proposing the thesis that the ideal of human existence is the life of game playing.[262]

True to the thesis entertained and in real Platonic fashion, *The Grasshopper* is not only an attempt to defend the game-playing thesis but also an illustration of it. The illustration is obvious enough in the silly, playful, imaginative, and most important, inefficient structure and style of the book. The defence of Suits' thesis, however, is tied to his use of a counterfactually grounded Utopia at the book's end.

The Grasshopper begins in Platonic fashion with a conversation between the Grasshopper and his disciples, Prudence and Skepticus. Using a fable of Aesop, Grasshopper, about to die with the oncoming of winter and his failure to prepare for it, offers an apology for his life devoted exclusively to play that is reminiscent of Socrates' own apology in *Apology* and *Crito*.

Firstly, because he was placed on earth to play and die, it would be wrong of Grasshopper to mock fate by doing otherwise.

Secondly, he offers the following hypothetical syllogism: if he is provident in summer (by working to store up food for winter), then he will live through winter but cease to be a grasshopper.

If he is improvident in summer, then he will die in winter. As he must be provident or improvident, he will either cease to be a grasshopper or die.

As did Socrates, Grasshopper chooses death in preference to a false life.[263]

Before leaving his disciples, Grasshopper gives them with a riddle in the form of a recurring dream.

Everyone alive, he relates, is actually playing elaborate games, though they go about their lives believing that they are pursuing most serious activities.

In a subsequent dream, Grasshopper tells everyone he meets that what has been revealed to him in the first dream is true.

As each person digests that truth, each ceases to exist so completely that it is as if he had never existed.

Truth not only sets one free; but it also annihilates![264]

Suits defines the ideal of human existence, pure autotelic activity, as that thing or those things whose only justification is that they justify everything else; or, as Aristotle put it, those things for the sake of which we do other things, but which are not themselves done for the sake of anything else.[265]

Suits' notion of autotelicity is derived from Plato, Aristotle, and other early teleologists. If some action, *ô*, is undertaken for the sake of some end, *w*, then it is *w* that one is really striving for, not *ô*. *It is w* that makes *ô* worth doing.

For Suits pure autotelic activity is game playing. He defines this as follows:

> To play a game is to attempt to achieve a specific state of affairs [prelusory goal] using only means permitted by rules [lusory means], where the rules prohibit use of more efficient in favour of less efficient means [constitutive rules], and where the rules are accepted just because they make possible such activity [lusory attitude].[266]

"Playing a game is the voluntary attempt to overcome unnecessary obstacles."[267]

Such voluntarily activity, Suits argues, is the only true, humanly meaningful autotelic activity, and he attempts to show this through his notion of Utopia.

Suits attempts to justify the claim that game playing is the ideal of human existence.

To justify this, he appeals to a counterfactual situation.

He has us imagine a Utopia where all instrumental human activities (i.e. types of work) are unneeded and have been eliminated.

Machines, activated by mental telepathy (presumably because any other means of activation would be a form of work), now do the work of humans.

So efficiently do they work that the number of goods they produce is plethoric, and there is a superabundance of each type of good.

Furthermore, psychotherapy and the social sciences have made such advances that all possible interpersonal problems have been solved.

People no longer need affection, approval, attention, and admiration.

The advances of psychotherapy even make moral principles superfluous.

Art, too, is unneeded and unpractised, since the motivation for its creation—human aspirations, frustrations, hopes, fears, triumphs, tragedies, and the like—do not exist in Utopia.[268]

Science, philosophy, and all other forms of investigative inquiry do not exist in any significant sense, for Utopia is a society where all the important questions have already been answered.

Even love, friendship, and sex disappear.[269]

In the end, Grasshopper argues that the only meaningful activity left is that of game playing.

Game playing is what remains as the human ideal when one abstracts away all instrumental activities.

As pure autotelic activity, "game playing makes it possible to retain enough effort in Utopia to make life worth living" since there is "nothing to strive for because everything else has already been achieved."[270]

When Skepticus asks whether that makes game playing "the whole of the ideal of existence"—presumably, whether game playing is a sufficient and necessary condition for the ideal of human existence in Utopia—Grasshopper replies that it appears so, at least at this stage in the investigation.[271]

Yet, that is not quite correct. Elsewhere, Grasshopper states that play is not sufficient but merely necessary for the ideal of human existence. He states "Game playing performs a crucial role in delineating that ideal—a role which cannot be performed by any other activity and without which an account of the idea is either incomplete or impossible."[272]

Grasshopper proceeds in reductio fashion.

He assumes that someone in Utopia wants to build a house. Since houses are in abundance in Utopia and readily available in every size, shape, ad form, the desire to build one is merely the desire to bring about some end through overcoming unneeded obstacles.

That, of course, is just to be playing a game.[273]

He also assumes that someone wants to solve a scientific problem. Since all scientific problems are solved in Utopia, the desire to solve a problem is, again, the desire to bring about some end through overcoming unneeded obstacles.

Therefore, like one who persists in a crossword puzzle without using the key answer, he too would be playing a game.[274]

Suits sums up this 'lusory attitude'.

> I am truly the Grasshopper that is, an adumbration of the ideal of existence, just as the games we play in our non-Utopian lives are intimations of things to come. For even now it is games which give us something to do when there is nothing to do. We thus call games 'pastimes', and regard them as trifling fillers of the interstices in our lives. But they are much more important than that. They are clues to the future. And their serious cultivation now is perhaps our only salvation. That, if you like, is the metaphysics of leisure time.[275]

Although there may be other ways of passing the time in Utopia, like loafing or travelling, game playing is the essence of Utopia—a condition sine qua non that gives meaning to human lives.[276]

Chapter 9
Genius, Talent and Exceptional Ability

A generous and elevated mind is distinguished by nothing more than an eminent degree of curiosity.
Samuel Johnson

Greatness knows itself.
Shakespeare, *Henry IV*

Men are wise in proportion, not to their experience but to their capacity for experience.
George Bernard Shaw

We first make our habits and then our habits make us.
John Dryden

The situation has provided a cue; this cue has given the expert access to information stored in memory and the information provides the answer. Intuition is nothing more and nothing less than recognition.
Herbert Simon

True genius is a mind of large general powers accidentally determined to some particular direction.
Samuel Johnson

Our everyday intuitive abilities are no less marvellous than the striking insights of an experienced firefighter or physician—only more common.
Daniel Kahneman

A moment's insight is sometimes worth life's experience.
Wendell Holmes

Doing easily what others find difficulty is talent, doing what is impossible is genius.
Henri-Frederic Amiel (1821–1881)

It always seems impossible, until it is done.
Nelson Mandela

Perfection is the child of time.
Bishop Joseph Hall (1574–1656)

The word genius isn't applicable to football. A genius is a guy like Norman Einstein!
Joe Theismann, American football player and coach

So I do believe…the works of genius are the first things in the world.
Keats

Lao Tzu commented "to see things in the seed that is genius." I think that what he meant is that a genius sees things that others don't 'beyond the horizon' of conventional thought and behaviour and acts accordingly. A genius is a person who has exceptional vision and who by thought, word or deed makes that vision manifest and reality. As did Plato, Aristotle, Epictetus, Heraclitus, Seneca, Mozart, Beethoven, Machiavelli, Descartes, John Locke, Spinoza, Leibniz, John Stuart Mill, Voltaire, Rousseau, Tolstoy, Rimbaud, David Hume, Immanuel Kant, Edmund Husserl, Ortega, Nietzsche, Bertrand Russell, Wittgenstein, Chaucer, Shakespeare, Moliere, Milton, Wordsworth, Keats, Shelley, Montaigne, Charles Dickens, D.H. Lawrence, James Joyce, Graham Greene, Jack London, Dylan Thomas, Steinbeck, Virginia Wolf, Hemingway, Dostoevsky, Giuseppe Tomasi di Lampedusa, Jean-Paul Sartre, Salman Rushdie, Iris Murdoch, J. K. Rowling, Einstein, Stephen Hawking, Niels Bohr, Darwin, Freud, Karl Marx, Hegel, da Vinci, Michelangelo, Rembrandt, Picasso and Salvador Dali. Such a person changes our vision and behaviour. For example Churchill, De Gaulle, Theodore Roosevelt, John F. Kennedy, Gandhi, Mandela and Martin Luther King shared the vision and changed behaviour. They identified collective frustration in the gap between current reality and where we want to be. In all domains including sport people who have exceptional vision stand head and shoulders above the norm witness Muhammad Ali, Pele, Don Bradman and Gary Sobers. They were able to convert the vision into the reality of their performance. They transcend the norm and display not just an exceptional but unique manifestation of originality, brilliance and virtuosity. That in my view is genius.

But that is only my view and there are many other approaches. It may be that the correct conclusion is that there is no exhaustive definition and that it is in the nature of genius that it is impossible to define. It can come in many shapes and forms and in many colours. All I can say is that when we see it we recognise it. I expand on that view later in this chapter.

There have been many attempts to define such terms as genius, creativity and talent. Most recently one philosopher defined "a genius as someone who communicates or creates new possibilities for experiencing the world or living in it."[277] Even then the same writer had to put it 'more precisely' as "genius is the ability, first to see things in new ways then to effectively create or communicate that vision (artists) or to put their new way of seeing into practice themselves (scientists and engineers)." Frederick the Great said "Genius is the transcendent capacity of taking trouble."

In my view, the definitional exercise is a futile one. Genius defies definition; that is the nature of genius. At best examples can be given of who might come into the category of genius. We are on safer ground attempting to do this but even then there will be differences of opinion. There is a difference between the concepts of being a genius and the concept of genius. As to the former the same philosopher put forward two examples of a genius in totally different domains— Picasso and David Bowie. His expressed reason for including them in the category of genius was in the case of Picasso that he greatly enlarged the possibilities of fine art, both beautifully and radically and that Bowie did the same for rock music and did it with panache. Another view is that talent is "impressive but genius is transcendental."[278]

The attempt to explain who is or was a genius is a pretty hopeless one. The best one can do is to make cryptic comment as many philosophers and others have done. I very much endorse Wittgenstein's, the master of cryptic and epigrammatic definition's view, that "Genius is courage in talent" or "Genius is talent in which character makes itself heard." Henry James described Kipling as "a man of genius as distinct from a man of fine intelligence." Nietzsche commented that a "man of genius is unbearable unless he possessed at least two things besides gratitude and purity." Nietzsche as usual was more sceptical when he observed "what a person is begins to betray itself when talent decreases when he ceases to show what he can do. Talent is also an adornment, an adornment is also a concealment."

Many would put Wittgenstein in the category of genius as an original thinker. He didn't think so himself having said in 1931 "I don't believe I have ever invented a line of thinking. I have always taken one over from someone else. I have simply straightaway seized on it with enthusiasm from my work of clarification. That is how Boltzmann, Hertz, Schopenhauer, Frege, Russell, Kraus, Loos, Weininger, Spengler, Sraffa, have influenced me."[279]

Maybe Wittgenstein was being self-effacing but I think not. As he said elsewhere—and as noted earlier in this book 'concept is a vague concept'. It follows that genius is a vague concept. Maybe it is in the category of not definitional precision but of a recognitional type. Like an elephant cannot be defined but I recognise one when I see one. My favourite observation on genius comes from sport and in light-hearted vein. Again I have noted this earlier in one of the quotations set out at the beginning of this chapter. It is by Joe Theismann, an American football player and coach. He observed "The word genius isn't applicable to football. A genius is a guy like Norman Einstein."

Genius cannot be precisely defined that is the nature of genius. However, I would loosely characterise it as being some strong quality in the mind and sometimes the 'body' that brings out some new and striking quality in nature. It is wholly different from the capacity to do something well or competently. Shakespeare was beyond genius – a man raised above any attempted definition of genius. His genius consisted in the faculty of transforming himself at will into whatever he chose; his originality was the power of seeing every object from the exact point of view in which others would see it. He was the Proteus of human intellect. Genius in Ordinary is a more obstinate and versatile thing. It is sufficiently exclusive and self-willed and peculiar. It does some one thing by virtue of doing nothing else: it excels in some one pursuit by being blind to all excellence but its own. Some sportsmen are in that category. It is the reverse of the chameleon; it does not borrow but lends its colour to all about it. Rembrandt saw things in nature that everyone had missed before him and gave others eyes with which to see them. This is the test and triumph of originality, not to show us what has never been and what we may therefore very easily never have dreamt of, but to point out to us what is before our eyes and under our feet, though we had no suspicion of its existence for want of sufficient strengths of intuition, of determined grasp of mind to see it and retain it.

A man does not affect to be original; he is so because he cannot help it and often without knowing it. Genius defies reason and logic. As Nietzsche pointed out genius can be madness and madness genius. It is beyond the norm – it follows no rules and knows no bounds. Doctor Bronowski said "a genius is a man who has two great ideas". In my view as Charles Darwin evidences one can be enough. If I had to choose the greatest idea emanating about man it would be Darwin's theory of evolution. It stands head and shoulders above all other ideas.

Creativity is another concept related to both genius and talent. It has been said that creativity is fundamentally the ability to come up with new ideas. It is related to talent which might be regarded as the ability to turn ideas into reality.

In other words having the technical ability to create a vehicle that will convey your creative vision. Sport may be a vehicle for doing this. In that respect creativity is in part a function of the degree to which you are able to see beyond your present ideas. It also means you are able to do things differently beyond the ability of others. Great sportsmen have the ability to do this. Charles Mingus observed "Creativity is more than just being different. Making the simple awesomely simple that's creativity." Great Sportsmen make things look simple. Auguste Rodin viewed his own work as very easy and simple. All I do, he said is to "choose a block of marble and chop off whatever I don't need."

There is no doubt that in life as well as in sport creativity is an elusive concept. Nietzsche saw creativity in the tragedy and poetry of ancient Greece as a marriage between 'Dionysian' outlook (spontaneity, irrationality, the rejection of discipline) and a more serious and ordered 'Apollonian' outlook. Einstein rightly saw it when he advised "Creativity is contagious, pass it on."

One of the problems with characterising sport as art is the existence of the rules. Hazlitt was of the view that "Rules and models destroy genius and art."

It is also said that a necessary condition is to keep trying or going. As Winston Churchill famously said "When you are going through hell keep going." And as Picasso reminded us "working won't make you a genius, but genius has to find you working." Philosophers have most recently focussed on the character of the unconscious mind. I have dealt extensively with the role of the unconscious throughout this book. It is of paramount importance.

But one cannot ignore the fact that one cannot create anything significant without conscious preparation. In sport and other domains it is necessary to exert a huge amount of conscious effort to learn the skills, concepts and other elements of your domain. As La Rochefoucauld perceptively noted "Ideas often flash across our minds more complex than we could make them after much labour." But conscious thought is invariably the precursor of unconscious and inspirational ideas and action. One may deliberately focus on a particular problem and any ideas that occur would emerge through recombining and altering elements through experience. That is why a period of passivity when your unconscious mind is still active may result in that flash of inspiration without knowing where it has come from. It is the Eureka moment.

It also seems to me that one of the hallmarks of individuals who have the touch of genius about them or the ability to display independence of spirit is not to go with the crowd. I do not often agree with Hegel but I do see much force in his opinion that to be independent of public opinion is the first condition of achieving anything great.

Romanticism was a sudden, violent explosion of emotions, enthusiasm and introspection that erupted in the second half of the eighteenth century destroying the classical belief in order, reason, symmetry, and calm and clearing the ground for the cult of personality by redefining the concept of genius. Nietzsche thought of himself as a genius "my genius is in my nostrils…I contradict as has never been contradicted and am nonetheless the opposite of a negative spirit. I am the bringer of good tidings, such as there has never been, I know tasks from such a height that any conception of them has hitherto been lacking, only after me is it possible to hope again."[280] Karl Vonnegut metaphorically said "We have to continually be jumping off cliffs and develop our wings on the way down."

Genius in the modern sense is defined by the Oxford English Dictionary "Instinctive and extraordinary capacity for imaginative creation, original thought, invention or discovery." It is an innovation of the late eighteenth century and it is not in Dr Johnson's dictionary.

Genius was a wholly romantic concept; not just a great artist but a great artist who has embarked on an inner journey and makes his own rules as he goes— Beethoven rather than Hayden, Rousseau rather than Dr Johnson, Rimbaud rather than Pope. Hume dismissed Romanticism as 'spilt religion' because it yearned for the infinite, wallowed in emotion, destroyed the intellectual, paid cursory attention to detail.

So what is Genius, talent and exceptional ability? The relationship between those concepts is a close one. Joseph Conrad saw this. He observed "only in men's imagination does every truth find an effective and undeniable existence. Imagination, not invention is the supreme master of art, as of life." And as Jean-Luc Goddard perceptively identified "It's not where you take things from, it's where you take them to."

Historically a number of approaches have been put forward.[281] The human information-processing approach, or the skills approach, has attempted to explain exceptional performance in terms of knowledge and skills acquired through experience. This approach, originally developed,[282] has tried to show that the basic information-processing system with its elementary information processes and basic capacities remains intact during skill acquisition and that outstanding performance in sport and other domains results from incremental increases in knowledge and skill due to the extended effects of experience. By constraining the changes to acquired knowledge and skill, this approach has been able to account for exceptional performance within existing general theories of human cognition. According to this approach the mechanisms identified in laboratory studies of learning can be extrapolated to account for expertise and expert knowledge and skill over a decade of intense experience in the domain. The long duration of the necessary period of experience and the presumed vast complexity of the accumulated knowledge has discouraged investigators from empirically studying the acquisition of expert performance. Similarly, individual differences in expert performance, when the amount of experience is controlled, have not been of major interest and have been typically assumed to reflect differences in the original structure of basic processes, capacities, and abilities.

The other major approach focuses on the individual differences of exceptional performers that would allow them to succeed in a specific domain. One of the most influential representatives of this approach is Howard Gardner, who in 1983 presented his theory of multiple intelligence in his book *Frames of Mind: The Theory of Multiple Intelligences* (hereinafter referred to as *Frames of Mind*),[283] drew on the recent advances in biology and brain physiology about neural mechanisms and localisation of brain activity to propose an account of the achievements of savants, prodigies, and geniuses in specific domains. He argued that exceptional performance results from a close match between the individual's intelligence profile and the demands of the particular domain. A major concern in this approach is the early identification and nurturing of children with high levels of the required intelligence for a specific domain. Findings within this approach have limited implications for the lives of the vast majority of children and adults of average abilities and talents.

Since the emergence of civilisation, philosophers have speculated about the origin of highly desirable individual attributes, such as poetic ability, physical beauty, strength, wisdom, and skill in handiwork.[284] It was generally believed that these attributes were gifts from the gods, and it was commonly recognised that "On the whole the gods do not bestow more than one gift on a person."[285] This view persisted in early Greek thought, although direct divine intervention was replaced by natural causes. Ever since, there has been a bias towards attributing high abilities to gifts rather than experience. As expressed by John Stuart Mill, there is "a common tendency among mankind to consider all power which is not visibly the effect of practice, all skill which is not capable of being reduced to mechanical rules, as the result of a particular gift."[286]

One important reason for this bias in attribution, is linked to immediate legitimatisation of various activities associated with the gifts. If the gods have bestowed a child with a special gift in a given art form, who would dare to oppose its development, and who would not facilitate its expression so everyone could enjoy its wonderful creations? This argument may appear strange today, but before the French Revolution the privileged status of kings and nobility and the birthright of their children were primarily based on such claims.

The first systematic development of this argument for gaining social recognition to artists can be found in the classic work on *The Lives of the Artist* by Vasari,[287] originally published in 1568. This book provided the first major biography of artists and is generally recognised as a major indirect influence on the layman's conceptions of artists even today.[288] Although Vasari's expressed goal was simply to provide a factual history of art, modern scholars argue that "the *Lives* were partly designed to propagate ideas of the artist as someone providentially born with a vocation from heaven, entitled to high recognition, remuneration and respect."[289] To support his claim, Vasari tried to identify early signs of talent and ability in the lives of the artists he described. When facts were missing, he is now known to have added or distorted material.[290] For example, Vasari dated his own first public demonstration of high ability to the age of 9, although historians now know that he was 13 years old at that event.[291] His evaluations of specific pieces of art expressed his beliefs in divine gifts. Michelangelo's famous painting in the Sistine Chapel, the *Final Judgment*, was described by Vasari as "the great example sent by God to men so that they can perceive what can be done when intellects of the highest grade descend upon the earth."[292] Vasari also tried to establish a link between the noble families and the families of outstanding artists by tracing the heritage and family trees of the artists of his time to the great families of antiquity and to earlier great artists. However, much of the reported evidence is now considered to have been invented by Vasari.[293] In the centuries following Vasari, our civilisation underwent major social changes leading to great social mobility through the development of a skilled middle class and major progress in the accumulation of scientific knowledge. It became increasingly clear that individuals could dramatically increase their performance through education and training, if they had the necessary drive and motivation. Speculation on the nature of talent started to distinguish achievements due to innate gifts from other achievements resulting from learning and training. In 1759 Edward Young published a famous book on the origin of creative products, in which he argued that "An *Original* may be said to be of *vegetable* nature: it rises spontaneously from the vital root of Genius; it *grows*, it is not *made*."[294] Hence, an important characteristic of genius and talent was the apparent absence of learning and training and thus talent and acquired skill became opposites.[295] A century later Galton presented a comprehensive scientific theory integrating talent and training that has continued to influence the conception of exceptional performance.

Sir Francis Galton was the first scientist to investigate empirically the possibility that excellence in diverse fields and domains has a common set of causes. On the basis of an analysis of eminent men in a wide range of domains and of their relatives, Galton argued that three factors had to be present: innate ability, eagerness to work, and "an adequate power of doing a great deal of very laborious work." Because the importance of the last two factors—motivation and effort—had already been recognised,[296] later investigators concentrated primarily on showing that innate abilities and capacities are necessary to attain the highest levels of performance.

He acknowledged a necessary but not sufficient role for instruction and practice in achieving exceptional performance. According to this view, performance increases monotonically as a function of practice towards an asymptote representing a fixed upper band on performance. Like Galton, contemporary researchers generally assume that training can affect some of the components mediating performance but cannot affect others. If performance achieved after extensive training is limited by components that cannot be modified, it is reasonable to assert that stable, genetically determined factors determine the ultimate level of performance. If all possible changes in performance related to training are attained after a fairly limited period of practice, this argument logically implies that individual differences in final performance must reflect innate talents and natural abilities.

The view that talent or giftedness for a given activity is necessary to attain the highest levels of performance in that activity is widely held among people in general. This view is particularly dominant in such domains of expertise as chess, sports, music, and visual arts, where millions of individuals are active but only a very small number reach the highest levels of performance.

One of the most prominent and influential scientists who draw on evidence from exceptional performance of artists, scientists, and athletes for a biological theory of talent is Howard Gardner to whom I have already referred. In *Frames of Mind*, Gardner proposed seven intelligences: linguistic, musical, spatial, logical-mathematical, bodily kinesthetics, and interpersonal and intrapersonal intelligence—each an independent system with its own biological bases. This theory is a refinement and development of ideas expressed in an earlier book,[297] in which the talent position was more explicitly articulated, especially in the case of music. Gardner wrote:

> "Further evidence of the strong hereditary basis of musical talent comes from a number of sources. Most outstanding musicians are discovered at an early stage, usually before 6 and often as early as 2 or 3, even in households where relatively little music is heard. Individual differences are tremendous among children, and training seems to have comparatively little effect in reducing these differences."[298]

He discussed possible mechanisms for talent in the context of music savants, who in spite of low intellectual functioning display impressive music ability as children: "it seems possible that the children are reflecting a rhythmic and melodic capacity that is primarily hereditary, and which needs as little external stimulation as does walking and talking in the normal child."[299] *Frames of Mind* contains a careful review of the then available research on the dramatic effects of training on performance. In particular, he reviewed the exceptional music performance of young children trained with the Suzuki method and noted that many of these children who began training without previous signs of musical talent attained levels comparable to music prodigies of earlier times and gained access to the best music teachers in the world. The salient aspect of talent, according to Gardner, is no longer the innate structure (gift) but rather the potential for achievement and the capacity to rapidly learn material relevant to one of the intelligences. Gardner's view is consistent with Suzuki's rejection of inborn talent in music and Suzuki's early belief in individual differences in innate general ability to learn, although Suzuki's innate abilities were not specific to a particular domain, such as music. However, in his later writings, Suzuki argued that "every child can be highly educated if he is given the proper training,"[300] and he blamed earlier training failures on incorrect training methods and their inability to induce enthusiasm and motivation in the children. The clearest explication of Gardner's view is found when he discussed his proposal for empirical assessments of individuals' profiles in terms of the seven intelligences. He proposed tests in which "individuals were given the opportunity to learn to recognise certain patterns [relevant to the particular domain] and were tested on their capacities to remember these from one day to the next."[301] On the basis of tests for each of the intelligences, "intellectual profiles could be drawn up in the first year or two of life,"[302] although reliable assessments may have to wait until the preschool years because of "early neural and functional plasticity."[303] Gardner's own hunch about strong intellectual abilities was that "an individual so blessed does not merely have an easy time learning new patterns; he learns them so readily that *it is virtually impossible for him to forget them.*"[304]

My reading of Gardner's[305] later books leads me to conclude that his ideas on talent did not fundamentally change. According to Gardner's influential view, the evidence for the talent view is based on two major sources of data on performance: the performance of prodigies and savants and the ability to predict future success of individuals on the basis of early test results. Given that our knowledge about the exceptional performance of savants and prodigies and the predictive validity of tests of basic abilities and talents have increased considerably in the past decade, I briefly review the evidence or rather the lack of evidence for innate abilities and talent. In his book *Creating Minds*, Gardner examined the lives of great innovators, such as Einstein, Picasso, Stravinsky, and Gandhi. Each was selected to exemplify outstanding achievements in one of seven different intelligences. His careful analysis reveals that the achievements of each individual required a long period of intense preparation and required the coincidence of many environmental factors. Striking evidence for traditional talent, such as prodigious achievements as a child, is notably absent, with the exception of Picasso. The best evidence for talent, according to Gardner, is their rapid progress once they made a commitment to a particular domain of expertise. These findings are not inconsistent with Gardner's views on talent because innovation and creation of new ideas are fundamentally different from high achievements in a domain due to talent. Gardner wrote, "in the case of a universally acclaimed prodigy, the prodigy's talents mesh perfectly with current structure of the domain and the current tastes of the field. Creativity, however, does not result from such perfect meshes."[306]

When the large collection of reports of amazing and inexplicable performance is surveyed, one finds that most of them cannot even be firmly substantiated and can only rarely be replicated under controlled laboratory conditions. Probably the best established phenomenon linked to talent in music is perfect pitch, or more accurately absolute pitch (AP). Only approximately 0.01% of the general population have AP and are able to correctly name each of the 64 different tones, whereas average musicians without AP can distinguish only approximately five or six categories of pitches when the pitches are presented in isolation.[307] Many outstanding musicians display AP, and they first reveal their ability in early childhood. With a few exceptions, adults appear to be unable to attain AP in spite of extended efforts. Hence the characteristics of absolute pitch would seem to meet all of the criteria of innate talent, although there is some controversy about how useful this ability is to the expert musicians. In 1993 a review of AP, the writers concluded that the best account of the extensive and varied evidence points towards a theory that "AP can be *acquired by anyone* [italics added], but only during a limited period of development."[308] They found that all individuals with AP had started with music instruction early—nearly always before age five or six—and that several studies have been successful in teaching AP to three- to six-year-old children. At older ages children perceive relations between pitches, which leads to accurate relative pitch, something all skilled musicians have. "Young children *prefer* to process absolute rather than the relative pitches of musical stimuli."[309] Similar developmental trends from individual features to relational attributes are found in other forms of perception during the same age period.[310] Rather than being a sign of innate talent, AP appears to be a natural consequence of appropriate instruction and of ample opportunities to interact with a musical instrument, such as a piano, at very young ages.

Other proposed evidence for innate talent comes from studies of prodigies in music and chess who are able to attain high levels of performance even as young children. In two influential books, Feldman showed that acquisition of skills in prodigies follows the same sequence of stages as in other individuals in the same domain. The primary difference is that prodigies attain higher levels faster and at younger ages. For example, an analysis of Picasso's early drawings as a child shows that he encountered and mastered problems in drawing in way similar to less gifted individuals.[311] Feldman also refuted the myth that prodigies acquire their skills irrespective of the environment. In fact, he found evidence for the exact opposite, namely that "the more powerful and specific the gift, the more need for active, sustained and specialised intervention"[312] from skilled teacher and parents. He described the classic view of gifts, in which parents are compelled to support their development, when he wrote, "When extreme talent shows itself it demands nothing less than the willingness of one or both of the parents to give up almost everything else to make sure that the talent is developed."[313] A nice case in point is the child art prodigy Yani,[314] whose father gave up his own painting career so as not to interfere with the novel style that his daughter was developing. Feldman[315] argued that prodigious performance is rare because extreme talent for a specific activity in a particular child and the necessary environmental support and instruction rarely coincide.

Contrary to common belief, most child prodigies never attain exceptional levels of performance as adults.[316] When Scheinfeld examined the reported basis of the initial talent assessment by parents of famous musicians, he found signs of interest in music rather than objective evidence of unusual capacity. For example, Fritz Kreisler was 'playing violin'[317] with two sticks at age four, and Yehudi Menuhin had a "response to violins at concerts"[318] at the age of one and a half years. Very early start of music instruction would then lead to the acquisition of absolute pitch. Furthermore, the vast majority of exceptional adult performers were never child prodigies, but instead they started instruction early and increased their performance due to a sustained high level of training.[319] The role of early instruction and maximal parental support appears to be much more important than innate talent, and there are many examples of parents of exceptional performers who successfully designed optimal environments for their children without any concern about innate talent.[320] For example, as part of an educational experiment, Laslo and Klara Polgar[321] raised one of their daughters to become the youngest international chess grand master ever—she was even younger than Bobby Fischer, who was the youngest male achieving that exceptional level of chess-playing skill. In 1992 the three Polgar daughters were ranked first, second, and sixth in the world among women chess players, respectively.

Although scientists and the popular press have been interested in the performance of prodigies, they have been especially intrigued by so-called savants. Savants are individuals with a low level of general intellectual functioning who are able to perform at high levels in some special tasks. In a few cases, the parents have reported that these abilities made their appearances suddenly, and they cited them as gifts from God.[322] More careful study of the emergence of these and other cases shows that their detection may in some cases have been sudden, but the opportunities, support, and encouragement for learning had preceded the original performance by years or even decades.[323] Subsequent laboratory studies of the performance of savants have shown them to reflect acquired skills. For example, savants who can name the day of the week of an arbitrary date (e.g., November 5, 1923) generate their answers using instructible methods that allow their performance to be reproduced by a college student after a month of training.[324] The only ability that cannot be reproduced after brief training concerns some savants reputed ability to play a piece of music after a single hearing.

In summary, the evidence from systematic laboratory research on prodigies and savants provides no evidence for giftedness or innate talent but shows that exceptional abilities are acquired often under optimal environmental conditions.

The importance of basic processes and capacities is central to many theorists in the human information-processing tradition. In conceptual analogies with computers, investigators often distinguish between hardware (the physical components of the computer) and software (computer programs and stored data). In models of human performance, 'software' corresponds to a knowledge and strategies that can be readily changed as a function of training and learning, and 'hardware' refers to the basic elements that cannot be changed through training. Even theorists who acknowledge that "practice is the major independent variable in the acquisition of skill,"[325] argue in favour of individual differences in talent that predispose people to be successful in different domains: "Although there clearly must be a set of specific aptitudes (e.g., aptitudes for handling spatial relations) that together comprise a talent for chess, individual differences in such aptitudes are largely overshadowed by immense differences in chess experience."[326] One writer went through many different domains to point out some necessary qualities that are likely to be mostly inborn, such as *"motor coordination, speed of reflexes* and *hand-eye coordination."*[327] These views were consistent with the available information at the time, such as high heritability for many of these characteristics. In the domain of sport psychology, it has been argued that the importance of fixed physiological traits for elite performance of athletes and that "there is good evidence that the limits of physiological capacity to become more efficient with training is determined by genetics."[328] They cited research reporting that percentage of muscle fibres and aerobic capacity "are more than 90% determined by heredity for both male and female."[329] However, more recent reviews have shown that heritability in random samples of twins are much lower and range between zero and 40%.[330]

154

It is curious how little empirical evidence supports the talent view of expert and exceptional performance. Ever since Galton, investigators have tried to measure individual differences in unmodifiable abilities and basic cognitive and perceptual capacities. To minimise any influence from prior experience, they typically base their tests on simple tasks. They measure simple reaction time and detection of sensory stimuli and present meaningless materials, such as nonsense syllable and lists of digits, in tests of memory capacity. A recent review[331] showed that efforts to measure talent with objective tests for basic cognitive and perceptual motor abilities have been remarkably unsuccessful in predicting final performance in specific domains. For example, elite athletes are able to react much faster and make better perceptual discriminations to representative situations in their respective domains, but their simple reaction times and perceptual acuity to simple stimuli during laboratory tests do not differ systematically from those of other athletes or control subjects.[332] Chess players' and other experts' superior memory for brief presentation of representative stimuli from their domains compared with that of novices is eliminated when the elements of the same stimuli are presented in a randomly arranged format.[333] The performance of elite chess players on standard tests of spatial ability is not reliably different from control subjects.[334] The domain specificity of superior performance is striking and is observed in many different domains of expertise.[335] Sport is a paradigm example.

This conclusion can be generalised with some qualifications to current tests of such general abilities as verbal and quantitative intelligence. These tests typically measure acquired knowledge of mathematics, vocabulary, and grammar by successful performance on items testing problem solving and comprehension. Performance during and immediately after training is correlated with IQ, but the correlations between this type of ability test and performance in the domain many months and years later is reduced (even after corrections for restriction of range) to such low values that some[336] questioned their usefulness and predictive validity. At the same time, the average IQ of expert performers, especially in domains of expertise requiring thinking, such as chess, has been found to be higher than the average of the normal population and corresponds roughly to that of college students. However, IQ does not reliably discriminate the best adult performers from less accomplished adult performers in the same domain.

Even physiological and anatomical attributes can change dramatically in response to physical training. Almost everyone recognises that regular endurance and strength training uniformly improves aerobic endurance and strength, respectively. As the amount of intensity or physical training is increased and maintained for long periods, far-reaching adaptations of the body result.[337] For example, the sizes of hearts and lungs, the flexibility of joints, and the strength of bones increase as the result of training, and the nature and extent of these changes appear to be magnified when training overlaps with physical development during childhood and adolescence. Furthermore, the number of capillaries supplying blood to trained muscles increases, and muscle fibres can change their metabolic properties from fast twitch to slow twitch. With the clear exception of height, a surprisingly large number of anatomical characteristics show specific changes and adaptations to the specific nature of extended intense training.

If one accepts the necessity of extended intense training for attaining expert performance then it follows that currently available estimates of heritability of human characteristics do not generalise to expert performance. An estimate of heritability is valid only for the range of environmental effects to which the studied subject have been exposed. With a few exceptions, studies of heritability have looked only at random samples of subjects in the general population and have not restricted their analyses to individuals exposed to extended training in a domain. The remaining data on exceptional and expert performers have not been able to demonstrate systematic genetic influences. Explanations based on selective access to instruction and early training in a domain provide a good or in some cases better accounts of familial relations of expert performers, such as the lineage of musicians in the Bach family.[338]

In summary, the traditional assumptions of basic abilities and capacities (talent) that may remain stable in studies of limited and short-term practice do not generalise to superior performance acquired over years and decades in a specific domain. Once the potential for change through practice is recognised, the view is that a search for individual differences that might be predictive of exceptional and expert performance should refocus on the factors advocated by Charles Darwin.[339] In a letter to Galton after reading the first part of Galton's (1869/1979) book: Darwin said "You have made a convert of an opponent in one sense, for I have always maintained that excepting fools, men did not differ much in intellect, only in zeal and hard work; I still think this is an *eminently* important difference."[340] In commenting on Darwin's remark, Galton agreed but argued that "character including the aptitude for work, is heritable."[341] On the basis of their review,[342] found that motivational factors are more likely to be the locus of heritable influences than is innate talent. We explicate the connection between these 'motivational' factors and the rate of improving performance in a specific domain in the last section of this chapter. But creativity needs to be worked on. As Jack London said "You can't wait for inspiration, you have to go after it with a club."

The conceptions of expert performance as primarily an acquired skill versus a reflection of innate talents influence how expert performance and expert performers are studied. When the goal is to identify critical talents and capacities, investigators have located experts and the compared measurements of their abilities with those of control subjects on standard laboratory tests. Tests involve simple stimuli and tasks in order to minimise any effects of previously acquired knowledge and skill. Given the lack of success of this line of research, a different approach is suggested a different approach that identifies the crucial aspects of experts' performance that these experts exhibit regularly at a superior level in their domain. If experts have acquired their superior performance by extended adaptation to the specific constraints in their domains, we need to identify representative tasks that incorporate these constraints to be able to reproduce the natural performance of experts under controlled conditions in the laboratory. Expert performance can be defined as consistently superior performance on a specific set of representative tasks for the domain that can be administered to any subject. The virtue of defining expert performance in this restricted sense is that the definition both meets all the criteria of laboratory studies of performance and comes close to meeting those for evaluating performance in many domains of expertise.

In many domains, rules have evolved and standardised conditions, and fair methods have been designed for measuring performance. The conditions of testing in many sports and other activities, such as typing competitions, are the same for all participating individuals. In other domains, the criteria for expert performance cannot be easily translated into a set of standardised tasks that captures and measures that performance. In some domains, expert performance is determined by judges or by the results of competitive tournaments. Psychometric methods based on tournament results, most notably in chess,[343] have successfully derived latent measures of performance on an interval scale. In the arts and sciences, selected individuals are awarded prizes and honours by their peers, typically on the basis of significant achievements such as published books and research articles and specific artistic performances.

Some type of metric is of course required to identify *superior performance*. The statistical term *outlier* may be a useful heuristic for judging superior performance. Usually, if someone is performing at least two standard deviations above the mean level in the population, that individual can be said to be performing at an expert level. In the domain of chess,[344] the term *expert* is defined as a range of chess ratings approximately two to three standard deviations (200 rating points) above the mean (1600 rating points) and five to six standard deviations above the mean of chess players starting to play in chess tournaments.

In most domains it is easier to identify individuals who are socially recognised as experts than it is to specify observable performance at which these individuals excel. The distinction between the perception of expertise and actual expert performance becomes increasingly important as research has shown that the performance of some individuals who are nominated as experts is not measurably superior. For example, studies have found that financial experts' stock investments yield returns that are not consistently better than the average of the stock market, that is, financial experts' performance does not differ from the result of essentially random selection of stocks. When successful investors are identified and their subsequent investments are tracked, there is no evidence for sustained superiority. A large body of evidence has been accumulated showing that experts frequently do not out-perform other people in many relevant tasks in their domains of expertise.[345] Experts may have much more knowledge and experience than others, yet their performance on critical tasks may not be reliably better than that of nonexperts. In summary, researchers cannot seek out experts and simply assume that their performance on relevant tasks is superior; they must instead demonstrate this superior performance. It is in Ryle's terminology not the knowledge of the 'that' but the 'how'.

For most domains of expertise, people have at least an intuitive conception of the kind of activities at which an expert should excel. In everyday life, however, these activities rarely have clearly defined starting and end points, nor do the exact external conditions of a specific activity reoccur. The main challenge is thus to identify particular well-defined tasks that frequently occur and that capture the essence of expert performance in a specific domain. It is then possible to determine the contexts in which each task naturally occurs and to present these tasks in a controlled context to a larger group of other experts.

Research on expertise in chess is generally considered the pioneering effort to capture expert performance. Ability in chess playing is determined by the outcomes of chess games between opponents competing in tournaments. Each game is different and is rarely repeated exactly except for the case of moves in the opening phase of the game. One who was himself a chess master, determined that the ability to play chess is best captured in the task of selecting the next move for a given chess position taken from the middle of the game between two chess masters. Consistently superior performance on this task for arbitrary chess positions logically implies a very high level of skill. Researchers can therefore elicit experts' superiority in performing a critical task by presenting the same unfamiliar chess position to any number of chess players and asking them to find the best next move. It was demonstrated that performance on this task discriminates well between chess players at different levels of skill and thus captures the essential phenomenon of ability to play this game.

In numerous subsequent studies, researchers have used a similar approach to study the highest levels of thinking in accepted experts in various domains of expertise.[346] If expert performance reflects extended adaptation to the demands of naturally occurring situations, it is important that researchers capture the structure of these situations in order to elicit maximal performance from the experts. Furthermore, if the tasks designed for research are sufficiently similar to normal situations, experts can rely on their existing skills, and no experiment-specific changes are necessary. How similar these situations have to be to real-life situations is an empirical question. In general, researchers should strive to define the simplest situation in which experts' superior performance can still be reliably reproduced.

Experts acquire skill in memory to meet specific demands of encoding and accessibility in specific activities in a given domain. For this reason their skill does not transfer from one domain to another. The demands for storage of intermediate products in mental calculation differ from the demands of blindfold chess, wherein the chess master must be able not simply to access the current position but also to plan and accurately select the best chess moves. The acquisition of memory skill in a domain is integrated with the acquisition of skill in organising acquired knowledge and refining of procedures and strategies, and it allows experts to circumvent limits on working memory imposed by the limited capacity of STM.

In many domains it is critical that experts respond not just accurately but also rapidly in dynamically changing situations. A skilled performer needs to be able to perceive and encode the current situation as well as to select and execute an action or a series of actions rapidly. In laboratory studies of skill acquisition, investigators have been able to demonstrate an increase in the speed of perceptual-motor reactions as a direct function of practice. With extensive amounts of practice, subjects are able to evoke automatically the correct reaction to familiar stimulus situations. This analysis of perceived situations and automatically evoked responses is central to our understanding of skilled performance, yet it seems to be insufficient to account for the speeds observed in many types of expert performance. The time it takes to respond to a stimulus even after extensive training is often between 0.5 and 1.0 seconds, which is too slow to account for a return of a hard tennis serve, a goalie's catching a hockey puck, a cricketer's slip catch and fluent motor activities in typing and music.

The standard paradigm in laboratory psychology relies on independent trials in which the occurrence of the presented stimulus, which the subject does not control, defines the beginning of a trial. In contrast, in the perceptual environment in everyday life, especially sport expert performance is continuous and changing, and experts must be able to recognise if and when a particular action is required. Most important, it is possible for the expert to analyse the current situation and thereby anticipate future events. Research on the return of a tennis serve shows that experts do not wait until they can see the ball approaching them. Instead they carefully study the action of the server's racquet and are able to predict approximately where in the service area the tennis ball will land even before the serve has hit the ball.

Abernethy (1991) reviewed the critical role of anticipation in expert performance in many racquet sports and cricket. (See the next chapter on Skill and Sport). Similarly, expert typists are looking well ahead at the text they are typing in any particular instant. The difference between the text visually fixated and the letter typed in a given instant (eye-hand span) increases with the typists' typing speed. High-speed filming of the moving of expert typists' fingers shows that their fingers are simultaneously moved towards the relevant keys well ahead of when they are actually struck. The largest differences in speed between expert and novice typists are found for successive keystrokes made with fingers of different hands because the corresponding movements can overlap completely after extended typing practice. When the typing situation is artificially changed to eliminate looking ahead at the text to be typed, the speed advantage of expert typists is virtually eliminated.[347] Similar findings relating to the amount of looking ahead and speed of performance apply to reading aloud[348] and sight-reading in music.[349]

Studies of expert performance have questioned the talent-based view that expert performance becomes increasingly dependent on unmodifiable innate components. Although these studies have revealed how beginners acquire complex cognitive structures and skills that circumvent the basic limits confronting them, researchers have not uncovered some simple strategies that would allow nonexperts to rapidly acquire expert performance, except in a few isolated cases, such as the sexing of chickens.[350] Analyses of exceptional performance, such as exceptional memory and absolute pitch, have shown how it differs from the performance of beginners and how beginners can acquire skill through instruction in the correct general strategy and corresponding training procedures.[351] However, to attain exceptional levels of performance, subjects must in addition undergo a very long period of active learning, during which they refine and improve their skill, ideally under the supervision of a teacher or coach. It is deliberate practice that appears to be necessary to attain these improvements.[352] In my view, Sir Joshua Reynolds got it right when he observed in his Discourse to Students of the Royal Academy on the 11th Dec 1769: "if you have great talents, industry will improve them: if you have moderate abilities, industry will supply those deficiencies."

By acquiring new methods and skills, expert performers are able to circumvent basic, most likely physiological limits imposed on serial reactions and working memory. The traditional distinction between physiological (unmodifiable physical) and cognitive (modifiable mental) factors that influence performance does not seem valid in studies of expert performance. For the purposes of the typical one-hour experiment in psychology, changes in physiological factors might be negligible; but once we consider extended activities, physiological adaptations and changes are not just likely but virtually inevitable. Hence we also consider the possibility that most of the physiological attributes that distinguish experts are not innately determined characteristics but rather the results of extended, intense practice.

A relatively uncontroversial assertion is that attaining an expert level of performance in a domain requires mastery of all the relevant knowledge and prerequisite skills. Our analysis has shown that the central mechanisms mediating the superior performance of experts are acquired; therefore acquisition of relevant knowledge and skills may be the major limiting factor in attaining expert performance. Some of the strongest evidence for this claim comes from a historical description of how domains of expertise evolved with increased specialisation within each domain. To measure the duration of the acquisition process, we analyse the length of time it takes for the best individuals to attain the highest levels of performance within a domain. Finally we specify the type of practice that seems to be necessary to acquire expert performance in a domain.

Most domains of expertise today have a fairly long history of continued development. The knowledge in natural science and calculus that represented the cutting edge of mathematics a few centuries ago and that only the experts of that time were able to master is today taught in high school and college.[353] Many experts today are struggling to master the developments in a small sub-area of one of the many natural sciences. Before the 20th century it was common for musicians to compose and play their own music; since then, distinct career patterns have emerged for composers, solo performers, accompanists, teachers, and conductors. When Tchaikovsky asked two of the greatest violinists of his day to play his violin concerto, they refused, deeming the score unplayable.[354] Today, elite violinists consider the concerto part of their standard repertory. The improvement in music training has been so considerable that according to Roth (1982), the virtuoso Paganini "would indeed cut a sorry figure if placed upon the modern concert stage."[355] Paganini's techniques and Tchaikovsky's concerto were deemed impossible until other musicians figured out how to master and describe them so that students could learn them as well. Almost 100 years ago the first Olympic Games were held, and results on standardised events were recorded. Since then records for events have been continuously broken and improved. For example, the winning time for the first Olympic Marathon is comparable to the current qualifying time for the Boston Marathon, attained by many thousands of amateur runners every year. Today amateur athletes cannot successfully compete with individuals training full time and training methods for specific events are continuously refined by professional coaches and trainers.

In all major domains there has been a steady accumulation of knowledge about the domain and about the skills and techniques that mediate superior performance.

In almost every domain, methods for instruction and efficient training have developed in parallel with the accumulation of relevant knowledge and techniques. For many sports and performance arts in particular, professional teachers and coaches monitor training programs tailored to the needs of individuals ranging from beginners to experts. The training activities are designed to improve specific aspects of performance through repetition and successive refinement known as deliberate practice.

From surveys of the kinds of activities individuals engage in for the popular domains, such as tennis and golf, it is clear that the vast majority of active individuals spend very little if any time on deliberate practice. Once amateurs have attained an acceptable level of performance, their primary goal becomes inherent enjoyment of the activity, and most of their time is spent on playful interaction. The most enjoyable states of play are characterised as flow,[356] when the individual is absorbed in effortless engagement in a continuously changing situation.

In Flow, the Psychology of Optimal Experience (1990) the Hungarian-American psychologist Mihal Csikszentmihalyi identified a type of highly focussed mental state conducive to productivity and creativity. He called it flow. He described flow as "being completely involved in an activity for its own sake." The ego flows away. Time flies. Every action, movement and thought follows inevitably from the previous one, like playing jazz. Your whole being is involved and you're using your skills to the utmost. Csikszentmihalyi advanced a model of creativity that stresses the importance of 'specific domain' to make a meaningful contribution within a culture, a time and place in history where many variables come together to form a situation where creativity can bloom. Renaissance Florence is sometimes cited as a domain where many such factors came together. The city was a financial and political powerhouse. The rich were encouraged to advertise their wealth and power through great art in which in turn attracted artists, sculptors, architects, all the seeds needed for a harvest of creativity. In his work Creativity in 2014 Csikszentmihalyi outlines the creative person as having "a sense that one's skills are adequate to cope with the challenges at hand, in a goal directed, rule bound action system." To achieve this sense the individual absorbs and develops cultural information. In some individuals this knowledge will reach such a level that they will be selected by the gatekeepers in the field for inclusion into the creative domain. His view is that the domains can be transmitted from one generation to the next through imitation and instruction. This cultural view of creativity goes against that of many thinkers such as Kant and Freud, who saw the spark of individual genius as the source of creativity. The processes of creativity are carried out in various domains including sport. Creativity need not be the exclusive preserve of those initiated in the higher forms of the relevant endeavour. In sport the athlete and spectator can have an input into the experience and appreciation of the sporting moment(s) and its significance. Likewise, those who read a novel, view a play or a film or look at a painting can also have an input into the meaning of a work of art and perhaps suggest new meanings which the creator might not have intended or seen. Examples of flow are referred to in detail later in Chapter 37 "Consciousness, Perception and the Senses in Sport." It is clear that creativity like genius is an elusive quality. No one theory can entirely encapsulate it. It has a touch of the magical and in Platonic terms 'inspired'—a break from the gods. They both represent and arise from a form of disinterested intellectual curiosity which G.M. Trevelyan, the distinguished historian characterised as the "life blood of real civilisation" (see Preface VIII English Social History). However, as John Dryden famously observed "Great Wits are sure to madness near allied, And thin partitions do their bounds divide."[357]

During play even individuals who desire to improve their performance do not encounter the same or similar situations on a frequent and predictable basis. For example, a tennis player wanting to improve a weakness, such as a backhand volley, might encounter a relevant situation only once per game. In contrast, a tennis coach would give that individual many hundreds of opportunities to improve and refine that type of shot during a training session.

Individualised training of students, who begin as very young children under the supervision of professional teachers and coaches, is a relatively recent trend in most major domains. It was only in 1756, for example, that Wolfgang Amadeus Mozart's father published the first book in German on teaching students to play the violin.

The differences in performance between experts and beginners are the largest that have been reliably reproduced with healthy, normal adults under controlled test conditions. From the life-long efforts of expert performers who continuously strive to improve and reach their best performance, one can infer that expert performance represents the highest performance possible, given current knowledge and training methods in the domain. Individuals' acquisition of expert performance is thus a naturally occurring experiment for identifying the limits of human performance. It is hard to imagine better empirical evidence on maximal performance except for one critical flaw. As children, future international-level performers are not randomly assigned to their training condition. Hence one cannot rule out the possibility that there is something different about those individuals who ultimately reach expert-level performance.

Nevertheless, the traditional view of talent, which concludes that successful individuals have special innate abilities and basic capacities, is not consistent with the reviewed evidence. Efforts to specify and measure characteristics of talent that allow early identification and successful prediction of adult performance have failed. Differences between expert and less accomplished performers reflects acquired knowledge and skills or physiological adaptations effected by training, with the only confirmed exception being height.

For a long time the study of exceptional and expert performance has been considered outside the scope of general psychology because such performance had been attributed to innate characteristics possessed by outstanding individuals. A better explanation is that expert performance reflects extreme adaptations, accomplished through life-long effort to demands in restricted, well-defined domains. By capturing and examining the performance of experts in a given domain, researchers have identified adaptive changes with physiological components as well as the acquisition of domain-specific skills that circumvent basic limits on speed and memory. Experts with different teachers and training histories attain their superior performance after many years of continued effort by acquiring skills and making adaptations with the same general structure. These findings imply that in each domain, there is only a limited number of ways in which individuals can make large improvements in performance. In my professional world, Justice Holmes once remarked "the hallmark of genius in a great lawyer or jurist, was his ability to cut through technicalities and go for the jugular." When mediating mechanisms of the same type are found in experts in very different domains that have evolved independently from each other, an account of this structure based on shared training methods is highly unlikely.

This chapter's heading links the concepts of Genius, Talent and Exceptional Ability. Is there a common conceptual thread running through them? They all share to a lesser or greater extent such common features as being the result of creativity, inspiration and imagination. But that only raises further questions about their meaning and significance in the scheme of things. Do they all essentially mean the same thing? What is the connection between creativity and inspiration? Where do inspirations come from? Do they all arise out of the workings of the unconscious mind?

This is a topic (the unconscious mind) which I have discussed in extenso throughout this book both in relation to sport and other domains. According to

Greek philosophy to be inspired is, as I have said earlier, to be the breath of God. That description coheres with the phenomenology of insight, the way an insight feels like it didn't come from you and it is mysterious to you how it arose. Philosophers now do not invoke divine inspiration but explain the phenomenon as the operation of the unconscious mind. But creativity is not simply an unconscious process. In Zen and the Art of Motorcycle Maintenance, Robert Pirsig wrote about the relationship between the mechanic's mind, hand and eye, as one of constantly assessing a problem and making the changes that seems to be called for and then reassessing and making further changes, in a continually unfolding creative process. This process, Pirsig claimed, was what united motorcycle mechanics with sculptors and other artists.

What is clear is that you can't create anything significant without conscious preparation. A person has to put in a huge amount of effort to learn the skills, concepts and other elements of the relevant domain.

In the next chapter, I look at the concepts of skill and perception which are relevant to sport, I also in later chapters consider the role of the conscious and the unconscious in sport and the time delays of consciousness and the embodied mind and the cognitive unconscious. It is a fascinating philosophical area of thought. There are so many aspects to it. Creativity emanates from the conscious and unconscious workings of the mind and body. In 2010, Professor Berys Gaut wrote a paper called 'The Philosophy of Creativity' in the journal Philosophy Compass and included a survey of issues such as whether the creative process is rational, whether creativity is a virtue and the relationship between creativity and knowledge. Gaut's view was that philosophers had so far paid little attention to psychology apart from the computational theory and the cognitive psychological approach.

That situation has now significantly changed as I have earlier illustrated in this chapter. Furthermore, in 2014 The Philosophy of Creativity was published by a whole range of thinkers including Gaut. Earlier than this in 1989 Professor Christene Battersby in a book entitled "Gender and Genius: Towards a Feminist Aesthetics," Battersby noted that historically the creative energies of the 'genius' were ascribed to sublimated male sexual energies and linked to a highly individualised male self. Women were commonly said to lack the individuality necessary for true creativity and also refused the ability to transcend or to sublimate their bodily instincts and reproductive capacities in the same way as the exceptional male 'geniuses'. Creative women credited with similar psychic powers were said not to be fully female. By the end of the nineteenth century, it had become a cliché to assert with Cesare Lombroso in the Man of Genius (1891) "there are no women of genius; the women of genius are men." In Philosophy Now, December 2022/January 2023, Issue 153, Battersby returned to this theme.

She said "Looking at the ways in which western and northern cultures were have over-valued an individualised mode of psychic creativity and under-valued the procreativity of female bodies, can provide us with resources for imagining creativity differently."

She argues that women in our culture are on the one hand, taught to think of themselves as not different from men. She further notes that the female subject's position is also, historically, more irredeemably bodily and less psychically isolated than that of a typical male-bound, through relationships of love, care, childbearing and childrearing, within interpersonal relationships to materiality and also to other embodied selves. She concludes: "Focusing on natality, as well as on social and maternal entanglements, can help us to reimagine creativity in a much more cooperative and dynamically interactive way."[358] Without entering the gender debate I remain convinced the interaction of the conscious and unconscious mind is the key to unlocking the door to understanding genius, talent and exceptional ability. They are a combination of the conscious and the unconscious working in tandem. The differences between the concepts, as Bertrand Russell said of mentality "it is all a matter of degree." They are not so much conceptual matters but largely empirical issues which philosophically need to be informed by findings from psychology, neuroscience and other cognitive sciences which I have discussed earlier in this chapter and elsewhere in this book. Maybe after all Thomas Edison will be proved "Genius is one percent inspiration and ninety-nine percent perspiration."

It is also to be remembered that a genius can be a complete pain. As Nietzsche noted: "A man of genius is unbearable, unless he possess at least two things besides: gratitude and purity."

In the context of a book about sport or beyond the concept of sport, can a sportsman be a genius? On the death of Pelé, the great Brazilian footballer, The Times amongst other commentators labelled him a genius.[359] He was described as a divine talent. He certainly had the beyond factors. Ferenc Puskas, the great Hungarian footballer said of him "The greatest player in history was Di Stefano. I refuse to classify Pelé as a player. He was above that." Bobby Moore took the view "Only 5ft 8in tall he seemed a giant of an athlete on the pitch—perfect balance and impossible vision." The coach of the Argentine 1978 World Cup winning side César Luis Menotti, said "When you talk about football don't bring Pelé into it because he is from another planet." The gift from the gods, the divine talent, being from another planet is a hallmark of genius generally and not simply in the sporting context. Someone with a talent or ability not just beyond the norm but beyond the horizon from another planet is arguably genius. I have referred earlier to the cartoonist Heblock's depiction of Einstein. He drew a circle of the world and then marked with an (x) a spot outside the circle with the words "Einstein lived here!"

Alex Bellos wrote in his book Futebol: "The Brazilian Way of Life," "Pelé was more than talismanic. As the most successful athlete in the world's most popular sport, he was also a powerful symbol of Black sporting excellence," The Times journalist Andrew Downie said "Pelé, like Muhammad Ali, his only sporting equal in the twentieth century, was a hero to people from ethnic minorities who had never before seen someone like them succeed on their own terms." Comparisons are odious. On his death arguments have ensued about his footballing ability compared with his Argentinian rivals Maradona and Messi. But does that matter. The question is can he be purely characterised as a genius in his own domain football. No one could possibly suggest that he nor indeed Muhammad Ali were any more than sporting genius. The concept of genius can be qualified and limited by the domain. I therefore agree with Bill Shankly that the word genius "is not applicable to football" in the sense of Michelangelo, Einstein, Wittgenstein or other great thinkers in our world. In that context, genius is in a different 'country'. People are different there. But in sport people like Muhammad Ali, Pelé, Michael Jordan, Maradona and Messi have taken the concept of sport to different levels. The concept of hero is different from genius. Sometimes both are demonstrated in one domain. Muhammed Ali, Pelé, Michael Jordan, Maradona and Messi are in that category and beyond the horizon.

Creativity and Constraints

Writing in the Sunday Times on February 19th 2023, cricket correspondent Ed Smith, a former National Cricket Selector, said creativity cannot be completely captured, defined, repeated and rolled out as a system. He then somewhat paradoxically attempts a definition: "…Creativity is intrinsically antisystemic and impossible to coerce. That is its definition. You can only create space for creativity, you cannot force it on people."

The use of 'create' and 'creativity' is somewhat tautological but the sense of creating space is meaningful in sport as well as in other domains. Great players create space in which to operate. I have referred to Barry John as running in a "different dimension of space and time." Great batsmen like Don Bradman, Peter May, Vivien Richards and Sachin Tendulkar created time, so did the footballers George Best, Pele, Maradona and Messi and tennis players Roger Federer and Djokovic.

In other domains like poetry in the form of the sonnet the poet has to operate within a "narrow plot of land." In the law, too, constraints in the form of rules create constraints under which the lawyer operates. In a later chapter, I look at John Elster's constraint theory of sport in particular boxing. To be fair to Smith he was expressly referring to bureaucratic and coaching constraints. However, talent and creativity manifest themselves notwithstanding these constraint. Indeed working within these constraints is the hallmark of sport and other creative pursuits. Creativity does not and cannot work within a vacuum. All definitions of sport emphasise the constraints it imposes and which are coercive. That is why Nietzsche's concept of overcoming is so apposite. Great sportsmen thrive and take advantage of those constraints and exhibit prodigious talent and creativity by overcoming or circumventing them. That is the fundamental nature of most creative people. They can break the rules and do things that other people less talented find impossible. In that sense I agree with Ed Smith that 'creativity' "antisystemic and impossible to coerce." That is why it takes courage to be creative and why Wittgenstein remarked that genius is 'courage in talent'. Constraints of any form do not necessarily inhibit creativity but often stimulate it. To break the shackles of convention and conformity are clear signs of the creative and talented performance Max Boyce sang:

"Disaster struck this morning when a fitter's mate named Ron cracked the mould of solid gold that once made Barry John."

Genius, talent and creativity are metaphorical and symbolic moulds of solid gold. They are to be treasured and stored in our memory forever and a day. These memories represent the true reality of our world.

De Tocqueville famously referred to the tyranny of the majority. I have extensively considered this elsewhere in relation to democracy (see my lecture on Democracy and Judicial Review given at The Law Society in London, 2018) In a different context the historian Daniel Bourstin, has written about the tyranny of common sense. It is a fascinating area of debate. Common sense is a paradoxical and ambivalent concept. One has to tread carefully about drawing any conclusions about its significance. Bertrand Russell, a hero Philosopher King to me, defined Aristotle "as Plato diluted by common sense". I do not know whether that was intended to be a compliment or a criticism but notwithstanding that observation I steadfastly remain a Platonist. Dilution detracts. It reduces impact and significance. Genius is uniqueness and by its very nature the antithesis of the common view. Aristotle wrote "What everyone believes is true". And is self-evidently nonsense and manifestly not 'common sense'.

Genius, Talent, Exceptional Ability and Common Sense

It is often said about highly intelligent, clever and talented people that they lack common sense. I heard it said 'ok he's a great, fantastic player but he is as daft as a brush'. Many examples come to mind but for obvious reasons I will not name any. I will leave readers to personalise observation! So, what is common sense? Can a so-called genius lack common sense? And is it a negative or positive characteristic? In my opinion like so many things in life, it all depends on the circumstances. I suppose the key to it is how people make decisions. People invariably decide things from feeling and not from reason. We react from the impression we form from a number of things on the mind, on the basis of the impression we form of what is true and well founded, though you may not be able to analyse or account for it in several particulars, in a gesture you use, in a look you see, in a tone you hear, not from reason or rules. You judge the expression, propriety, and the meaning from habit, not from reason or rules. Reason is the interpreter and critic of nature and genius, not their law giver or judge. For instance, in my world of advocacy there is a clear distinction between eloquence and wisdom, between ingenuity and common sense. A man may be articulate and able in explaining the grounds of his opinion and yet may be a mere sophist because he sees only half of the subject. Another may feel the whole weight of a question, nothing relating to it may be lost on him. That is the wise man. Edmund Burke expressed the opinion that there is nothing so true as habit and, in some respects, I would say that the same could be said of common sense. But there can be bad habits and inadequate common sense. Common sense is tacit reason that it can also be bad reasoning depending on the circumstances. Conscience likewise is the tacit sense of right and wrong, or the impression of our moral experience and moral apprehension. It works unseen and it is regarded as instinctive. Common sense and conscience are important instinctive reactions to what is perceived to be truth and nature. But in order to be enduring and applicable must bear the test and abide the scrutiny of the most severe and patient reasoning. If they do not, they result in poor conclusions. Depending on the circumstances common sense can be good and bad. Common sense should act as a check weight on sophistry and suspend our rash and superficial judgements. It is only a judge of things that fall under common observation. It rests upon the simple process of feeling; it anchors in experience. It is not nor can it be the test of abstract, speculative opinions and manifestations. Genius can defy common sense and does so often in science, especially physics and mathematics. Quantum physics is a good example. The scientific, artistic and creative minds are not shackled by it. They are counter-intuitive. Which superficially can be attacked as contrary to common sense, they are often the hall mark of genius, talent and exceptional ability in thought and ability. At first flush they might even be perceived as daft or mad confirming Dryden's conclusion that 'great wits are sure to madness near allied. And thin partitions do their bounds divide'. In some circumstances common sense is inimical to genius, talent and exceptional ability and vice versa. We form our impressions from intuition, counter – intuition and unconsciously formed opinions. Common sense is the result of the sum-total of such unconscious impressions. Genius and taste depend much upon the same principle exercised on loftier grounds and in more unusual combinations. Sir Joshua Reynolds in his Discourses said that all the arts which we have 'addressed themselves only to two faculties of mind, its imagination and sensibility'. Capacity is not the same thing as genius or indeed common sense. Capacity may be described as to relate to the quantity of knowledge, however maybe acquired;

genius as to its quality and the mode of acquiring it. Capacity is a power over given ideas or a combination of ideas. Genius is the power over those which are not given, and for which no obvious or precise rule can be laid down. Capacity is the power of any sort; genius is power of a different sort from which as yet been shown. Retentive memory, a clear understanding is capacity but not genius. It requires good capacity to play well at chess but chess is a game of skill as is sport generally and not of genius. Know what you will of it, the understanding still moves in certain tracks in which others have stood before it, quicker or slower with more or less comprehension and presence of mind. The greatest skill strikes out nothing for itself, from its own peculiar resources. The nature of the game is the thing determinative and fixed. There is no royal or poetical road to check-mate and adversary. There is no place for genius but in the indefinite and unknown. The discovery of the Binomial Theorem was an effort of genius; but there was none shown in Jedediah Buxtons being able to multiply a figure 9 in his head. He was simply a man of capacity who possessed considerable intellectual riches. A man of genius finds out a vein of new core. Originality is seeing nature differently from others, and yet as it is in itself. It is not singularity nor affectation but the discovery and display of a new and valuable truth. All the world do not see the whole meaning of the world they have been looking at. Habit blinds them to some things; It is short sightedness to others. Every mind is not a gauge and measure of truth. Nature has her surface and dark recesses. She is deep, obscure and infinite. It is only minds on whom she makes her fullest impressions that can penetrate her shrine. It is only those whom she has filled with her spirit that have the boldness or the power to reveal her mysteries to others. But nature has a thousand aspects, and one man can only draw out one of them. In my view who ever does this is a man of genius. A little originality is more esteemed and sought for than the greatest acquired talent because it throws a new light upon things, and is peculiar to the individual. The other is common and maybe had for the asking to any amount. I have written earlier of Gilbert Ryle's magnificent philosophical work The Concept of Mind in which he expresses his views on the concept of multiple intelligence. I glean from that there is no such thing as intelligence in all things. The truth is that there are many forms of intelligence. There are very few of us, if any who have the ability to do all things well in this life. The supreme accomplishment is to find one thing you do well and do it. Genius is the ability to do one thing uniquely and better than anyone else. In Shakespearean terms why should a man do more than his part? In my opinion there is no greater impertinence than to ask if a man is clever out of his profession. For example, no man can play all the harlequin tricks of sport let alone philosophy, science, law and art. In conclusion I am reminded of William Hazlitt who concluded in Table Talk (1959) "Good nature and common sense are required from all people; but one proud distinction is enough for any one individual to possess and aspire to".

Chapter 10
Sporting Skill and Perception

What'er he did was, done with so much ease, in him alone it wasn't natural to please.
Absalom and Achitophel-Dryden

The most powerful drive in the ascent of man is his pleasure in his own skill. He loves to do what he does well, and having done it well, he loves to do it better.
Jacob Bronowski (1973), *The Ascent of Man*

Few things are impossible to diligence and skill.
Samuel Johnson

Every habit and faculty is preserved and increased by correspondent tactics—as the habit of running by running.
Epictetus

Skill is knowledge of any art together with expert ability to put that knowledge to use.[360]

"It is the embodied capability of a particular human subject."[361]

It is at once a form of knowledge and a form of practice. It is both practical knowledge and knowledgeable practice.[362]

Sporting activities may be considered as skills or as being comprised of skills, and the degree of proficiency attained by the individual reflects his skill level.[363]

Though the exercise of skill depends on perceptual knowledge, such knowledge is obtained directly, in the course of active engagement with other persons and things—the environment. It does not presuppose the processing into images, by some devices known as 'intellect', of 'sense-data' delivered by receptor organs of the body.[364]

Skill is a relative quality and is not to be defined in absolute terms.

Ingold gives the example of playing the cello.[365]

His experience is that when he sits down to play, everything falls naturally into place—the bow in his hand, the body of the instrument between his knees—so that he can launch directly, and with the whole of his being, into the music.

He dives in, like a swimmer into water, and loses himself in the surrounding ambience of sound. This is not to say that he ceases to be aware or that his playing becomes simply mechanical or automatic; quite the contrary, he says he experiences "a heightened sense of awareness." The awareness "is not of my playing, it is my playing."[366]

As with speech and song, the performance embodies both intentionality and feeling. The intention is carried forward in the activity itself. It does not consist in an internal mental representation formed in advance and lined up for instrumentally assisted, bodily execution.

The novice becomes skilled not through the acquisition of rules and representations, but at the point where he or she is able to dispense with them. They are like the map of unfamiliar territory, which can be discarded once you have learnt to attend to the features of the landscape and can place yourself in relation to them. The map can be a help in beginning to know the country, but the aim is to learn the country, not the map.[367]

The cello teacher may place marks on the finger board to show the novice where to put his fingers in order to obtain different notes. The novice is thereby enabled to feel for himself the particular muscular tensions in the left hand, and to hear the resulting intervals of pitch.

Having learnt to attend to these things, his fingers will find their own place, and the marks which serve no further purpose, can be removed. The same can apply to any other branch of apprenticeship in which the learner is placed, with the requisite equipment, in a practical situation, and is told to pay attention to how 'this' feels or how 'that' looks or sounds—to notice those subtleties of texture that are all important to good judgment and the successful practice of a craft.

This kind of learning exemplifies what has been called "understanding in practice."[368]

The latter phrase denotes the theory of learning long favoured by cognitive science, according to which effective action in the world depends on the practitioner's first having acquired a body of knowledge in the form of rules and schemata for constructing it. Learning the process of acquisition, is thus separated from doing, the application of acquired knowledge.

Understanding in practice, by contrast is a process of enskillment, in which learning is inseparable from doing, and in which both are embedded in the context of a practical engagement in the world, that is, in dwelling. The kind of know-how thus gained, "constituted in the settings of practice, based on rich expectations generated over time about its shape, is the site of the most powerful knowledgeability of people in the lived-in world."[369]

Skilled movements tend to break down when they are consciously thought about. The learner should be spending his time rather on increasing his understanding of the needs of a situation by more specific and more precise interpretation of the sense impressions he receives.

Perhaps the acuity we should strive for is not enhanced general body awareness but rather a more sharply defined and specific sensitivity to what is happening in those key manoeuvres upon which the success or failure of a complex movement pattern may depend.[370]

But skill is also affected by the amount of insight which in turn depends on what the individual perceives.

At any time, there are a very large number of stimuli present both outside and inside the individual. An individual cannot perceive them all and so he selects those which he considers important and groups them so that they make sense to him. The stimuli arrive through many different sensory channels which may change in importance as learning proceeds.

The stimuli from the visual, auditory, tactile, kinaesthetic and other sense organs are all organised to form a perceived whole. Frequently there are gaps in the actually observed stimuli and these he fills up on the basis of what he has experienced before.

The footballer may move into a space at the appropriate moment. Two hockey players may look the same when hitting for goal, but the goalkeeper may react differently through previous knowledge of them.

The unskilled performer may notice a number of stimuli but he will be unable to perceive which are the important ones or what the response should be. He will tend not to perceive any pattern to the stimuli and since the capacity to take in information is limited the number of stimuli to which he can pay attention will be relatively few.

The skilled person on the other hand possesses a mental framework which takes into account a large number of the stimuli which have occurred before. He notices small changes from the expected display and is therefore able to react to them quickly.

First, the skilled man perceives the change in the pattern of stimuli and secondly he takes appropriate action to meet the change.

Prior experience greatly affects how and what is perceived because individuals are continually trying to fit new information into some general framework. This framework which is usually based on previous experience is both spatial and temporal. When a game is being played, the other players and the ball are perceived as located in space and their positions are related to the lines of the field and to each other.

The events which occur are also noted as being related in time, and series of events can be appreciated as forming sequences or rhythms. It is on the basis of such a framework that anticipation of future events can be made.

Timing and anticipation have been put forward then as criteria of skill.

High degrees of skill coincide with high degrees of spatial precision and timing. During skilled performances, responses to stimuli are set in appropriate sequential order.

In javelin-throwing, for instance, timing refers to the duration of the constituent actions in a pattern of movement. A javelin-thrower with good timing so orders the duration of the constituent actions in the throw that maximal force is produced at the appropriate moment. The muscle fibres required for the action respond in such a way as to produce maximum force at the best moment to result in maximum effect.

In this case, then, good timing is only attained when the technique is in accord with the mechanical requirements of the skill as related to each individual's body.

Similarly in skills such as high and long-jumping, shot-putting and discus-throwing and also in sprinting or hurdling short distances.

So far as these types of skill are concerned any smoothness results from the fact that mechanical efficiency is greater when movements flow into one another and no excess muscular contractions are executed. No mechanism is perfect; noise (amount of uncertainty) may be present anywhere—in the encoder, in the decoder or in the transmission where processing takes place. Therefore it is a rarity when the response equals the stimuli.

If however the movements have to be performed over such a period of time that decisions on pacing have to be made then timing may refer to the ordering of a series of responses so that the best results in the long run may be secured.

Thus in a long-distance running race good timing does not refer so much to the technique of the winner but to his ability to arrange his efforts over the distance to the best advantage. Here he has to respond to a certain extent to the actions of his opponents and so external stimuli are of some importance.

There is a further complication in many activities such as racquet games, fencing, judo and boxing in that "the subject himself partly determines the temporal conditions in which the next response will have to be made."[371]

This partial control not only arises from the techniques used but also to a large extent from the ability to perceive and attend to the important changes in the display. Timing in these circumstances can be defined as creating the most favourable temporal conditions for response.

Conditions for response are considered to be favourable when the response can be made with the least hurry, at the best moment, while leaving the organism in a satisfactory post-response state.

Bad timing in the present makes timing in the immediate future more difficult whereas good timing in fact provides the optimum temporal conditions for response.

In cases of this kind good timing is not directly observed but it is inferred from the fact that the performer is able to make unhurried and smooth movements.

The good batsman seems to have all the time in the world to make his strokes and do what he wishes with the ball.

Good timing so far as the performer is concerned normally results in a feeling of ease and lack of effort relative to the results achieved.

There are limits, upper and lower, within which the timing can rapidly fluctuate and the essential adaptive spacing of actions still be maintained. Outside those limits only the skill breaks down.[372]

The tennis player can keep his timing even, on occasions, when losing. But if the pressure becomes too great, the person has not time to pay attention to the important signals in the display. This results in his having less time in which to act and so, at a certain point, his technique breaks down and any feeling of good timing is lost.

The same type of thing can also happen to a whole team of football or hockey. Pressure can be produced not only by the speed of the ball but also by the expectedness of the situation.

Frequently movement takes less time than deciding on the action and so, if the ball is put by the opponent in an unexpected place, the response is slowed down because of the need to make a sudden decision and pressure is increased.

Anticipation, the ability to look forward and judge correctly what is going to happen next, is therefore an aid to good timing in many circumstances. The person who can anticipate well can start to respond early to a situation which is arising and he therefore truly has more time in which to make the necessary movements.

Anticipation may occur in at least two ways.

When a footballer kicks a ball in the air its path is determined at the moment of impact. The unskilled performer may have to wait until the ball has almost landed before he acts. The skilled man on the other hand will estimate the future position of the ball on the basis of a small number of signals occurring very early on.

Similarly in cricket, the path of the ball is determined when it leaves the bowler's hand and, if the wicket can be relied on, the batsman can begin to prepare his stroke on the basis of signals arising from the bowler's action and on his knowledge of the bowler's abilities.

In these cases then anticipation is based on the ability to predict future events from the first few signals in an invariant sequence.

It is difficult for people to learn to anticipate in this way where conditions are bad.[373]

For instance, if cricket is practised on bad surfaces the batsman will not be dealing with an invariant sequence until after the ball has hit the ground and will therefore not learn to pay due attention to the bowler's behaviour. The tennis player who learns on poor courts will build up techniques which stand him in good stead in his local club, but which prove of little use in the championships of the world.

Poor conditions are notoriously great levellers of performance too, for the skilled performer can no longer use his superior anticipatory abilities to advantage and indeed they may even prove disadvantageous.

For example, a good tennis player knows how balls with top-spin react and therefore adjusts his movements accordingly when his opponent uses it. But on a bad surface, any bound may result and therefore the expert's response may be worse than that of the individual who normally waits to see what the ball will do before responding.

Anticipation may also occur where an invariant sequence is not involved. The skilled games player has to anticipate the actions of his opponents and of his team-mates. In this case his prediction cannot be certain, but must be based on the probability of certain events occurring. He has to be able to weigh up the odds. He is more likely to be accurate in his prediction than the unskilled because he has learnt the relative significance of various signals and therefore knows the ones to which he should pay attention.

Decision-taking is of major importance. Some games players never seem to be able to do the right thing at the right moment whereas others seem to have developed the ability of keeping some unorthodox move up their sleeve and producing it when the occasion demands it.

One of the most impressive scholars in the area of skilled performance for many years was A.T. Welford. He researched the human mechanisms that operate between sensory input and motor output, primarily using relatively simple laboratory tasks involving reaction time or the tracking of moving targets. His concept of these mechanisms as composing a "communication channel of limited capacity" indicates a desire to determine which mechanisms function, how and with what restrictions during human performance.

Mathematical formula and logic, along with extensive data, enabled Welford to propose 'laws' about a person's channel capacities. One of the most famous of his proposals is the single-channel operation of the human organism. An examination of an individual's ability to process information presented simultaneously and to perform dual tasks led Welford to postulate limitations in the channel capacity to attend to cues, process information, and respond effectively in a number of on-going acts. Welford also suggested a formula for predicting speed of arm movement as reflected by the nature and distance of the target and other variables.[374]

Welford theorises that movement time is determined more by central processes controlling movement than by factors of muscular effort involved. Choice reaction time is primarily affected by the translation mechanism. Performance is limited by the phasing and co-ordinating movement of the central mechanism.

The writing of Welford and other scholars concerned with skill acquisition and motor performance tend to be in agreement on the fact that the learner responds to both external or situational demands and internal (self-controlling and regulating) mechanisms that operate as response-produced stimuli. An understanding of how a person processes situational or response-induced information with what capacities and rapidity, suggests instructional techniques that would be favourable to the learner. Redundant information or too much information could be a waste of time or overtax the channel capacity. Too little information might result in inadequate cues and poor performance. It also might indicate that the channel capacity is not being used to its fullest advantage.

Assuming that man, like a computer, functions with higher order (executive) programs or routines and subroutines or subprograms, it would appear logical that executive routines function with higher-order subroutines, at early levels of learning, but they 'delegate' their authority to lower-order routines in later skill development, thus freeing the system to attend to other matters. As an example, the boy learning to play football for the first time concentrates very hard on dribbling the ball. There is little choice or chance for him to do anything else. Yet once the skill has been mastered, he apparently dribbles with very little conscious control over his activity and attends to such matters as previewing the game situation, thinking ahead about alternatives, and making decisions. The subroutines of control for dribbling have been freed, and other lower-order subroutines are activities for this role. Additional subroutines for other movements are operative. The executive programme can be broadened, that is, contain a greater goal.

Some scholars have gone beyond the descriptive terminology of man-computer comparisons and have demonstrated the hierarchical functional organisation of the nervous system. Jacques Paillard, [375] for instance, talks in terms of two levels of nervous systems structures as he analyses skilled movements, the lower motor-neuron keyboard and the upper motor-neuron keyboard. This better reflects the current view as to neuronal activity.

Skilled movement patterns depend upon the role of the corticomotor-neural tracts in conveying messages. The lower motor-neuron keyboard consists of the medulla and spinal cord, and directly controls peripheral activity. The upper motor-neuron keyboard is associated with critical involvement and represents highest-level control. The activities within and between these keyboards are integrated in a manner depending upon functional requirements. Less complex or better learnt responses will be performed in a nonvolitional manner and be under control of the lower keyboard. Complex routines are under the control of the upper keyboard. Complex routines are under the control of the upper keyboard. The ability to perform highly skilled acts will depend upon the internal organisation of the entire motor arrangement. The lower keyboard will respond accurately to the signals of the upper keyboard as a function of the upper keyboard as well as the close connection between the two boards.

Using neurological terminology and concepts, Paillard's work (carried out in 1960) attempted to identify executive programs and subroutines, discussing ways in which they work dependently and more or less independently.

But notwithstanding this research his findings were inconclusive. The science (the physical analysis) could not scientifically explain the metaphysical. It was beyond (meta) science. It could not identify why some people are more or less skilful or perceptive than others. Therein lies the mystery and wonder of sport.

Chapter 11
The Concept of Deliberate Practice[376]

Deliberate practice is above all an effort of focus and concentration. It isn't work and it isn't play but it is something entirely unto itself. – Geoff Colvin

In Talent is Overrated (2008) Geoff Colvin argues that top performers in sport, business, in science and in the arts owe their success to the intense practice of their skills rather than to some unique genius.

I am not so sure that he is right about that and it is certainly not as simple as that. In chapter 9 of my book Genius, Talent and Exceptional Ability I argue that success results from a rich tapestry and amalgam of factors combining in one individual at a given point in time of a particular domain. But deliberate practice is invariably a component part of it.

It isn't work and isn't play, but is something entirely unto itself.

We commonly use the term 'practice' when talking about two domains, sports and music, but that habit can lead us astray.

Our habitual use of the term in sports and music may stop us from thinking of how deliberate practice can be applied in other domains, such as business or science, in which we almost never think about practising.

Deliberate practice requires that one identifies certain sharply defined elements of performance that need to be improved we then work entirely on these.

The great soprano Joan Sutherland devoted countless hours to practising her trill—and not just the basic trill but the many different types (whole-tone, semitone, baroque). Tiger Woods has been seen to drop golf balls into a sand trap and step on them, the practice shots from the near impossible lie. The great performers isolate remarkably specific aspects of what they do and focus on just those things until they are improved then it's on to the next aspect.

Choosing these aspects of performance in itself an important skill. Noel Tichy, a professor at the University of Michigan business school and former Chief of General Electric's famous Crotonville management development centre, illustrates the point by drawing three concentric circles. He labels the inner circle 'comfort zone', the middle one 'earning zone' and the outer one 'panic zone'.

Only by choosing activities in the learning zone can we make progress. That is the location of skills and abilities that are just out of reach. We can never make progress in the comfort zone because those are the activities we can already do easily, while panic zone activities are so hard that we don't even know how to approach them.

Identifying the learning zone, which is not simple, and then forcing oneself to stay continuously in it as it changes, which is even harder—these are the first and most important characteristics of deliberate practice.

It can be repeated a lot. High repetition is the most important difference between deliberate practice of a task of performing the task for real when it counts. Tiger Woods may face that buried line in the sand only two or three times in a season and if those were his only opportunities to work on hitting that shot, he certainly wouldn't be able to hit it very well.

Repeating a specific activity over and over is what most of us mean by practice, yet for most of us it isn't especially effective. Two points distinguish deliberate practice from what most of us actually do.

One is the choice of a properly demanding activity in the learning zone. The other is the amount of repetition.

Top performers repeat their practice activities to stultifying extent.

Ted Williams, baseball's greatest hitter, would practise hitting until his hand bled.

Feedback on results is certainly available.

There are situations in which a teacher, coach or mentor is vital for providing crucial feedback.

It is highly demanding mentally.

Deliberate practice is above all an effort of focus and concentration. That is what that makes it deliberate as distinct from the mindless playing of scale or hitting of tennis balls that most people engage in.

The work is so great that it seems no one can sustain it for very long. Elite athletes say the factor that limits their practice time is their ability to sustain concentration.

The best practice in doing something is doing it. In all domains doing what you know you can do well is the key. And doing it time and time again is best practice. As a barrister I do not practice but I do it time and time again. Practice makes perfect and perfection requires practice. Gary Player, the great golfer, famously said "the more I practise the luckier I get."

Deliberate practice does not create talent and exceptional ability and certainly not genius. It simply accentuates and exploits it. In that regard I note the observation of Edward Lytton in Last Words "genius does what it must, talent does what it can,".

Furthermore, in this context Publius Syrus in his Maxims said "Practice is the best of all instructors". (Maxims 439)

The imprimatur should be for everybody "if you can do it just do it and the more you do it the better you get".

However, in the same way as you cannot get blood out of a stone you cannot create a genius through practice deliberate or not. In that respect Colvin in my view is wrong, it is practice that is overrated.

Chapter 12
Knowledge and Sport

There are two ways of acquiring knowledge, one through reason, the other by experiment.
Roger Bacon

Knowledge is the only fountain both of love and the principles of human liberty.
Daniel Webster (1782–1851)

Knowledge is power.
Disraeli

The central fact of our epoch is that knowledge has grown; man's brain has not.
Geoffrey Pyke (1894–1948)

All I know is how little I know.
Socrates

The great end of life is not knowledge but action.
Thomas Huxley

Knowledge is more than equivalent to force.
Samuel Johnson

Knowledge in truth is the great sun in the firmament life and power are scattered with all its beams.
ibid Webster

Imagination is more important than knowledge.
Einstein

Knowledge has killed the sun, making it a ball of gas with spots. The world of reason and science...is the dry and sterile world the abstracted mind inhabits.
D.H. Lawrence

Once and for all, there is a great deal I do not want to know—wisdom sets bounds even to knowledge.
Nietzsche

Science is a procedure for testing hypothesis, not a compendium of certain knowledge.
Steven Jay Gould

The purpose of Polanyi's work *Meaning*[377] is to examine the development of individual knowledge—specifically, how people come to know things. He explores the development of 'personal knowledge' through 'from-to' knowing and the notions of subsidiary and focal awareness. Focal awareness is simply that which is consciously focussed upon. Polanyi uses hammering a nail into the wall to exemplify this idea. He writes, "When I use a hammer to drive a nail…I watch the effects of my strokes on the nail as I wield the hammer" and he feels the hammer's head strike the nail. His *focal awareness* is at the end of the hammer hitting the head of the nail and its relationship to the intended result of driving the nail into the wall. According to Polanyi, it is our focal awareness that we *attend* to in order to achieve the primary goal of our actions—in this case, the hammer striking the nail.

In contrast, *subsidiary awareness*, or tacit knowing, constitutes "items or particulars that we are aware of in the act of focusing our attention on something else, away from them."[378] It is a reliance on "the trained delicacy of eye, ear, and touch" that inform the activity currently being undertaken and focussed upon directly.[379] The 'subsidiaries' that comprise our tacit knowledge include a potentially massive body of personal knowledge learnt and embodied information—that we use or sense while performing specific focussed upon actions.

Polanyi's discussion of focal and tacit awareness significantly adds to the current analysis by supplementing and extending Moe's interpretation of the Background. Simply, Moe speaks in terms of abilities and dispositions—a 'readiness' to certain types of skills and behaviours.[380] This 'readiness' and 'tendency' to action implies higher level abilities. From Polanyi's perspective, those abilities and dispositions must be constructed from more fundamental bits of knowledge—those minutiae of personal knowledge that include felt sensations and simple movements.

He expands the notion of the Background by including not only higher-level learnt abilities but also the fundamental fragments of our knowledge—the available material that can be used in a variety of potential movement solutions which ultimately allows us to create those abilities.

He uses the idea of 'dwelling' to describe how subsidiaries merge into the knower's focal awareness and further to the development of skilled performances.

He explains now, "We can paralyse the performance of a skill by turning our attention away from its performance and concentrating instead on the several motions that compose the performance."[381]

Polanyi takes a step back from 'readiness' or 'tendencies' to act. It more broadly includes fundamental felt sensations and embodied simple movements that can potentially constitute many different types of tendencies and readiness needed in potentially diverse movement settings. For example, the subsidiary feeling of the hammer's handle in the hand can be similar to the feelings of a bat in a cricket player's hand or racquet in a tennis player's hand—bodily sensations much more fundamental than a gestalt 'readiness' to strike a ball.

His description of dwelling more clearly describes how the interactions among the triad of subsidiaries, focal targets and an integrating knower produce skilful performances. Polanyi's triad provides a useful model to explain why some performances are segmented, fragmented and unpolished while others are smooth, effortless and easy.

Simply, it more cleanly shows why the golfer who focuses on the smaller fragments of the swing is more likely to produce a fragmented and inaccurate shot. In addition, it also shows how the golfer, who dwells within those subsidiaries by focusing on the complete swing and not its individual parts, will most likely drive a more accurate ball.

He believes that a purpose or a goal motivates and guides the process of performing and mastering a skill.[382]

Moving to higher levels of skilful performance requires one to move beyond dwelling within the triad of subsidiary awareness, focal targets, and personal integration. Conscious effort is needed to acquire knowledge and skilfully apply it.

Polanyi provides a path that connects subsidiary knowledge and purposeful action. This does not occur in the manner described by classical cognitivism. "The imagination does not work like a computer," Polanyi writes, "surveying millions of possible useless alternatives; rather it works by producing ideas that are guided by a fine sense of their plausibility, ideas which contain aspects of the solution from the start."[383] There is spontaneity here—a smooth and intuitive move to the right solutions that are often felt implicitly rather than developed explicitly.

Purposeful or goal-directed action requires an active mind.[384]

Purposeful action arises from dwelling in subsidiary knowledge. The process begins by focally attending to the learning of skills. As we learn them, we acquire and embody these skills as subsidiaries for use in future problem-solving situations. From this experience and from dwelling within these subsidiaries, we find meaning in our lives. We integrate these subsidiaries constantly to produce our focal awareness—that which secures our conscious attention.

It is here that the interaction of intuition, imagination and ideas comes into play. Conscious and goal directed striving remains important to skilful performance. When confronted with a task, our intuition comes into play. We implicitly select plausible and useful subsidiaries given the situation. We then integrate the subsidiaries to form speculative solutions or at least "reduce the vagueness of the problem and offer a firmer guidance for the next push towards a possible solution."[385]

Skilled motor performance appears to build from learning and embodying simple skills and sensations. In turn, this creates a background of subsidiaries that generates possibilities. This background of tacit knowing creates one's perception of meaning in the world. Through intentional striving by performers, the imagination actively casts forth potential ideas and solutions to confronted challenges. This latter mechanism is catalysed through one's intuition which spontaneously integrates subsidiaries to create explicit solutions.[386] The rapid and automatic judgments and choices of sportsmen illustrate those skilled intuitions in which a solution comes to mind quickly because familiar cues are recognised. Intuitive judgments can be made with high confidence even where they are based on non-regressive assessments of weak evidence. Of course many judgments are a combination of analysis and intuition.

Mark Twain provided a useful description of the balance between non-conscious and conscious negotiation of the bike-riding practice.[387] As a beginner cyclist, Twain explained how he had to consciously focus his attention on balancing the bike by embodying an important cycling truth: when losing balance, one must turn in the direction of the fall, not away from it as beginners intuitively tend to do. Twain explains: "It is hard to believe this [rule] (turning into the fall), when you are told it…it is opposed to all your notions. And it is just as hard to do it, after you come to believe it. Believing it, and knowing by the most convincing proof that it is true, does not help it: you can't any more DO it than you could before; you can neither force nor persuade yourself to do it at first. *The intellect has to come to the front now.* It has to teach the limbs to discard their old education and adopt the new."[388] Twain's comments specifically note that he had to consciously focus on turning into the fall in order to learn and embody this necessary skill. Once learnt, this skill became only implicitly attended from and was, from that point forward, rarely a subject of the skilled bike rider's focal awareness.

After honing his cycling skills, Twain declared himself a skilled rider able to "steer as well as he wants," although he still needed to consciously "keep his attention on his business."[389] However the 'business' he was attending to was not any one specific skill or rule. Those were now part of his background—subsidiaries to be dwelt in. Instead, his new conscious attention to 'business' informed him of such challenges as the grade of the street, the consistency of the road, and the distance between himself and the kerb or passing carriages while he was riding a bike. Further, his new focal awareness, informed from his embodied, tacit knowing, guided him towards successfully negotiating those challenges. Such conscious attention is confirmed by a modern-day skilled cyclist. He recalls similar conscious attention to such things as "watch that pothole, keep a nice rhythm up this slight hill, enjoy the descent on the other side."[390] These riders, while highlighting Breivik's goal-directed thinking, also described how dwelling in Polanyi's subsidiary abilities manifested in higher skilled performance.

Chapter 13
The Mastery of Movement

There is a well-known Chinese story of the centipede which, becoming immobilised, died of starvation because it was ordered always to move first with its seventy-eight feet, and then to use its other legs in a particular numerical order. This story is often quoted as a warning against the presumption of attempting a rational explanation of movement.

Man's inner urge to movement has to be assimilated to the acquisition of external skill in movement.[391]

According to Laban[392] there exists an almost mathematical relationship between the inner motivation of movement and the functions of the body.

"Man moves in order to satisfy a need. He aims by his movement at something of value to him. It is easy to perceive the aim of a person's movement if it is directed to some tangible object—yet there exist intangible values that inspire movement."

Movement reveals many different things. It is the result of the striving after an object deemed valuable, or of a state of mind. Its stops and rhythms show the moving person's attitude in a particular situation. It can characterise momentary mood and reaction as well as constant features of personality.

Human movement, with all its physical, emotional and mental implications, is the common denominator of the dynamic actions of sport.

Sport is dynamic because phase of the activity fades away immediately after it has appeared. Nothing remains static and the leisurely inspection of details is impossible.

Laban describes the inner impulses from which movement originates as 'effort'.[393]

The components making up the different effort qualities result from an inner attitude (conscious or unconscious) towards the factors of movement, weight, space, time and flow.

It has been said that the plays and dances of primitive tribes originate in an endeavour to make themselves aware of certain selected effort combinations.[394]

Sport may well have had the same origin.

Apart from the awareness or rather combined with it, is the fixation of the selected effort combination in memory and movement habit. It is a peculiar kind of building up of ideas about movement qualities and their use. Laban introduced the idea of thinking in terms of movement as contrasted with thinking in words.[395]

Movement and thinking in his view were a gathering of impressions of happenings in one's own mind, for which nomenclature is lacking.

This thinking does not, as thinking in words does, serve orientation in the external world, but rather it perfects man's orientation in his inner world in which impulses continually surge and seek an outlet in sport, acting and dancing.[396]

Man's desire to orientate himself in the maze of his drives results in definite effort rhythms.

Tribal or national dances were created through the repetition of such effort configurations as are characteristic of the community. These dances often show the effort range cultivate by social groups living in a definite milieu.

The languid, dreamlike dance of the orient, the proud, fierce dance of a Spaniard, the temperamental dance of the southern Italian, the well-measured round dance of the Anglo-Saxon are examples of the effort manifestations selected and fostered during long periods. They have become expressive of the mentality of particular social groups.[397]

Laban suggests that an observer of tribal and national dances can gain information about the states of mind or traits of character cherished and desired within the particular community. Formerly, he says, these dances were one of the main means of schooling the young to adapt themselves to the habits and customs of their forbears. They are in this way as much connected with education as with ancestor worship and religion.[398]

"The urge to play and dance has thus developed into an astonishing variety of movement traditions in all fields of human activity."[399]

A great deal of significance is attached to dance in the development of man. Dance became an adjunct to fighting, hunting, loving and much else.

Again according to Laban in dancing, or movement thinking, man first became aware of a certain order in his higher aspirations towards spiritual life. He unconsciously learnt of both the contradictory and the balancing factors within his actions, but he did not know how to use and to control them.

Religious dances were fundamental to man's understanding of his environment. They represented those superhuman powers which, as he conceived, directed the happenings of nature and determined his personal and the tribal fate. He gave physical expression to certain qualities he noted in the actions of the superhuman powers.

Laban says

"In such personification of effort action, primitive man learnt the reconciling trend of events and in his movement-thinking he pictured the power behind it all as a god with gliding gestures."[400]

Gods as conceived by primitive man were the initiators and instigators of effort in all its configurations. They were symbols of the various effort actions.

"The strange poetry of movement that has found expression in sacred dance enabled man to build up an order of his effort actions, which, in essence, is valuable and understandable to this day."[401]

Laban believed this to be the fundamental basis of civilisation.

"Man alone has become aware of the gods. That is to say, man is the only living being who is aware of and responsible for his actions; and he has thus become king of creatures and lord of the earth. From effort awareness in ritual and national dances arise the conventions of the various forms of economic and political order in human society. In the teaching of children and the initiation of adolescents, primitive man endeavoured to convey moral and ethical standards through the development of effort thinking in dancing. The introduction to Humane effort was in these ancient times the basis of all civilisation."[402]

The flow of movement is strongly influenced by the order in which the parts of the body are set in motion and its control is intimately connected with the control of the movements of the parts of the body.

Body movements can roughly be divided into steps, gestures of arms and hands, and facial expressions.[403]

The astonishing structure of the body, and the amazing actions it can perform are some of the greatest miracles of existence. Each phase of movement, every small transference of weight, every single gesture of any part of the body reveals some intrinsic feature of our life.

Rationalistic explanations of the movements of the human body insist on the fact that it is subject to the laws of inanimate motion.

The weight of the body follows the laws of gravitation.

The skeleton of the body can be compared to a system of levers by which distances and directions in *space* are reached.

These levers are set in motion by nerves and muscles which furnish the strength needed to overcome the weight of the parts of the body that are moved.

The *flow* of motion is controlled by nerve-centres reaching to external and internal stimuli.

Movement take a degree of *time*, which can be exactly measured.

The driving force of movement is the *energy* developed by a process of combustion with the organs of the body.

The fuel consumed in the process is food.

There is no doubt about the purely physical character of the production of energy and its transformation into movement.[404]

The switching on and off of the energy and the regulation of the flow of movement according to the intensity of the life sustaining instinct might also be purely mechanical, but here occurs a break in our rationalistic explanation.

The motion of a falling stone is arrested when it reaches the ground or other support. The acceleration of the speed of its fall and path in space are constant. Both can be exactly calculated.

The movement of a dropping arm, however, can be arrested at any time by the control mechanism of the body. The stop is brought about by purely mechanical means, namely, by calling into play of an antagonistic muscle which suspends the arm in the air. Yet the cause of the stop is less easy to explain. The falling arm might be stopped because the moving person has realised that there is a dangerous object in the way and instinctive reflection demands the avoidance of injury, and, therefore, the cessation of movement. The individual has accumulated certain experiences of situations and objects which are bound to cause pain or injury, and, therefore, he tries to avoid them.

It is difficult to ascribe the remembrance of such experiences and the prompt reaction to them to a purely physical or even psychological mechanism. Mechanical theories have been advanced to explain these peculiarities of the behaviour of living beings, but they are not convincing.[405]

It is a mechanical fact that the *weight* of the body, or any of its parts, can be lifted and carried into a certain direction of *space*, and this process takes a certain amount of *time*, depending on the ratio of *speed*. The same mechanical conditions can also be observed in any counter-pull which regulates the *flow* of movement.

Man has a power which enables him to choose between a resisting, constricting, withholding, fighting attitude, or one of yielding, enduring, accepting, indulging in relation to the '*Motion Factors*' of weight, space and time, to which, being natural accidents, inanimate objects are subjects.[406]

This freedom of choice is not always consciously or voluntarily exercised; it is often applied automatically without any contribution of conscious willing.

But we can observe consciously the function of choosing movements appropriate to situations; that means that we can become conscious of our choice and can investigate why we so choose.

It can be observed whether people yield to the accidental forces of weight, space and time, as well as to the natural flow of movement in the sense of having a bodily feeling of them, or whether they fight against one or more of these factors by actively resisting them.

Mastery of the former is what makes sportsmen.

Every human movement is indissolubly linked with an 'effort'[407] which is its origin and inner aspect. Effort and its resulting action may be both unconscious and involuntary, but they are always present in any bodily movement; otherwise they could not be perceived by others or become effectual in the external surroundings of the moving person. Effort is visible in the action movement of the sportsman or dancer and it is audible in song or speech. If one hears a laugh, or a cry of despair, one can visualise in imagination the movement accompanying the audible effort.

The fact that effort and its various shadings cannot only be seen and heard, but also imagined, is of great importance for their representation, both visible and audible, by the sportsman and the dancer. He derives a certain inspiration from descriptions of movements that awaken his imagination.

The mode or style of movement used sheds a light on a particular aspect of bodily expression.

As with the dancer the sportsman's body follows definite directions in space—Directions form stops or patterns.

Visible movements in sport and dance engender in the spectator the reaction of feeling. The spectator's participation is aroused in action and reaction and in the outcome of the conflict. It is important not only to become aware of the various articulations in the body of their use in creating rhythmical and spatial patterns, but also of the mood and attitude produced by bodily action.[408]

Bodily actions produce alterations of the positions of the body, or of parts of it, in the space surrounding the body. Each of these alterations takes a certain time and requires a certain amount of muscular energy.

Laban suggested that it was possible to determine and describe any bodily action by answering four simple questions.

(1) Which part of the body moves?
(2) In which direction or directions of space is the movement exerted?
(3) At what speed does the movement progress?
(4) What degree of muscular energy is spent on the movement?[409]

Laban analyses the component parts of a thrusting kick of the right leg in a forward direction (say as in kicking a football).

The answers to the four questions would be

(a) The moving part of the body is the right leg.

(b) The region of space to which the movement is directed is forwards. The movement is straight.

(c) The muscular energy spent on the movement is relatively great. The movement is strong.

(d) The speed of movement is rapid. The relative pace is quick.

I would like to identify the essential component parts of movements in cricket batting.

Stance

First of all, there is the *stance* which is the spot where one or both legs supporting the weight of the body rest on the ground.

Transference of weight or steps

Each *step* creates a new *stance*.

Directions and levels of steps

The spatial *direction* of a step is relative to the immediately preceding *stance*.

Gestures with successive actions

Gestures are actions of the extremities which do not involve transference or support of weight. They can occur towards, away or around the body, and can be done with successive actions of the various parts of a limb.

Gestures with simultaneous actions

Movements of the extremities towards the body and away from it can also be performed with *simultaneous* actions of their various articulations. In these gestures the movements of hand, forearm, upper arm, shoulder-girdle or those of foot, foreleg, upper leg, pelvis, are started and completed at the same time.

Leg gestures may precede a step

Gestures may be done as a preparation to a transference of weight.

Leg gestures may follow a step

These gestures are felt as a result of the step, as, for instance, after a left step backwards, the right leg performs a folding movement across the left knee.

Gestures via several directions constitute definite steps of movement.

Among the spatial forms of movement, we can distinguish round, angular and twisted shapes as being basic.

Directions and levels of gestures

Directions and levels of arm and leg gestures are relative to the joint in which movement takes place.

Medium level of a leg gesture is at hip height, medium level of an arm gesture is at shoulder height. Movements reaching above these joints are high, those reaching below are deep. Gestures leading into the vertical line below, at, or above the moving joint are deep, medium, or high respectively.

Combined arm and leg movements

The body is our instrument of expression, through movement.[410]

"The body acts like an orchestra in which each section is related to any other and is part of the whole. Its various parts can combine in concerted action, or one part may perform alone as a 'soloist' while others pause. It is also possible that one or several parts take the lead and others accompany. Each action of a particular part of the body has to be understood in relation to the whole which should always be affected, either by participating harmoniously or by deliberately counteracting or by pausing."[411]

Extension of Gestures

The normal reach of our limbs, when they stretch away from our body without changing stance, determines the natural boundaries of the personal space or 'kinesphere' in which we move. This kinesphere remains constant in relation to the body, even when we move away from the original stance; it travels with the body in the general space.

The very great number of combinations of the sub-divisions of the body, time, space and muscular energy corresponds to the possible acts of movement which can be registered in a logical manner. The order shown in such combinations will be best observed by the movements used in dancing, since they are relatively large and clear, and therefore more easily recognised.

The analysis of bodily actions in sport, play, work and everyday behaviour is based on the same "thinking in terms of movement" as is applied for that of dance movements.

Bodily attitudes during movement are determined by two main action forms. One of these forms flows from the centre of the body outwards, while the other flows from the periphery of the space surrounding the body inwards to the centre. The two actions underlying these movements are those of 'gathering' and 'scattering'. Gathering can be seen in bringing something towards the centre of the body while scattering can be observed in pushing something away from the centre of the body. Many combinations of these two actions can be performed by our limbs each working independently.

The most natural way to move is the style of harmonious movement which was common in ancient Grecian and Roman cultures. Small diversions from the harmonious ideal never deform the movement to the extent that the attitude become disharmonious.

The movement habits of primitive tribes, however, appear to the European mind and feeling as grotesque or disharmonious.

The grotesque forms of Moorish dances were introduced at the time of the Crusades, but their asymmetric tortuousness was surpassed by movement fashions which can still be seen in the sculptures ornamenting our Gothic cathedrals.[412]

Movement styles have been in some way useful in a particular period or in a particular country, according to the main needs of the civilisation.

Such partial aesthetical and utilitarian conceptions which lead people to say that this or that tennis-player, skater, batsman or film star 'has style' descend frequently to tiny details of movement habits. The small difference of a gathering or scattering position of a tool, for instance, might provoke favourable or unfavourable opinion of a champion athlete or a film star. This subconscious evaluation of people's movement is practised by almost everybody.

Rhythm has been used as a language apart and rhythmic language conveys measures without words. It may be that language has grown out of man's own rhythmic instinct.[413]

We have lost this language of the body. Our approach to life differs all too fundamentally from that of our ancestors who first moved to the rhythm of the drum. Sport may be in part a recrudescence in our time of the sense of rhythm and mime in early dance.

Chapter 14
The Gunslinger's Lament—
When Reaction Beats Intention

Attack is the reaction I never think I have hit hard unless it rebounds.
Samuel Johnson

Get your retaliation in first.
Carwyn James

It's not what happens to you but how you react to it that matters.
Epictetus

Nobel laureate Niels Bohr considered why during a gunfight one who drew first was the one to get shot.[414]

He suggested that the intentional act of drawing and shooting is slower to execute than the reactive action in response, an idea grounded in the everyday trade-off between stimulus-driven behaviour and intentional, planned actions.

The distinction between different classes of action is not merely semantic: evidence for differential neural bases for intentional as opposed to reactive movements is provided by neurophysiology.[415] Further behavioural evidence points to a distinction between different types of movement and switching between the two modes of operation can result in a cost.[416] Modern research is generally supportive of the benefits associated with reactive movement consistent with Bohr's intuition and the gunslinger legend.

The relevant research demonstrates that reactive movements are associated with faster execution times and that this quickening of movement does not appear to relate to having another human as a model for one's own action. It is suggested that different cortical processing routes for the control reactive versus intentional movements, and that faster movement dynamics may constitute a basic property of reactive movement production. The suggestion of a distinction between reactive versus intentional movements is consistent with a range of previous studies that report changes in the balance of the involvement of a number of cortical and subcortical areas during the production of different classes of action.[417]

Previous behavioural work also suggests this distinction. For instance, countermanding the production of a intended movement to react to an external trigger can have a cost suggesting a delay imposed by switching between different modes of movement triggering.[418]

Under this paradigm, participants could be provoked to move sooner than they intended by seeing their opponents actions.

Reactive movement can be advantageous in producing faster execution times (albeit with increased error rates).

As a general survival strategy the evolution of a movement system of producing quick movements that support faster responses to the environment seems reasonable. However, within the context of a gunfight, a strategy based purely on reaction seems unlikely to increase evolutionary fitness as the advantage produced by reacting is far outweighed by the time taken to react to the opponent.

Anecdotal reports suggest that Bohr tested his original idea with colleague George Gamow with toy pistols, with the reaction Bohr apparently winning every duel. It may be that those victories can be ascribed to the benefits associated with reaction. But it is also possible that they suggest that Bohr was a crack shot, in addition to being a brilliant physicist!

But I doubt it.

What is the broader philosophical value of this phenomenon.

What significance should we afford to the above hypothesis as this is no more than what it is?

In general terms, the argument is that every-day behaviour involves a trade-off between planned actions and reaction to environmental events. The evidence from neurophysiology, neurology and functional brain imaging suggests different neural bases for the control of different movements types. The evidence is broadly to the effect that there is a reactive advantage in movement execution, whereby the same action is executed faster in reaction to an opponent. With subject analysis of movement times revealed a 10 per cent benefit for reactive actions. This was maintained when opponents performed dissimilar actions and when participants competed against a computer, suggesting that the effect is not related to facilitation produced by action observation. Rather faster ballistic movements may be a general property of reactive motor control, potentially providing a useful means of promoting survival.

However, reactions to stimuli often produce, albeit quicker than actions, unsatisfactory responses. In the courtroom responding too quickly to an argument often results in ill-considered comments and knee-jerk reactions to a well-prepared submission. It is not all about speed but the quality of the response that counts. I know this to my own forensic cost!

But quick reactions are invariably a characteristic of success in sport especially ball games like tennis, cricket, football, baseball and boxing. But even there it is more than just speed of reaction. Many other factors are present— timing, spatial awareness, focus, concentration, intuition, counterintuition, determination, strength, courage and imagination to mention but a few.

Speaking for myself on this theme 'et grosso modo' most of the serious mistakes I have made in my life have been my reactions to things rather than my planned intentions. The speedy reactions in our modern world using texts and emails are a vivid example of that. I suspect I am not the only person who feels like that mais c'est la vie.

Good reaction is not only the hall mark of the gunfighter, but most sportsmen. Most sports especially ball games require fast reaction and as we have seen in this chapter reaction beats action. Reaction is an evolutionary concept. Historically man has adapted to his environment and to the circumstances he finds himself in at any given time. Man is what he has become and is what he is because of the way in which he has reacted to the circumstances.

Chapter 15
Movement and Cognition

It is the mark of an educated mind to entertain a thought without accepting it.
Aristotle

There can be no transforming of darkness into light and of apathy into movement with emotion.
Carl Jung

All truly great thoughts are conceived by walking.
Nietzsche

With curious art the brain, too finely wrought, prays on herself, and is destroyed by thought.
Charles Churchill (1731–1764)

What we think or what we know or what we believe is, in the end of little consequence. The only consequence is what we do.
John Ruskin

Fritjof Capra wrote in *The Web of Life* "The new concept of cognition[419] the process of knowing, is thus much broader than that of thinking. It involves perception, emotion and action the entire process of life. In the human realm cognition also includes language, conceptual thinking and all the other attributes of human consciousness. The Santiago theory [of Maturana and Varela] provides, in my view, the first coherent scientific framework that really overcomes the Cartesian split. Mind and matter no longer appear to belong to this separate category but are seen as representing merely different aspects, or dimensions of the same phenomenon of life." Sport in its many forms reflects those different dimensions in so many ways.

In an article 'Neurology and the Soul'[420], Oliver Sacks, the neurologist, stated "There has always, seemingly, been a split between science and life, between the apparent poverty of scientific formulation and the manifest richness of phenomenal experience."

Later, he continues:

"Implied in all this is the necessity for an adequate concept of the individual and of mind, a concept of how individual persons grow and become, and how their growing and becoming are correlated with their physical bodies."

The infant immediately starts exploring the world, looking, feeling, touching, smelling, as all higher animals do, from the moment of birth. Sensation alone is not enough; it must be combined with movement, with emotion, with action. Movement and sensation together become the antecedent of meaning. This evolution of self, this active growth and learning and becoming of the individual, is made possible by 'selection', the strengthening of connections within neuronal groups in accordance with the individual's experiences (and needs and beliefs and desires). This process of selection cannot arise, cannot even start, unless there is movement—it is movement that makes possible all perceptual categorisation.

As Moshe Feldenkrais has said movement is a mystery.

"The execution of action by no means proves that we know, even superficially, what we are doing or how we are doing it. If we attempt to carry out an action with awareness—that is, to follow it in detail—we soon discover that even the simplest and most common of actions, such as getting up from a chair, is a mystery, and that we have no idea at all how it is done."[421]

Moshe Feldenkrais was an Israeli physicist interested in the mechanics and control of bodily movement.

The underlying idea had occurred to Feldenkrais over a period of time just by observing people, watching them with a physicist's eye and thinking how difficult it would be to get a motorised vehicle to move as the human body does. The more he thought about walking, the more he marvelled at the sophistication of the brain as a controller of movement. Somehow athletes could perform, dancers could dance fluently, sometimes magnificently and the incomprehensibly complicated physics, biomechanics and physiology of the body—the underlying ballet of bones and joints and muscles—would take off itself completely outside of awareness, confidently, silently and reliably orchestrated by the brain. But some people are not like that. Why should that be? Was there something wrong with their brain? He suspected that people just lose contact with their bodies.

Professor Shaun Gallagher in his seminal work "How the Body Shapes the Mind" believes that movement and the registration of that movement in a developing proprioceptive system (that is, a system that registers its own self-movement) contributes to the self-organising development of neuronal structures responsible not only for motor action, but for the way we come to be conscious of ourselves, to communicate with others, and to live in the surrounding world. Across the Cartesian divide, movement prefigures the lines of intentionality, gesture formulates the contours of social cognition, and, in both the most general and most specific ways, embodiment shapes the mind.[422]

There are two basic sets of questions that concern the structure of experience.

The first set consists of questions about the *phenomenal* aspects of that structure, and specifically the relatively regular and constant phenomenal features that we find in the content of our experience. In regard to embodiment it is necessary to explore to what extent and in what way an awareness of my body enters into the content of my conscious experience? To what degree and in what situations am I, as an experiencing subject, aware or unaware of my own body? Does intentional action, for example, involve an explicit or implicit awareness of the body?

The second set of questions focuses on aspects of the structure of consciousness that are more hidden.

These may be more difficult to get at because they happen *before we know it*. They do not normally enter into the phenomenal content of experience in an explicit way and are often inaccessible to reflective consciousness.

Gallagher uses the term *prenoetic* to signify these hidden aspects. The basic question can be phrased in this general way: To what extent, and in what ways, are consciousness and cognitive (noetic or mental) processes, which include experiences related to perception, memory, imagination, belief, judgment, and so forth, shaped or structured *prenoetically* by the fact that they are embodied?

This question gets at a different issue from questions about phenomenal aspects of consciousness, which may or may not include an awareness or sense of the body. To ask about the prenoetic effects of embodiment is to ask about what happens behind the scenes of awareness, and about how the body anticipates and sets the stage for consciousness.

The answers to these questions not only provide important insights into the nature of experience, but they also tell us something about the nature of the human person, sometimes more abstractly referred to as the *self*. In some fashion, quite obviously, the human person is embodied in human form and matter. The human body, and the way it structures human experience, also shapes the human experience of self, and perhaps the very possibility of developing a sense of self. If the self is anything more than this, it is nonetheless and first of all this, an embodied self.

There is no one methodology that will provide a full picture. As a result Professor Gallagher borrows from already established insights provided by a number of different disciplines: developmental psychology, neuropsychology, the neurosciences, cognitive linguistics, phenomenology, and philosophy of mind.

His approach involves the interpretation of a large amount of empirical data.

A distinction divides up a conceptual space in a certain way; it does not necessarily divide up reality in the same way. The questions about phenomenal and hidden (prenoetic) structures signify a theoretical distinction that from a behavioural or existential perspective may seem too abstract. The distinction between phenomenal and prenoetic structures, however, is a relative one, and is not meant to signify two different ontological categories or classes of things that pertain to embodied consciousness. Furthermore, if one takes this distinction as signifying something along the lines of manifest versus latent, or conscious versus unconscious, one would be led in the wrong direction. It may be best to think of it as a functional distinction, so that the lines drawn between phenomenal structure and prenoetic structuring may differ from one context to the next.

Gallagher is cautious about how distinctions work between conceptual and practical contexts because he adopts one that is controversial but he says productive—the distinction between *body image* and *body schema*.

He argues that if the clear and proper distinction is made, these concepts carve up the conceptual space in a way that leads to a productive understanding of embodied consciousness. This distinction helps to clarify not only perceptual experience and action, but also a variety of specific phenomena such as neonate imitation, phantom limbs, deafferentation, unilateral neglect, and the linguistic nature of gesture.

Proprioception signifies one of the specific areas where the distinction between phenomenal consciousness and physical body gets redefined. Proprioception is itself a complex phenomenon that is articulated in slightly different ways in different disciplines. On the one hand, neuroscientists may treat somatic proprioception as an entirely sub-personal, non-conscious function—the unconscious registration in the central nervous system of the body's own limb position. In this sense, it results in information about body posture and limb position, generated in physiological (mechanical) proprioception located throughout the body, reaching various parts of the brain, enabling control of movement without the subject being consciously aware of that information.[423] On the other hand, psychologists and philosophers sometimes treat somatic proprioception as a form of consciousness. One is said to be proprioceptively aware of one's own body, to consciously know where one's limbs are at any particular time as one moves through the world.[424] Thus proprioception can mean either non-conscious *information* or a form of conscious *awareness*. Conceptually, Gallagher tries to keep these different senses apart by maintaining the distinction between proprioceptive *information* and proprioceptive *awareness*, respectively.[425] On the embodied experiential level, however, these two aspects of proprioception are fully integrated.

There is an essential relation between movement and cognition. Although movement is involved, at almost every turn, in considerations about perception and other acts of cognition, it is only rarely made the theme of philosophical investigation.[426] It is true that philosophers from Aristotle to Locke and Berkeley have indicated that a certain movement in the soul, a successive flow structure in the mind, is responsible for structuring thought and for gaining regular connections among our experiences. But this is not the movement Gallagher means. Rather, *bodily* movement is closely tied in various ways to perception and to other forms of cognition and emotion.[427]

There is now a large amount of evidence from a variety of studies and disciplines to show that the body, through its motor abilities, its actual movements, and its posture, informs and shapes cognition.

- Visual perception involves the constant task of keeping the world relatively stable when many different features of the perceiving body may be in constant motion. From eye saccades and interocular motor adjustments to overall bodily balance, visual stability necessarily depends on motor control. For example, micro-saccadic eye movement helps to prevent the loss of visual object perception.[428]

- The shape and size of objects are perceived not simply in phenomenal terms (phenomenal size of an object depending on distance from the perceiver), but in pragmatic terms (as something I can grasp or manipulate). Phenomenally, the size and shape of an object may vary perspectively or across distance; pragmatic ally, they remain invariant— the perceived object is something I am capable of picking up or it is not.[429]

- There is good evidence to suggest that in certain cases, when we observe an object, even when we are not specifically required to reach for it or pick it up, 'canonical neurons' in the ventral premotor cortex (area F5) responsible for the motoric encoding of actions such as reaching and grasping are selectively activated (see Gallese 2000;[430] Murata *et al*; Rizzolatti, Fogaii, and Gallese 2000).[431] The visual observation of such objects automatically evokes the most suitable motor program required to interact with them, and the activation of motor preparation areas in the brain form part of what it means to perceive such objects.

- The development of perceptual and cognitive abilities is enhanced in correlation to a greater amount of crawling and mobility in infancy (Campos, Bertenthal, and Kermoian 1992;[432] see Thelen 1995),[433] and more specific perceptual strategies develop only when infants are able to execute certain motor abilities.[434][435] Studies in biodynamics show that movement and proprioception are intrinsically related to perception (Lockman and Thelen 1993).[436]

- Both ontogenetic and phylogenetic studies suggest that motor-related neuronal processes and structures are integrally linked to sensory and emotive processes, and that much of this integration is organised by motor representations of the body (Panksepp 1998).[437]

- Body posture can affect attention and certain kinds of judgment (Kinsbourne 1975).[438] If subjects turn head and eyes to one side just prior to making a judgment, the direction of turning influences cognitive performance (ibid). When subjects listen to a sentence with head and eyes turned right; their performance in cued recall is better than when they listened with head turned towards the left (Lempert and Kinsbourne 1982).[439] In addition, trunk orientation incudes directional biases in the ability to shift attention. The rotation of a subject's trunk to the left increases their response times to cued targets on the right and decreases their response times to cued targets on the left (Grubb and Reed 2002).[440]

- The discovery of 'mirror neurons' in the premotor cortex, neurons that are activated either by the subject's own motor behaviour *or* by the subject's visual observation of someone else's behaviour, shows a direct and active link between the motor and sensory systems and has important implications for explaining how we understand other people (di Pellegrino *et al* 1992,[441] Gallese *et al* 1996;[442] Rizzolatti *et al* 1996).[443]

This is not an exhaustive list; but it provides a quick sense of what is in fact a small part of the empirical evidence available, and a brief indication of how bodily movement and the motor system influence cognitive performance—how the body shapes the mind. Gallagher believes that it may even be possible to say that bodily movement, transformed onto the level of action, is the very thing that constitutes the self.[444]

In that sense life is the movement of thought. In the next chapter I look more closely at that idea, the primacy of movement and animation.

Chapter 16
The Primacy of Movement

You can never step twice in the same river.
Heraclitus

Experience shows the problem of the mind cannot be solved by attacking the citadel itself—the mind is function of body—we must bring some stable foundation to argue from.
Charles Darwin[445]

We must begin our examination with movement.
Aristotle[446]

'Animation' designated the way in which mind acquires a locality in the spatial world, its spatialisation, as it were, and together with its corporal support, acquires 'reality'.
Edmund Husserl[447]

True ease in writing.
As those move easiest comes from art, not chance who have learnt to dance.
Pope

It is the special quality of...animation which accounts for the fact that what is Bodily and ultimately everything Bodily from no matter what point of view can assume psychic significance, therefore even where at the outset it is not phenomenally the bearer of a soul.
Edmund Husserl[448]

Paul Klee characterised a drawing as "simply a line going for a walk". Movement is at the heart of life and its essence.

The seminal work on this topic and its significance is Maxine Sheets-Johnstone's wonderful book The Primacy of Movement.[449]

It is not a new idea. Heraclitus, 513BC said "Everything flows and nothing stays."

The necessity of incorporating movement in our epistemological and metaphysical investigations of the animate world from the very beginning, and in our scientific and historical investigations of the animate world is essential to our understanding of sport.

This necessity derives from corporeal matters of fact that define our lives from infancy onward and in an evolutionary sense, define the lives of all animate forms including man.

Movement is at the root of our sense of agency and it is the generative source of our notions of space and time.

Self-movement structures knowledge of the world—moving is a way of knowing and thinking in movement is foundational to the lives of animate forms.

It is philosophically instructive to examine in a carefully critical manner those cognitivist accounts of mind—or consciousness—that bypass an understanding of actual living bodies.

Not uncommonly, these accounts bypass living bodies for much the same reason that they reduce minds to matter.

In general, present-day philosophers and scientists begin their studies of mind, consciousness, and related topics from the viewpoint of perception, especially visual perception, movement being seldom accorded equal time or viewed with equal seriousness.

The purpose of *The Primacy of Movement* is essentially to reverse direction, to shift the perspective from which both epistemological and metaphysical—and scientific and historical—studies commonly proceed.[450]

It is to demonstrate that movement offers us the possibility not only of formulating an epistemology true to the truths of experience, but of articulating a metaphysics true to the dynamic nature of the world and to the foundationally animated nature of life.

Ecological paleoanthropologist Steven Kuhn's critical observation that "Too often, research is framed in terms of 'inherited' questions"[451] is exactingly relevant to this endeavour.

When it comes to assessing our capacity for thinking, the terms of the inquiry and subsequent discussion quickly gravitate to language—or to kindred forms of what is designated 'symbolic behaviour'.

Bodies are hardly at the forefront of thought about thought; neither for that matter is movement.

Not only this but language is not infrequently conceived a solely human phenomenon with no significant historical antecedents, that is, a phenomenon with no substantive evolutionary linkages whatsoever, whether on the basis of deficient anatomies (Lieberman 1983[452], 1972, [453]Laitman 1983[454]), of deficiency in linguistic design features,[455] of a deficiency in rational behaviour[456] of a deficiency in communicative repertoires,[457] or of a deficiency with respect to a Centre of Narrative Gravity.[458]

While there is no question but that human language and other so designated forms of 'symbolic behaviour' such as the creation of art objects are *culturally* unprecedented phenomena and ones that bring with them an untold richness and unending capacity for knowledge, there is every reason to question that such forms arose *de novo*, that they have *no* evolutionary ties. Language, for example, sprang full-blown from the mouths of waiting hominids. Present-day humans are on that account thoroughly unique products of evolution.[459]

The concern here is less directly with showing how that pre-eminence is unfounded and how sustaining it is myopically self-serving[460] than it is with showing how an immediate and thoughtless turn towards language and other so designed 'symbolic behaviours' is precipitous: it deflects us from a recognition and understanding of a phylogenetically and ontogenetically more basic phenomenon, the phenomenon of movement. Indeed, those intricate and subtle everyday gestures whose once invented and now learnt articulations constitute human speech are consistently taken for granted or ignored.

This calls to mind philosopher Maurice Merleau-Ponty's essay, 'The Primacy of Perception' (1964). The title of this chapter does not signal a declaration of war.

In fact, there is no question of a contest of any sort between movement and perception, and this is for two reasons: creaturely movement is the very condition of all forms of creaturely perception; and creaturely movement, being itself a creature-perceived phenomenon, is in and of itself a source of knowledge.

Indeed animation is the originating ground of knowledge. Not only is our own perception of the world everywhere and always animated, but our movement is everywhere and always kinaesthetically informed.

The foundational significance of movement should in consequence be doubly apparent to anyone concerned to investigate the nature of animate life.[461]

Because this significance has been largely ignored in contemporary Western science and philosophy, because perception—most especially visual perception—language, information-processing, computational modelling, and other such topics are at the focal point of contemporary attention, the primacy of movement has in fact gone unrecognised and unexamined.

The purpose of this chapter is to correct the omission in the most basic possible way, by going back to actual experience, to the things themselves—or more precisely, to us ourselves—thereby showing first how movement is the generative source of our primal sense of aliveness and of our primal capacity for sense-making, and second how a descriptive account of the phenomenon of self-movement elucidates cardinal epistemological structures inherent in kinaesthetic consciousness.[462]

To bring these kinetic and kinaesthetic understandings and structure historically and resonantly to the fore, it is necessary to frame them in the context of philosopher Edmund Husserl's notion of animate organism. Not only to show Husserl's consistent concern with, and insights into, animation, but to call attention to his non-species-specific sense of animation. By his very use of the term 'animate organism', Husserl was clearly rendering an account of something not exclusive to humans, but something broader and more fundamental than human animate organism.

He regarded nonhuman creatures as animate organisms along with humans and included them in an account of reality and nature,[463] never referring to them, for example, in a demeaning way as 'brutes' in the manner of Descartes and other philosophers, even present-day ones.[464]

This non-exclusive conception is not the result of a love of animals, or of a particular familial or cultural upbringing. It is the result of regarding the world, and in particular, nature, within the phenomenological attitude; that is, when one brackets one's every day, natural attitude towards the world, which attitude of course includes a certain attitude towards nonhuman animals, and in turn perceives nonhuman animals in a neutral way without the values—whether social, religious, or even scientific-medical—that one ordinarily brings to one's perception of them, then one of course observes them as *animate organisms*, i.e. live things, beings that move.

This is the way they appear, this is the original way in which we experience them. Indeed, this is the way infants and very young children experience them prior to ingesting any particular familial or cultural attitude; they experience them simply as things that move, as *animate forms*. It is noteworthy to point out that perceiving nonhuman animals in the phenomenological attitude is conceptually concordant with an evolutionary viewpoint.

In each case, one sees animate organisms *as living, moving things that by their very animate nature are continuous in kind, there being no fundamental break between nonhumans and humans.*

In other words, while an understanding of pan-animate aspects of animate life are required, so also are understandings of animate organisms in their uniqueness.

This dual understanding recalls the challenge of knowing Neandertals "in their own terms." What the latter knowingly requires is something philosopher Eugen Fink in another but quite pertinent context termed a 'constructive phenomenology'.[465] This dual understanding is implicit in what follows: a fleshing out of the phenomenological distinctiveness of the animate organism that is human against the background of what is phenomenologically pan-animate.

Husserl used the phrase 'animate organism' not only many times over but with a progressively greater and greater range of meaning in referring to living beings.

In *Ideas I*, for example, after saying, "let us imagine that…the whole of nature…is 'annihilated'." (i.e. that our experiences of the world do not add up harmoniously and are in fact totally refractory to harmonisation), he goes on to remark that "Then there would be no more animate organisms and therefore no more human beings. I should no longer exist as a human being: and, a fortiori, no fellow human beings would exist for me."[466] Clearly, Husserl initially ties the phenomenon 'animate organism' to nature as a coherent whole.

In *Ideas II*, he states that the sensuous and the psychic "are given as *belonging* to the [man or animal] Body in question, and it is precisely because of them that it is called Body or organism, i.e. an *'organ' for a soul* or for a spirit."[467] Of such bodies, he writes, for example, "I see a playing cat and I regard it now as something of nature, just as is done in zoology. I see it as a physical organism but also as a sensing and animated Body i.e., I see it precisely as a cat."[468] Here, Husserl clearly ties the phenomenon 'animate organism' not only to living nature, but to living creatures in the full sense of their livingness, i.e. of their carrying on activities in the world, of their being dynamically engaged as in playing, and the like.

In *Ideas III*, he writes of animate organism from the very beginning, focusing in particular on the way in which we perceive an animate organism and on what he terms the science of 'somatology': "We perceive the animate organism" he says, "but along with it also the things that are perceived 'by means of' the animate organism in the modes of their appearance in each case, and along with this we are also conscious of ourselves as human beings and as Egos that perceive such things by means of the animate organism."

In short and in sum, 'animate organism' refers in more and more refined ways to living beings whose animateness is the foundation of their perceptual world, including the perceptual world of their own bodies. It implicitly called attention first and foremost to creaturely movement, the term 'animate organism' underscores the originary significance of movement to creaturely life.

In this Husserlian context there are epistemological dimensions of this originariness.[469]

To begin with, Husserl makes the point (as does Ludwig Landgrebe more extensively in later commentaries) that "Originally, the 'I move', 'I do', precedes the 'I can do'."[470]

In *The Roots of Thinking* (Maxine Sheets-Johnstone) elaborated on this precedence noting that "an awareness of corporeal powers [the awareness of 'I cans'] does not (and could not) arise *ex nihilo*. It arises from [everyday] tactile-kinaesthetic activity: chewing, reaching, grasping, kicking, etc."

The awareness of corporeal powers is thus not the result of reflecting musings, whether with or without language…[and hence is] not a matter of wondering. What can I do?

"On the contrary, the sense of corporeal powers is the result either of moving or of already having moved." She gives as example the tactile-kinaesthetic act of chewing: in that act, a creature "catches itself in the act of *grinding something to pieces*."[471]

In such acts corporeal powers give rise to corporeal concepts, fundamental human concepts such as grinding, sharpness, hardness, and so on.

Now if we take seriously that the (experience) 'I move' precedes the (conceptual realisation) 'I can do', and if we take with equal seriousness the fact that specific perceptual awareness of ourselves arising in everyday tactile-kinaesthetic acts of doing something are the touchstone and bedrock of our discovery of 'I cans' and in turn of corporeal concepts, then it is clear that movement is absolutely foundational not only to perceptual realisations of ourselves as doing or accomplishing certain things or making certain things happen—such as "grinding something to pieces"—and to correlative cognitive realisations of ourselves as capable of just such acts or activities, but to perceptual-cognitive realisations of ourselves as alive, i.e. as living creatures, animate organisms, or animate forms.

Aliveness is thus a concept as grounded in movement as the concept '*I can*'.[472] Indeed, we intuitively grasp the coincidence of aliveness and animation from the very beginning. With no prior tutoring whatsoever, we take what is living to be that which moves itself and to apprehend what is not moving and has never moved to be precisely inanimate. Experimental studies and observations of infants readily document this intuitive knowledge.[473] They document as well our fascination with movement. What moves straightaway captures our attention; it is consistently at the focal point over what is not moving.[474]

This focal tethering to movement is no less first-nature to other creatures than it is to ourselves. We are all of us attuned to the animate over the inanimate; we are all alive to movement from the start. Indeed, animation is at the core of every creature's engagement with the world because it is in and through movement that the life of every creature—to borrow Husserl's phrase in the first epigraph—'acquires reality'.

Given the fact that we intuitively equate aliveness with movement, it is difficult to explain why philosophers would overlook the primacy of movement in their renditions of what it is to be human, taking instead a textual model which reduces movement to mere visual and/or manual gestures coincident with reading and writing;[475] a computer model which reduces movement to mere 'output', the necessary but comparatively dull aftermath of a vastly more interesting and prestigious 'input'; an objective model which either typically disregards movement by considering only objects in motion and, in effect, ignores self-movement, or typically instrumentalises movement by de-cognising it, making it no more than a means, a necessary but purely serviceable accoutrement of perception (or knowledge); or, finally, taking no model at all, simply trivialises it.

Most importantly and pointedly in terms of experience—that is, given that we humans all begin life by wiggling, stretching, opening our mouths, swallowing, kicking, crying and so on—it is odd that philosophers would overlook the *sui generic* character of movement and fail to explore its significance. In the beginning, after all, we do not *try* to move, *think* about movement possibilities, or put ourselves to *the task* of moving. We come straightaway moving into the world; we are precisely not *stillborn*.

In this respect, primal movement is like primal sensibility: "it is simply there," Husserl says.[476]

Moreover in the beginning, we are not surprised by our movements, disappointed by them, or wish that they were different.[477] In the beginning, we are simply infused with movement—not merely with a *propensity* to move, but with the real thing.

This primal animateness, this original kinetic spontaneity that infuses our being and defines our aliveness, is out point of departure for living in the world and making sense of it.

It is the epistemological foundation of our learning to move ourselves with respect to objects, and thus the foundation of a developing repertoire of 'I cans' with respect to both the natural and artefactual array of objects that happen to surround us as individuals in our particular worlds. It is in effect the foundation of our sense of ourselves as agents within a surrounding world. But it is even more basically the epistemological foundation of our sense of who and what we are.

We literally discover ourselves in movement. We grow kinetically into our bodies. In particular, we grow into those distinctive ways of moving that come with our being the bodies we are.[478]

In our spontaneity of movement, we discover arms that extend, spines that bend, knees that flex, mouths that shut, and so on. We make sense of ourselves in the course of moving.

We discover ourselves as animate organisms. These kinetic-kinaesthetic self-discoveries constitute their own specific repertoire of 'I cans'; that is, quite apart from our 'I cans' relative to a world of objects, we discover a realm of sheer kinetic 'I cans': I can stretch, I can twist, I can reach, I can turn over and so on.

This realm is in truth an open-ended realm of possibilities. That it is so means that our individual repertoires are ultimately a measure of how far we grow into the bodies we are, a measure of both the extent to which we give ourselves over to the spontaneity of movement and the extent to which we explore the kinetic dimensions of our animate nature.

In discovering ourselves in movement and in turn expanding our kinetic repertoire of 'I cans', we embark on a lifelong journey of sense-making.

Our capacity to make sense of ourselves, to grow kinetically into the bodies we are, is in other words the beginning of cognition. In making sense of the dynamic interplay of forces and configurations inherent in our on-going spontaneity of movement, we arrive at corporeal concepts.

On the basis of these concepts, we forge fundamental understandings both of ourselves and of the world.

We discover opening and closing in the opening and closing of our eyes, mouths and hands; we discover that certain things go together such as a certain constellation of buccal movements and certain feelings of warmth—as in the act of nursing; we discover a differential heaviness in lifting our head and lifting our arm and a differential over-all bodily tension in the two movements as well.

In making kinetic sense of ourselves, we progressively attain complex conceptual understandings having to do with *containment*, with *consequential relationships*, with *weight*, with *effort*, and with myriad other bodily-anchored happenings and phenomena. In effect, our first cognitive steps are taken by way of our own movement.

With these steps we begin to discover the nature of our being in the double sense of finding a coherency of experiences and of articulating a particular form of life.

Insofar as our primal animateness is the bedrock of just such kinetically and kinaesthetically-rooted conceptual understandings, our primal animateness is, to borrow (and singularise) a phrase from Husserl, "the mother of all cognition."

A remarkable analogy in fact exists between the originariness of movement and the originariness sought by phenomenology, the context in which Husserl actually used the phrase.

The analogy is adumbrated in Husserl's remark that "Phenomenology in our sense is the science of 'origins', of the 'mothers' of all cognition; and it is the maternal-ground of all philosophical method: to this ground and to the work in it, everything leads back."[479]

211

Everything cognitive leads back equally to movement, to animate nature. Clearly, our first consciousness is a tactile-kinaesthetic consciousness that arises on the ground of movement that comes to us spontaneously.

Indeed, on the ground of fundamental and invariant species-specific kinetic acts that we simply 'do' in coming into the world, acts such as kicking, stretching, sucking, swallowing and so on.

Kinaesthetic consciousness in turn defines an emergent, progressively expanding consciousness whose structures can be thematised, i.e. analysed phenomenologically.

In particular, kinaesthetic consciousness unfolds on the ground of spontaneous movement and in its initial unfolding reveals not only corporeal concepts on the order of those described above, but spatio-temporal concepts that are basically qualitative in nature and that emanate from what we discover to be the creative, i.e. freely variable, character of our movement.

I can, for example, lift my head abruptly or in a sustained manner; I can open my mouth minimally or widely; I can kick my legs rhythmically or at random; and so on.[480]

Any movement we make has certain degrees of freedom. That it does—that our movement is freely variable—is a measure of the qualitative nature of movement and potential conceptual richness of our unfolding kinaesthetic consciousness.

It is furthermore suggestive of how spatialities and temporalities are kinetically created—and even of how space and time are fundamentally constituted in and through our experience of self-movement.

In sum, our primal animateness is of profound epistemological significance. In the beginning is movement. Our very emergence as cognising subjects is grounds in our original kinetic spontaneity.

In effect, what is already there—but not by any means already *"all there"* as Merleau-Ponty would have it[481]—is not the world and the body. What is already there is movement, movement in and through which the perceptible world and acting subject come to be constituted, which is to say movement in and through which we make sense of both the world and ourselves. As Henry Vaughan (1622–1695) poetically put it in Silex Scintillans:

> "Man is the shuttle, to whose winding quest
> And passage through these looms
> God order'd motion, but ordained no rest."

That 'I move' arises on the ground of our primal animateness is of equally profound epistemological significance, for it means that movement is the ground on which transcendental subjectivity—in a broad sense, our sense-making or constituting faculty—arises.

Movement awakens transcendental subjectivity in the form of kinaesthetic consciousness.[482] To see this relationship is to corroborate and extend Landgrebe's account of "[a] prelinguistic acquaintance with oneself as the centre of a spontaneous ability to move."

In the context of his account, Landgrebe writes that "kinaesthetic *motions…*are the most fundamental dimension of transcendental subjectivity, the genuinely original sphere, so that even the body *(Leib)*, as functioning body, is not just something constituted *but is itself constituting as the transcendental condition of the possibility of each higher level of consciousness and of its reflexive character.*"[483]

The kinaesthetic correlates of perception—what Husserl calls 'the kinesthesis'—are hence not simply practical perceptual affordances (to use a Gibsonian term[484]), necessary *"functions of spontaneity* belong[ing] to every perception."[485]

They are, in their own right, *perceptual experiences*, the most fundamental of perceptual experiences, and as such are at the very core of the constituting I, that is, of transcendental subjectivity.

As might be evident, kinetic free variations disclose four primary qualitative structures of movement having to do with force or effort, with space, and with time.

These qualitative aspects of movement are of course separable only reflectively, that is, analytically, after the fact; experimentally, they are all of piece in the global qualitatively felt dynamic phenomenon of self-movement. Any movement has a certain felt tensional quality, linear quality, amplitudinal quality, and projectional quality.[486]

In a very general sense, the felt tensional quality has to do with our sense of effort; the linear quality with both the felt linear contour of our moving body and the linear paths we sense ourselves describing in the process of moving; the amplitudinal quality with both the felt expansiveness or contractiveness of our moving body and the spatial extensiveness or constriction of our movement; the felt projectional quality with the way in which we release force or energy.

Linear and amplitudinal qualities obviously describe spatial aspects of movement; tensional and projectional qualities obviously describe temporal aspects of movement, what we recognise as the felt intensity of our moving bodily energies and the felt manner in which we project those bodily energies— in a sustained manner, for example, in an explosive manner, in a punctuated manner, in a ballistic manner, and so on.[487]

Temporal aspects of movement are the result of the way in which tensional and projectional qualities combine; that is, the temporal quality of any movement derives from the manner in which any particular intensity (or combined intensities) is kinetically expressed.

On the way to spelling out the nature of these qualities more precisely, it is necessary to draw specific attention to the fact that movement *creates* the qualities that it embodies and that we experience; thus it is erroneous to think that movement simply takes place *in* space, for example.

On the contrary, we formally create space in the process of moving; we qualitatively create a certain spatial character by the very nature of our movement—a large, open space, or a tight, resistant space, for example. Great football and rugby players do that. I have referred to the description of Barry John as running in a different dimension of time and space. His good friend George Best did the same in football as did the late lamented Brazilian footballer Pelé and more recently Diego Maradona, Lionel Messi: all arguably sporting geniuses.

In sum, we learn our bodies by moving and in moving both create and constitute our movement as a spatio-temporal dynamic. If we look more deeply into the matter, we discover that movement is the originating ground of our sense-makings, in phenomenological terms, the originating ground of transcendental subjectivity; we constitute space and time originally in our kinaesthetic consciousness of movement.[488]

Flux, flow, a streaming present, a stream of thought, consciousness, or subjective life, a style of change—all such descriptive terms are in both a temporal and spatial sense rooted in originary self-movement: they are all primordially present not in the constitution of objects but in our original spontaneity of self-movement, in our original experience and sense of our dynamically moving bodies. For many examples of this, see my chapter 39 'Consciousness, Perception and the Senses in Sport'.

To think the reverse is to overlook precisely that in the beginning was movement: we all of us came into the world moving and at the same time had to learn our bodies and to move ourselves. In effect, to think the reverse is to overlook *animation*, and the fact that that animation is the very bedrock of our coming to know the world.

It is ultimately to ignore the transcendental clues Husserl himself provides in his consistent references to and descriptions of, both animation and animate organism. His phenomenological insights into the fundamental meanings of animation and of animate organism are in fact a validation of his methodological use of intentional objects as 'transcendental clues'.[489]

In particular, Husserl, took animation and animate organism as transcendental clues to understanding how we come to make sense of the world.

However incomplete his phenomenological analyses of animation and animate organism, his insights are springboards to understanding how, in self-movement, we come to constitute ourselves as spatio-temporal forms of life including dance and sport—how, in a broad sense, we make sense of ourselves—and how we derive our very concept of a spatio-temporal world on the basis of our own moving bodies. In this respect, his insights are themselves clearly transcendental clues to the cardinal epistemological structures of kinaesthetic consciousness.

Chapter 17
Thinking Is Movement

Winged words.
Homer

Mind is the great lever of all things; human thought is the process by which human ends are ultimately answered.
Daniel Webster

[Joe Louis] the great boxer was a human replica of Rodin's thinker. In the ring he associates ideas and responds with lightening-like rapier thrusts about as rapidly through the medium of mind and muscle as an Einstein calculates cause and effect in cosmic theory.
Edwin Bancroft Henderson

All thoughts and actions emanate from the body. Therefore the description of a thought or action—however abstruse it may be—can be beaten home by bringing it onto a physical level. Every idea intuitive or intellectual, can be imaged and translated in terms of the body, its flesh, skin, blood, sinews, veins, glands, organs, cells or senses.
Dylan Thomas[490]

In lazy apathy let Stoics boast their virtue fix'd: it is fixed as in a frost; contracted all, retiring to the breast
But strength of mind is exercise, not rest.
Alexander Pope

Sport is the movement of thought conscious and unconscious. Thinking in movement is clearly not the work of a symbol-making body, a body that mediates its way about the world by means of language, for example, it is the work of an existentially resonant body.[491]

An existentially resonant body creates a particular dynamic world without intermediary.

An improvisational dance, the world it creates is neither a part of the everyday given world nor a temporary fictitious world, but a protean world created moment by moment. Sport in many ways is a form of improvisational dance.

Experienced as an elongated or ongoing present, it is a world in which there are no befores or hereafters, no sooner-or-laters, no definitely expected endings or places of arrival. That is a 'beyond' which this book explains exponentially.

For just such reasons, the dance or sport being created is not a dance or sport that the dancer or sportsman might acknowledge as being 'about' something, unless that something were movement itself.

To appreciate and to understand such a phenomenon is akin to appreciating and understanding what Gertrude Stein meant when she said "a rose is a rose is a rose." Clearly a rose is not about something. Neither is it a jumble of petals.

The same may be said of dance or sporting improvisation. The kinetic intelligence that creates the dance or sport informs the dance and sport itself. No more than the dancing and sporting body must movement stand necessarily for or refer to something beyond itself in order for the phenomenon to be dance or sport. But sport does have a meaning beyond itself which is the philosophical subject matter of this book. But that is not deny the significance of movement metaphysically or physically in the scheme of things.

To have meaning is not necessarily to refer and neither is it necessarily to have a verbal label. Movement—animation—can be in and of itself meaningful.

To appreciate—and indeed, to fathom—such non-linguistic strata of experience, we turn towards that which is animate; we find in our highly symbol-laden human world patches where thinking in movement comes to light.

In so doing, we discover that fundamental creative patterning of thought that is founded upon a kinetic bodily logos; we discover mindful bodies, thinking bodies, bodies that, in improvisational dance, break forth continuously into movement and into this dance, bodies that moment by moment fulfil a kinetic destiny and so create kinetic meanings. The same is true of sporting movement.

When we reflect upon our experience of moving in just such ways, examining the experience from a phenomenological perspective and discovering the phenomenon of thinking in movement, we are in turn propelled to rethink our notion of thinking—and in the process, to realise that insights gleaned from a descriptive account of Improvisational dance and sport have consequences for epistemology and evolutionary accounts of animate life as well as for aesthetics.[492]

It will be helpful to consider two assumptions about thinking, assumptions that, the preceding descriptive account notwithstanding, might otherwise impede a clear and unprejudiced grasp of what it is to think in movement. The first assumption has to do with thinking itself and has several layers. To begin with, it is commonly assumed that thinking is tied to language and that it takes place only via language. It is furthermore commonly assumed that thinking and language are tied in an exclusive way to rationality.

The basis for these assumptions seems itself to be an assumption: that thinking, language and rationality form a holy, albeit human, triumvirate, a congealed sacred hallmark of pre-eminently *human* existence. To link thinking, language, and rationality in this manner, however, is to claim a necessary and inherent interdependence before examining the evidence from experience itself and prematurely to declare impossible something that may not be impossible at all, and perhaps, on the contrary, quite common, i.e. thinking in movement.[493]

Moreover to deny peremptorily the possibility of thinking in movement on the basis of the foregoing assumption(s) may readily involve a further assumption, namely, that thinking takes place only by means of something, in particular, a symbolic system of some sort—e.g. mathematical, linguistic, logical—a system having the capacity to mediate or carry thought referentially. As the previous descriptive account has demonstrated, however, to affirm the possibility of thinking in movement is to regard movement neither as a vehicle for thinking nor as a symbolic system through which reference is made to something else.

Indeed, steadfast and serious reflection on the phenomenon of improvisational dance and sport shows that movement is neither a medium through which a dancer's or sportsman's thoughts emerge nor a kinetic system of counters for mediating his or her thoughts; movement constitutes the thoughts themselves.

One might in this context paraphrase Maurice Merleau-Ponty's remarks upon language and say that, in order to understand what it means to think in movement, "*movement* must somehow cease to be a way of designating things or thoughts, and become the presence of that thought in the phenomenal world, and moreover, not its clothing but its token or its body."[494]

Similarly, one might paraphrase neurologist Kurt Goldstein's remarks upon language and say that, "As soon as man uses *movement* to establish a living relation with his fellows, movement is no longer an instrument, *no longer a means: it is a manifestation, a revelation of intimate being and of the psychic link which unites us to the world and our fellow men.*"[495] I certainly believe that sport provides that fundamental link which unifies us to the world and our fellow men.

The assumption rooted in a reification of thinking and a substantive metaphysics may be accompanied by a parallel assumption rooted in a Cartesian separation of mind and body.

To assume that thinking is something only a mind does, and doing or moving are something only a body does is, in effect, to deny the possibility of thinking in movement.[496] Sport is a phenomenon of that kind as is dance from which it may have originated.

If thinking is furthermore assumed to be always separate from its expression—a thought in one's head always existing prior to its corporeal expression—then thinking must necessary be transcribed—or, given a strictly linguistic conception of thinking, *transliterated*—into movement. When the mind formulates a thought, for example, the tongue and lips move to express it; when the mind thinks of going to the store, the body complies by walking or driving it there, likewise kicking a ball or jumping over a hurdle.

The notion that thoughts must corporeally transliterated, that they exist separately from and prior to their expression, has been justly criticised by philosophers such as Wittgenstein and Merleau-Ponty, "When I think in language," Wittgenstein points out, "there aren't meanings going through my mind in addition to the verbal expressions."[497] Merleau-Ponty similarly points out that "speech is not the 'sign' of thought, if by this we understand a phenomenon which heralds another as smoke betrays fire...Nor can we concede...that [speech] is the envelope and clothing of thought."[498]

Although in these examples, it is a question of *language* and not of movement, the same critical insights into the phenomenon of *thinking* apply.

What the descriptive account of improvisational dance and sport challenges is not the possibility that thinking, or a single thought such as an image, never occurs prior to its overt expression in some form, that is prior to a movement or an action of some kind. When one thinks in general terms about what one will say prior to expressing the thought verbally to others, verbal thinking clearly occurs prior to its active expression. What the descriptive account challenges is the notion that thinking always and necessarily takes place in this way, thus that the mind is always one thoughtful step ahead of the body, always three beforehand to mobilise it into action. Sport, especially, evidences the fundamental flaw in that proposition. It is a fallacy.

There is an aspect of this assumption that we would do well to clarify in some detail.[499] Though typically so regarded, movement is hardly given its due when presumptively conceived merely as the medium of a body's everyday transactions with the world.

Movement is, on the contrary, first and foremost the natural mode of being a body—a ready a perpetual kinetic susceptibility and effusion, as it were, of animate life. Serious reflection on this fact readily leads one to the realisation that animate forms readily inhabit movement in the literal sense of living in it and that thinking in movement is foundational to being a body, as much an epistemological dimension of bodily life as a biological built-in that makes sense.

One aspect of this naturally kinetic manner of being—this spontaneous thinking in, and opening up into movement—is implicit in Merleau-Ponty's remark that Cezanne's description of himself as 'thinking in painting' is a description of a process in which 'vision becomes gesture'.[500] On this issue see my detailed chapter 27 on Gesture.

His remark is clearly not intended to mean that movement follows perception, i.e. doing follows seeing, but that perception is interlaced with movement, and to the point, we might add, where it is impossible to separate out where perception begins and movement ends or where movement begins and perception ends. The one informs the other—inextricably, and all the more inextricably when it is a question not of *vision* becoming gesture, but of *movement* becoming movement.[501]

Consider, for example, the two basic ways in which thinking in movement may enter into the creation of a dance and playing sport. One can readily distinguish between thinking in movement in and of itself and a kind of thinking in movement that is analogous to Cezanne's 'thinking in painting'.

The distinction is in fact integral to an understanding of the difference between improvisational dance and sport—what we might characterise as the creation of dance and playing sport as artistic process—and non-improvisational dance—the creation of dance as artistic product and sporting endeavour. In creating the latter kind of dance, a choreographer obviously thinks in movement as she creates the dance, precisely in a way similar to the way in which Cezanne 'thinks in painting'. In a similar way, the sportsman plays as a choreographer and thinks in movement as Cezanne 'thinks in painting'. Thought becomes subsumed in movement and play.

The old division between instinctive and learnt behaviour is a spurious one, as most biologists have come to realise, an oppositional way of thinking that does not accord with facts of life. In their classroom text *Biological Science*, William Keeton and James Gould, for example, state that "[I]it is extremely unlikely that any behaviour can be classified as strictly innate or strictly learnt: even the most rigidly automatic behaviour depends on the environmental conditions for which it evolved, while most learning, flexible as it seems, appears to be guided by innate mechanisms." They conclude that "*Instincts*…can be defined as the heritable, genetically specific neural circuitry that organises and guides behaviour," and that "behaviour that is thereby produced can reasonably be said to be at least partially innate."[502] Instructive cases in point that confirm this conception of behaviour are paths and shelters. Sport is a classic example of this form of behaviour.

Intelligence in action is instinctive. *All* animals—humans included—could hardly survive much less reproduce if intelligence in action were not instinctive. In just this sense, a kinetic bodily logos is at the heart of thinking in movement and sport. It is what makes such thinking spontaneous and contextually appropriate to the situation at hand. It is what ties thinking not to *behaviour* but to *movement*, that is, to kinetic meanings, to a *spatio-temporal-energic semantics*.

Intuitive and instinctive behaviour are malleable precisely because they are fundamentally kinetically dynamic patterns and not chunks of behaviourally labelled 'doings'.

To think in movement is not to think in monolithic compartmental wholes: eating, mating, courting, defending, aggressing, threatening, and so on; it is to think in dynamic terms—in terms of speed, postural orientation, range of movement, force, direction and so on. That is what sport and dance fundamentally embody.

Behavioural variations exist precisely because *kinetically dynamic possibilities* exist. It is just such kinetically dynamic possibilities that distinguish one creature from another: one creature runs faster than another, is more agile over a rough terrain than another, is more awkward in climbing than another, is less easily aroused or startled than another, is quicker to withdraw than another, and so on.

From this essentially kinetic vantage point, the malleability of what are called instinctive behaviours, indeed, their *evolution*, is a matter of movement. Instincts have their genesis in animation—prima animation.[503]

When circumstances change, ways of living change, and these changes in the most basic sense are a matter of movement possibilities. A kinetic bodily logos is not some kind of adaptive mechanism; it is a real-life dimension of animate forms. An intelligence of action is a built-in of animate life. Thinking in movement is the natural expression of this elemental biological character of life.[504]

Sport is a paradigm and rich example of thinking in the movement and the natural expression of this elemental character of life. The relationship between dance and sport is an interesting one both in terms of their roots and their essence. Both are creative manifestations of the human mind and body in action and reaction and in the aesthetic and non-aesthetic and the gratuitous and non-gratuitous. That complex relationship is outside the ambit of this book although I have dealt briefly with that comparison in the next chapter. However, what I would say anecdotally that I would add to the most charismatic persons I have ever met namely Muhammad Ali and El Cordobes, a figure from dance, Rudolph Nureyev who in terms of body image and schema was in the same category of physical and mental intensity—beyond the horizon.

Chapter 18
Movement and Perception

*If the doors of perception were cleansed everything would appear as it is,
infinite.*
William Blake

Where there is no vision, the people perish.
Proverbs 29-18

Moving and perceiving are two sides of the same coin.[505] I am indebted in
this chapter to David Morris's illuminating book The Sense of Space (2004).

Tennis and cricket aren't just a matter of swinging a racquet or bat but of
perceiving a 'gamey' geometry of the court or field, of learning the angles,
vectors and way of seeing that give advantage in tennis and cricket as a moving
game of chess.[506]

Bergson gives an insightful account of learning a skill in *Matter and
Memory*.[507] Learning a skill is a process in which movements are recomposed by
repeating their decomposition, what Morris calls a synthesis that proceeds by
repeating an analysis.

The tennis student does not see the tennis instructor doing a backhand and
then immediately reproduce the movement as a whole. To do that, he would
already need to be able to do the backhand. But that is precisely what he cannot
do, what he is trying to learn. So the instructor breaks the backhand into
fragments, and he learns how to perform the fragments, by deploying or
modifying habits, folds, that I have already have.

Still, he has not learnt the backhand if he performs it as a series of movement
fragments. That sort of fragmentary, chopping movement is characteristic of the
learner, or of the comic—like Jacques Tati as M. Hulot—who draws out humour
in human movement by unhinging it from within.

It is not until fragments slide into a smooth whole that I have learnt the
backhand. How does he achieve the smooth whole? Precisely by doggedly
repeating the chopping fragments in a sequence that follows the smooth whole
modelled by the instructor, until the fragments start flowing into one another in
my own movements.

I am a being who gets to play tennis only by *learning* how to. I cannot directly or instantaneously copy the moving figure manifest in the instructor's moving body. That figure must be unfolded into a network of folds, analysed, either by the instructor, or by me trying to break things down.

But the folds revealed by analysis are fragments of a whole, folds implicated in one another in the instructor's body. The folds retain something of their co-implication. As I repeat fragmentary folds in my body, these implications are activated, and the folds gradually fold back into a whole, are synthesised, in the way that someone playing with an unfolded piece of origami, by following folds hinged into one another in the paper, may fold it back up (although it won't come out exactly the same).

Learning a skill is not a direct transfer of moving wholes from instructor to learner, but a 'synthesis by analysis': it is only by repeating, over and over, analysed movement in my body that I synthesise the complex movement of the instructor's body.[508]

Crucially, the synthesis by analysis is conducted within the movement of bodies. (Being able to learn or re-jig movements just by watching or talking is characteristic of experts well past the simple stage of acquiring skills.)

Skills acquisition depends on synthesis by analysis and on the interrelation of the bodies in which alone this synthesis by analysis occurs.

What is transferred in skill acquisition is not just a way of moving, but a *sens* of the world. In learning the backhand, I learn how to approach the ball, court, and world, I learn a whole attitude and orientation to the world, a *sens* that I did not yet have.

I do not gain that new *sens* directly, as if handed it ready-made and entire.

When I first step onto the court, I glimpse a *sens* in the instructor's way of moving, but I do not yet have it.

But in learning, an aspect of the instructor's moving schema of perception is unfolded along the lines of moving bodies, analysed; and when I play and replay that analysed movement in my moving body, I refold it, synthesise it, and begin to acquire a new *sens*.

I do not already need to have the *sens* to learn it, because I never seize *sens* entirely from the instructor. In some sense, I learn the *sens* from *myself*, from moving in a certain choppy way and having it fall together in my own movement.

When people first learn to play a sport, they draw upon conscious reflection of what they ought to be doing with their bodies, but mainly they draw unconsciously upon the already learnt skills that we all have in using our bodies such as running.

As the training of the brain and body proceeds, that conscious reflection on the manipulation of the body falls away and they can take the plunge through having what we commonly call a strong 'feel' for the game.

For many people the greatest fulfilment and enjoyment comes with total immersion into the activity, so that self-awareness is scattered to the winds, and they become wholly what they desire in body and spirit without reservation.

The brain and body anticipate inputs, perceive and make movements without need for reflection. It is precisely this kind of unconscious, but directed skill in the exercise of perception that the concept of intentionality must include.

The examples of the athlete and dancer demonstrate what Walter J. Freeman considers to be the three main properties of intentionality.[509]

Unity

The first is unity. Our brains and bodies are entirely committed to the action of projecting ourselves corporeally into the world and our perceptions are unified across all our senses at rates faster than we perceive. Here he distinguishes between the self, which is unified, and the awareness of self that we experience as the ego, which is not unified but can he says be "splintered like sunlight on the waves."

Wholeness

The second property is wholeness: the entirety of life's experience is brought to each moment of action. The experiences of games and dancing are generalised and continually built upon. It includes an effort, described by Aristotle and again by Goethe two centuries ago, as a blind, organic striving towards realising our full potential with the constraints of heredity and environment.

Purpose or Intentionality

The third property of intentionality is purpose or intent because whether athletes and dancers are aware of it or not, their actions are directed to some end.

Freeman's view is that:

> "Perception is a continuous and mostly unconscious process that is sampled and marked intermittently by awareness, and what we remember are the samples, not the process. The fact that consciousness need not enter into the description of intentionality opens up a new vista. Consciousness is not a good place to start a theory of brain function, because there is no biological test to prove whether consciousness is present in a supine subject, other than to ask 'Are you now or were you ever awake?' Animals cannot answer, not because they can't remember or make representations in their own ways, but because they can't create

and represent abstractions equal to the required level of sophistication of communication."[510]

Evolutionary biologists have shown that complex operations of brains and bodies originate in simpler animals and have evolved into human capacities. Therefore, we can, says Freeman, infer that animals have intentions by observing their behaviours, even if we do not know whether they are conscious of what they do.

Chapter 19
Sport as Absorbed Coping and Readiness to Hand

Though man a thinking being is defined
Few use the grand prerogative of mind
Jane Taylor (1733–1824)

Experience is the Child of Thought, and Thought is the child of action. We cannot
learn men from books.
Disraeli

Experience is never limited; and it is never complete, it is an immense sensibility;
a kind of huge spiderweb of the finest silken thread suspended in the chamber of
consciousness and catching every air-borne particle in its tissue.
Henry James.

Dreyfus contends that at higher levels of skilled behaviour we use 'absorbed coping' in our dealings with the world. His writings are seminal.

Absorbed coping is based on the direct interaction of our bodies with the physical environment.

> "In absorbed coping the agent's body is led to move so as to reduce a sense of deviation from a satisfactory gestalt without the agent knowing what a satisfactory gestalt will be like in advance of achieving it. Thus in absorbed coping, rather than a sense of trying to achieve success, one has a sense of being drawn towards an equilibrium."[511]

This leaves little room for conscious attention and control at higher skill levels. Consciousness recedes into the background.[512] Dreyfus coined this expressed by reference to Heidegger's analysis of our environing world and has extended this analysis to sport. He maintains that skilled motor behaviour at expert levels is characterised by nonthematic and absorbed coping.

His views are predominantly based on Heidegger's *Sein un Zeit* (Being and Time), originally published in 1927.[513]

Heidegger calls the specific human mode of being 'existence'.[514]

He discusses several deep and fundamental modes of being related to human existence.

The most important mode of being is disclosed when we use things as equipment ('Zeug') in our daily dealings with the world.

His notion of a fundamental 'equipmentality' of the world comes close to ideas of the fundamental 'affordances' of our environing world.

Gibson, another outstanding thinker uses the concept to characterise our environment in its ability to present (afford) possibilities for use or exploration. A stone invites being thrown, a tree climbing, and so on.

Both Heidegger and Gibson try to overcome the subject-object dichotomy by showing how humans and the surrounding world interact with each other, impacting and shaping each other in important ways.

Humans are geared into the world, and the things in the world are functionally tied together in equipmental wholes. Heidegger maintains that an item of equipment never really exists in isolation, because it is dependent on the "totality of equipment, in which it can be this equipment that it is."[515]

This means that that equipment is "essentially something-in-order-to." Such totalities of equipment are constituted by their serviceability, conduciveness, usability, and manipulability. Because an item never functions alone, there is an unbuilt assignment or reference in the equipmental structure. Items or equipment refer to the other equipment in the context, like "ink-stand, pen, ink, paper blotting pad, table, lamp, furniture, windows, doors, room."[516] The room presents itself not here in its geometric spatial sense but as a place to live and where each thing in the room is already disclosed as part of an equipmental structure.

Each piece of equipment can genuinely show itself in practice, like a hammer in hammering, but in practice this is not thematically grasped as an objective thing or as a piece of equipment. We just use the hammer in a suitable nonthematic way. Heidegger says that the hammering "has appropriated this equipment in a way which could not possibly be more suitable."[517]

When things are encountered as ready-to-hand for use, we understand them in a mode very different from a theoretical and scientific outlook. It is not possible to get from one outlook to another without a complete attitudinal shift.

> "No matter how sharply we just *look* at the 'outward appearance' of Things in whatever form this takes, we cannot discover anything ready-to-hand."[518]

But readiness to hand has its own outlook. For Heidegger, the theoretical outlook is paralleled by a practical look or sight. Practical behaviour is not

BEYOND *the* CONCEPT *of* SPORT

W ROBERT GRIFFITHS

BEYOND THE CONCEPT OF SPORT

W Robert Griffiths

SPORTS & RECREATION/ Cultural & Social Aspects

PB £17.99   9781035845101
HB £24.99   9781035845118
EB £3.50   9781035845125

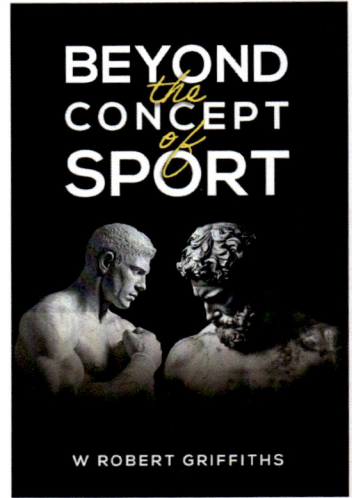

Beyond the Concept of Sport challenges tired assumptions about athletics to unveil sports' underestimated yet far-reaching social impact and philosophical significance. While governments downplay its influence, author Robert Griffiths recognizes that sports – especially cricket – deeply channels the human spirit for participants and fans alike. He eschews cliches to deliver fresh insight into sports' resonance.

This is no dry academic tome, but rather a lively examination blending scholarship with accessibility. Griffiths brings iconoclastic yet sage opinions to unpacking how athletic pursuits shape culture and consciousness. He illuminates the under-appreciated role sports play in forging identity, purpose, inspiration, escapism and more for millions globally. Grappling with issues often overlooked when discussing athletics, Griffiths' unconventional analysis explores fandom, nationalism, arts, business, and the very meaning woven through sports' rituals.

Written with passionate intellect, wry wit, and a distaste for the dull, *Beyond the Concept of Sport* cries out to be read by both die-hard fans and curious sceptics. After all, few human realms spark such fervour and unity across the world's divides like that of sports. This book captures that emotional impact while elevating sports as a subject worthy of serious yet spirited consideration.

Please send me ...... copy/ies of

Beyond the Concept of Sport
W Robert Griffiths

Please add the following postage per book:
United Kingdom £3.00 / Europe £7.50 /
Rest of World £12.00

Delivery and Payment Details

Format	Price	Qty	Total
Paperback ☐ Hardback ☐			
Subtotal			
Postage			
Total			

Full name: ....

Street Address ....

City:.. County:..

Postcode: .... Country: ....

Phone number (inc. area code): .... Email: ....

I enclose a cheque for £.................. payable to Austin Macauley Publishers LTD.

Please send to : Austin Macauley Publishers Ltd®, 1 Canada Square, Canary Wharf, London, E14 5AA

Tel: +44 (0)20 7038 8212, +44 (0)20 3515 0352

orders@austinmacauley.com
www.austinmacauley.com

AUSTIN MACAULEY PUBLISHERS
LONDON · CAMBRIDGE · NEW YORK · SHARJAH

atheoretical in the sense of 'sightlessness'. It has its own kind of sight, which Heidegger calls circumspection (Umsicht). It is this practical circumspection that guides our lives in daily situation.

In our circumspective dealings with concentrate not on the equipment as such but on the work to be done. The equipment must withdraw in order to be efficient in its availability...[519]

In *Being and Time* Heidegger reorganised and discussed just three modes of being: Dasein, readiness-to-hand, and presence-at-hand. In his later works he also discussed modes of being we find in art, in language, and in the thing. A stone can be seen as something to be used as equipment or studied as an object that is present-to-hand, but it could also be seen as something to play with or as something to look at in its beauty or holiness.

Even if there is a similar referentiality and a similar equipmental structure in the workshop and on the playing field, the purposes of sport and work are different. Therefore instrumentality and functionality will be guided by different goals.

Sport has an element of play, exploration, and experimentation that instrumental relations in other contexts lack.

The ontology of play is different from the ontology of work.

Furthermore, Heidegger does not focus on the quality of performance. In sport, though not always in play, this is important right from the start. Even in football among children one must play to win for the game to function at all. Sport has an element of competition and winning even in everyday settings that makes the quality of performance important in a sense that is different from the work world.[520]

Whereas Heidegger focuses on the ontology of dealing with things as equipment or objects, Dreyfus goes further into discussions of our bodily skills.

Heidegger had already paved the way for this, even if he never focussed on the body as such. In a sense, the body is absent from Heidegger's analysis of equipmentality in *Being in Time*. The hammer is there, but the hand that holds the hammer is not in focus, and the body that controls the arm disappears altogether.

This lacuna in Heidegger's thinking was filled by Merleau-Ponty, Polanyi, and others who strongly focussed on the embodied nature of the self's relation to the world.

Dreyfus uses Merleau-Ponty to underline the bodily directed nature of our dealing with the world. We have a direct access to the world through a motor intentionality that nonthematically and nonsconsciously helps us to navigate in the daily environment.[521]

Heidegger talks about the practical circumspection of our everyday orientation in the environing world.

Heidegger says:

"When we enter here through the door, we do not apprehend the seats, and the same holds for the doorknob. Nevertheless they are there in this peculiar way: we go about them circumspectively, avoid them circumspectively and the like."[522]

He then goes on to mention an example of flow experience in climbing as an example of such:

"nonthematic, non-self-referential awareness." One rock climber remarks, "you are so involved in what you are doing, you don't see yourself as separate from the immediate activity...You don't see yourself as separate from what you are doing."[523]

Dreyfus here elucidates our daily absorption in the environing world with what goes on in flow situations in sport. Breivik thinks this is to conflate two types of activities into one category. Dreyfus is right in interpreting manoeuvring through doors, using doorknobs, making use of seats, and so on in its nature of nonthematic activities.

But he thinks Dreyfus is on a wrong path so far as sport is concerned.

States of flow are reached only under special circumstances, like when our minds are focussed, when we perform well and experience a sort of emotional elation.

As an interpretation of Heidegger, it is also problematic because it shifts the Heideggerian focus from ontology to mind states.

Breivik would reserve the concept of flow to efficient coping under challenging circumstances and not as a characterisation of everyday dealing with our surroundings.[524]

Dreyfus claims that in flow states we are masterful in our dealing with the world. He introduces the concept of 'masterful coping', citing a passage from Aron Gurwitsch about being involved in situations where we allow ourselves to be immersed in the situation and subordinate to it. The move from masterful nonthematic dealings to flow experiences to masterful coping is problematical in two ways.

First, it conflates into one broad category what should be distinguished and analysed more precisely as increasing levels of performance, accompanied by specific views.

Second, it is problematic as a reasonable interpretation of Heidegger's views in *Being and Time* and other early works.[525]

Dreyfus thinks that 'absorbed coping' is a good and concentrated expression of the meaning of the readiness to hand condition.

However, the notion of absorbed coping is problematic.

Coping involves an element of skill or mastery. Heidegger never uses concepts like masterful coping, expertise and the like. There is only one place in *Being and Time* where a certain idea of expertise comes in. "The hammering does not simply have knowledge about (um) the hammer's character as equipment, but it has appropriated this equipment in a way which could not possibly be more suitable."[526]

Heidegger talks here of the individual Dasein's adjustment to the equipment but not of the expertise in using it. I may have my best grip on the hammer but still be relatively poor in using it skilfully to hit the nails. The possibility of poor circumspective dealing is obvious when Heidegger says that we also stumble into furniture in a room.

> When we enter here through the door, we not apprehend the seats as such, and the same holds for the doorknob. Nevertheless they are there in this peculiar way: we go by them circumspectively, avoid them circumspectively, stumble against them, and the like.[527]

According to Heidegger, entities can be encountered in a condition of readiness-to-hand or a condition of presence-at-hand.

Dreyfus equates these two conditions with respectively absorbed coping and conscious representational intentionality. This means that what happens in sport must be included in one of these two conditions or modes of being. Dreyfus has to accommodate this two category model with is five-stage skill model that has a more gradual evolving logic.

The model describes how skills develop through stages called novice, advanced beginner, competence, proficiency, and expertise. When adults learn a new skill they start following decontextualised and explicit rules. They use conscious mental representations and deliberate action. At the expert level they use intuitive and mindless nonrepresentational coping.

This means that from Level 4 to Level 5 there is a shift from conscious deliberation to nonconscious absorption.

Dreyfus underestimates the need for planning, preparation, and improving both in everyday life and sport. He places us in the middle of action where we are mindlessly performing our daily duties, but he forgets the vast amount of conscious activity that surrounds the mindless coping.

Heidegger sees readiness-to-hand as the ontologically primary condition. It is the condition where we are and where we remain for the most part. Dreyfus underlines his in his own version:

"We should try to impress on ourselves what a huge amount of our lives—dressing, working, getting around, talking, eating etc.—is spent in this stage, and what small part is spent in deliberate, effortful, subject/object mode, which is, of course, the mode we tend to notice, and which has therefore been studied in detail by philosophers."[528]

But this almost turns us into mindless zombies. Dreyfus is aware of this and enumerates the various ways in which daily comportment differs from the behaviour of robots. Among other things, humans have awareness, are adaptable, are startled, and switch to deliberate action when the going gets difficult. But it is really not enough.

According to Dreyfus we seem not to be consciously aware of what we are doing for the most part of our daily lives. But this is surely too simple. An interrogation of our own experiences shows that we are much more flexible switching between absorption and deliberation, mindless coping and conscious improvement on a regular basis.[529]

Breivik maintains that we do not use equipment in such an absorbed and mindless way as Dreyfus, pace Heidegger seems to think. Even the Heideggerian carpenters are not totally absorbed in their activities where the equipment disappears into the equipmental context and only the work to be done is in focus. The carpenter needs to design the house, find a suitable place to build it, think about materials, and collect the necessary tools and equipment. All this is very thematic, very conscious, very deliberate, and even self-referential.[530]

Bernard Heinrich's times for the 50 mile and 100km are world records at master level (over age 40). Heinrich has commented:

Even the tiniest inefficiencies of movement can make a huge difference over a long distance. I often noticed that muscle tenseness could be relaxed by conscious effort. I then focussed attention on my calves, thighs, arms, trying to relax them even during training runs, so that the most essential running muscles would be exercised. For a mile or so, I would monitor and hence try to control the kick of my arm swings, to make sure no energy was wasted in side-to-side motion.[531]

But Heinrich also describes the absorbed unconscious state and how he switched between conscious and unconscious mind states:

During running I feel my breathing. Like my heartbeat, the breathing rhythm is usually also unconscious. It is timed to the same unconscious metronome that times the footsteps. I like the feeling of the strong, steady rhythm with everything in sync. At times I listen to it—in just one instant I can bring it up on my screen of consciousness. Three steps with one long inspiration, a fourth step and a quick expiration. Over and over again. My mantra. My mind goes blank.[532]

We see here the mind wandering in and out between conscious effort and control and automatic subconscious state. The movements of a trained athlete are to a large extent, automatic. Yet they can be consciously improved during training and even during competition like table tennis and boxing. And this seems to be the case not just in slow-paced events like marathon running but also in sports with high speeds and fast decisions.[533]

Conscious here means a strong presence and awareness of what goes on. And second, it seems to be a form of thinking that is different from what we find among beginners.[534]

Heidegger's analysis is very relevant for the context of sport. Dasein's 'world' is connected to Dasein's ultimate projects. The purpose of playing on the playing fields or competing in the sports arenas is different from the purposes of, for instance, the world of work. The ontology of sport needs to be further analysed.

In a way, the ontology of *Being and Time* is too narrow—only the ontology of Dasein, readiness-to-hand, and presence-at-hand are discussed.

Later Heidegger opened up to the richer ontological field in which additional ways of Being elucidated (the thing, art, language, technology). Breivik says this needs to be further extended and applied to play and sport.[535]

Dreyfus maintains that our daily coping is mindless or unconscious in character. When equipmental breakdown occurs, however, there is a shift in our representational intentionality. But this is an exception; for the most part we are in a mindless state.

However, reflecting on situations from sport, we see that deliberation and conscious attending play an important part. Dreyfus in his skill model has tried to show how the highest skill levels show increasingly unconscious and automatic form of behaviour. Experts just act. They do not need to think reflectively and thematically. The body knows best. Parallel networks in the brain do the job. Even Dreyfus admits, however, that when the going gets tough or when things need to be maintained or repaired, consciousness intrudes. But Dreyfus underestimates conscious and deliberate goal-directed striving. Building on some examples from sport, it is significant how an elite athlete uses conscious attention to direct his performance not just during training but also during competition. In the rush to celebrate 'flow' experiences, we must not forget the important role played by reflection and thematization.[536]

Using Michael Polanyi's work *Meaning*,[537] Hopsicker supplements and extends the analysis of skilled-motor behaviour presented by Moe and Breivik. From these three perspectives, he examines kinds of knowing and how our intellect operates at the tacit and focal levels during the learning and performance of complex motor activities.[538]

Moe concluded that classic cognitivism provided little help in understanding physical activity. He claimed that "the cognitivist's belief in human information processing gained support neither at the experiential level of consciousness nor at the biological level of the brain."[539] Specifically, it neither held up to the "phenomenological description of how we experience everyday life" presented by Dreyfus nor supported the "logical analysis of information processing related to the biological brain" presented by Searle.[540]

He was convinced that classical cognitivism "produced an incomplete or erroneous understanding of intentional movement in sport."[541]

However, he came to very different conclusions about Dreyfus and Searle's alternative approach to intentional movements—specifically the notion of "background knowledge of the body."[542] This 'Background' consists of the embodied learnt abilities and practical 'know how' that inform intentional action.

Both Dreyfus and Searle had similar takes on the 'Background'.

Dreyfus viewed such knowledge as a 'readiness' or a 'tendency' to get a 'maximal grip' on the situation and surroundings. It is the body's "implicit and practical ability—a bodily schema or readiness—for moving towards an optimal relationship with the environment."[543]

Searle believed that the 'Background' consists of "intentional phenomena such as meanings, understandings, interpretations, beliefs, and desires."[544] He described such background knowledge as 'neurophysiological capacities' manifesting in abilities, dispositions, tendencies, and causal structures generally.[545] They are 'preintentional stances' towards the world that allow linguistic and perceptual interpretations to take place. They also "structure consciousness, constitute motivational dispositions, facilitate certain kinds of readiness, and dispose us to certain sorts of behaviour."[546]

While Dreyfus and Searle conceded that "early in the process of learning we often follow rules," they also believed that skilled motor behaviour relies much less on such obligation.[547]

These critics of classical cognitivism used the notion of the 'Background' to avoid the problems of infinite rule-following regress.

From Dreyfus' perspective, the skilled performer's body is simply "drawn to the appropriate position" without being explicitly aware of doing so.[548]

From Searle's perspective, the body is simply governed by developed "neuronal capacities, dispositions, and knowing how" that are "functionally equivalent to a system of rules." Therefore, Searle argued, explicit rules simply become irrelevant.[549]

There is value in Dreyfus and Searle's notion of background knowledge as an alternative to classical cognitivism and to the traditional way of thinking about intentional movements in sport.[550] Background capacities "are intended to provide intentional states in general with a sound foundation."[551] Further, "it explains how differences in Background capacities produce differences in actual performances."[552]

The Background, from the accounts of both Searle and Dreyfus, tends to eliminate the need to consciously focus on explicit rules in guiding performance.

The acquisition of Background knowledge does not automatically result in highly skilled athletes acting in a non-conscious manner a majority of the time. This is Dreyfus' notion of 'absorbed coping'—the idea that at higher levels of skilled behaviour we do not act by following discrete representations and rules. Rather, our bodies interact in a 'nonconscious zombie-like' manner with the physical environment.[553]

Absorbed coping "is not based on discrete representations and rule-following but on the direct interaction of our bodies with the physical environment."[554] Breivik proceeded to dispute the characteristic outcome of this thesis—that skilful performance is largely "mindless or unconscious in character."

Comparing Dreyfus' notion of absorbed coping to states of flow, such states "are reached only under special circumstances, like when our minds are focussed, when we perform well and experience a sort of emotional elation."

While these states of flow "may be observed in the workshop and on the playing field, they are not typically but relatively rare mental states, even among expert performers."[555]

It underestimates conscious and deliberate goal-directed striving.

Non-conscious behaviour may be present during specific moments of performance, but this is not the norm. Conscious decision-making is still present and functioning even at the highest levels of skilful performance.

Chapter 20
Husserl and Passive Synthesis

Whether it be the heart to conceive, the understanding to direct, or the hand to execute.
Junius

Passive synthesis is the prepredicative and prejudicative experience, which takes place ontologically before active judgments.[556]

According to Edmund Husserl, an exceptional thinker in his field, external perception constantly pretends to do what, by its very nature, it cannot.[557] He argues that somehow we create an objective sense, which endues as a unity throughout continuing unfolding manifolds of appearances; yet we never perceive that objective unity.

Our perception is always spatially and temporally partial, but that which we intend through perception is whole.

In external perception, the spatial object only presents itself from one side, and the pretention of external perception is to transcend our perspectival limitation on the object by constituting a transcendent, objective unity or sense.

My perception of the front of a cricket ball points to its back and the continuity of colour, shape and composition with the front.[558]

Passive synthesis is that "realm of bodily habits that were once actively acquired but subsequently have become sedimented into a style of comportment and yet are accessible reflectively."[559]

"Everything habitual belongs to passivity, even the activity that has become habitual."[560]

Learning a sport is a continuous activity with multiple layers of sedimented habituation contribution to the ability to play without active thinking.

He defines pretentions as "presentations directed towards an object by virtue of genetic synthesis."[561] This means that our forward learning intentions are presentations in consciousness.

A ball can be bowled in a number of directions. Because batsmen see through repetition where the ball often goes, his perception of the logically indeterminate action can awaken in retention that presentation directed at the ball in its anticipated place.

Passive association generates the phenomenon of expectation and all intentions that anticipate.[562]

It is a phenomenology of association or an 'inductive association' which results in the generation of these anticipatory intentions.

Prominent data emerge for themselves in conscious experience.[563] Some of these prominent qualities concresce or go together, with others, whereas some discrese, or emerge as discrete and different.[564] This associative process gives rise to that which can be awakened through repetition.

Those awakenings give rise to the phenomenon of expectation.

This expectation as anticipation is the key to successful sporting ability. Note the composure of the great batsman who is in position in anticipation of the arrival of the ball. The coaching aphorism is "to beat the ball not meet the ball."

Husserl's enigmatic phrase, 'active passivity' applies directly to playing because it characterises the passive sense, which contains an intentionality of drive and desire, but which does not originate from "an activity proceeding from the ego."[565]

These passive syntheses pass continuously from object to object and excite anticipations and possible determinations, but they are not judgments in the Kantian sense.

Playing and passivity run together in a continuous process.

Reaction is not the explanation. It either requires that a stimulus response analogous to a knee jerk is at work, or it considers the action a function of active judgment, which our experience refutes. It has been suggested that it is active judgment but just a very fast one. This cannot be right as so many examples of the burden of conscious, active judgment in sports refute this contention.

Active intentions can cohere in the continuity of these passively generated synthesis. For example, when one swings the golf club in an attempt to hit a low hooking shot from around a tree, one might actively emphasise turning the hands over abruptly at impact; but the ability to succeed in this shot depends on the passivity at work in all the other aspects of the swing.

Husserl's phrase 'active passivity', resonates in the numerous publications on the Tao of sports, and these speak to the truth concerning active passivity.

The Taoist concept of Wu Wei is one of creative inaction or effortless action. Taoism's language explains the experience of this sense of active passivity.

Hyland's explanation of play, at least as it concerns 'peak experiences', refers not to Wu Wei, but to the Zen Buddhist concept of 'self-forgetfulness'.

We commonly refer to the athlete in this zone as 'unconscious'. This refers to a realm accessible by a different mode of cognition than the analytic, active mode described by Kant. Husserl's analysis phenomenologically gets to the heart of this mode.

Husserl's overall contention: that meaning is given in the perception itself, before any judgment or predication about the perceptual experience, attempted to ground scientific knowledge on a transcendental logic. In doing so, he gives us a language and an analysis of lived experience with tremendous explanatory power.

The role played by the objects of perception in the aesthetic undermine the language of positive science that imports an action-reaction mechanism into features of playing that are motivational and not stimulus responses.

Playing in both its thoughtlessness and in its internal relation to anticipation demands an internal descriptive account.

"Husserl's explication of passive synthesis yields powerful explanations of the otherwise ineffable, perhaps mystical experience of playing and its parallel phenomena of acting out of passivity in an anticipation of otherwise indeterminacy."[566]

It is a fundamental sensory, conceptual and symbolic phenomena of the mystical nature of how we think and do things. No complete explanation has been given as to its origin and workings in the human brain. It is one of life's great mysteries.

Chapter 21
The Hand

Bless the hand that gave the blow.
Dryden

Whatever the hand findeth to do, do it with thy might.
St Luke Chpt.10

Great is the hand that holds dominion over man by a scribbled name.
Dylan Thomas

The significance of the hand cannot be overstated. In a wonderful book by Géza Révész first published in 1943 the author says:

> "The hand created our entire civilisation and culture. We owe to the hand all the tools of labour employed in the service of social life...By work and gesture we express our thoughts, our feelings, and aspirations and enter into communication with our fellow men. Even the realisation of our ideas, intentions and aims presupposes a materially controlled and intellectually directed hand...We come into the world with moving hands; with folded hands we are laid in the grave!"[567]

From the biological standpoint it is often emphasised that the improvement of the hand and its many-sided usefulness were of great significance for the morphological and functional development of the brain, and thereby that the intellectual progress of mankind.[568]

Révész indicated the relation between the activity of the human hand and anatomical and physiological suppositions about speech whereby the locomotor function of the hand and arm is brought into localised relationship with the region of the brain serving speech.

Well ahead of his time he appreciated that the relationship is a mutual one. The function of speech also, in its form, exercises an influence on the morphological and functional development of the hand.

Form and function are closely tied with the extension of needs. Between hand and need there exists a reciprocity of action; needs shape the hand and the hand creates new needs and discharges those tasks which arise from those needs. So long as the hand only has to master tasks in the vital sphere of life; it remains animal-like, confined to its pure biological functions, morphologically primitive and incapable of development. If, however, needs of social, cultural and civilised life arise, then the hand acquires its *human character*, through its working, forming and expressive functions. These mental functions endow the hand with its human character.[569]

In German the notion of '*Handeln*' embraces all meaningful and goal-directed human activities. It characterises unequivocally the total personality of man.

This idea is not limited to external manipulation, that is, actions which affect changes in the outer world. It also includes inner action, the purposeful activities of the mind.

In his mode of manipulation man experiences his real 'I'.[570]

The general significance of the human hand lies therefore in its manipulative function. Man is induced to various forms of manipulation by his conscious and unconscious strivings, by his desires and by the goals towards which he aspires. Impulses, aspirations, wishes, decisions press for realisation, and this takes place chiefly through the mediation of the hand.[571]

There is no occupation which can dispense with the hand, and we can scarcely think of a situation in the waking state or in normal circumstances in which man refrains completely from any purposive movement of the hand. Speaking itself seldom occurs without some movement of the hand.

Gestures (as we shall see) support and supplement spoken language.

The ability to act in a practice fashion is one of the chief criteria of a life governed by the will.

All manifestations of a child's willpower are recognised in his movements; in the first stage and at the earliest stage in gestures and actions. The first and most important habits are connected with the mechanism of grasping.

The hands great adaptability and resourcefulness exert an extraordinary challenging effect on the activity of the mind and vice versa.[572]

All handicaps or injuries to the capacity for manual movement have an unfavourable influence, directly or indirectly, on the mental faculties and consequently on the entire personality.

Everywhere we see the significance of manipulation and everywhere it is the hand that manipulates.

The whole of man's life is filled with 'action-movement', the activity of the hand is always prominent.

It is therefore only natural that most ideas of action are borrowed from the achievements of the hand in touching and grasping.

The metaphorical use of these concepts leads to the idea of mental 'manipulations', that is, to the activity of a mental and moral personality.

The metaphorical application of designations and ideas connected with the activity of the hand are encountered in the most diverse spheres of human life. If we wish to convey that we have acquired something mental, we say, that we have 'grasped' it.

The general notion of mental acquisition we express by the word 'comprehend'. The words 'comprehension' and 'comprehensible' point most clearly to the activity of the hand.

Even the word 'impression', in the sense of an influence on the mind, and the word 'expression', in the sense of a reaction to an impression, are borrowed from the tactile motor sphere.

Likewise, words like 'dismember', 'dissect', 'maintain', 'display', 'turn over' may denote actual activities of the hand as much as they may relate, in the transferred sense, to mental activities and achievements.[573]

No expressions are more apt to describe inner mental states than those borrowed from the activity of the hand: 'something moves me', 'I am gripped', 'I am held'.

In the moral sphere we find, among other instances, the following images: 'to lay the hand on the heart', 'the helping hand', 'to wash the hand in innocence'.

Duties are designated by similar expressions: 'a handshake', 'an earnest', 'placing the hand upon'.

There are similar expressions referring to social relationships: 'my life rests in his hands', 'to offer the hand', 'to give the hand', "He is his 'right hand'." We even find that a 'hand' is identified with a person, as in the nautical expression: "there are seventeen hands on board."

It has been pointed out that knowledge mediated by the sense of touch is most capable of bearing a reality character. According to one view, the sense of touch carries a greater force of conviction about the reality of the external world than the other senses.[574]

This led Locke and Berkeley to believe that tactile space is the *original* and that the visual space is derived from it. This has led to the theory of the primacy of the sense of touch. I do not wish to go into the detail of such a hypothesis. It is beyond the scope of this book.

As far as the biological place of the hand including the arm is concerned, the hand represents an organ which includes two interrelated systems, the sensory and motor.

The hand, as a sensitive apparatus, receives manifold sensory stimuli which release differentiated sensations: it yields the sensations of pressure, touch, vibration, temperature, pain, movement, localises tactile stimuli, perceives changes of location in the hand-arm system, and takes note of the spatial quality of objects that are touched.

The hand is the only organ distinguished by such manifold forms of sensation and perception.[575]

Besides having these sensitive experiences, the hand constitutes the centre of impulses of movement, of giving form, of willing, and at the same time is the executive organ of these impulses.

Because of these functions the hand occupies a specially unique place in the biological system.

There is no question that the sphere which is controlled by the hand and in which the hand raises its full power is the sphere of work. We must account for the great part played by the hand in all working activity by its instrumental character.

"The hand is a universal instrument, the manifold character of which cannot be ignored."[576]

When we realise the rich variety of working movements and working performances which the hand can carry out together with the arm and motor system of the body, we are justified in declaring that in the activity of the hand, in the capacity for performance of this unique instrument, all that tools can achieve is potentially included.

When we consider what kind of working movements the most diverse tools, implements and machines individually or in combination with one another can execute, it seems that all these work processes can be readily covered in the activity of the hand.

The hand grips, pulls, pushes, presses, holds, throws, flings, tears, beats, turns, folds, ties, wraps, kneads, threads, raises, fills, breaks, bends, puts into order.

It thus carried out working movements which belong to the most important functions of the majority of implements and tools.

The principal tools of mankind owe their *origin* to the hand.

They have come into existence by imitation and transference of the position and working movements of the human hand.

The hammer is a copy of the clenched fist, and the thumb, in the pressure of opposition to fingers, is a model for the tongs.

The most important first grips have promoted the construction of different kinds of hook, ring and tongs; and the precision grip of the hand has suggested the finest precision instruments and meanings equipment.

Tools are created by the hand. Only the hand that creates tools and uses implements sensibly possesses an instrumental function.

The human hand is precisely an instrument of this kind, a universal instrument worthy of admiration. It became such a thing in the first place by participating in work of the most diverse kinds.

However, the human hand was able to take part in work and make working implements, and use instruments sensibly, just because it became an oran of *thinking* and *speaking* man.[577]

No one has characterised the human hand better than Aristotle.

In his *Metaphysics*[578] he writes:

"It is not a hand in any state that is part of man, but the hand which can fulfil its work, which therefore must be alive, if it is not alive it is not a part."

The human mind first transformed the hand into an instrument; out of the original grasping hand the mind shaped a hand that works.[579]

Chapter 22
The Brain and Hand

I am a brain my dear Watson, and the rest of me is mere appendage.
Sherlock Holmes

Whose little body lodg'd a mighty mind.
The Iliad of Homer

What hand and brain went ever paired.
Robert Browning

The futility of maintaining a separation of mind and body is self-evident when one looks at high levels of achievement in purely 'physical' skills like sport. It depends on a mastery of both procedural and declarative knowledge, and akin to the creative skills displayed by highly successful mathematicians, sculptors and research scientists. Success is dependent on the unity of mind and body.

Indeed, intelligence, the capacity for innovative response to the world is an aspect of the entire organism.

At the forefront of this is the relationship between hand and brain.

The hand speaks to the brain as much as the brain speaks to the hand. Self-generated movement is the foundation of thought and willed action of the underlying mechanism by which the physical and psychological co-ordinates of the self-come into being.[580]

This may result in the entity which metaphorically is called 'mind'. But we not only have no real understanding of it but difficulty in seeing mechanisms that give rise to it, how any mechanism can give rise to it. This is the problem Schopenhauer called the 'world knot'. Neurophysiology and neuroanatomy have been the main avenues of approach to knowledge about the physical brain. However, more recently, other sciences, including physics and mathematics, have joined a broader front of disciplines going under the name of neuroscience.

The machine has been a metaphor when studying the functioning of various parts of the body. In the brain, too, we speak of mechanisms by which certain functions are accomplished. But the machine-body analogy breaks down in certain significant respects that concern the individuality of humans. This includes such features as intrinsic unpredictability and the impossibility to duplicate or reset. Unlike a computer, which can be made to yield all its stored information (by printing and transferring it to some other memory device or to another computer), no human can willingly or unwillingly be drained of all his knowledge.

Norbert Wiener wrote in 1963 in the introduction to a Symposium on Nerve Brain; and Memory Models:

> "In a comparative study of human performance and machine performance, it must be realised that the human being does some things much better than the machine and some things worse. The human system is not as precise nor as quick as a computing machine. On the other hand, the computing machine tends to go to pieces unless all details of its programming are strictly determined. The human being has a great capacity for achieving results while working with imperfect programming. We can do a tremendous amount with vague ideas, but to most existing machines vague ideas are of absolutely no use."[581]

The difficulty is that the connections between neurons in the real brain are still not known, and if they were known, their complexity would make it all but impossible to calculate what might happen in such a system. The only real model of a brain, it has been pointed out, is another brain.

"The brain does not live inside the head, even though that is its formal habitat. It reaches out to the body, and with the body it reaches out to the world. We can say that the brain 'ends' at the spinal cord, and that the spinal cord 'ends' at the peripheral nerve, and the peripheral nerve 'ends' at the neuromuscular junction, and on and on down to the quarks, but brain is hand and hand is brain, and their interdependence includes everything else right down to the quarks."[582]

In 1833, Sir Charles Bell completed and published the Fourth Bridgewater Treatise The Hand, Its Mechanism and Vital Endowments as Evincing Design.

His analysis of the behavioural consequences of variation in anatomic structure, and his insights into the relationship between movement, perception, and learning, were revolutionary and seminal. Its singular message, that no serious account of human life can ignore the central importance of the human hand remains as valid as when it was first published. It is as "as admonition to cognitive science."[583]

From the perspective of biomechanical anatomy, the hand is an integral part of the entire arm, in effect a specialised termination of a cranelike structure suspended from the neck and upper chest.

Should those parts of the brain that regulate hand function be considered part of the hand? Wilson[584] says the perspective of physiological or functional anatomy suggests that the answer is yes. The precise definition of hand may be beyond us. We understand what is meant conventionally by the simple anatomic term, but can we say with certainty where the hand itself, or its controls or influence, begins or ends in the body.

Bodily movement and brain activity are functionally interdependent and their synergy is so powerfully formulated that no single science can independently explain human skill or behaviour. In fact it is not clear that what we have asked is a scientific question. The hand is so widely represented in the brain, that hand's neurological and biomechanical elements are so prone to spontaneous interaction and reorganisation, and the motivations and efforts which give rise to individual use of the hand are so deeply and widely rooted, that we must admit, says Wilson, "we are trying to explain a basic imperative of human life."[585]

The shoulder contributes to movements that not only transport but also orient the hand. And in shifting our attention to its role in supporting the function of the hand per se, there is no such division or segregation of these functions in the body itself; the musculoskeletal system functions in a fully integrated way, so that the overall movement is unitary and fluent. In reality, shoulder, arm and hand functions are exquisitely tuned and responsible to one another in both neuromuscular and biomechanical terms. The kinetic and informational processes take place simultaneously from the body outward to the hand, and from the hand inward to the body.[586]

Jugglers performed at the 40,000 seat Hippodrome in Constantinople during the reign of Justinian I, some fifteen hundred years ago, and contests of juggling skill were a popular Native American Sport long before the arrival of Europeans.[587]

Nothing could be intuitively more clear than the critical dependence of the skill of juggling on hand-eye co-ordination; what is not so obvious is the degree to which almost all physical skill flows from the maturation of motor skills under the guidance of both visual and kinaesthetic monitoring.

Charles Bell said:

> "This faculty of searching for the object is slowly acquired in the child; and in truth, the motions of the eye are made perfect, like those of the hand, in slow degrees. In both organs there is a compound operation; the impression on the nerve of sense is accompanied with an effort of the will to accommodate the muscular action to it."[588]

He asserted that both the hand and the eye develop as sense organs through practice, which means that the brain teaches itself to synthesise visual and tactile perceptions by making the hand and eye learn to work together. According to Bell, the learning process must involve the correlation of sensory information from retinal (light) and cutaneous (tactile) receptors with what he calls a 'muscle sense'. In other words, the brain activity orients the receptors in the eye or the hand towards a target of interest, and then moves them precisely during a process of exploration. The resulting image constructed by the brain must of necessity be based both on the messages from retinal and/or skin receptors and on the record of guided eye or limb movements occurring during the collection of sensory data. This is an extremely sophisticated concept, and contemporary research into the functional organisation of perception has strengthened it.

Howard Austin said:

> "The most powerful technique for making global programme changes it to call on a specialist who can give expert advice on the problem at hand. The expert can be the coach, teacher, or some other external (to the learner) source or other interval programme. The ability to find, or give yourself good advice then becomes a critical part of the motor programming process and hence of athletic ability."[589]

Advice techniques, skill models and various other high level editor features play the dominant role in the construction of new motor programmes. These activities are inherently intellectual in nature and hence lead to the claim that so-called physical skill is largely mental activity.

The features that distinguish the human hand from the hand of apes are nearly imperceptible and in fact were recognised by anatomists relatively recently; while there has been no dispute that human use of the hand is unmatched in the animal world, no one thought there was anything special about its design. In a serious discussion about evolution the term 'design' is a battle cry. Bell, writing—before Darwin, meant quite specifically that the condition of the human hand was evidence of an intelligence outside of man who was, in his words, the 'anonymous author' of the design. Richard Dawkins argues the case against design in The Blind Watchmaker (1987) and Daniel Dennett does the same— Darwin's Dangerous Ideas (1995).

Chapter 23
The Hand in Sport

Voltaire said of Sir Isaac Newton that with all his science he knew not how his hand moved.[590]

> "…the hand acquires knowledge through action; its knowledge guides its action; and its acts acquire more knowledge through communication. In this respect, the hand is perfectly suited to its special relationship with the brain."[591]

The hand communicates with it in various ways. In manipulation, especially precision manipulation, it has continually to report to itself in order to initiate actions and to keep those actions on track. The conversation may be virtual or subconscious, as the feedback from muscles and joints that is said to enable the cerebellum to 'calculate' (and recalculate) trajectories during manipulative movements; or just above the threshold of consciousness, as when hand adopts and maintains a certain posture. This mixture of unconscious and subconscious feedback is essential for the sensorimotor integration that underpins co-ordinated exploratory or manipulative activity.[592]

One study[593] investigated the respective roles of motor programmes and perceptual guidance in two relatively simple but voluntary tasks—bimanual finger oscillation and bimanual four-finger tapping. The authors argue on the basis of their findings that the co-ordination of the two hands is on the basis of visual perception and not on the basis of 'motoric' representation of the pattern of actions. They speculate that 'voluntary movements' are, in general, organised by way of a simple representation of perceptual goals, whereas the corresponding motor activity of, sometimes extreme, formal complexity is spontaneously tuned in. This may be the kind of movement organisation that makes the richness and complexity of human voluntary movements possible, be it sports and dance, skilful tool use or language. In short, the deliberate voluntary activity, however simple depends upon seeing (in the widest sense) what needs to be done), seeing that it is being done, and seeing that it has been done. The requisition of motor programmes requires knowledge of what one is about. Almost how this translates into action when no one knows how to mobilise motor programmes to order remains a mystery.[594]

The dexterous hand is not isolated. It soon interacts with an intelligence eventually co-driven by other factors, with language and sociability. Nor is it isolated in the literal sense of being separated from the arm: intelligence and wakefulness can creep up the upper limb. The genius of reaching may learn much from the dazzle of dexterity. And, as William Calvin has suggested, throwing may have a central role in driving the growth of intelligence and (in our terms, the sense of self and agency).[595]

Raymond Tallis says:

"The hand…took humans over the threshold dividing consciousness from self-consciousness, unreflective instinctive behaviour from true agency and sentience, which is shared with other animals, from knowledge and feeling for truth, which is not."[596]

The variety of hand games—or games that involve the hand—seems limitless. The unadorned hand is involved in throwing. This has many modes: simple throwing, baseball chucking, the exquisite intricacies of spin bowling and the power of discus hurling and shot putting and hammer throwing. The hand deploys its power in grappling (wrestling), in lifting (weights), in pulling (carts) and tugging (of war). The sporting hand may be supported by special gloves— as in boxing, cricket, baseball, goalkeeping. It may be armed with bats, racquets, clubs, cues, lariats, firearms. Catching is best seen as the genius of reaching writ large plus the genius of grasping.

In order to catch a ball that is sent in your direction without any warning— you have to move smartly to a place where a ball is within reach of your hand. The whole body has to be transported to the right place at the right time and there to assume an appropriate posture. This may be a 'relatively' simple matter of positioning yourself so that the ball falls into your hands. Even this is not quite as simple as might appear. How do you know where to go in order to intercept the ball?

Experiments that involve simulation of 'catching' on 'trained' neural networks suggest that when you catch a ball, you run so that as you watch the ball your angle of gaze above the horizontal increases continuously but at a decreasing rate.[597]

It is remarkable, however, that this is not how we believe that we succeed in arriving at the right place at the right time. According to experimental psychologist Peter McLeod, if you ask people:

"to describe what happens to their angle of gaze as they run to catch a ball, they say that they look up as the ball rises through the air, and then down as it falls. While this is a good description of what happens if you look at a ball in flight from the side, it is completely incorrect if the ball is coming towards you."

McLeod's observation on this discrepancy is of particular interest.

"It's almost as if the subconscious self knows it can solve the problem and doesn't want to have interference from the conscious self. The conscious is allowed to have some theory to keep it quiet, but the subconscious just gets on with solving the problem."

He says the knot is drawn even more tightly when we consider the routine virtuosity of the slip fielders whose diving catches require the body to be flung across the field and whose intercepting hand is unfolded from a moving reference point. So there has to be an instantaneous selection of the end-posture in order that the flying body's target position, and its trajectory to that position may be determined and the arm undergo the right amount of unfolding. Unless, moreover, the flying body's journey and the brachial unfolding are correctly paced, the desired intersection with the anticipated position of the ball will not occur. Modelling will have to take account not only of the ball's flight, but also of the current position and speed and direction of movement of the arm.

There is a good deal to sort out in advance about what should take place— happen or be enacted—when the hand makes contact with the ball. Mere collision is not enough, that would result only in a sore hand, a dropped catch and an angry bowler. The posture of the hand needs to be predetermined. Too small an aperture will mean that the ball never gets bedded in and bounces off (leaving a sore fist in its wake). Too large an aperture will mean that it slips out (resulting in a frustrated fielder with a stinging palm). Assumption of the right aperture and posture of the hand is not enough: the open door of welcome has, at the very moment of contact, to turn into a barricade or cage of fingers, with fingers and thumb moving in precise formation to make the arrival into a capture. The work of the snatch squad has to be simplified by arranging that the stiffness or impedance of the arm measured at the hand is precisely matched to the ball at the time of the impact. If the impedance were too high—and the ball ran into a brick hand—it would bounce out again before the fingers had a chance to hug it; and if the impedance were too low, it would knock the hand out of the way. There is evidence that reflex pathways are switched on and off, hereby continuously modulating the impedance just before and just after the impact.

The genius of catching is merely a variant on the genius of reaching and is riddled with passivity; when we reach out for an object we are dependent upon a host of mechanisms—for example, those that regulate the movement of the arm in the ballistic phases of its trajectory—of which we have little understanding. It's important, however, not to draw the wrong conclusion from the fact that catching incorporates many layers of automatism; to suggest that the cricketer is mere the 'site' of the catching and to replace 'Jones catches the ball' with 'It—Jones's body—catches the ball'. We should not make this error, and incorrectly conclude that cricketing stars are automata, if we recall the full context of the spectacular catch. This will include the game of cricket which Jones took the trouble to attend; the many hours that Jones spent practising catches; the special arrangements he made to ensure that he could attend the nets; and so on. Remembering this will ensure that we reinsert the self—Jones—into the moment that seems quite void of Jones' consciousness.[598]

The moment of dazzling virtuosity is the product of hours, days, weeks and months of dedication.

Sport is full of these virtuoso versions of the manifold dexterities of daily life. And a sporting career will clock up many thousands of documented and undocumented acts of bodily genius.

Chapter 24
Pulsation and Coordination

'The Ways of the Hand'[599] by David Sudnow describes the ways embodied beings acquire the skills of giving order to, or better, finding order in our temporally unfolding experience.

"It is a phenomenology of how we come to find our way about in the world, whether it be the world of jazz, discourse, typing, tennis or getting on and off the bus."[600]

The aim is not explanatory but descriptive. It is a phenomenological account of handwork as known to a performing musician.

Like Merleau-Ponty's Phenomenology of Perception, Sudnow's work has implications for those who want to understand the nature of skilful performance. His details description of his acquisition of the skilled hands of a jazz pianist shows the limitations of a cognitivism that thinks that having a skill consists in interiorising the theory of a domain.

Sudnow starts in 'Beginnings', by hunting for particular features, in his case the notes on the piano keyboard, and practising following rules, such as typical jazz scales, until they become second nature.

After much experience such a novice progresses to the stage where he finds himself able to reach for gestalts, like chords or scales as a whole, without having to think about them, and then to begin to apply maxims, such as "repeat this melodic cluster," as in his 'Going for the Sounds'.

Next at a level one might call intermittent competence, the student has to form a strategy to get from one situation to the next, as Sudnow begins to do in the first part of 'Going for Jazz'.

Finally, this too becomes something the hand can do, so that now there is a strategy without a strategist, although such proficiency is still interrupted by the occasional need to thematise aspects of the performance.

After years of accumulating specific experiences of many thousands of ways to move, he gradually masters the essence of improvisational play with the development of a finely shaped rhythmic co-ordination that synthesises such movements into true jazz sentences.

As 'Going for Jazz' reaches the climax, there is finally no longer an I that plans, not even a mind that aims ahead, but a jazz hand that knows at each moment how to reach for the music. The hand takes over.

In the course of his detailed phenomenology, Sudnow implicitly corrects a sensible but surprising error in the phenomenology of perception.

Merleau-Ponty occasionally characterises the lived body as an 'I can', whereas Sudnow is clear that it is not he but his hand that reaches for the jazz. Just as in the Odyssey, Homer says of his heroes that, 'when they sat down to a banquet, their hands went out to the food in front of them.'

Sudnow's detailed description suggests that cognitivist theory of skill acquisition, taken for granted from Socrates to Descartes to Kant to Husserl to Piaget, has the phenomenon upside down.

Rather than moving from specific cases to abstract principles, skill acquisition seems to move in the opposite direction, from principles followed until they are interiorised, to the possession of so many types of concrete cases paired with types of responses that each situation leads fluidly to the next.[601]

Empiricists who think of skills in terms of associations of experiences or the formation of linear neural connections (what Merleau-Ponty's contemporaries called 'the reflex arc'), would have to defend their view in the face of the phenomenon observed by both Merleau-Ponty and Sudnow that one can transfer one's skill from what one hand has learnt to the other hand.

Sudnow's work moves in the opposite direction from Merleau-Ponty's.

Like any philosopher, Merleau-Ponty provides only enough detail in his description of action and perception to motive his move to generality and ultimately to ontology, whereas Sudnow purposefully restricts himself, in what he calls a 'production', to reveal only the concreteness of situated relevant detail.

In articulating one of the most subtle, rich, intricate, and inarticulate skills human beings have developed, he provides new insights into how the body takes over a domain and, most particularly, how it uses varying styles of pulsation to co-ordinate the temporal unfolding of skilled activity, whether it be music, speech or sport.

Research that comes from another direction, from such broad details as that the body moves forward more easily than backwards and has to balance in a gravitational field—can also lead to new understanding of what Merleau-Ponty calls 'motor intentionality' and thus of the body as a way of being that is neither subject or object, but the discloser of the spatio-temporal world.[602]

Sudnow's and Todes's[603] work carry forward and go beyond Merleau-Ponty's phenomenology of the active body. Together they are uniquely at the forefront in doing Merleau-Ponty—inspired research on embodiment, and not, as so many others do, merely interpreting Merleau-Ponty's philosophy.

Sudnow makes an interesting comparison with speech.

"I came to see my passable first phenomenology of aspects of jazz piano performance as a suggestive preface for the phenomenological description of articulated gestures of all sorts, talking included."[604]

"The report is about jazz piano playing. But he also saw it as a sort of prolegomenon to the study of talking. There is so much in common between ordinary speaking and musical improvisation that, at the least, not to expect descriptions of experience at producing one to inform approaches to the other is plainly unreasonable."[605]

He described the body as making rapid and finely articulated moves from one place to the next on time, proper places and timings very closely defined by cohorts of fellow speakers. The body finds its way from place to place in the course of moving and certainly in general not by figuring out places to go in advance. It takes years to become a mature speaker and listener in each domain.[606]

He had the most vivid impression that his fingers seemed to be making the music by themselves.[607]

This hand chooses where to go as much as 'I' do.[608]

Sudnow considered that while there are a vast number of physiological and experimental psychological studies of timing in human and other animal behaviour, there were none documenting the specific and absolutely critical role that pulsation plays in co-ordinating bodily movements.[609]

He referred to the difficulty of co-ordinating the movements of his right and left hand playing jazz and he drew an analogy with workers lifting a weight.

He said that despite his intentions, the left hand's reachings were often less than really rhythmically solid, the right hand's movings having their ways of upsetting the lefts (and vice versa).

Sudnow defines the essence of the pulse.[610]

The example of the workers suggests the interbody work of pulsing. Each worker adjusts the thrust of his movements by aiming towards an upcoming time of arrival, established by a preceding count, appraising the speed required to manage the weight and the distance to be traversed, adjusting the force and extent of the move accordingly, holding these variables in delicate bodily balance.

Their joint pulsing joins their respective moves in accord with the same phasing structure that unites the drawing of two lines of differing lengths, although the particular patterns of acceleration and deceleration always vary with respect to the tasks at hand. A fortiori with cricket and timing.

When two or more moves are co-ordinated towards various destinations for simultaneous arrival, turnaround phases for acceleration to deceleration are shared in common by the variously relevant solo and/ or multibodied movements.[611]

Pulse is a sequential course of articulations which is paced by reference to a prospective time of arrival at the completion of the phase. All reaches are given their smoothness and internal pacing layouts by reference to prospective arrival times.

The ubiquity of rhythmic pulsation for finely organising gesturing of every variety, perhaps most highly refined in the unfolding articulations that constitute music and speech, is profound.

Chapter 25
Sport and the Human Condition

Great dreamers' dreams are never fulfilled, they are always transcended.
Alfred Lord Whitehead

Don Quixote is a madman, but seen from within his madness has a transcendent content.
Ortega.

Everyone is born a genius, but the process of living de-geniuses them.
Buckminster Fuller

Ortega, the great Spanish philosopher, who saw sport as an essential component of what is truly 'human' about human being, points to a more profound statement above the nature of human existence. Ortega's philosophy is essentially a philosophy of praxis, of human actions in the world. Indeed human actions are seen as constitutive of human being per se. Man is what he does.

Echoing Heidegger's and other existentialists' theme of the human being's 'thrownness' in the world, Ortega's vision of human existence stresses that "life is given [to] us, since we find ourselves in it, without our knowing how or why[612]."

Thus no matter how much a particular human being is embedded in a particular cultural context, he or she still at some basic level feels adrift in an alien environment.

This is a further reason why humans and animals are different: "Animals are adapted" to their environments "while man is essentially unadapted. Man is everywhere a foreigner."[613]

Human life, however, is in general terms not one characterised by despair and submission to this eternal condition of alienation. Instead, in the fact of the "frightening reality that life ultimately and constitutionally is[614]," humans struggle in the face of adversity. It is another aspect of human life, beyond that of existential angst, that makes this possible.

"The life that is given us is not give ready-made, and instead we have to forge it for ourselves, each one his own life. Life is something we *must do*, and it gives us a lot *to do*...life is not a fact, not a *factum*, but a *faciendum,* something that has to be accomplished, it is not a substantive but a gerund[615]."

Each person's life is an open-ended project, and existence confronts him or her with the challenge of fulfilling that project to the best of his or her ability, given the circumstance s/he finds him/herself in.

For human beings, existence is "a poetic task, like the playwright's or the novelist's: that of inventing a plot for his existence, giving it a character which will make it both suggestive and appealing[616]." Ortega's response to a philosophy that equates human existence solely with the more despairing aspects of it is this: "Life as anguish, Mr Heidegger? Agreed! But, it is also this: an enterprise[617]."

Ortega sees the human being metaphorically not just as a poet writing the story of his or her own life, he also sees him or her through the figure of the sportsperson. Each individual is engaged in the difficult business of continuing to play the games that make up his or her life, despite the vicissitudes that existence throws at him or her.

In Ortega's understanding, the figure of the sportsperson is quite as noble and profound as that of the artist. It is by seeing each individual life as a project within the vast and unfriendly environment that is the cosmos that compels Ortega to see sports as profoundly symbolic of the human condition.

The latter is seen as being made up of both despair (at the fundamentally alien nature of the cosmos from the point of view of humankind) and aspiration, the desire to make of oneself what one can.

Whereas Christianity sees the brief span of individual human being as 'a vale of tears', paganism "makes a *Stadion* for sports of that vale of tears[618]." This is because continuing to be oneself, even against the most extreme form of adversity fate can throw at one, and struggling to become more than one currently is despite the unfavourable conditions life dictates, both involve the "willing acceptance of painful effort," and this latter "is the very definition of athletic striving[619]."

In other words, what keeps each human soul going in the face of even the bitterest setbacks and torments is the same form of will that underpins sportive striving. This involves the undertaking of a task that is recognised as arbitrary (literally as unnatural, because no natural reasons compel, authorise, or underwrite it) but which is taken *as if meant something*. To that degree, the project of one's own life is taken seriously, just as sports are taken seriously by those engaged in playing them.

Ortega sees sports as occupying a middle ground between the grinding seriousness of necessary labour and the wholly capricious nature of play. Sports are serious enterprises but also involve the playing out of a role, of a mode of being that is not wholly synonymous with 'oneself', and this 'play' element tempers the seriousness of the game because one's being is never totally contiguous with the sporting role.

If sports are a mixture of earnestness and playfulness, then they are exactly homologous to the human condition itself. This is because the latter involves both an engagement with the serious business of living and the auto-creation of a self, which is undetermined by nature, which can gaze upon itself, and which is thus—at least potentially—open to being changed by its own will to change.

Ortega's argument is that the characteristically human (rather than animal) response to life is at its root fundamentally 'sportive' in character. Sports are a microcosm of the existential situation that is human being. On this view, the human is an existential and metaphysical sportsperson engaged in a game called human life. Human activities of all sorts are games undertaken in order to give the human being some direction and meaning in an otherwise directionless universe. Actual sports are a manifestation and expression of the nature of humanity per se, a species that expresses itself by submitting to freely chosen yokes, in addition to those forced upon its members by a natural world characterised by exigency and necessity.

Ortega's ideas bear some apparently similar to those in Bernard Suits. Suits developed his ideas as to the nature of play and games, and their roles, as important constituent features of human life, without reference to the work of Ortega although they predate Suits' ideas by up to four decades.

There are four key similarities between each set of ideas.

First, Ortega's claim as to sports as yokes that are freely chosen by participants in them is akin to Suits'[620] argument that the central feature of sportive rules is that in each sport the players choose to renounce "the simplest, easiest and most direct approach to achieving…a goal is always ruled out in favour of a more complex, more difficult and more indirect approach." Ortega would agree with Suits' contention that choosing more 'difficult' rather than the most 'efficient' means for achieving a particular aim is an elemental feature of all 'sports' worth the name.

Second, Ortega regarded the deliberately 'handicapping' nature of sportive rules as an important feature of the human condition in that it is the desire to submit to freely chosen rules that in part makes humans truly human and thus beings who transcend the limitations otherwise set on them by the forces of nature. There is some correspondence here with Suits'[621] argument that playing rule-driven games is at the heart of properly 'human' forms of life.

Third, Suits[622] argues, in some ways in like fashion to the later Wittgenstein[623], that all forms of human practice are more or less 'elaborate games' although this is mostly unrecognised by the participants in the games. Ortega, writing not only before Suits but also the later Wittgenstein, concurs that each individual is like a sportsperson, striving to play the games that together make up the entirety of human life.

Fourth, both Suits[624] and Ortega seem to concur that for someone to play one of the games of life without despairing that it is pointless because it is "only a game," they must perceive the game "as if it meant something;" the rules of the game must be perceived by the participant not as arbitrary constructions but as binding imperatives, the following of which is a worthwhile exercise.

For both Ortega and Suits, only if games are played in a particular spirit, which involves feeling that their pursuit is rewarding in and of itself, will they provide the players with a certain sense of ontological security and prevent feelings of ennui and anomie from coming to haunt the consciousness of each participant.

For both thinkers, sportive and game-like activities are central features of human existence, precisely insofar as they can furnish meaning and purpose to individual lives.

Chapter 26
The Divided Brain

Make it thy business to know thyself, which is the most difficult lesson in the world.
Miguel de Cervantes (1546–1616)

To know thyself may be the most difficult of lessons; to know the brain is surely one of the most challenging of the sciences, one that does not lend itself to easy labels.
W.H. Calvin[625]

Even in antiquity, there was a view that separate portions of the brain that could carry out different functions. An anterior-posterior 'localisation' of perception, reasoning and memory into anterior, middle and posterior parts of the brain, respectively, was the prevalent theory for 2000 years.[626]

Furthermore, crossed innervation of the hemispheres was clearly demonstrated in the days of Hippocrates, but was not widely known or utilised. It was not until the early 18th century that definitive demonstrations of localised brain functions were first presented. A century later it was demonstrated that the anterior-posterior distribution of motor and sensory functions was in the cerebral cortex.[627]

The idea that each hemisphere of the brain carries out a mental life of its own is relatively recent and controversial. Although Broca[628] and his contemporaries established the existence of separately functioning hemispheres as early as 1861, the phenomenon was considered merely a clinically useful curiosity. Subsequent investigations perpetuated the centuries old usage of the 'mind' as a unified entity, and that concept still prevails as serious religious and philosophical discussions as well as popular imagination. Current researchers, however, have amplified the implications of hemispheric asymmetry developing a dual brain theory for understanding human mental function. According to this theory each hemisphere embodies a separate representation of reality and the self. Human behaviour results from their complementary interaction.

There has been a reluctance to accept the cerebral hemispheres as functionally discrete entities.

There has always been a strong religious/academic tendency to speak of the mind rather than the brain when discussing human mental function.[629]

Since the time of Gall, the father of phrenology, whose major treatise on the localisation of functions within portions of the brain was published in 1810,[630] there has been a schism between those brain scientists who seek cerebral localisation for function and another group who consider that the most important brain functions reflect activity of the entire brain.

The biggest problem in accepting that a major mental function can be carried out by one human cerebral hemisphere but not the other, concerns the uniqueness of this finding in biology. The right hemisphere appears to be the mirror image of the left hemisphere. In fact, there are many dual organs in the human body in which the degree of symmetry is considerably less.

The right and left lungs are readily distinguishable grossly and do not even have the same number of major divisions.

The two kidneys, two thyroids, two parathyroids, and even the two testicles are routinely asymmetrical. The difference between analogous areas of right and left cerebral hemispheres are far less striking than the similarities. Throughout animal biology there are countless examples of bilateral organs that are mirror images of each other and carry out identical functions.

Thus the human body has two lungs, two kidneys, a pair of thyroid glands, paired reproductive glands and multiple double appendages (including eyes, ears, arms, legs etc); and almost without exception, the right and left sided paired organs.

The demonstration by Broca and his contemporaries that language was a function of the left hemisphere was accepted as a clinical fact without serious attempts to seek or understand other asymmetrics of the human cerebral hemispheres.[631]

Evidence from hemispherectomy and split-brain surgery has not ranked high in discussions of mental function. Holistic approaches have held sway in approaches to mental activity.

However, it is now becoming clear that recognition of hemispheric asymmetry of function is essential to an understanding of human mental function.

"Among milestones of thought, Broca's demonstration of hemispheric specialisation may well rank with Darwin's theory of evolution and Freud's demonstration of the unconscious as the major intellectual discoveries of the 19th century."[632]

In 1981 Roger Sperry won the Nobel Prize in Physiology and Medicine, shared with David Hubel and Torsten Wiesel from Harvard for his work on hemispheric specialisation in the split brain.[633]

In the latter years of his work he concentrated his creative efforts on the so-called big three: consciousness, free will and values—"three long-standing thorns in the hide of science." According to Sperry:

> "Materialistic science could not cope with any of them, even in principle. It is not just that they are difficult problems, they are in direct conflict with the basic premises and models. Physical science has had to renounce them to deny their existence, or to say that they are key and the domain of science."[634]

As an alternative to materialist reductionism, Sperry offered his own concept of mentalism. His key point was that in his view the higher levels of brain activity control the lower. The higher cerebral properties of mind and consciousness are in command.

When asked why we shouldn't perceive ourselves as a pair of separate left and right persons instead of the single, apparently unified mind and self that we feel we are, his answer was on these lines.

The normal bilateral consciousness can be viewed as a higher emergent entity that is more than just the sum of the awareness in the separate right and left hemispheres, and supersedes these as a directive force in our thoughts and actions. The two hemispheres normally function together as an integrated whole, and the mind as a bilateral unit then supervenes and integrates the activities within each hemisphere.[635]

In 1977 Sperry's view was:

> "It has now become a familiar story in neuroscience that, when you divide the brain surgically by midline section of the cerebral commissures, the mind also is correspondingly divided. Each of the disconnected hemispheres continues to function at high level, but most conscious experience generated within one hemisphere becomes inaccessible to the conscious awareness of the other. The parallel mental functions of the separated hemispheres are found to differ further in important ways, the most conspicuous being that the disconnected left hemisphere retains the ability to speak its mind, much as before, whereas the right hemisphere, for most practical purposes, is unable to express itself either in speech or in writing."[636]

Sperry asked some important questions. Are there in the brain divided two separately conscious minds—two co-conscious senses sharing the one cranium?

And if so, what does this signify regarding the nature and substrate of mind and unity of the conscious self in the normal intact brain?[637]

The results of his experiments led him to the view that both disconnected hemispheres retrain mental function at a rather higher level, but are no longer cognisant of most mental functions of the partner hemisphere. The two disconnected hemispheres can further be shown to function concurrently but independently in parallel, by presenting different stimulus items simultaneously to the two hands or to the two visual fields.[638]

His finding was that:

> "Under these conditions, each of the two hemispheres is found to process concurrently in its own separate perceptual-cognitive-mnemonic functions; these may be grossly incompatible or even mutually contradictory without either hemisphere noticing that anything is wrong, so separate are the inner experiences of the disconnected hemispheres."[639]

Sperry realised that this was contrary to conventional wisdom concerned for the essential unity of the conscious self. There was a reluctance not to accept that the mind and self remain unified within the language hemisphere or centred in the intact brain stem or in the person as a whole, and that the non-speaking, subordinate hemisphere operates only as a computer like unconscious automation. There was in Sperry's view no real justification according to his test findings for denying consciousness to the disconnected mute hemisphere. He concluded on this point:

> "Everything we have observed in many kinds of task performances over many years of testing reinforces the conclusion that the mute hemisphere has an inner experience of much the same order as that of the speaking hemisphere, though differing in quality and cognitive faculties. Clearly, the right hemisphere perceives, thinks, learns and remembers, all at a very human level. It also reasons non-verbally, makes studied cognitive decisions and carries out novel volitional actions. Further, it can be shown to generate typical human emotional responses when confronted with affect-laden stimuli and situations."[640]

Sperry's tests cast doubt on the earlier view that the right hemisphere was subordinate to that of the left hemisphere.

They showed that the mental performance of this hemisphere after commissurotomy were found repeatedly to be superior and dormant to that of the left hemisphere in a growing series of non-verbal, largely spatial tasks.

The tasks involved were of the kind where a single spatial image processed on a whole proved to be more effective than a detailed verbal or mathematical description.[641] His experiments, in his view, justified the concept of a complementary evolution of both hemispheres as opposed to the older classic view of one sided dominance.

It was difficult now to accept that the right hemisphere was an automaton lacking in conscious awareness.

He considered the right hemisphere to have the clear ability to learn from experience, remembering test items it has seen or felt on prior testing.

There was mounting evidence for higher cognitive faculties and a complementary specialisation in the right hemisphere. In particular, the right hemisphere has a well-developed seemingly normal conscious self with a basic personality and a social self-awareness.[642] It was no longer possible to support the argument that a person is correlated one-to-one with a body. It was necessary to refine the concept in terms of the critical brain states and neural systems involved.

Some have held the extreme position that each hemisphere must have a separate mind of its own,[643] that we are all a compound of right and left minds, or persons. Sperry's view was that the conscious mind is normally single and unified, mediated by brain activity that spans and involves both hemispheres e.g. bilateral motor controls are extensively present in both hemispheres. We are also heavily dependent on input from the hands and from the half-fields of vision.

In Sperry's opinion:

"Consciousness is defined rather simply to be a holistic or emergent, functional property of high-order brain activity."[644]

He warns that this is not an implication that mind is separate from matter in the dualistic sense. Mentalism is not equivalent to dualism. It does not bolster belief in the mystical, the paranormal or the supernatural.

But it opposes prior materialist doctrine that "Man is nothing but a material object having none but physical properties" and that "science can give a complete account of man in purely physio-chemical terms."[645]

A modern neuroscientist and clinical psychologist has attempted to categorise the relevant functions of each hemisphere albeit in a fairly rough and ready fashion[646] as follows:

Functions of the Left and Right Hemisphere

The left hemisphere controls	The right hemisphere controls:
Linguistic consciousness	Unconscious awareness
The right half of the body	The left half of the body

The right hand	The left hand
Perceives right half of body	Perceives faces and both half of body and maintains the body image
Talking, reading, writing, spelling	Emotional and melodic speech
Speech comprehension	Singing and swearing
Linguistic and verbal thinking	Comprehension of music, emotion body language, and environmental sounds (chirping birds, buzzing bee, babbling brook, thunderstorm)
Verbal intelligence	
Verbal memories	
Dreaming in words, thought dreams	
Rhythm, temporal and sequential	
Information processing	Visual, emotional, musical, creative and geometric thinking
Keeping score of football game	Emotional and childhood memories
Marching	Insight and intuitive reasoning
Math	Seeing the forest
Typing	Reading between the lines
Grammar	Perceiving the overall 'big picture'
Logical and analytical reasoning	Visual-spatial processing
Confabulation	Throwing and catching a football
Perception of details	Riding a bicycle
Seeing the trees	Dancing
Broca's expressive aphasia: Loss of Speaking ability	Visual closure
Wernicke's receptive aphasia: Loss of Language comprehension	Gestalt formation
Agraphia: Loss of writing ability	Initial self-concept
Alexia: Loss of reading ability	Dreaming in visual-emotional images
Acalculia: Loss of math ability	Amusia: Loss of facial recognition
Apraxia: Loss of ability to perform Skilled temporal-sequential movements	Prosopagnosia: Loss of facial recognition
	Anosognosia: Left-sided neglect and denial, disturbances of the body image

There is a considerable amount of evidence that the right hemisphere subserves a type of awareness that is considerably more ancient than, as well as qualitatively different from that manifested by the left.[647]

It has been suggested that this was not always the case and that the right hemisphere developed this character over thousands of years. Originally the human brain and the original mind of man increasingly became more proficient in the hunting and stalking of prey, exploitation of the environment and socialising as bands of tribes increased in sizes and complexity. The suggestion is that with the exception of the left hemisphere for handedness, both halves of the brain performed quite similar functions for many millions of years.

However, it is argued that with the development of language and writing, the more ancient speciality associated with the original mind became increasingly crowded out of the left half of the cerebrum and came to be represented within the right hemisphere. However, the evolution and elaboration of right hemisphere functions continued unabated.[648]

The right hemisphere is now thought responsible for discerning distance, depth, and movement; for recognising environmental factors; and for controlling most aspects of emotion, social behaviour and body language, as well as the capacity to sing, dance, chase or throw something with accuracy and run without falling or bumping into things. It is associated with non-verbal environmental awareness.

Most significantly so far as sport is concerned, the right hemisphere is dominant in the perception of movement, speed, distance, and depth as well as the geometric analysis of gestalts, angles and visual relationships.[649]

The right hemisphere enables us to manoeuvre in space, to detect, analyse and react to the movements of others, and to determine how these actions and motions are interrelated.

The visual-spatial capabilities of the sportsman are significantly dependent on the functions of the right hemisphere.[650]

The right hemisphere can visualise and fully comprehend and perceive a complex visual array as an interrelated whole (or gestalt).

Furthermore, the right hemisphere perceives touch and physical sensations that occur on both sides of the body. Whereas the left hemisphere is limited to receiving information from only half of the body and its surrounding environment, the right brain is able to perceive sensations regardless of where they arise so as to map them to the body image. The right hemisphere is much more sensitive and aware of the body and its environment.[651] This is again of vital significance to the attribution of the skills of the sportsman.

It is important to emphasise that the right and left hemispheres are not completely dichotomous and that they share many of the same abilities. Some functions are simply processed much more efficiently and effectively in one than in the other hemisphere.[652]

So far as visual-spatial matters are concerned, it appears that the brain possessed some rudimentary temporary-sequential capabilities as long ago as two million years and that both the right and left cerebral hemispheres were probably equally adapted.[653]

This adaptation would have been reflected in the activity that ancient men and women probably engaged in fairly equally, that is wandering over long distances, scavenging, hunting and chasing small game, as well as gathering.

However, the capacity of the human brain to process visual-spatial distance, depth, movement and other environmental variables probably became embodied even further with the onset of big game hunting about 300,000 years ago.

The Left Hemisphere and Motor Control

Over 80 per cent of our population is right handed. One reason that most of us become right-hand-dominated is during prenatal development, nerve cells and fibres in the left hemisphere that control hand and arm movement began to mature and grow more quickly than those in the right hemisphere.[654] It is said that this is genetically programmed.

We become right-hand-dominant because the left brain gains a competitive advantage and obtains a head start in regard to motor control and development.

Language especially its grammatical and syntactical components is directly related to handedness and motor control. It has been said that the right hand operates as a kind of motoric extension of language and thought in that it acts at the behest of linguistic impulses.[655]

When one is speaking, the areas of the brain controlling right hand use become activated in part because of the spread of neural excitation from the speech area to the immediately adjacent cortical regions that control hand movement.

A number of theories have been proposed to explain the evolution of handedness and language. Handedness and temporal-sequential motor control probably preceded the development of language specialised nerve cells.[656]

There is considerable evidence that over the course of evolution the predominant use of the right hand enabled the left hemisphere to develop nerve cells specialised for counting, naming, temporal-sequential processing and thus far the meditation of grammatical-syntactical speech.

Once the right hand preference became established, the left hemisphere continued to evolve and adapt as it increased its proficiencies in right handed motor control.

Tool making also evolved, as did the ability to make more efficient weapons and hunting implements. The left hemisphere not only became dominant over hand and motor control but was beginning to become increasingly proficient in temporal and sequential processing.[657]

The original mind has not been discarded. Rather, as the left hemisphere became increasingly associated with language and linguistic consciousness, the original mind appears also to have evolved and to have become more intimately associated with the right cerebral hemisphere. It is probably for these reasons that, among modern humans, the right hemisphere is associated with presumably the more primitive unconscious, whereas the left hemisphere maintains the more recently evolved language-dependent conscious mind.

The Unconscious Mind

The conscious mind is linked to the functional integrity of the left hemisphere. It is not however a rigid divided. It has been said that "the left side is involved with conscious response and the right with the unconscious mind."[658] This distinction must ultimately come down to what these concepts mean. Conscious and unconscious are broad and loose categories. They are not mutually exclusive. As Bertrand Russell said of mentality they are matters of degree. They are not capable of being precisely defined. Indeed depending on the context they may have different meanings. As has been pointed out, consciousness might be the 'waking state', an 'experience' or a 'state of mind'.[659] It has other meanings too.

In any linguistically defined way the concepts are therefore largely inadequate to convey fine ontological distinctions. That is why Ortega, James and others were so hostile to the concepts. Nonetheless, subject to the caveats I have expressed and for literary convenience, I will not abandon their usage as an aid to understanding the cerebral processes they are generally accepted to represent.

Ian McGilchrist in the most recent work on the hemispheric differences in the brain refers to the "primacy of the unconscious will."[660] Drawing on evidence from the experiments of Benjamin Libet and others which I have described as showing the time delays of consciousness, he cites the view that they seem to "deny to consciousness any major role in the conduct of our day to day affairs."[661] This is only a problem, says McGilchrist, if one imagines that for me to decide something I have to have willed it with the conscious mind.[662]

In particular he makes reference to Jaynes' view that very little brain activity is in fact conscious and that it seems that there is some current justification for the view that this is less than 5 percent and probably less than 1 per cent.

Speaking for myself, I am not sure there is any probative evidence available to this effect. And as I have earlier said, it really depends upon what you mean by conscious and unconscious.

In his cerebral magisterial work *The Master and His Emissary. The Divided Brain* and the *Making of the Western World* (2009), McGilchrist focuses at length on the relationship between the right hemisphere and the unconscious mind. He notes that there is evidence that the right hemisphere experiences material that the left hemisphere cannot be aware of.[663]

He says

"If what we mean by consciousness is part of the mind that brings the world into focus, makes it explicit, allows it to be formulated in language, and is aware of its own awareness, it is reasonable to link the conscious mind to activity almost all of which lies ultimately in the left hemisphere. One could think of such consciousness as a tree growing on one side of a fence, but with a root system that goes deep into the ground on both sides of the fence. This type of consciousness is a minute part of brain activity, and must take place at the highest level of integration of brain function…"[664]

In his opinion the right will has a will, can intend, mean, will and choose, just as the left hemisphere can.

He quotes from Hans Vaihinger who wrote:

"The organic function of thought is carried on for the most part unconsciously. Should the product finally enter consciousness also, or should consciousness momentarily accompany the process of logical thought, this light only penetrates to the shallows, and the actual fundamental processes are carried on in the darkness of the unconscious. The specifically purposeful operations are chiefly, and in any case at the beginning, wholly instructive and unconscious, even if they later press forward into the luminous circle of consciousness."[665]

McGilchrist says that research findings on the study of gesture[666] show:

"these intentions arise from the right hemisphere and are prior, in every sense, temporally, logically and ontologically."[667]

In McGilchrist's view, McNeill's work is an illuminating companion to Libet's work on the will. In both cases, he says, the conscious left hemisphere believes that it is an originator, whereas in fact it is a receiver of something that comes in from elsewhere.[668]

He suggests that the conscious left hemisphere thinks that it is in control while the reality is it is selecting from a broader world that has already been brought into being for it by the right hemisphere and often it is not even doing that, since far more than it realises, its choices have already been made for it.[669]

However, it would be a mistake to limit the right hemisphere's activities to what has been labelled the subconscious and the unconscious. The right hemisphere does a great deal. Right brain activity and unconscious mental activities are not synonymous.

As I have earlier commented, the right hemisphere predominates in the perception and identification of environmental and non-verbal sounds, the maintenance of body image, the comprehension of melodic and emotional speech and the perception of most aspects of musical stimuli.

The right hemisphere is also dominant in the analysis of geometric and visual space, including depth perception, position, distance, movement, and stereopsis.[670] It is these mental capacities which are essential for success at sport. The role of the right hemisphere in this regard does not feature significantly in the learning on the two hemispheres. Save for Joseph, very little mention is made of this linkage at all. I will return to this topic later.

One of the points about right hemisphere activity is that it performs so much of its activity naturally and thus seemingly surreptitiously.[671]

However, from a conscious left hemisphere perspective much of what occurs within the domain of the right hemisphere is not susceptible to conscious scrutiny or analysis, partly because it is non-linguistic, cannot be labelled or categorised verbally, and is not temporal-sequential.

As has been said:

> "In this regard, right brain processing can be an unconscious phenomenon because it is non-verbal. It is unconscious because the left brain cannot verbally recognise it and may not know about these events until after they have occurred or until after it has lost control and the consequences have been experienced."[672]

Does this therefore mean that right brain-mediated events and our associated feelings are unconscious?

This is not necessarily the case. What seems to originate unconsciously is often amenable to conscious scrutiny and examination. We are capable of becoming fully aware of what is occurring in the right hemisphere, although it is non-linguistic and non-conscious.

We can judge distance when throwing or catching a ball and we are aware of where our legs and arms are in space when we run; and we can do all those things without thinking about them. If we choose to, we can also become consciously aware of all of them as they occur.

There are thus a myriad of activities at which the right brain excels and in which the left brain plays a minor supporting role.[673]

It has been suggested that our educational system stifles right brain development by stressing and rewarding left-brain activities such as reading, writing and arithmetic. Many right brain activities, such as drawing, painting, sculpting, music, dancing, sports and gymnastics are relegated to recreational status.[674]

Because the two halves of the brain cannot fully communicate and process, store in memory and recall different aspects of what appears to be the same event, considerable conflicts may occur between the right and left hemispheres. When this happens, the left brain becomes alerted to the possibility of forces that are acting outside its control.

Right brain activities are only unconscious in so far as the left brain does not know about them, cannot label them, or is not actively paying conscious attention.[675]

The right brain is associated with a highly developed form of mental awareness that can be referred to loosely as an unconscious awareness.[676]

The unconscious mind consists of two levels. The more accessible aspect of the unconscious consists of the highly refined social-emotional, melodic, environmental, visual-spatial, pectoral and non-linguistic aspects of the right brain.[677]

There is, however, a relatively inaccessible aspect of the unconscious that is associated with the very primitive, very ancient region of the brain, the limbic system.

The limbic system is a 'system' of nervous tissues and cell assemblies located and buried within the depths of the central most portion of the brain.[678] The most ancient structure within the limbic system is the hypothalamus. It serves the body tissues by attempting to maintain internal homeostasis and by providing the immediate discharge of tensions, almost reflexively.

This is the mental region most deserving of the label unconscious, as its functions originate and often occur for the most part independently and outside of conscious awareness. The limbic level of the mind could be referr3ed to as the primary unconscious as its appearance precedes all other aspects of mental functioning. It has been said that:

"The limbic unconscious is the bedrock on which the foundations of the mind will be laid."[679]

Impulses, desires and related feelings and motivational states originating in the limbic system often occur completely outside conscious awareness. It is very difficult for the conscious mind to scrutinise or understand this aspects of the psyche. A great deal of what transpires in this region of the mind is biological or reflexive and never enters conscious awareness at all.

Half the limbic system is located within the left, the other half within the depths of the right cerebral hemisphere. There are two amygdala, two hippocampus and so on. Whereas, these impulses are also almost completely foreign to the language—dependent conscious mind, and the left hemisphere, the right hemisphere, being more involved in emotional functioning, is often (but not always) able to discern and recognise these limbically induced feeling states and desires for what they are.

Right and Left Handedness

In the Western world at the moment about 89 per cent of people are broadly right-handed, and the vast majority of these have speech and semantic language centres in the left hemisphere. In the other 11 per cent, which are broadly left-handed, 75% still have their speech centres in the left hemisphere. It is therefore only about 3 per cent of the population overall who are known not to lateralise for speech in the left hemisphere.[680] Of these some might have a simple inversion of the hemispheres, with everything that normally happens in the right hemisphere happening in the left and vice versa. But there may be a subset of left handers who may have a partial inversion of the standard pattern, leading to brain functions being lateralised in unconventional combinations. For them, the normal partitioning of functions breaks down. In McGilchrist's view, "this may confer special benefits, or lead to disadvantages, in the carrying out of different activities."[681] This is particularly true of sports such as cricket, tennis and boxing. Cricket is an especially good illustration of this phenomenon. So many outstanding batsmen and bowlers have been left handed. In most recent times one can think of Gary Sobers, Brian Lara, Tony Lock, David Gower and Derek Underwood to name but a few.

Chapter 27
Gesture

Grace was in all her steps.
Heaven in her eye, in every gesture dignity and love.
John Milton, *Paradise Lost*

Gesture is a revealing topic. The Times in a leading article on 21st March 2023, referred to a Canadian judge's recent ruling that someone who raised their middle finger at a neighbour in a Montreal suburb didn't break the law. The judge said "It may not be polite or gentlemanly but it does not trigger criminal liabilities." He described gesture as "a God given enshrined right that belongs to every red blooded Canadian." The Times editorial commented "the right to be rude in word and gesture is an integral part of our freedom and expression."

Gesture at times is a most effective and graphic means of communicating feeling and meaning. It is a form of bodily metaphor and symbolism. In this chapter I explore the relationship between gesture, mind and body and its relevance to sport.

Gestures are an integral part of language as much as our words, phrases and sentences—gestures and language are one system.[682] Language is a modality of the human body. It generates out of movement. As Merleau-Ponty expressed it, the body converts a certain motor essence into vocal form.[683]

Gestures have attracted the attention of writers for at least two millennium.[684] However, the original interest was mainly in rhetoric. Quintilian in the first century A.D., specified in detail the gestures that orators should use during their speeches. But these gestures are very different from spontaneous gesture.

The word 'gesture' comes from the Latin for action, for carrying out activity, and for performing. Sport is a form of symbolic and metaphorical gesture reflected in the sportsman's performance related to body schema image and language. I have dealt elsewhere with these concepts and how they are at the heart of sporting activity. The act of performance itself is gestural. The scoring of a goal, the running in of a try, the taking of a wicket amount to gestural statements in the act of performance. They are manifestations of a highly symbolic achievement in the contest. They are sporting metaphors which give meaning and value beyond the concept of the game.

According to McNeill, in his masterly work Hand and Mind (1992), gestures are closely linked to speech, yet present meaning in a form fundamentally different from that of speech. His own hypothesis is that speech and gesture are elements of a single integrated process of utterance formation in which there is a synthesis of opposite modes of thought-global-synthetic and instantaneous imagery with linear-segmented temporally extended verbalisation.[685]

McNeill focuses on the spontaneous and idiosyncratic gestures that occur while one speaks. An example is the hand rising upwards while the speaker says "and he climbs up the pipe."

"Language-like gestures" are similar in form and appearance to gesticulation but differ in that they are grammatically integrated into the utterance; an example "the parents were all right, but the kids were [gesture]," where the gesture fills the grammatical slot of an adjective.

The issue of how gestures and speech relate in time is crucial for understanding the system that includes gesture and speech as two parts. Gestures slightly anticipate speech; gestures and speech have a constant relationship in time.[686]

In McNeill's view, the anticipation of speech by gesture is important evidence for the argument that gestures reveal utterances in their primitive form, there is a global-synthetic image taking form at the moment the preparation phase begins, but there is no yet a linguistic structure with which it can integrate.[687]

Gesture expressions regularly anticipate their co-expressive speech.[688]

Gestures both anticipate and synchronise with speech. This is not the paradox it may seem. Anticipation and synchronisation refer to different phases of the gesture. The synchrony rules refer to the stroke phase: anticipation refers to the preparation phase. It is only the stroke of the gesture that is integrated with speech into a single smooth performance, but the preparation for the stroke slightly leads the co-expressive speech.

Language has the effect of segmenting and linearising meaning. What might be an instantaneous thought is divided up and strung out through time. A single event, say, somebody sitting down on a chair, is analysed into segments: the person, the chair, the movement, the direction and so forth. These segments are organised into a hierarchically structured string of words. The total effect is to present what had been a single instantaneous picture in the form of a string of segments. Segmentation and linearisation are part of all linguistic systems.

It has been explained that the linear-segmented character of language is a property that arises because language is unidimensional while meanings are multidimensional. Language can only vary along the single dimensions of time—phonemes, words, phrases, sentences, discourse: at all levels. Language depends on variations along this one axis of time.

This restriction forces language to break meaning complexes into segments and to reconstruct multidimensional meanings by combining the segments in time.[689]

Gestures are different in every way. This is because they are themselves multidimensional and present meaning complexes without undergoing segmentation or linearisation. Gestures are *global* and *synthetic* and never hierarchical.[690]

Gestures and language thus differ from each other on a number of fundamental dimensions. Yet they are also closely linked. Such linkages imply that gestures and speech should be viewed as aspects of a single underlying process.[691]

They have the following similar characteristics. They only occur during speech, are semantically and pragmatically co-expressive, are synchronous, develop together in children, break down in aphasia.[692]

Speech and gesture are therefore co-expressive manifestations of a single underlying process. The underlying process is equally speech and gesture, and there is a subsequent evolution of expressive action with outputs in both channels concurrently. The channels, moreover, have a constant relationship in time, with the gesture manifestation the primitive stage of the shared process and speech its final socially presentable stage.[693]

Gestures Affect Thoughts

Gestures do not just reflect thought but have an impact on thought. Gestures, together with language, help constitute thought.[694]

It is necessary to return to the earlier remarks made as to the anticipation of speech by gesture. McNeill calls this interval *deep time*. Gesture is evolving over this same interval.

The gesture is, at the beginning stage in the growth point, an image. It is schematic, reflective of context and a psychological predicate at the moment of speaking, but lacks an outer kinesic form.[695]

At the final stage, when there is also speech, the gesture takes on its kinesic form—the movements we usually take to be the gesture.

Thus the two channels evolve together.

At the final stage, the gesture stroke and speech are integrated into a single performance in which there is a synthesis of gesture and a specific utterance form.[696]

If, says McNeill, we make the assumption that thought is multi-dimensional, a dialectic of speech and gesture means that some dimensions of thought are presented in the gesture and others in linguistic form. There is a synthesis and at the moment of synthesis, language and gesture are combined into one unified presentation of meaning.[697]

This is an act of communication, but also an act of thought. Not only the listener but the speaker is affected. That is, the speaker realises his or her meaning only at the final moment of synthesis.

Only at the final synthesis is there the joining of a linear-segmented, analysed representation with a global-synthetic and holistic representation. The synthesis—its analytic and holistic qualities—is a single mental representation for the speaker which did not exist until the instant fusion at the rhythmical pulse.

An underlying rhythmical pulse is essential for synthesis.[698] It provides the point of convergence that completes the evolution of the utterance. At this moment, there is a synthesis of the image with the linguistic structure integrated into one performance of speech and gesture.

> "The kinesic base which underlines both gesture and the gesture-like component of speech (i.e. the intonation) is represented as a rhythmic pulse, if expressed gesturally, the pulse peak corresponds to the stroke position of the gesture. If expressly vocally, the pulse peak is represented by an intonational peak, which presumably because of the more elaborate processing required by speech production [viz. self-organisation of the utterance] may be somewhat delayed relative to the gestural stroke."[699]

This synthesis is achieved at a specific rhythmical moment. The synthesis is the integration into the single performance of the gesture stroke and the co-expressive linguistic segment, both presenting the same meaning in combined forms.[700]

While gestures may be performed by either hand, or both together, most speakers tens to use one hand rather than the other during extended stretches of speech.

Gestures are manual actions, and thus might be lateralised to the dominant hand like other manual actions.

But they are also part of the symbolic process, and thus may be lateralised depending on functional factors such as the role of the gesture in the discourse. So observing gesture hand preferences can provide insights into the cerebral control of gesture and the neurological underpinnings of gestures in relation to speech and discourse.

Studies of gesture performance generally show gestures being made with the dominant hand but find no such lateralisation for non-gestive movements like smoothing the hair or fiddling with objects.

The dominant hand for most complex motor movements is controlled by the language-dominant hemisphere in right-handed people, although left-handers show more complicated relationships.[701]

Thus the half of the brain that is dominant for language also appears to be significant locus for the production of the gestures that accompany speech.

It has been suggested that the specific quality of the language dominant hemisphere that is responsible for this lateralisation is the programming of complex movement sequences. Complex movements—either oral or manual—would engage the left hemisphere of right-handed people (the right hemisphere of left-handed people) and this is the case when the speaker may have learnt to write with the other hand.[702]

But is appears that more than motoric complexity engages the speech-dominant hemisphere in the production of gestures, and that the specific symbolic character of the movement also must be taken into account.[703]

'Split-brain' patients are individuals who have had their two cerebral hemispheres disconnected in a surgical treatment for epilepsy in which the corpus callosum and the other smaller interhemispheric connections are severed, leaving only brain stem connections. Each hemisphere remains intact and functions in a normal way apart from its inability to interact with the other hemispheres plus any abnormality associated with the patients' epilepsy.

It is thus possible to study the special functions and limitations of the two hemispheres in isolation in such patients.

The effects of the operation are subtle but profound. There is a complete division of the patient's awareness into two halves.[704]

Neither half is directly aware of the other—indirect awareness is possible—and this is achieved through one hemisphere's observing the effects of the other hemisphere on 'its' body, as if observing a second person.

For example, in one of the classic texts a female split-brain patient was unexpectedly shown a pin-up, in such a way that it was confined to her right hemisphere.[705]

The patient laughed but when the experimenter asked why she had laughed she answered (the left hemisphere speaking) that she didn't know, that the "machine was turning or something."

The left hemisphere was trying to make sense of why there had been this laughter. When the pin-up was flashed to the left hemisphere there was again laughter but now the laughter was spontaneous accompanied by a report that among the apples and the other ordinary objects that were being shown there had been a nude woman.[706]

Separate tests of the two hemispheres in these kinds of experiments show a sharp division of linguist and visual-spatial capabilities.

Drawings shown to the left hemisphere can be described verbally in the normal way, but the same drawings shown to the right hemisphere usually evoke no response at all.

If the patient is urged to respond he will say (the left hemisphere speaking) that there was some kind of vague event or he will simply make up some story.

On the other and; the right hemisphere is far superior to the left in perceiving through palpation oddly shaped objects that do and have recognisable components.

Thus the right hemisphere seems better at taking in global holistic perceptual patterns. The left hemisphere tries to build up complex objects out of components rather than taken as a whole.[707]

But can that really be true? I do not accept that I am a person of divided brain, I am me and my brain. I am who I am like a man of many parts I may be a split personality, I do what I do and behave as I am without knowing why. That is the mystery and complexity of me and my personality. I am what I am and much more than that. I am "beyond the horizon."

Chapter 28
The Voice, Communication and Sport

Sweet is every sound
Sweeter thy voice, but every sound is sweet.
Alfred Lord Tennyson

'Tis not enough no harshness gives offence, the sound must seem an echo to the
sense.
Alexander Pope

The relevance of this chapter to sport and beyond is the role of the voice not just the commentators who in their own way create the atmosphere of the sporting event and moment. In cricket, for example, the voice of John Arlott was synonymous with the game itself. The Hampshire burr like that of Lord Denning in the courts was almost a metaphor for the domain they graced themselves. The voice tells people who we are and where we came from. From hearing a single sentence most people can say whether a speaker comes from America, Australia, Scotland, Wales or Cornwall. George Bernard Shaw said: "It is impossible for an Englishman to open his mouth without making some other Englishman despise him."

The spectators too through their voices create an atmosphere by vocalising their support for their teams and enhance the sense of the occasion and the living in the moment which sport provides. Singing is an essential component of any major sporting event. It is not just the ceremonial singing of the National Anthems but the spontaneous resort to song that heightens the competitive spirit of the sporting rivalry. Songs like 'I'm forever blowing bubbles', 'You'll never walk alone', 'Delilah', 'Swing low sweet chariot' and most recently at the football World Cup 'Sweet Caroline' demonstrate the close association sport has with song. They are almost part and parcel of each other—mind and body (voice) in unison. Cheering your team on is the hallmark of the sporting occasion. Chanting too emphasises the tribal and primeval nature of sport and its roots.

I want to explore further the view I expressed earlier about the body's ability through the voice to communicate in some instances more effectively than words. The control of pitch, intonation and volume enable us to pass on the finest shades of meaning. I have referred to the significance of gesture and the hand in communication and the role they play in the art of persuasion and oratory. Shakespeare dealt with it through Hamlet's soliloquy on speech.[708]

I have earlier looked at Shaun Gallagher's seminal work "How the Body Moves the Mind" which explains the relationship the body has with the mind and thinking. The purpose of this Chapter is to touch on the relationship between one aspect of the body—the larynx, the voice and communication orators, singers, actors, broadcasters and sports commentators, heavily rely upon the voice to convey meaning and emotion and excitement. If actions speak louder than words certainly the voice does. The following names are just a few examples—Enrico Caruso, Luciano Pavarotti, Joan Sutherland, Elvis Presley, Frank Sinatra, Bob Dylan, John Lennon, Shirley Bassey, Roy Orbison, Tom Jones, Laurence Olivier, John Gielgud, Ralph Richardson, Richard Burton, Dylan Thomas, Anthony Hopkins, Glenda Jackson, Maggie Smith, Catherine Zeta Jones, Betty Boothroyd, Barbara Windsor, Theodore Roosevelt, Winston Churchill, David Lloyd George, Aneurin Bevan, John Kennedy, Ian Paisley, Richard and David Dimbleby, David Attenborough, John Humphreys, Huw Edwards, John Arlott, Murray Walker, Harry Carpenter, Peter Alliss, Cliff Morgan, Eddie Butler, Ron Pickering, David Coleman, John Motson, Bill McLaren Dan Maskell, Kenneth Wolstenholme, Peter Dimmock, Ted Lowe and Sid Waddell.

It was something about their voices and articulation and not simply about the words and content. The pitch, the cadences, the modulation, the accent, the spacing and timing are the hallmarks of what is pleasing to the ear and brain. It is in that sense that the voice communicates better than the words. The bible speaks of the tongues of angels; more correctly it should have been the larynxes. Sir Anthony Hopkins was once asked what was his technique in acting; he modestly answered: "I learn my lines and say my words." We all know from listening to his performances that it is a bit more than "saying my words" it is about how you say them. In the words of the song "It's not what you do but the way that you do it." Orchestral music is obviously about sound but on reflection so is choral and vocal singing. Philosophically what does that tell us? Maybe it is that our mind and our bodies are more in tune with the resonance of sound than with the meaning of words in that it relates to the primeval roots of how man developed in evolution. Communication being the use of sound was the basic instinctive way of communicating feelings, emotions, ideas and ultimately thought. Dance, music and singing were early man's most profound way of expressing himself and communicating with others. In modern man they remain as his essence. That perhaps was why Hamlet said:

"It offends me to the soul to hear a robustious periwig-pated fellow tear a passion to tatters, to split the ears of the groundlings, who, for the most part, are capable of nothing but inexplicable dumb-shows and noise; would have such a fellow whipped; it out herods Herod; pray you avoid it."

Physically when we speak muscles in the larynx bring the vocal cords together. As the air rushes from our lungs out through the larynx, the front portion of the vocal cords vibrate, producing sound. The upper portion of our throat then modifies this sound to produce speech.

Finally, as a South Walian, I cannot resist referring to a line from the 1941 Oscar-winning adaptation of Richard Llewellyn's novel "How Green Was My Valley," a story about a Welsh mining community at the turn of the 20th century "Singing is in my people as sight is in the eye." He acknowledges that "the consonantal nature of the Welsh language, like Italian and German, tends itself to clearly articulate expression." In fact Welsh was apparently used as the basis for the Elvish language in J.R.R. Tolkien's The Hobbit and the Lord of the Rings, partly due to its melodic, ethereal nature. Unlike the English accent, the Welsh is characterised by its tendency for the pitch to jump down to a stressed syllable and then rise from it.

Wales is known as the land of song. Its language and accent is musical and melodic.

The musical tones of the likes of Oscar-winning Sir Anthony Hopkins clearly demonstrated that melody.[709] Think also of Richard Burton, of Dylan Thomas and of Catherine Zeta Jones.

A stereotypical Welsh accent is set apart from others in the UK by how its pitch changes. They can take the language gallantly by the hand and dance with it, leading it to and fro, up and down, and relishing each word.

Milton Shulman's 1953 review of the film Valley of Song in the Evening Standard with perhaps tongue in cheek summed it up pithily when he said:

"There are many things you can say about a Welshman, but never be rude about his larynx."

One writer has said that "if you want to hear the finest English that has ever been spoken then you have to listen to a Welshman."[710]

I certainly agree with that sentiment. But as a Welshman, I am biased! Many famous Welshmen are rightly known for rhetoric, the forensic command of language that ignites those bonded in common cause. Aneurin Bevan had "the gift of simplifying an issue with a deadly clarity in his voice and manner combining to wield sentences with scalpels."[711]

So the conclusion for me on this topic is that the sound/the voice is often more important than the words in terms of communication. In modern contemporary songs, the reality is that we don't discern the words; we are entranced and captivated by the sound. It's the voice that counts. And this is not a modern phenomenon.

We have enjoyed and appreciated classical opera in Italian, French, German and even English without understanding the words sung. It's the sound not the words—stupid!

Chapter 29
Sport and War

The supreme art of war is to subdue the enemy without fighting.
Sun-Tzu

There is such a thing as legitimate warfare; war has its laws, there are things which may fairly be done and things which may not be done.
Cardinal Newman

Everlasting peace is a dream and not even a pleasant one, and war is a necessary part of God's arrangement of the world. Without war, the world would slide dissolutely into materialism.
Helmut von Moltke

As long as war is regarded as wicked, it will always have its fascination. When it is looked up as vulgar, it will cease to be popular.
Oscar Wilde

Serious sport is war without the shooting.
George Orwell

Without a sign his sword the brave man draws,
And asks no omen but his country's cause.
Pope, The Iliad of Homer

My argument is that war makes rattling good reading but peace is poor reading.
Thomas Hardy

The German people is no warlike nation. It is a soldierly one, which means it does not want a war but does not fear it. It loves peace but it also loves its honour and freedom.
Adolf Hitler

There is a tendency in life to want to make things simple. Even the literary great George Orwell tried to do so when he characterised sport as 'serious sport is war without the shooting'. The idea that everything can be reduced to simplistic and comparative form is absurd. Complexity is the essence of being and living. As The Who sang in that great work Substitute 'the simple things you see are all complicated'. In parenthesis one of my greatest experiences as a young barrister was to represent Peter Townshend in a relatively minor court case. The sequela was a wonderful metaphysical and reflective evening with Peter Townshend, Roger Daltrey and the road manager who gave me a wonderful insight into their creative talents and composition. It was akin to the experiences with Muhammad Ali and El Cordobès to which I have referred to elsewhere in this book. Why do we need to strive to keep things simple? Konrad Lorenz in his readable yet somewhat presumptuous psychological treatise On Aggression (1996 New York Bantam, pg 271), was of the view that 'the main function of sport today lies in the cathartic discharge of the aggressive urge'. It is not so simple as that. Reductive philosophy is an inadequate form of intellectual thinking. In particular, as far as this chapter is concerned, the philosophical relationship between war and sport is much more complex and profound.

Throughout recorded history Sport and Warfare have been to varying degrees inseparable. War has been regarded as sport – the Sport of Kings according to Huizinga. For hundreds of years, sports were characterised as a prelude to or preparation for actual war. The first sports called 'blood sports', were merely a mirror image of the sort of encounters that took place on the battle fields, and the Ancient Greek word agon – meaning a test of valour, match or contest – has survived in English form as agonistic or perhaps more poignantly as 'agony'.

Even when things had evolved to the stage of the Greek Olympiads the sorts of competition emphasised were still skills like speed or javelin throwing which would be useful in battle.

Similarly, the jousting tournaments of medieval knights were a kind of mock warfare and the training that preceded them was regarded, more or less explicitly as training for battle. And we have the Duke of Wellington's testimony that "Waterloo was won on the playing fields of Eton". Likewise, American military history is replete with examples of close connectivity between militarism and the concept of sport. Generals Eisenhower and Patton were explicit in their references to that close relationship. Eisenhower's view was that "the true mission of American sports is to prepare young people for war".

Vince Lombardi was often compared with General Patton in his sporting methodology. Furthermore, it is significant that the Vietnam War had gone hand in hand with a tremendous surge in the popularity of American football; a game which Lombardi when he was coaching, militarized mastodons of the Green Bay Packers described as "nothing but discipline".

A symbol is a concept typifying or representing something which possesses analogous qualities or character by association in fact or in thought. Sport and war have a symbolic philosophical relationship. Thucydides said, "history is philosophy teaching by example." No better example of the relationships between sport and war is what was characterised as the Football Wars between El Salvador and Honduras in 1969 in the build up to the 1970 World Cup. This was a competition which neither country had ever competed in before. In the final deciding match between the two countries in the 11th minute of extra time, El Salvador's Mauricio 'Pipo' Rodriguez scored the winning goal. El Salvador had won. Although the players hugged, shook hands and left the pitch together, within three weeks their countries were at war.

The sporting contest had a significance beyond itself. Fifty years later at the age of 73 Rodriguez said, "we felt we had a patriotic duty to win for El Salvador. I think we were all afraid of losing because in these circumstances it would have been a dishonour that followed us for the rest of our lives. What we didn't know was the significance of that win and the historical importance of the goal—that it would be used as a symbol of war." (quoted in—*The Hundred Hour War* by Dan Hagedorn). A matter of honour had been converted by a sporting contest into a symbol of war. But the sportsman still had respect for his opponent. Rodriguez concluded, "El Salvador retained an immense appreciation and respect for the Honduras players nor were the games between enemies but between sports rivals." Therein possibly lies the philosophical difference between war and sport.

Vladimir Putin's invasion of Ukraine has highlighted the analogies and disanalogies between sport and war. It has been characterised as an unjust and unlawful war, a battle between right and wrong and where there can be only one victor—Ukraine. Russia is an aggressive totalitarian state which does not respect the rules of engagement and although it has might on its side but not right. Ukraine is not only the underdog but has the moral high ground. In that context winning is not the only thing but the only thing that can be accepted ethically is that Ukraine rebuts Russia's aggressive invasion and wins. Arguably, that is a fundamental conceptual difference between war and sport—the ethical component of the competition has to be factored in war but not in sport. But is that right? Thomas Hobbes thought that "every creature lives in a state of war by nature." If that is right, sport is no exception.

This chapter looks at the apparent comparative and contrasting features of sport and war.

Isaac Asimov said "Violence is the last refuge of the incompetent." Sport is the antithesis.

I am indebted throughout this chapter to two outstanding works of history and philosophy of John Keegan's A History of Warfare (1993) and Norman Fischer's essay Competitive Sport's Imitation of War (2002).

"In war," Napoleon observed "three-quarter turns on personal character and relations; the balance on manpower and materials." Maybe sport is the same, depending on your metaphysical definition of materials.

Keegan suggests the relationship between war and sport may well be traced back to the charioteers who were the first great aggressors in human history.[712]

In his view, "The adoption of the war chariot and the imposition of the power of war charioteers throughout the centres of Eurasian civilisation in the space of some 300 years is one of the most extraordinary episodes in world history."

The association with contemporary horse racing is obvious. It arises out of the domestication and improvement in physique of the wild horse. Even today when mankind everywhere expects to travel by internal combustion engine, horse flesh engages passions and mobilises money on a vast and universal scale. The richest men and women (including our late lamented Queen Elizabeth II) compete to display their wealth and power through the ownership of thoroughbred horses. Horseracing is 'the sport of kings' on which republica-multimillionaires rejoice to expend their fortunes, but few kings or millionaires ever risk as much of what they have as the common man who believes he knows a winner in the world of the horse, the poorest feel themselves the potential equals of the wealthiest in the land, as the sayings go "sport is a great leveller" and "animals can make fools of us all."[713]

The horse however pampered, whatever its bloodlines, may choose to repay an owner's expectation in hypochondria or ill-will; contrarily the horse may stay the course against all the odds make its rider, trainer, breeder and owner men of stature overnight, bring joy to the hearts of a thousand humble punters and send bookmakers home with lighter pockets than they set out with. The modern thoroughbred is a force to be reckoned with, and the great thoroughbred may end his days more famous than most statemen of his lifetime. The greatest of thoroughbreds acquire regal and dynastic status; pilgrimages are made simply to see them run, while the descent of their genes into subsequent generations is catalogued with all the care taken to establish the legitimacy of a Bourbon or Habsburg. A great horse, in a sense, becomes a king. It is therefore not surprising that kings were made by the first great horses.[714]

As Fischer has identified there is arguably a close symbolic association between war and sport. When Bill Shankly became manager of Liverpool Football Club he said he wanted to make the club a 'bastion of invincibility'.

Certainly, the language of competitive sport is ubiquitously infused with metaphors of war and battle.[715] But is this just a trite and obvious metaphorical analogy without any real substance philosophically?

Some critics of sport's role in society seek to lower its status upon these grounds. Competitive sport, it is said, both encourages and masks the sort of violence that finds its most complete expression in war or in warlike attitudes within society. An apparently obvious reply is that competitive sport and war have significant differences. These differences seem centred around the most significant—in war, there is *essentially* a risk of life. Violence in sport has a clear limit, and this limit would seem to be transformative of the purposes of the activity. In particular, sport possesses a clearly more liberal or generous view of the purposes of competition.

But this transformation of purposes does not mean that the purposes and motives of the two activities are entirely unrelated. The origins of sport seem to be in some sort of *imitation* of war.[716] The disagreements that arise about sport's moral and social value come out of differing interpretations of the effect of this imitation upon both spectators and competitors. Such disagreement comes down to differing understandings of the nature of this imitation. If sport is in some way an imitation of war, an investigation into the nature of this imitation and the manner in which sport transforms and modifies the values of war would be of help in understanding what well-played competitive sport can contribute to man society and education.

The issue of imitation is one of both love and knowledge. Aristotle points out both senses of imitation; in the *Poetics*, he says that imitation is the human being's most native mode of knowing.

By this, the human manifests in any poetic or making activity its nature as 'rational animal'.[717]

Imitation is not merely the reproduction of a facsimile. Imitation 'distorts'; but this distortion is a revelation, precisely because the essence of a being is not simply revealed on its surface, but must be represented imaginatively in order to be completely grasped. A proper account of imitation, therefore, requires an account, not just of the essence revealed but of the distortion of its appearance. An account of an imitation must account for both analogies and disanalogies between the imitation and the imitated. An imitation will reveal something more essential or complete than the thing imitated can do on its own. Homer's portrayal of Achilles, or Plato's of Socrates, shows us more than mere acquaintance with these men could. If sport is an imitation of war, then sport will reveal the essence of the warlike to be more than the warrior could account for qua warrior. From the perspective of the warlike, this will seem a loss; from the perspective of a whole human life, this distortion is both a revelation and a gain.

Further, this gain is one *implicit in the self-understanding of the warrior himself.* This is what underlies the ubiquity of the metaphors of war in the discussion of sport.

Sport is a sign of the recognition of the self-sufficiency of a virtue within the very context of a contest that is interpreted by the warlike as the acquisition of self-sufficiency. Sport can reveal the virtuosity of the virtue of courage. This is an issue because courage's apparently primitive and foundational nature is unclearly related to the completion of virtue. The issue is whether a character that appears to be a pre-condition of excellence can itself manifest the presence of full excellence. This is an issue relating to courage in war because of war's apparent concern with the necessary conditions of activity and their acquisition; war seems particularly illiberal, and we have a sense that the completion of virtue is more liberal. The meaning of contest in war appears fundamentally acquisitive. Sport, by transformation of the context of contest, reveals what is noble in the warlike and strips it of its appearance of ignobility. Sport does idealise competition. But this idealisation is a revelation of competition's larger context and contribution to excellence. It helps us to recognise that, even in fields which seem dominated by a context of scarcity, the pre-eminent human concern is liberal and self-sufficient.

Given the complexity of warlike courage and its relation to a more liberal and comprehensive conception of virtue, images of war and battle are both inevitable and to some degree appropriate when speaking of the competition of sport. The warlike in us is the ground where virtue or excellence first reveals itself. This is not to say that the warlike is the entirety of virtue, but it is to say that the self-sacrifice involved in battle is a spirit that is more liberally rewarded and revealed in sport. But this does not mean that the warlike is entirely eliminated; rather, it is recognised and reinterpreted by the revelation of its own satisfaction.

There are two areas of important disanalogy between war and sporting activities. One has to do with the sort of discipline in the light of the goal of victory; the other has to do with differing attitudes regarding one's opponent. Underlying these disanalogies, which are crucial for the understanding of the comparison, is an analogous attitude and self-interpretative stance; a willingness to project oneself beyond the limitations of a concern with mere physical well-being and of others who wish to achieve the same goal. The psychic self-interpretations of the excellence of the lover of war and lover of sport, see the latter as a development of an initiative present in the former, one that leaves behind the context of war in order to artfully express itself. It is in this sense that sport is an imitation of the warlike.

That the spirit of sport is disanalogous to that of war seems clear from its non-seriousness; it is a certain adaptation of our playful or imitative instinct. War, by contrast, is most serious, and what it is serious about is victory. Sport seems serious about victory, too. As famed professional football coach Vince Lombardi said "Winning isn't everything, it is the only thing." However, the difference between victory in the two arenas can be discerned comically. For example, Lombardi's own words have been used by the satiric newspaper *The Onion* in a faux article in which Lyndon Johnson deploys Lombardi's Green Bay Packers to Vietnam to help turn the tide of the war. One of Lombardi's quotations, using the language of war, is given a darkly humorous undertone: "As I have always said, a man's finest hour is that moment when he has worked his heart out in a good cause and lies exhausted on the field of battle victorious."[718] In war, 'exhaustion' in victory is terminal in a way that it ordinarily is not in sport.

However, the Lombardi's quotation does touch on an apparent likeness between sport and war. Both participate in a kind of physical discipline that is for the sake of victory in context. It is the mode of this participation that differs. As Santayana put it, sport is "the liberation of martial energies from the stimulus of necessity," which is to say that in athletic competition, physical discipline for the sake of victory is for its own sake, not in the service of other, less liberal purposes. Victory in competitive sport is only for what Santayana calls 'the honour of success'.[719] Lombardi's famous saying, then, that victory is the 'only thing' rings strangely true for sport in a way that it does not for war. Victory in war is never finally for victory but for something else. If final victory is only found in the possession of an excellent activity, war is not finally about victory.

This liberation of martial energy from the constraint of necessity is precisely what makes sport a *playful* imitation of war. For any serious imitation of war's profound heart must take account of the grimmest necessity of all—death. All truly profound and interesting treatments of war in art and literature address the significance of this necessity. While there is defeat and risk in sport, it is not a risk of one's entire existence. Victory in sport is not tainted the way it is in war by the sacrifice of things that no one in his right mind would wish to lose. But as I have commented in the Prologue sport is a form of sacrifice in the sense of representing Terry Eagleton's view in his work Radical Sacrifice a kind of self-dispossession which is the condition of a renewed life.

I now consider the philosophical background to the claim that sport is victory-seeking battle freed from the constraint of necessity, and what effect that has on the self-interpretation of both competitors and their spectators. The divergent modes of victory in sport and war are revealed in the sort of psychophysical discipline appropriate to each. In both cases, we are concerned with a physical striving that concerns itself with victory. But the physical striving of sport is purely internal to the act of performing, which is in itself pleasing, while the act of war is more intimately connected with that which is essentially unpleasant and toilsome. This meaning of sport's freedom from necessity can be illustrated by reference to two comments found in philosophy.

The first occurs in Plato's *Republic*, where Socrates and Glaucon are discussing the type of gymnastic education appropriate to the guardians of the city in speech. Although he begins the discussion with the language of the games ("the men are champions in the greatest contest, aren't they?"), Socrates explicitly contrasts the psychic and physical training of the athlete with that of the guardian. The guardian must not worry about injury and deprivation of food and sleep, but must carry on in spite of these things. He must particularly *not* act like an athlete, who possesses a too 'high-tuned' yet sleepy psychophysical habituation; the athlete has a strict regimen of food and rest, and becomes unfit or injured the moment that he strays from such a habit. The warrior must be more endurant and flexible than this.[720] This flexibility is necessitated by a difference in purpose between a soldier in war and an athlete in contest, which reveals a characteristically different stance towards the necessities of embodiment. Sport transforms the body for the sake of a particular bodily excellence, whereas the psychophysical stance of the warrior is of the willingness to *forsake* his embodiment and its necessities, not merely transform them; this necessarily has an effect upon the psychophysical transformation that soldierly training demands.

These issues are given fuller development in Aristotle's discussion of the virtue of courage in the *Nicomachean Ethics*. It would not be fully unjust to say that for Aristotle courage is in some sense a foundation for all the other virtues; it is the virtue that is treated first in his catalogue of moral virtues. It is a foundation in that it is what is first for us, not necessarily what is first in itself; it appears to be a foundation of virtue, as a nativity and preservation of rather than as the full flowering of nature. As the development and becoming of nature is an enterprise of struggle, courage intimately involves pain, and it hence appears to be in some way not for its own sake. Courage is the only virtue for which Aristotle takes up explicitly the question of whether it is inherently pleasant, indicating how strong the customary disposition is to regard it as painful.

In his discussion of courage, Aristotle deals with the relationship between pain and pleasure in the paradigmatic act of courage in war and draws an analogy between war and competitive sport.[721] The fact that the end of courage in war, the courageous action, is pleasant is observed in much the same way that the pleasure of the honour of victory that the boxer receives seems obscured by the blows and physical exertion that the boxer endures in order to obtain the honour.

Aristotle's analogy is clearly not meant to be taken as an equivalence; the very next sentence implies that the likeness implied is conditional (*el dê toiouton ésti kaì tó perí tên andreían*: if this is also the case concerning courage).[722] Taking into consideration what makes the situations disanalogous, we can read Aristotle's conclusion even more strongly. Even in the case of the boxers, the end is small in comparison with the cost in the midst of the fight, but not so when victory is won. Even more so, then, is the end small for Aristotle in comparison with the cost of war, the prospect of death means the end to a good life for the best of human beings, and the best human beings *do not make the best soldiers*.[723] This should indicate the incompleteness of courage *if one interprets the willingness to engage in war as its highest manifestation*.

The paradigmatic act of courage for Aristotle is facing one's own death in war. It is the only act of virtue that appears inherently and not merely accidentally to involve resistance to pain; it is an end, not merely of life but, for the best of human beings, activity according to virtue and thus happiness. For the virtuous human being, the contemplation of his end is truly painful in a way it is not for less virtuous human beings. If facing death in war is the paradigm of courageous action, it appears to be the choice of making oneself the instrument of the good's actuality yet, as involving one's own death, it is the choice to no longer be the human good's full actuality. The former is something that all the virtues in a sense participate in; this is how courage is the root of virtue and the pleasure one can take in contemplating it. By contrast, the latter is that which none of the other virtue appears to participate in, and which seems to be the root of pain's being essentially part of courage.

While it is true, that sport seems to require something like this courage (insofar as it requires instrumentalising the body for the sake of an excellence it is not yet), and that defeat in some way might be described as something resembling a death (insofar as it is the recognition of a lack of substantiality in the realm of athletic excellence), it would seem to be an exaggeration to identify the two sorts of courage and defeat. Sport, like all other virtue, requires a warlike courage in its coming to be, but in its actuality warlike courage seems to come to be it—it resembles those excellences that are most pleasant in their full actuality. This gives a different meaning to Aristotle's boxing analogy in an important way. If the pleasure in courage is likened to the pleasure in sport, then it may be by the virtue envisioned in sport that we can measure the place of the virtue present in battle. For all the instrumental hard work and discipline that goes into sport, performing the activities well is in itself enjoyable and pleasant to. Heroism in war involves the contemplation of the painful, heroism in sport the pleasant. This is why heroism in war seems to touch us more deeply and, as Aristotle recognises, is most honoured.[724]

What sport does imitate in battle is a certain joy in contest. Contest, the physical and psychic struggle to overcome both body and opponent as a measure of oneself, itself becomes a pleasure, and is freed from the ultimate pain or master. Nietzsche, perhaps more than any other thinker, emphasises this aspect of contest as the essence of authentic human existence. For him, what made the Greek civilisation great was its incorporation of the idea of contest into every aspect of the community's existence, including art oratory and religion as well as sport.[725] Yet it is precisely this wrestling of the impulse for contest from the context of war into other realms that causes concern amongst sport's critics. As Nietzsche notes, the contestants in Greece despised one another, and it is the fear of setting loose this warlike attitude in society freed from the potentially curbing costs of war (in pain and death) that concerns critics of sport. Sport seems to promote competition without cost.

This critique, however, is based upon a double illusion. First, it exaggerates the effect that the fear of death has as a curb on the warlike. The warlike within us is not curbed by these costs, but is in fact exhilarated by them. Nietzsche again puts this finely when he celebrates tragedy; tragedy celebrates the warlike in our soul, which is fulfilled in opposition to pain that *cannot be overcome*.[726] Death does not deter the warlike but in fact is its destiny that is embraced by its very virtue; the *Iliad* is this story. If sport frees the martial impulse from necessity, this is precisely the way in which competition is released from a preoccupation with a proof of oneself by opposition to the greatest necessities. In fact, it is the warlike individual's preoccupation with the necessity that makes his desires unlimited.

In fact, by freeing the martial impulse from the realm of the necessary, sport only arouses the martial impulses to transform them by re-constituting them in a different setting. The truly sporting attitude puts limits upon the warlike in the very act of imitating the contest of war. For victory in sport, unlike in war, is judged by the judges or officials in competitive sport, and their standard is properly respected by all; and even when challenged, it is only in light of the rule that the official is to embody. Victory in this sense is the only thing in sport, and this is a victory judged not by the mere subjection of the victim by any means, but by an excellence in action enforced by the judges. The demonstration of excellence subordinates contest to itself, and this is why the activity of sport is essentially pleasant despite its obvious pains. The competitive spirit is indeed animated by sport, as the critics of sport recognise; however, it is animated in the context of a liberating physical excellence (which in the end produces pleasure), while war is fought primarily under constraint of necessity (which makes it painful), even in its liberating moments.[727]

One of the most telling consequences of the differing emphases in war and sport is the different nature of the social relationship between competitors. While it is true that on the field of battle, there are and have been various codes of conduct which show some measure of respect for the fallen or defeated enemy, this practice is both inconstant and largely ineffective in an important respect: The victorious and the defeated, when both survive, often despite each other. Victory in war—*conquest*—is hardly ever total in an important respect. The only way that it could seem so would be to literally eliminate one's enemy, which deprives one of the meaning of conquest in an important sense,[728] of different passions, from envy and spite to outright hated, whether of self or conqueror, due to some perception of injustice done to them by war and to the loss of a way of life judged by them to be noble and free. At best, what a conqueror can expect is grudging respect for her power, which is not simply respect for her excellence, although over time it may become identified with such.[729]

Victory in war is incomplete, and enmity goes on by other means even after conquest. Even if war achieves the total submission or destruction of the enemy, it is hard and not satisfying, for Hegelian reasons. The famous 'master/slave dialectic' begins in war and ends in slavery, the latter putting greater limits existentially on the victor than on the defeated. One way of putting this is that the victor in war may win, but she cannot 'win over' her opponent, and is actually prevented from doing so because there is no equality presumed between victor and vanquished as a result of there being no good held in common. The victor despise the vanquished, and hence the recognition gained by the victor from the vanquished is spoiled in its victory.[730] To the victor in war might indeed go the spoils, but they are spoiled spoils.

Victory in war is partial. It is partial because it is by necessity partisan, and the partisan by his own admission takes a part for the whole. It is partial in the end because it deals with scarce resources or what we might call the necessary. Now the necessary is the aspect of war that makes it 'serious'; and we should actually say more accurately that war deals with that which is necessary. The necessary, as opposed to the serious, picks out something objective regarding the intended activity in question, whereas 'serious' is what we might call a subjective mode of comportment towards any one set of activities. This distinction is needed because one can be serious towards those things that are not in the realm of the necessary. Sport, insofar as it moves us to transcend questions of scarcity and the necessary, seems to be this sort of activity.

Of course, any life of action must in the end deal with the necessary or the scarce; this is another side of saying, with Aristotle, that the life of politics, the realm of action, is oftentimes a life of toil.[731] However, it seems that there are two basically different attitudes regarding the external goods necessary for complete actions. One deals with the necessary only as it is related to the accomplishment of particular actions, which actions are not merely instrumental but rather essential to the self-sufficient life. For Aristotle, this is the primary and truthful sense of dealing with the necessary, which is the concern of the beginning of community. However, this concern with the necessary is not a preoccupation with it, and Aristotle remarks that it truly is not the proper, but only a subordinate, concern of the free citizen or head of a household.[732] A preoccupation with the necessary, however, is opposed to and not necessitated by what we might entitle the noble use of the necessary. This preoccupation with the necessary, we might, with Aristotle, entitle the acquisitive attitude (*chrematistike*).

The warlike attitude differs from the sporting attitude primarily in differing interpretations of honour. By saying that the warlike has to do with the necessary, I do not mean to trivialise its purposes as if they merely had to do with material preconditions of life. War seems, more often than not, to have to do with the meaning of the distribution of these resources and their uses, which involves the questions of purpose and recognition, of value and honour. When I say that war has to do primarily with the necessary, I mean that it emphasises the honourable and valuable as, by nature, a scarce resource. Honour in war comes in conquest interpreted as the subjugation of the necessary. The warlike honours the opposition to the necessary, and this opposition is unbridgeable by the self of the warrior because honour is interpreted as a scarcity needing to always be re-won. Again, as Nietzsche says, the warlike is celebrated in opposition to a pain that *cannot* be overcome. This existential conflict is at the heart of war and significantly manifests itself in one's relation to an opponent.

Perhaps the greatest achievement of sport is the substitution of defeat for death and the detachment of one from the other. The victor in sport does, of course, acquire something from the defeated, and this, just as in war, is honour. Likewise, the defeated does lose something; in some sense, the defeat in sport is more substantial than in war, yet this substantiality is precisely what makes defeat in sport less 'serious' than death. What the victor acquires and the defeated relinquishes is the respect and honour of one who has lived for the same purpose for which she has lived. By contrast, the defeated in war who loses his life to another can go to his grave without recognising the defeat of his substantial way of life, even if the means to its continuance have been taken. The victor in war, that is, does not in principle receive the recognition of his excellence in the very grasping from the defeated in his life; the substance of a man's life is not destroyed in taking his life. This is why the appropriate recognition of the enemy in war, usually by religious burial, takes on the significance it does. By allowing the dead enemy to be honoured as worthy in their own self-sacrifice, the victor makes his victory worth having. In burial, victor and defeated are seen to have a common ground in their self-sacrificing excellence. In it, both victor and defeated gain the recognition that the battle has stripped from them. These sorts of 'rules of battle' are, in this regard, quite 'sporting'.

In sport, the defeated gives over the recognition of her substance, the purpose of her activity, to the victor; in this manner, defeat in sport is a greater loss of substance than that in war. This loss of substance, however, is in another sense not as great, because in the contest in which one is defeated, one participates in the actuality of the psychophysical excellence that one desires and loves.[733] Competitors in sport are tied to a commonality of purpose in their competition that war does not appear to possess. In a sporting contest, the exhibition of this excellence subordinates the warlike perspective upon honour-seeking to the desire for an excellence that the contest makes possible. Honour becomes freely given, not forcibly taken as a scarce resource; it becomes truly honourable because it is freely given by those who know—judges, competitors, and spectators.

This is why it is best said that victory in sport, by contrast with war, is complete. The completeness is the result of sport's essential non-partisanship. The aim of competition in sport is not the destruction of the opponent but rather the defeat. This defeat is one that is recognised by the defeated in her love for the excellence involved in playing the game well. Those of us who have participated in or even watched an ill played game find the satisfaction of the victory of our side diminished when victory is either too easy or too ugly. In fact, the satisfaction of the victor increases with the excellence of the defeated; both athletes and spectators become bored or distracted when the game is not skilfully competitive, which effects the excellence of the game played and the satisfaction in victory. It would be difficult to say the same thing in war. Whereas the defeated in war are overcome by a force that they need not recognise as their own, in sport the defeated are defeated by themselves in the guise of a more magnificent other. We might say that in sport, the defeated are ennobled by their participation in the greatness of the victors; thus can victory be "the only thing." The defeated's recognition of such victory, in fact, contributes to the victory and its visible completeness.[734]

Thus, sport is not fundamentally about enmity. Rather, enmity is subjugated and transformed to a higher purpose, one that sport shares with art. Sport, like art, is about the transformation of some material into an idealised form, for the final purpose of revealing an ultimate significance or interpretation of our activity.[735] The material here is one's own body and with it the particularised aspect of one's soul, which one often experiences as a recalcitrant other.[736] Through training and exercise, the body and its associated passions and thoughts are geared towards or equipped with the appropriate actions, and come to perform those acts more or less well. Competition and contest is both a stage and a test for that excellence, a matter of public display and self-measurement. Competition is crucial to self-measurement in sport because through it one is not merely recognised by oneself, but is tested and recognised by those who know, both the other competitors and those spectators who might appropriately be called judges. It transforms mere private opinion into a splendid athletic self-knowledge.

This idealisation is also a revelation and thereby a transformation. Sport's imitation of war brings out and isolates that nobility which is present in war. Underlying the warrior's spirit is the sporting spirit. A sign of this is given in the *Iliad*, where the death of Achilles' friend Patroklus is followed logically by funeral and by games.[737] The burial reconciles victor and defeated; sport follows in the wake of this reconciliation as an activity performed in the spirit constituted by burial.

Sport, we could say, is an activity of peacetime; hence, the historically incorrect assertion that sport could appear only in times of peace in Greece has this much truth to it—the competition in sport is not one of enmity, but one in the service of a common good, and hence one of peace. The competition in sport, as possessing a common good in bodily excellence of some sort, is a kind of friendship. When a sporting contest is taken seriously by its competitors, a mutual recognition of superiority and inferiority (but not of insignificance) in regards to the same standard that both have strived to embody overcomes the alienation that is characteristic of most other forms of competition.

Sport differs from those competitive pursuits that we ordinarily deem serious precisely in its abstraction from the viewpoint of necessity and scarcity. There is defeat in sport, but it is one that ennobles the community of sport and one acquiesced to by the defeated in his very efforts. War has neither of these two features, at least on first look. The disanalogies in what has preceded seem to vastly outweigh the analogies. However, one must hesitate before concluding that analogies of war and sport are senseless. In exhibiting the disanalogies, I have noticed two types of phenomena that contravene a simple separation of these two activities.

One is the presence of a serious struggle for honour and its accompanying psychophysical self-overcoming. Accompanying this is the fact that the activities of sporting friendship are not easy and relaxed, but coexist with the enmity-producing self-overcoming spirit. Second, it is not clear that the warlike is so simply subject to a grave preoccupation with the necessary.

A sign of this is the practices of burial and 'fair play' that, although perhaps foolish from one perspective on war, seem to be genuinely part of the spirit of the warrior, and are practised (if inconstantly) even in the most gruesome and inhuman of wars. This is related to a more essential objection to the portrait of war painted thus far. One could say that war is the paradigm, in some sense, for ennobling experience; what could be more ennobling, could garner one of the greatest recognition, than death in war? This is a death in which one proves one's dedication to a way of life and political community that one judges noble and so sanctifies that way of life as a whole. If nobility is the sacrifice of one's mere living for the sake of its significance, or of mere living for what it means to live well, then willingness to engage in battle does seem the paradigm. It is to the incorporation of this objection, and to discerning the appropriate psychic ground of the analogies of sport and war that I now turn.

The contrast between the warlike and the sporting has been characterised thus far by opposing concerns for, respectively, acquisition and freedom/ nobility. The considerations above have made it clear that although sport does clearly imitate war in its manifestation of the desire for the acquisition of honour and victory, it puts such a concern within a context which is more friendly and liberal than that of war. Contest, seen as fundamentally acquisitive, is subordinated to a common desire for excellence that is both at issue between competitors and guides the individual athlete's psychic self-interpretation.

The contrast between acquisitive and noble or free concerns I have adapted from Aristotle's *Politics*. When speaking of the activity of *oikonomia* (household management), Aristotle distinguishes between *oikonomia* proper and what he calls *chrematistike*—the art of acquisition for its own sake. *Oikonomia* proper is part of the art of property management, which is grounded in and guided by an excellence of a community, whether that of the household or the city. By contrast, *chrematistike* is not grounded by a community and its proper conception of an excellent activity, but is rather what is properly a means taken as an end in itself.[738]

War, like acquisitive activity of all types, by nature or properly is bounded by concerns for excellent activity proper to a community. However, like that of the economic type, it can be practised outside of the bounds of what is by nature proper.[739] The basic point is that in neither the case of war nor in the case of unbridled economic acquisition is there the notion of a common good and public virtue that can mediate claims of acquisition. Without such a notion, acquisition is in principle limitless, whether 'peacefully' in terms of finance or 'violently' in terms of war.

Aristotle's analysis penetrates more deeply. The reason that humans interpret their lives as naturally under the guidance of *chrematistike* is that "they are zealous for living, but not for living well." These human beings are concerned primarily not with the quality of life, we might say, but mistake such a concern with the quantity of life under their jurisdiction, because ultimately the fear of death—the ultimate sign of scarcity—is their guiding concern, and they pursue the means to preserving the sight of themselves everywhere. It is important not to misunderstand this as a mere desire for physical self-preservation, even if this is at the root of the problem. Every animal faces death, but it seems the curse or blessing of the animal that has intellect to take its death as something extraordinary, and to be infinitely avoided. The desire for acquisition is primarily a concern with honour, with a seeing of oneself reflected back in all things. One manner of doing so is to own all things, or at least as many as possible. Aristotle rightly identifies this as a limitless task; even if accomplished in space, time marches on, and it must be constantly reacquired.[740]

War, or perhaps more properly the conquering attitude of war, is a mere arm of this preference of living to living well. By demanding that all be subject to one by conquest, one is trying to preserve one's existence even beyond one's life; *chrematistike* is just the logical extension of the desire for self-preservation ruling the soul in an animal whose horizons are not merely limited to its finite existence. Maybe this explains Putin's desire to conquer Ukraine.

Living well, by contrast, implies a limit to desire. This limit is provided by a standard of excellence that is proper to the animal in question. This limit can be a limit because in its own sense, it is not limited to the finite possession of a particular living thing, but can be participated in commonly by those living things when the circumstances provide such a possibility. The end of the communities of virtue, that is, can be achieved, since it is not dependent upon the desire to see the individual everywhere recognised, which is an unlimited task, but by a set of finite activities. It is precisely these activities; freedom from the questions of self-preservation that makes them ends or limits. If their being depended wholly on their embodiment, then they would be trapped in the infinite cycle of self-preserving behaviour exhibited by the practitioners of *chrematistike*.

The first comprehensive sign of the virtuous self-limitation of acquisition paradoxically occurs *within* what seems to be merely acquisitive actions; it occurs, in fact, in war. This is the complexity of the phenomenon of battle and its virtue, courage.[741] Participation in battle is not simply reducible to an acquisitive motive, although it does partake of it. The warlike in us cannot be done full justice by a reduction to the acquisitive impulse. For war, as opposed to mere acquisition that is motivated by self-preservation, involves in the most obvious manner a risk of mere life. As in the Hegelian master/slave dialectic, a being raises itself from mere animal desire to a human desire by a risk of its animal life for the sake of its recognitive or ideal life.[742] War is the very beginning (although not the end) of a life of the freedom from mere animal acquisitiveness.[743] The human being in battle-subjugates the animal necessities hunger, thirst, sex, and in the extreme case, self-preservation, for the sake of something that is not simply reducible to those necessities, whether this be the honour of family, friendship, city or virtue itself. This sort of freedom from the merely necessary desires is what can be termed the noble.

There is something actually more profound about defeat in battle, meaning death, than in sport, and it seems to have nothing to do with acquisition or self-preservation in the crude sense, rather, it is that impulse which grounds the possibility of all virtue. In the soldier's choice of his death, he affirms that his life is not simply reducible to the desire for his own continued existence. This is what is inherently pleasant about the courage in battle, that which is obscured by the pains of death, and actually occurs within the very pains of death. By acting in this manner, the courageous human being reveals that *he already has a self-possession that transcends his finite existence*. It is not Achilles the man who is remembered in the stories of Homer; it is the memory of Achilles that is remembered, and has a higher degree of actuality than does Achilles himself; young Greek men were instructed to wish to 'die *like* Achilles', not die Achilles' death. Sport, like the poetry of Homer, abstracts from the immediate presence of the pains of death and is thus more able to make visible than can war the presence of self-possessed virtue. Before this point is explicated, further clarification is needed regarding the character of courage as it is inflected in the warlike.

The virtue that a soldier needs reveals the presence of a self-possession not limited by finite existence, which we might call the noble or free. However, we can still strictly distinguish the warlike from the courageous. This is not to say that the courageous human being will not have occasion to go to war, it has more to do with the motives by which the courageous human being will desire to exhibit his courageous independence. There is something incomplete about the *appearance* of the virtue of courage in situations of war. This may be why Aristotle uses sport as a way of explicating what is eudaemonic and hence inherently pleasant about the excellence particular to battle. Recall that I discussed the uniqueness of courage in war as a virtue in the Aristotelian scheme; it *seems* both painful to contemplate and to perform, especially for human beings of complete virtue. In the formula used above, the virtuous human being feels himself to be the instrument of virtue and not its continuous actuality, this despite a transcendence of finite concern already accomplished in the virtuous act of courage. The full exercise of the virtue of courage contemplates the cessation of a good life, whereas the full exercise of the other virtues are a living actualisation of excellence.

This fact about the nobility found in war leads us back to a reconsideration of the acquisitive aspect of this nobility in the warlike character, or what we might term its mixed nobility. It can be seen as an emphasis upon a virtue that has to do with the acquisition of nobility rather than its full actuality. This is not to deny the spectacle of this courage as an object of wonder and depth. In fact, it is precisely its wondrousness that correlates with its incompleteness.[744] Even in looking at sport, we tend to dramatize our perception of excellence in this direction; we find dramatically interesting and profound the story of obstacles overcome, without which the story of excellence achieved seems cold. But this in some way a distortion; what we need is to recast our opinion of our wonder. The human being, like any other animal, tends to look upon the appearance and acquisition of its good as more grand and profound than the fully actualised exercise thereof, and to reinterpret the former as the latter by means of this amazement. In this wondrous acquisition is warlike courage of the greatest significance and, hence, inspires the greatest awe. Courage in battle, at first glance, appears to tread on the ground where one's freedom or nobility is opposed to one's existence and hence must *yet be won*.

This is precisely, however, where the warlike diverges from the courageous simply. Courage, interpreted simply as the warlike although a sign of the already implied transcendence of this opposition, and hence of virtue entirely possessed, puts an emphasis on the motion of acquiring the excellent, and when detached from more gentle or liberal activity, has a tendency to glory in the experience of the opposition and its overcoming for its own sake. This is a sign of such a character's preoccupation with the necessary even under the auspices of the noble. This constant desire for such a challenge, if it takes control of the centre of one's life or soul, promotes an essential restlessness that is the opposite of the self-possession of the virtuous character. This is the final meaning of Aristotle's claim that those who are interested in acquisition are more interested in living than in living well. Mere life is a restlessness without end; however, there never is a 'mere' life but species of kinds of life, which allow us the judgment of living better or worse, and the aforementioned possibility of a self-possession that is not subject to this infinite restlessness of acquisition.[745]

Competition is secondary, as a display of and necessity for the possessed excellence in action. This is why Aristotle claims that the best human beings do not make the best soldiers; the soldier is interested in *making* her life worthy, whereas the best human being already lives worthily. The best soldier is she who is both lesser with regard to the virtue of courage itself and possesses little in the way of other virtue.[746]

Now that the virtue of courage proper has been distinguished from the character of the warlike, it is necessary to say how the sporting event's imitation of war effects and reflects the distinction between these two different characters. What distinguishes the courageous form the warlike is the emphasis upon a self-possession already achieved as opposed to an emphasis upon an excitement over the motion of acquisition or the becoming of self-possession. This is what lies at the heart of Aristotle's statement that the most courageous and virtuous human being is not likely to be the best soldier.

In war, however, this distinction is often easily obscured. What obscures it is like that which obscures the honour of victory in the case of the boxer in Aristotle's example. It is the blows and injuries suffered—and in war, death; the context of necessity and scarcity obscure the true honour in battle. One has only one body, which is assaulted, and to the sight of most human beings, this understandable preoccupation with the bodily or self-preservative side of war's honour obscures the true self-possession that the soul can and does achieve within the acquisitive toil of war. War's preoccupation with necessity and the body infects the appearance of the virtue that is appropriate to it. The very same motions that can be performed from the virtue of self-possessed courage can also often be done from the striving for self-possession in the violent death of or subjugation of an opponent, although it is possible that sooner or later an act will reveal itself which can if not entirely by nature, inclined towards self-preservation, it is possible that many will mistake true nobility within battle for the acquisitive motive, and the self-possession of virtue will remain unseen.

But is perhaps sport that is the most direct reflection of the birth of nobility in the very arena of competition and defeat that characterises war. While religion and poetry can comment upon this subject in speech, sport presents us with a live spectacle: an act whose outcome, like that in war, is the gaining of honour. Yet unlike war, where the tolls and bodily harm and their accompanying focus upon acquisition of one's self-possession against one's finitude, obscure the self-possession already won in the courageous act, in sport the primary emphasis is upon playing the game well on a self-possession already achieved and defined, even if in the contest the emphasis appears to be upon the gaining and manifesting of that self-possession. Sport, that is, is a display of excellence to which gain in competition is subservient; this is its liberality, a liberality revealed deep in the heart of the warrior.

Yet it could also be said that this is a subordinate concern in the case of sport. The mode of this subordination, indicated above, is what is unique to sport and what distinguishes sport from war and other competitive 'arenas'. In sport, what the athlete competes for and what the spectators, when they are true, come out to see, is the showing of a self-possessed excellence. This is revealed in the language that both athletes and spectators use in describing the performance of an excellent athlete: "Tiger Woods seemed to be playing by himself today." In such language we reveal that the excellence proper to sport is not simply dependent upon the contest, but rather that the excellence of the contest is due to the excellence of the play, an excellence that the contest can reveal but not fully determine. Sport is friendly[747] and serious competition, which takes nothing from the contest, but adds to it a standard to which competitors strive and are mutually devoted, the character of which devotion and standard is the self-possession that already exists.

There is a standard which already exists that mediates the competition of competitors in sport.

This seems to be true because in sport the question of who ought to be honoured is already settled by the rules,[748] whereas in war, this is precisely what is to be determined and is subject to force. This lack of determination regarding the excellent means that there is no set of activities, no natural end, to which the actions of contest could be referred to for judgment. In war, rule exists wholly as a result of victory; in sport, victory exists as a result of rule. Because in sport rule is always already determined, the activity of a good sportsman properly resembles the character of self-possession that virtue has essentially; it has been noted above that in a sense of an excellent human beings do not go beyond themselves in their acts. In other words, the sporting human's actions will manifest the stability of one who acts well knowing that it is well, while the warrior appears to be in an unstable situation with regards to rule, no matter how things may stand in his heart. Like in the majority of our lives, the arena in which the warrior perseveres is not the arena for which he longs.

Sport allows for the manifestation of self-possession more clearly than in war because its abstraction from honour is considered as scarce. Competitors are allowed to focus upon their devotion to the excellent because they are not constantly faced with the prospect of the necessary. As such, sport can reveal more clearly what war tends to conceal with its preoccupation with the necessary, pain and death. What makes sport unique is that it accomplishes this revelation within the very realm of contest which the warlike claims as its own.

Sport manifests the separation of self-possession from the need to prove it to oneself and others by contest within the very act of proving it through contest. As such, sport can reveal the self-possessed courage which is present in war in a more clear manner than it can be revealed in war itself. This is not to deny the presence of such virtue in war, but rather its visibility. This visibility in sport is potentially educative; it can help preserve the presence of self-possession in all situations that involve conflict and competition, including war itself.

There is a virtue which is appropriate to sport that comes closer to the self-possession manifested in true courage than the self-opposed character that first appears in the warlike character. This virtue is called 'sportsmanship'. Sportsmanship is not merely reducible to the victory in contest, for we attribute it to both victor and defeated alike. Rather, it has to do with the mode of one's devotion to the athletic excellence involved.

Sport, because it abstracts from the question of death and necessity, is able to reveal that behind the quest for honour is a quest for excellence in activity. Serious human beings, says Aristotle, desire honour to assure themselves of their worth and so honour is subordinated to virtuous activity.[749] As Santayana says, sport's contest is for the 'honour of success'. Success is not meant here as it is in the everyday 'business of the world'; in the same article, Santayana quickly disabuses us of this notion.[750] What is meant is an excellent achievement not a mere acquisition of necessities. Sport is an artful aid to remind us that behind the various contests for honour or power is a longing for self-possessed excellence. In this manner, it is a break from the 'business of the world', which explains our own feeling of refreshment and revitalisation when we see and participate in sport well played. It takes us to a world 'beyond'.

In this, sport is truly artful, insofar as art is concerned with "the purification of spirit from its thraldom."[751]

One of the most profound insights of cognitive science is that the slower, conscious brain more often post facto "rationalises the faster decisions that the subconscious brain has already taken."[752] I have dealt earlier with this issue generally. But I now want to look at it in the context of war and sport. Mike Martin in his excellent book 'Why We Fight' argues persuasively[753] that when it comes to war, humans are driven to fight because of their subconscious motivation to seek status and to belong. He says "When we are subconsciously driven to fight for status and belonging we consciously interpret this and justify it, as fighting for moral reasons, religious reasons or ideological reasons."[754]

I agree with Martin, ideology, no matter how 'twisted' does not cause people to kill themselves and others. Religious differences do not cause wars. They are merely the conscious framework that we use to justify our intuitive desires for status and belonging to others and to ourselves. Martin argues that leaders who start wars are pursuing status, just the same as people who kill in gangs and in bar fights and that these different forms of violence share the same evolutionary shaped drives[755].

He refers to the ideas of Thucydides who began writing his History of the Peloponnesian War in 431 BCE who wrote the war was caused by three motives namely: fear (of others), honour (pride in your group) and interest (possible resource gain).[756]

Martin's central thesis is that power is analogous to status, belonging is analogous to identity. In my view sport, like war relates to status and identity and the pursuit thereof. To that extent the desire to play sport shares the same psychological evolutionary roots as war. It is significant that as men get older and can no longer play sport they seek other means of gratification in the pursuit of power and identity, politics, wealth, honours and literary, artistic and sporting achievement and admiration or association with the admiration and achievements of others. The most recent illustration of this concept is the football World Cup in Qatar in 2022. The Times Newspaper pointed[757] out that foreign nations have been invaded by the English dressed in crusader uniforms for more than 1,000 years. It notes that these days their swords are made of plastic but they are no more welcome and that has resulted in England football supporters being banned from attending the World Cup match against the United States from dressing up as Christian invaders after it was ruled offensive to Muslims in these various domains.

This would include being a supporter of sporting teams, their icons and share in their achievements and success in doing that they strengthen their sense of identity and belonging to a group and sharing common experiences including respect for and the love of the game. This might possibly be explained by biological considerations such as testosterone or the loss of it but as has been pointed out by others. Nothing in biology makes sense except in the light of evolution.

And that example of this phenomenon is the tale told by the TV series SAS Rogue Heroes which is based on Ben Macintyre's book of the same title. The book has as one of its heroes Blair 'Paddy Mayne'. The book compares and contrasts Bulldog Drummond's bravado and raw machismo with the psychological complexities of these men. They were brutal killers as well as a band of brothers. Mayne was a founding member of the SAS during the Second World War and also represented the British and Irish Lions on their South African tour of 1936. Mayne was born in Newtownlands in County Down who showed talent as a cricketer, rifle shooter and golfer—he had an eight handicap as well as becoming the Irish Universities boxing champion. After three caps for Ireland, he was selected for the British Isles touring team of 1938. He made 20 appearances for Ireland. He was described by a member of the South African Morkel family who wrote "He is the finest all-round forward I have ever seen. His staying power has to be seen to be believed." As Rich Broadbent observed[758] "The same could be said for his off-field behaviour." Ben Macintyre discovered that Mayne befriended Bill Travers, the Welsh hooker and that they went looking for bar room brawls. In Behind, the Lions contemporary Harry Bowcott said the pair would go to the Cape Town docks and wait until someone said rude things about them and then demolish them. Bowcott said of Mayne: "Magnificent physique and a very quiet fellow you thought wouldn't hurt a fly—until you saw him roused. Mad as a hatter."

Before the Lions played the Springboks in Johannesburg, Mayne and Travers came across a gang of shackled convicts who were sleeping beneath scaffolding as they erected a temporary stand. Finding out one of them had been jailed for seven years for stealing a chicken, Mayne returned with a set of bolt croppers and freed 'Rooster' and at least one other prisoner. Sadly, the prisoners were recaptured with 'Rooster' still wearing Mayne's jacket. On the pitch Mayne was indefatigable and against a team considered the unofficial world champions, played a part in a landmark 21-116 win at Newlands. After the tour he played three more times for Ireland before Germany invaded Poland and his sporting career ended. He went on to become one of the British Army's most decorated soldiers. He supposedly was recruited for the fledgling SAS by David Stirling. His bravery for the handpicked unit was legion. He is said to have destroyed more than 100 aeroplanes and was part of the raiding party that stormed the Sidi Airfield in Egypt in 1942. This daring mission involved a daylight attack in jeeps. Stirling was captured ending up in Colditz Castle. Mayne took over command of the SAS and led campaigns in Italy, France and Germany. Recommended for the Victoria Cross, with a citation approved by Field Marshall Montgomery, he never got it and had to make do with the D.S.O. He was also made an officer of the Legion of Honour.

He had a successful post-war law career, but injuries sustained from parachute landings made it difficult to even sit and watch sport. He died in a car crash after a night of poker and booze in 1955 aged 40.

I mentioned earlier the most charismatic people I have ever met. Muhammad Ali and El Cordobes. Obviously, I never met Paddy Mayne but I so wish I had. If history as Thucydides said is philosophy by example. Paddy Mayne was such an example. The brave man, the courageous man who exemplified the man of war and the sportsman. The same virtues and vices which both activities display could be used and abused for good and bad causes. He was the epitome of Roosevelt's Man in the Arena. He was Dylan Thomas' no good boyo who didn't know who was up there and didn't care. A man whose virtues were his disguised vices.

Nietzsche said of himself in "Why I am so wise" (Gotzen-Demmerung 1889), "War is another thing. I am by nature warlike. To attack is among my instincts. To be able to be an enemy, to be an enemy—that perhaps presupposes a strong nature, it is in any event a condition of every strong nature. It needs resistance consequently it seeks resistance." As a philosopher, he said, he challenged problems to a duel ("so does the sportsman").

Chinese philosophy, War and Sport

Little is known for definite about Sun-Tzu (544-496 BC) and his life during the Warring States period after the decline of the Zhoo Dynasty but his classic, The Art of War has been one of the central works of Chinese literature for 2500 years. His wisdom has survived time. His opinions such as, "no nation has ever benefitted from a protracted war" and "Without a full understanding of the harm caused by war it is impossible to understand the most profitable way of conducting it "are as pertinent and relevant now as they were when stated more than two thousand years ago.

The subject matter of the book is based on the emphasis he places on the importance of psychological warfare. His philosophy was that ultimate excellence lies not in winning every battle but defeating the enemy without ever fighting. To him the skilful strategist defeats the enemy without doing battle, captures the city without laying siege and overthrows the enemy state without protracted war. It was the psychological strategy that counts to win the war. He concluded know the enemy, know yourself, and victory is never in doubt not in a hundred battles. The key was to know oneself and the enemy because if you do not you will fail in every battle. In his assessment the victorious army is victorious first and seeks battle later; the defeated army does battle first and seeks victory later. The skilful strategist cultivates the way and preserves the law and he is thereby master of victory and defeat.

What he says about war in my view could equally be said in the modern world of sport. Both activities have a shared vision of success, defeat and achievement. In my opinion therein lies the true analogy between sport and war. Both exemplify the ancient Chinese philosophy of Sun-Tzu in The Art of War.

Chapter 30
Sport and Race

Morality knows nothing of geographical boundaries or distinctions of race.
Herbert Spencer

This is a hugely controversial and topical subject evidenced by the different views expressed about such things as "taking the knee." But I do not want to avoid it on the grounds of controversy or topicality. A consideration of the significance of the relationship between sport and race is an essential component of any comprehensive philosophical analysis of the concept of sport and beyond which is the subject matter of this book. Philosophy deals with the intrinsic and extrinsic nature of things and mankind. As I said in the Prologue a complete account of sport depends on a philosophical examination of it. It drapes over other matters such as the biological, sociological and historical. It is only on that basis and only on that basis that its other essential features can be made wholly intelligible.

Sport and race and the relationship between them cannot be avoided in any true analysis and investigation of the intrinsic and extrinsic nature of sport or indeed race.

There is a view that sport has damaged race relations and preserved the rights of race.[759]

In 1997, Darwin's Athletes by John M. Hoberman was published in the United States. Hoberman was Professor of Germanic Languages at the University of Texas at Austin. It is a learned courageous and original work and essential reading for those interested in the study of sport and its racial connotations. Hoberman characterises his book as an exercise in racial dialogue.

It is a call to mount a campaign against the 'damage' caused by the pathology of white racism.[760]

The thesis of *Darwin's Athletes* is provocative: sport has done more harm than good to African-Americans. Hoberman argues that sport has played an important role in leading African-Americans to embrace the damaging idea that physical self-expression is the essence of being black. He also argues that sport has helped to preserve deep seated racial myths among whites, while at the same time fostering an illusory view of racial integration in America. For Hoberman, these factors outweigh the contributions that sport has made to racial integration, to promoting friendships among blacks and whites, to teaching values such as deferred gratification and fair play, and to paving the way to higher education. Although one might ultimately disagree with one or more of the positions that Hoberman takes in his book, it is unlikely that one can read *Darwin's Athletes* without a serious reassessment of one's own views on race and sport.[761]

Darwin's Athletes is divided into three parts. Part I, entitled "Shooting Hoops Under the Bell Curve," is the most provocative and controversial section of the book, as it asserts that pride in black athletic achievement "is damaging black American in ways that African-Americans in particular find hard to acknowledge."

In Part I, Hoberman argues that Western racism has inflicted upon African-Americans a 'physicalised' and thus 'primitive' identity from which they have yet to escape. This physicalised identity has become 'athleticized' through African-American engagement in sport. Hoberman characterises Jesse Jackson as 'breathing new life' into the industrial education movement of Booker T. Washington when Jackson asserts that young black athletes "create a tremendous industrial base for black America."[762] Similarly, Hoberman quotes Charles S. Farrell, national director of Jackson's Rainbow Coalition for Fairness in Athletics, who states, "Athletics is to the black community what technology is to the Japanese and what oil is to the Arabs."[763] Hoberman concludes:

> The black athletes today who refine their athletic skills and little else at American universities are thus the damaged inheritors of an educational philosophy that once promoted manual training as the highest cultural achievement to which black youngsters should aspire.[764]

This physicalised identity of African-Americans became athleticized, Hoberman argues, as African-Americans became increasingly entrapped in a 'sports fixation'. He continues:

> The sports fixation is a direct result of the exclusion of blacks from every cognitive elite of the past century and the resulting starvation for 'race heroes'; it has always been a defensive response to the assault on black intelligence, which continues to this day. That is why the sports syndrome has made athleticism the signature achievement of black America, the reigning symbol of black 'genius'.[765]

As evidence of the power of this sports fixation, Hoberman not only puts forth the athleticized prose of writers such as Amiri Baraka and Ishmael Reed, he points out the intellectual vacillations of academics such as Harry Edwards.

> Even if Harry Edwards, who has spoken out against the fixation for many years, has declared that the highest form of human genius is athleticism: "If I were charged with introducing an alien life form to the epitome of human potential, creativity, perseverance and spirit, I would introduce that alien life form to Michael Jordan."[766]

This sports fixation, which disposes African-Americans to identify 'blackness' with 'physical prowess' and 'physical expression' has had profound consequences for the African-American community. First, it has engendered a physicalised form of anti-intellectualism. This physicalised anti-intellectualism, asserts Hoberman, damages black children "by discouraging academic achievement in favour of physical self-expression, which is widely considered a racial trait." Studying academics in school, then, becomes a form of 'acting white'.

This physicalised anti-intellectualism, Hoberman argues, has promoted 'compensatory efforts' by African-American writers and academics "to convert black physicality into intelligence." He quotes the black sportswriter Edwin Bancroft Henderson who wrote the following in a 1936 article on Joe Louis entitled 'The Negro Athlete and Race Prejudice':

> The great boxer was a human replica of Rodin's 'Thinker'. In the ring he associates ideas and responds with lightning-like rapier thrusts about as rapidly through the medium of mind and muscle as an Einstein calculates cause and effect in cosmic theory.[767]

Hoberman also discusses the compensatory efforts of Afrocentrists who respond to doubts about black intelligence by positing a theory of cognitive 'styles'. Among the work he cites is that of Alice M. Scales who, in a 1987 article in the *Negro Educational Review*, distinguishes between 'reflective' cognition, which is a white style of cognition, and 'impulsive' cognition, which is a black style of cognition. In the last analysis, Hoberman asserts, these compensatory efforts simply reinforce the white racist folklore that "has always made the body the essence of black humanity and a sign of its inferior status."

Hoberman argues, even African-American writers and thinkers who consciously reject the identification of blackness with physicality can be entrapped by a fixation on sport that undercuts their rejection. Kariamu Asante, an Afrocentric professor at Temple University is a case in point. It is Asante's position, stated in his own words, that "when you see Michael Jordan going to the hoop...you're seeing the African-American approach to things."[768] Now Asante explicitly states that Jordan's going to the hoop is not a 'natural' phenomenon but a 'cultural' phenomenon rooted in the role that rhythm plays in traditional African culture. Yet what all too often gets communicated by Asante's use of Michael Jordan as a cultural paradigm is the message that to be black is to be physical. It is as if African-Americans had heeded the advice of Charles Murray and Richard Hernstein, authors of *The Bell Curve*, who argue that the 'wise' cultural response to residence at the bottom of the mental ability scale would be to develop a 'clannish self-esteem' based on the demonstrated aptitudes of one's group.[769] Hoberman concludes:

> A black middle class (and its intelligentsia) that remains infatuated with sports cannot campaign effectively against racial stereotyping that preserves the black man's physicality as a sign of his inherent limitations.[770]

This physicalised anti-intellectualism, Hoberman continues, has subverted "more productive developmental strategies founded on academic and professional development," and it has discouraged "more productive cultural and intellectual interests" that are driven by 'theory', 'argument' and 'concept-creating'. It is for these reasons that Hoberman dedicates *Darwin's Athletes* to Ralph Ellison:

> Ellison saw the physicality of black self-expression as an unsatisfactory substitute for unhampered intellectual development. Here, as elsewhere, he opposed a notion of racial essence that is conducive to both black and white separatism.[771]

A consequence of the African-American sports fixation is that it has fostered an illusion of racial progress or, as Hoberman puts it, "a virtual integration" that serves as a racial coping strategy for blacks as well as whites. "The presence of large numbers of black athletes in the major sports," Hoberman writes, "appears to have persuaded almost everyone that the process of integration has been a success." But this "sense of closure," he goes on to argue, is 'wishful thinking' that is rooted in 'black apathy' and 'white auto-intoxication'. Thus he writes of the Reverend Jesse Jackson:

"We believe," Jackson writes in 1993, "that sports can help change the despair in our communities into hope, replace low self-esteem with confidence and rebuild a true sense of community that transcends neighbourhood and racial boundaries." For all its noble intentions, this declaration revealed a stunning lack of historical perspective. Could Jesse Jackson not have known that he was invoking the millennial hopes for sport that the NAACP had proclaimed back in the 1920s and 1930s? Had the passage of most of a century taught him nothing about what the African-American engagement in sports could and could not do for his people? The recycling of noble rhetoric is, in fact, a constant by-product of the black sports fixation precisely because it has produced so little of permanent value for most black Americans.[772]

Hoberman, then, comes down squarely on the side of the 'justly famous' 1968 *Sports Illustrated* series on the black athlete, which concluded that the social utility of integrated sport has proven to be largely fraudulent and illusory. Although he acknowledges that the importance of pursuing integration within the sports world "varies by historical period," he concludes that the integration of sport "has long served as a distraction" that has obscured the racial struggles directly affecting the rights and dignity of the far larger numbers of black Americans who do not dwell within the sports world.

Hoberman, however, is not done. He argues that there is a "dirty little secret" associated with the integration of sport. Within the integrated sports world itself, the subordinate racial status of the black athlete has been 'preserved'. It has been preserved in two ways.

First, integration of the world of sport has not fundamentally changed the attitudes of whites towards black athletic performers once these performers leave their stage.

Second, there is a racial imbalance of power within college and professional sport that is the product of a "genuine colonial arrangement" that has preserved a traditional white hierarchy of owners, general managers, coaches, trainers, writers, and broadcasters.

The one positive role that Hoberman can see for virtually integrated sport today is to 'keep alive' the idea that racial integration can actually work. "If integrated sport boosts the morale of a multiracial society in this way," he writes, then it may 'buy time' for more significant bridge building measures to take effect.

Hoberman calls for a deep and comprehensive critique of the entire relationship between African-Americans and sport, and in Parts II and III he attempts to lay the historical and anthropological foundations for such a critique.

Part II of *Darwin's Athletes* is entitled 'Prospero and Caliban' after the characters in Shakespeare's *The Tempest*, and it presents the past century of sport as an "arena of racial competition." Hoberman writes:

> The ascendancy of the black athlete and the growing belief in his biological superiority represents a historic reversal of roles in the encounter between Africans and the West. White European pre-eminence during the 19th century included the presumption of physical as well as intellectual and characterological superiority over other races, and athletic ability played a significant role in establishing white male authority in colonial societies.[773]

This historic 'reversal of roles' in the arena of sport, however, did not undermine the myth of white superiority. Instead it 'preserved' it. How can this be? Hoberman's answer is prefigured in Part I when he writes about the 'physicalisation' of black identity through the 'athleticizing' of the black essence. As long as African-Americans think of themselves first and foremost in terms of their bodies, the myth of white racial superiority will be preserved. The brutish Caliban, who is so closely identified with the physical as to be subhuman, will always be inferior to the scholarly Prospero who used his mind to discover the secrets of white magic that makes him all-powerful.

Part III of *Darwin's Athletes* is entitled 'Dissecting John Henry' and is subtitled 'The Search for Racial Athletic Aptitude'. In this final section of his book, Hoberman shows how ideas about black athletic superiority belong to a "more comprehensive racial folklore that has long imagined black people to be a hardier, physically stronger, and biologically more robust human subspecies than other races." Today, this racial folklore is often put forth in the form of "tabloid science." Unlike "genuine science," tabloid science "streamlines its presentations" and "feigns omniscience" in order to excite its readers.

As an example of tabloid science, Hoberman cites Amby Burfoot's cover story in the August 1992 issue of *Runner's World* magazine entitled, "White Men Can't Run." This article claimed to scientifically demonstrate the physiological superiority of East Africans in distance events and West Africans in the sprints. The most significant evidence presented in the article was unpublished muscle fibre research done by the Swedish physiologist Bengi Saltin. The research results, as published by Burfoot in his article, were "purged of uncertainty," asserts Hoberman, whereas the research results eventually published by Saltin "raised more questions than they answered."

Hoberman goes on to argue that Saltin subsequently cast a shadow on the whole muscle fibre theory by calling into question the representative character of any given muscle biopsy sample. This did not, however, dampen the splash made by Burfoot's article, which has subsequently been featured by right wing proponents of differential evolution and innate racial differences such as the Canadian psychologist J. Phillip Rushton.

"One of the cleanest tests" of differential evolution, writes Rushton, "comes in the realm of athletic competition." If we can demonstrate that the races have evolved differently in this realm, then it is more likely that they have evolved differently in other realms as well. As it stands, of course, this is a relatively weak inductive argument.

Hoberman points out, however, that racial folklorists also fashion a deductive argument using the premise of differential athletic aptitude in conjunction with the Darwinian "law of compensation," which postulates an inverse evolutionary relationship between brain and brawn. This deductive argument would take the following form:

Premise 1: If Africans have evolved with superior physical aptitude, then they have evolved with inferior intellectual aptitude.

Premise 2: Africans have evolved with superior physical aptitude.

Conclusion: Africans have evolved with inferior intellectual aptitude.

Hoberman then examines the scientific literature on nerve fibres, muscle fibres, bone density, human growth hormone, testosterone, hip width, leg length, calf size, and so on, and their possible links to athletic performance and intelligence. On the basis of his examination of the literature, he concludes that we should reject the folklorists' argument.

In short, it is Hoberman's position in Part III of *Darwin's Athletes* that genuine science has not yet demonstrated that biological differences in racial athletic aptitude exist and that it has provided no evidence to support a Darwinian Law of Compensation. He writes:

> This is not to say that bioracial differences of athletic significance do not exist. It is possible that there is a population of West African origin that is endowed with an unusual proportion of fast-twitch muscle fibres, and it is somewhat more likely that there are East Africans whose resistance to fatigue, for both genetic and cultural reasons, exceeds that of other racial groups. But these hypotheses are not even close to scientific confirmation, and there is not scientifically justified reason to tie such plausible athletic traits to mental aptitudes, despite the promptings of the racist heritage that says we should.[774]

When discussing the Darwinian Law of Compensation, Hoberman refers to the "classic muscle/mind trade-off Darwin presented in the *Descent of Man*."[775]

The only mention of a law of compensation on the part of Darwin that I have been able to find occurs in *The Origin of Species*, where Darwin takes pains to argue that "there is hardly any way of distinguishing" between the effects of a purported law of compensation and the effects of natural selection, whereby one part of an organism develops over time through use and another reduces over time through disuse.[776]

In Part III, Hoberman argues that we must overcome our fear of racial biology if we are to combat racism, and he states in a footnote that he is "inclined to agree" with the view that biological race can be studied independently of ethnicity.

Although Hoberman is careful to argue for the superiority of the life of intellectual expression ('theory', 'argument', and 'concept-creating') over the life of physical expression ('sport') in terms of its utility in a knowledge based society, *Darwin's Athletes* in fact represents intellectual expression as *intrinsically* superior to physical expression. This is manifest in Hoberman's assumption that the value placed upon physical expression in African-American life is to be viewed first and foremost as a compensation for the limitations placed upon blacks by slavery, Jim Crow, and racism. It is further manifest in Hoberman's assumption that African-American assertions of a non-intellectual life of the mind of 'genius' in sport are simply compensatory efforts to "convert black physicality into intelligence."[777] Finally, it is manifest in Hoberman's willingness to accept the traditional identification of the physical with the primitive.[778] Hoberman at no point argues for the intrinsic superiority of intellectual expression as a way of being in the world, he simply assumes it in much the same way that Freud assumes that sublimating the instinctual domains of the id through art and science is higher and finer than sublimating them through business and commerce.

In my view, this assumption of the intrinsic superiority of intellectual expression on the part of Hoberman is a source of major resistance to his work on the part of those for whom physical expression, as a way of being in the world, is as intrinsically valuable as intellectual expression. How would Jewish people, for example, countenance the assumption that the value they place upon intellectual expression is to be viewed first and foremost as a compensation for the limitations placed upon Jews by the diaspora? Until we confront the issue of the relationship between intellectual and the physical expression in human life, we cannot fully address the issues of race and sport that John Hoberman so courageously and insightfully raises for us in *Darwin's Athletes*.

Chapter 31
Sport and the Good Life

We know too much and feel too little. At least we feel too little of those creative emotions from which a good life springs.
Bertrand Russell

We are shaped and fashioned by what we love.
Goethe

Hadley Cantril, in his book 'The Why and How of Man's Experience', quotes a passage from A.A. Milne's 'The House at Pooh Corner' in which Christopher Robin asks Pooh:
"What do you like doing best in the world, Pooh?"

> "Well," said Pooh, "what I like best"—and then he had to stop and think. Because although Eating Honey was a very good thing to do there was a moment just before you began to eat it which was better than when you were, but he didn't know what it was called. And then he thought that being with Christopher Robin was a very good thing to do, and having Piglet near was a very friendly thing to have; and so when he had thought it all out, he said "What I like best in the whole world is Me and Piglet going to see you and You saying 'What about a little something' and Me saying 'Well, I shouldn't mind a little something, should you, Piglet?' and it being a hummy sort of day outside, and birds singing."

Cantril quoted this to illustrate his theme that one at least of the essential components of satisfaction in life is social, which is the enjoyment of other good things in good company. It is the enjoyment of the taste of honey in good company.

"To cultivate the pleasures of my senses was throughout my life my main preoccupation. I have never had any more important objective."

If this had not been said by Giovanni Jacobe Casanova de Seingalt, it could have been the expression of a quite reputable philosophy and psychology of aesthetics.

Much of interest to the philosopher of sport is apparent even on a cursory reading of Ortega's "Man the Technician."[779]

According to Ortega, despite the many ways in which different technologies embody different practical and intellectual interests, that there is an essence of technology that underpins all its various and multiple manifestations. Technology in its essential form is always part of an attempt by human beings to free themselves from the tyranny of labour aimed at meeting the basic necessities of human existence and reproduction.[780]

It is based on reducing the time and effort spent in activities of this sort in favour of living what a particular group of people perceive as the good life. In each period, the good life involves what humans within a given cultural context feel to be not only the most pleasurable way to live but also the way of life that seems to them to involve a reaching towards what they see as the most sublime elements of what is meant to be a human being.

Technology aims at achieving a condition in which human pleasure and aspiration can be made possible by freeing up time and energy for 'nonessential' activities. But if this condition is met, even minimally, a problem arises for the social group in question. Ortega asks, "What is man to do after he has eliminated [at least to some degree] what nature compels him to do? What fills his life? For doing nothing means to empty life to not live."

The answer Ortega gives to these questions as to the existential dilemma of *filling one's time up* is that although there are many ways of achieving this, some of the most telling from the philosopher's point of view are sportive activities. These are simultaneously both means of achieving individual enjoyment and expressions of the group's highest aspirations as to what human life should be like. Despite the many different ways in which diverse communities can conceive of the good life, Ortega argues that a common thread runs through all conceptions. They are all based on some sense of life lived free from external constraints, an autonomous rather than heteronomous form of existence in Kant's terms.

Such autonomy takes the form of a game like attitude towards life, wherein humans freely decide to adopt and to live by what are, in effect, the conventional and arbitrary rules of a game. The good life, whatever its particular definition in a particular society, is always glimpsed in those

"Moments when the pressures and responsibilities of life are shuffled off and man indulges in the diversion of a game. This...shows strikingly to what degree the human programme of life can be extra-natural. For games and their rules are sheer invention in comparison with life as it comes from nature's own hands [where the latter comprises the] enforced existence of struggle with his [man's] environment. [The good life is glimpsed when the individual is involved] in the unreal and purely imaginative orbit of his games and sports...when people are in the mood to play we may assume that they feel comparatively safe regarding the elemental needs of life...In this state of mind man delights in his own magnanimity and gratifies himself with playing fair...He will not cheat, for cheating means to give up the attitude of play: it is 'not cricket'. The game, it is true, is an effort but an effort which is at rest in itself, free from the uneasiness that hovers about every kind of compulsory work[781]."

Humans become most human when at play, because they have *chosen* to adopt the rules under which sportive activity occurs, rather than operating wholly under the demands of life conditions not of their choosing. By entering into sportive activities, humans give their "mundane destiny the grace of a game[782]."

This grace, however, is not characterised by unfettered enthusiasm or Dionysian revels. Instead, Ortega makes a distinction between the rule-governed nature of sports and the capricious nature of pure play. Sports are a *via media* between the "overbearing seriousness of life" and the "strenuous effort of living" on the one hand, and the pleasure involved in the "irresponsible triviality of play" on the other hand[783]. This distinction Ortega takes as further evidence for the assertion that humans are in part 'extranatural' because although both humans and animals can engage in play, the latter cannot engage in rule-governed sports or at least cannot *fully* engage in such activities. There is an echo here of the work of the later Wittgenstein[784]; see also [785] where it is precisely the rule-governed nature of human life that makes it explicitly *human*.

Thus far our reconstruction of Ortega's argument may not seem to have progressed much beyond a fairly standard claim, based on an assumed divide between culture and nature, that argues that sports are part of culture rather than nature because they are rule-governed, and rules are cultural because they are the 'arbitrary' constructions of a particular group in a specific time and place. However, Ortega's viewpoint goes beyond the banal idea just outlined in that it explicitly links technological innovation to sportive activities, with the former being seen both as the 'material' condition of possibility of the latter and also, in an indirect way, as being motivated by a human desire to allow the latter to occur. This position could be said to be a significant elaboration and refinement of the trivial argument that sports are cultural rather than natural phenomena.

Chapter 32
Sport and the State

No man is free who is not a master of himself.
Epictetus

All his life long, man is imprisoned by our institutions.
Rousseau

The Republican form of government is the highest form of government; but because of this it requires the highest type of human nature—a type nowhere existing.
Herbert Spencer

The state was made for man, not man for the state.
D.H. Lawrence

Government is the problem not the solution.
Matt Ridley

Politics (patriotism) is the last refuge of the scoundrel.
Samuel Johnson

The life of a nation is secure only while the nation is honest, truthful and virtuous.
Frederick Douglas

   Sports are central to Ortega's general phenomenology of human existence. They also play a crucial role in his conceptualisation of concrete social institutions. In particular, on the nature of the generic mechanism of political organisation, the State, and man as a political animal.
   He sees this institution as having historical roots that are fundamentally sportive in origin. His treatment of the nature of political life is centred around a consideration of the hidden sportive beginnings of the state.

Ortega's understanding of the state rests on a distinction between mundane, necessary labour on the one hand, and the activities pursued freely in the time opened up by technologies on the other.

Ortega[786], drawing on biological notions of random mutations, argues that all organic life, both human and animal, is characterised by "two great classes of activity, one original, creative, vital par excellence—that is, spontaneous and disinterested; the other of utilitarian character, in which the first is put to use and mechanised."

In other words, spontaneous and joyful creativity and the sober, instrumental guiding of particular activities towards particular ends are the two poles of being.

Whereas work and technology were seen by Ortega to be the bases for, and to make possible, leisure, in the mediation on political life, the casual order is reversed. "Utility does not create and invent; it simply employs and stabilises what has been created without it."

Innovations and novelties, unguided and unplanned, are the primary elements of organic life, with instrumental utilisations of these being seen to be purely secondary and derivative.

As Ortega puts in, "in the history of every living entity we shall always find that life at first is prodigal invention and that it then selects among the possibility thus created, some of which consolidate in the form of useful habits[787]."

It is from this basis that Ortega argues for the centrality of sportive practices in human life, for the latter are equated with the unfettered, creative dispositions inherent within the human version of organic existence:

> "If the classic instance of the obligatory effort which strictly satisfies a need is to be found in what man calls work, the other, the effort *ex abundantia cordis* [from the overflowing abundance of the human heart], becomes most manifest in sport…We thus feel induced to invert the inveterate hierarchy [which places instrumental labour over 'leisure' activities]. Sportive activity seems to us the foremost and creative, the most exalted, serious, and important part of life, while labour ranks second as its derivative…life, properly speaking, resides in the first [i.e. sportive activity] alone; the rest is relatively mechanic and a mere functioning[788]."

It is Ortega's rejoinder to Marx's notion of homo faber—the human being as essentially the fabricator of goods.

For the early Marx[789], it is useful labour that, under the right social conditions, gives joy to the human heart.

Ortega, by contrast, makes two arguments opposed to the position of Marx. First, labour is for man a necessary evil and the human seeks always to relinquish it in favour of the more pleasurable pursuits of leisure and sports. Second, homo faber is always but the impoverished and unhappy cousin of homo ludens, man as player of games. It is in the pursuit of sportive activities, not instrumentally guided labour, that man truly finds contentment.

There is a sharp division between their respective accounts of the origin of the state and political life.

For Marx, the state arises after the period of primitive communism, when there appears a class of people who control the labour of the wider population of producers. The state is a regularised mechanism of domination by the elite group over the rest of the population.

But for Ortega, the state does not appear for reasons associated with labour and control thereof, but for reasons "precisely the opposite of a reaction to imposed necessities."

The state does not, as on the Marxian account, arise for utilitarian and rationally purposeful reasons (such as the domination of one class by another) but for irrational reasons, which are themselves part of the key principle of life, the constantly flowing superabundance of energies—in this case, human energies—wilfully expressing themselves. "In the beginning there is vigour, not utility[790]."

This vigorous outflow of energy Ortega locates in the groups of young men that are seen as a key part of primitive societies. "It was not the worker, the intellectual, the priest...or the businessman who started the great political process, but youth (in the role of) the warrior, the athlete[791]."

According to Ortega's account, the first form of organisation proper in human society is comprised not of social classes but of age groups: the young, the mature, and the elderly. The group of young men is the most powerful group in such a society, for they have bonded together for the pursuit of common interests.

Ortega sees this as a 'natural' function of the transition from boyhood into adolescence: "The isolation of infancy breaks down, and the boy's personality flows out into the coeval group...he is absorbed by the anonymous personality of the group which feels and wishes for him...youth is the season of friendship[792]."

Members of the group of youth "decide to unite and to live together— obviously not for the purpose of remaining idle" but for achieving what they see as certain pleasurable aims, the primary of which is the abduction of women from other tribal groups[793].

It is here that the roots of the State lie, for in order to be successful in their aims, the young men must fight the men of the other tribes.

Political organisation arises because "war calls for a leader and necessitates discipline, thus bringing into being authority, law and social structure." Warfare also requires discipline and Ortega regards *ascesis*, a condition described by a Greek word that refers to the self-imposed hardships undergone by athletes in training, as being simultaneously sportive and military. Asceticism is seen as having its roots in primitive youth associations, in which play, sports, the erotic drives of heterosexuality, and the exercise of martial skills are all mixed up together.

Overall, it is the effervescence of youth, rather than directly utilitarian reasons, that begets the state, and thus its most basic form should be seen as "an association of the young for the purpose...of performing all sorts of...exploits. Rather than a parliament or a cabinet of bigwigs, it resembles an athletic club."[794]

It is not the empirical validity or otherwise of Ortega's views that is at issue here.

Instead, the point is to demonstrate how Ortega's metaphysics places so much emphasis on the 'sportive' aspects of human existence. He deployed it to explain more concrete human phenomena.

Ortega, himself regarded his meditations on the beginnings of man as a political animal "as an instance of the creative power inherent in the activity of sport."[795]

His focus on the explicitly sportive origins of a phenomenon such as political life is a unique one among major European thinkers.[796]

What is interesting about his opinion that politics originated as play, is that its origin reflects its current status it is no more than a game and not just in the minds of the cynical few. Morality is virtue in action and has its own morality however, we have to bear in mind that self-interest is a predominant characteristic in man. In a speech a hundred years ago, Lord Birkenhead said "Self-interest is and ought to be the mainspring of human conduct." He continued "The world continues to offer glittering prizes to this have stout hearts and strong swords." As FE Smith he was a brilliant advocate who won glittering forensic prizes. It is not surprising that he thought this was the hallmark of virtue and the mainspring of human conduct. But what he said I have no doubt is the hallmark of outstanding athletes.

Chapter 33
Communicative and Strategic
Action in Sport

Without knowing the force of words, it is impossible to know men.
Confucius

In the beginning was the word.
John 1.1

For words are wise men's counters they do but reckon by them; but they are the money of fools.
Thomas Hobbes, Leviathan

A powerful agent is the right word. Whenever we come upon one of those intensely right words...the resulting effect is physical as well as spiritual and electrically prompt.
Mark Twain

Solon used to say that speech was the image of action.
Diogenes Laertius, circa 200 AD

Language is the dress of thought.
Samuel Johnson

We know more than we can tell.
M. Polanyi

Habermas distinguishes between two forms of social action. Communicative action demands that the social agents come to an agreement over the nature and legitimacy of their respective actions. The alternative to communicative is strategic action.[797] While communicative action presupposes that I respect my fellow agents as competent and autonomous beings, and we mutually rely upon each other's competence in order to interact successfully, in strategic action 1 objectify others. The other becomes a means to the attempt to manipulate the other, if necessarily against their will. The other becomes a resource that I strive to exploit in order to further the achievement of my goals. Thus if I use either acts of violence or deceptions to coerce the other to act as I want, then I am acting strategically. In either case, I am undermining his/her autonomy, by restricting the possibility of free choice of action. I prefer the view of Pope in his Essay on Criticism "words are like leaves and where they abound much fruit of sense beneath is rarely found."

Edgar says:[798]

"Participation in sport would appear at first glance, to demand strategic rather than communicative action. I do not communicate with my opponents; I attempt to better them through greater physical force or deception. The ultimate objective of the game is to undermine their authority and sporting competence to the degree necessary for my sporting competence to triumph. Sport manifests both overt strategic acts of (legitimate) violence and covert acts of deception."

Habermas offers a critical interpretation of Wittgenstein's use of the image of a language game in exploring the entwining of language and action.[799]

The rules of a game, on Habermas's account, stood proxy for the grammatical rules that govern the meaningful use of words to actions, in an utterance. While Habermas recognises the importance of Wittgenstein's analysis of language games, he criticises him for, in effect, distorting language to turn it into a strategic as opposed to a genuinely communicative competence.

Habermas claims that a game cannot refer to anything outside itself. I do not agree e.g. the boxing matches between Joe Louis and Max Schmelling in 1936 and 1938 clearly carried a significance that went beyond a mere sporting context. There is no doubt that while a particular contest, such as a particular boxing match need not have any significance beyond being a contest exemplary of that sport, some contests can legitimately be ascribed with a significance that goes beyond the intrinsic meaning of the sport. Certain sports open themselves up to the possibility of particular contests acquiring this significance. Furthermore, certain sports acquire the status of national sports, precisely because they allow their exponents to exercise certain competences that are of significance to the wider social culture, be this rugby union in Wales or New Zealand or cricket in Australia or in the West Indies or bullfighting in Spain or ice hockey in Canada or baseball in the USA.

A model of such interaction (or dialectics) might be provided in Adorno's philosophy of music.[800] A formalist account of music would suggest that meaning within music is purely intrinsic, which is to suggest that a particular event within the performance has significance not through its relevance to some external event (or even through the evocation of an emotion) but rather through its relationship to other events within the performance. A sequence of chords, for example, repeats or develops upon an earlier sequence. Thus an event in music has meaning much as an event within a game, for its meaning is determined by convention and constitutive rules of the game. In my view, the body is more communicative than words. The larynx is more important than the tongue. I will return to this here later.

Adorno's point, though, is that these purely formal elements of music are typically historically derived from socially significant activities or patterns of thought (such as dances, marches and religious celebrations). His contention is that certain approaches to composition (in what typically might be termed 'art' music), because they demand a rigorous reflection on and pursuit of logic inherent in these forms, come to comment on those primal social events that are, as it were, embedded in forms. In contrast to the formalist music critic, who focuses exclusively on meaning. Inherent to the musical form, the socially informed or dialectical critic will seek to make explicit the social comment that is ingrained in that form. A parallel to sport could be constructed, by suggesting that certain contests focus the social history embedded in the rules of the sport, allowing either the participants or commentators to recognise the extrinsic significance. Boxing is by no means unique as a sport in carrying with it a complex history of ideological and class struggle. CLR James's 'Beyond a Boundary' to which I have already referred is a philosophical, political and historical analysis of the significance of cricket in the West Indies and the impact of colonialism on the development of the West Indian's identity and culture.

Chapter 34
Constraint Theory

The greatest fault of a penetrating wit is to go beyond the mark.
La Rochefoucauld maxim 377

Most people live whether physically, intellectually or morally in a very restricted circle of their potential being. We all have reservoirs of life to draw upon of which we do not dream.
William James

I once admitted to my shame that football was a brutal game.
Alfred Cochrane

Constraint theory is Jon Elster's unique attempt to develop, from within methodological individualism, a conception of rational action that is thick enough to capture the relationship between such constraints and preferences.[801] Although it has many dimensions and applications, the seeds of constraint theory can be found in Elster's *Ulysses and the Sirens*, wherein he argues that all human action is the result of "two successive filtering devices:"

> The first is defined by the set of structural constraints which cuts down the set of abstractly possible courses of action and reduces it to the vastly smaller subset of feasible actions. The second filtering process is the mechanism that singles out which member of the feasible set shall be realised.[802]

A thorough examination of the nature and functions of both of these 'filtering devices' or constraint mechanisms in human action is what distinguishes Elster's work from more atomistic accounts of rational-choice-guided action. Elster is interested not only in the rationality of individual preferences and choice-maximising action but also in (a) the role that constraints play in shaping those preferences and choices; (b) the fact that many constraints on human action are not merely structurally prefiltered but can, themselves, be reflexively *selected* by individual actors; and (c) the forms of choice and action, such as creative action, wherein these two filters are not successive but interactive.

The mythological hero Ulysses is Elster's model of a constraint theorist at work. For in his encounter with the Sirens, Ulysses deliberately has himself constrained by his crew.[803] In this way, Ulysses' own intentional reduction of his feasible set (a choice of constraints) frees him to hear the call of the Sirens (a choice within those constraints). In Elster's study, *Ulysses Unbound*, he attempts to characterise more fully this interaction of choice of constraints and choice within constraints[804] in an analysis of creativity and constraints in the arts, wherein artists bind (and often seek to unbind) themselves in various ways. In this study, as well, Elster provocatively suggests but does not develop the relevance of constraint theory for the philosophy of sport.[805]

Sport as Constrained Maximisation

In his constraint-theoretical account of art, Elster argues that artistic creation is "guided by the aim of maximising aesthetic value under constraints[806]." In art, the philosophically interesting constraints are not so much objectively given or 'hard' constraints such as technical or physical limitations but, rather, 'soft' constraints or conventions—those "restrictions that *constitute* a specific genre[807]."[808] On Elster's account, soft constraints or conventions in the arts are constitutive rules.[809] Adhering to them, like adhering to the rules of chess, does not merely normatively regulate artistic endeavours; it defines such behaviour as an artistic endeavour. Indeed, despite the dream of high and postmodernism, Elster demonstrates that in the absence of soft but constitutive constraints no coherent account of artistic creation can be given. Sonnet writers are, at least at one decisive level, simply artists who have chosen to bind themselves by working within the conventions of the sonnet form. The absence of the constraints of the sonnet would mean, quite literally, the impossibility of sonnet writing. "Expression needs form," as Elster rightly insists.[810]

Yet, Elster is no mere formalist. His analysis concedes that sonnet writers, if they are to be composers who achieve anything of aesthetic value or excellence, cannot simply adhere to or unintentionally 'embody' their chosen constraints. There is always more to writing a sonnet than simply following the rules of sonnet writing. Sonneteers must intentionally seek, through reflexively monitored revisions, continued variations, and deliberate experimentations, to maximise their creativity and skills of written expression within their elected constraints; they must, that is to say, practise the rational art of what Elster calls 'constrained maximisation'[811] to excel in their creative production.

Such is the case with competitive sportsmen as well. They too, are engaged in a more or less complex practice aimed at constitutively constrained maximisation. In playing a particular sport, they, like artists, have chosen their soft constraints and deliberately adopt the constitutive conventions or rules of a specific game. Following the rules of tennis is what it means to play tennis and be a tennis player, just as following the conventions of the sonnet is constitutive of sonnet writing and being a sonnet writer. Of course, *how* one plays and competes—the quality of the choices and the skilfulness and creativity of the moves one makes—within those rules is what makes one a better or worse tennis player. In this respect, Roger Federer is Shakespearean in his athletic realisation of constrained maximisation in the sport of tennis.

The basic argument by analogy being introduced here is simply that, at least at one level, competitive sports are athletic genres in which individuals strive to achieve constrained maximisation. The inverse of such a striving would be something like engaging in sport (or art) as a leisure activity or recreational hobby wherein the only objective is constrained diversion. Maintaining a distinction between constrained maximisation and constrained diversion does not necessarily imply that there are any essential differences between a 'serious athlete' and 'recreational player' or between athletic action and merely play.[812] Instead, from the rational-choice perspective, such a distinction simply makes explicit the relative potential for increasing individual skill and creativity in each case. Like the accomplished sonneteer, the competitive athlete's committed rational (utility-maximising) orientation towards his or her constraints characteristically enables skill levels and forms of creative action that are difficult if not impossible to realise in the pleasure-maximising activity of constrained diversion wherein individual relaxation and recreation, and not excellence and innovation, is sought.[813]

There is no need to push this analogy between constrained maximisation in art and constrained maximisation in sport too far here. There are many distinctions to be made, but two in particular seem especially relevant in this context. First, the orientation athletes adopt in relation to the rules of their chosen sport is not identical to that of artists in relation to the conventions of their chosen genre. For example, artists often self-consciously seek to violate or reject the conventions of their genre by subverting or exceeding them. This is especially the case in much contemporary art. Athletes, by contrast, must endeavour to maximise their creativity and skill levels within existing constraints. Indeed, the 'deconstruction' of soft constraints or the establishment of new ones is not, generally speaking, one of their aims.[814]

Second, the relative constitutive quality of soft constraints (rules) in sport and conventions in art are quite different. For example, today almost anything and everything counts as 'poetry'. But what counts as tennis remains playing by the rules of tennis. And what counts as playing tennis as well is an individual's ability to maximise his or her choices and actions from within the constitutive constraints of tennis. The conclusion to be drawn from this difference in the nature of soft constraints in sports and the arts is not that the elimination of all (or even most) soft constraints further enables creativity and skill. Rather, the opposite is typically the case: In eliminating constraints rather than choosing them, creativity and skill are undermined rather than potentially optimised. What sport, unlike contemporary poetics, makes clear is that where a certain number of soft constraints obtain, more possibilities to tighten or loosen the bounds of action exist.

The binding that sportsmen regularly engage in is, thus, at once more demanding and potentially more enabling than the conventional bounds of artists, at least those of a high or postmodern sensibility. Consequently, it is understandable that sportsmen rarely attempt constrained maximisation of their chosen athletic genre on their own, but rather routinely seek out additional constrainers—coaches, trainers, advisors—to help them optimise their constraints.[815] This important use of additional constrainers in sports also helps sharpen the distinction between constrained maximisation and constrained diversion introduced previously. Individuals seeking constrained diversion in sport do not require additional constrainers to realise their goals. Indeed, it would be odd to argue that sport hobbyists *need* coaches or trainers simply to enjoy their athletic diversions; the term 'hobby coach' rings counterintuitive.

In the light of the important role of constraints and constrainers in athletics, there is a constraint theory of sport. Such a theory holds that athletic endeavours involve the interplay of three types of individual choices designed to achieve constrained maximisation. These are (a) choice of constraints (electively self-binding), (b) choice of constrainers (electing others—coaches, trainers, etc.—to bind one's self), and (c) choice within constraints ('a') and constrainers' constraints ('b').

The central task of a constraint theory of sport should be to make explicit and describe how this dynamic matrix of choices and constraints works in case studies of particular sports.

This theory is not without important antecedents in the philosophy of sport. In fact, two core elements of game playing singled out by Bernard Suits in his "The Elements of Sport" are consistent with aspects of a constraint theory of sport. Specifically, constraint theory shares with Suits a conception of the rules of sport as constitutive rules. It also shares with Suits a general sense that game playing entails that game players adopt a certain attitude qua game players towards those rules—or what Suits describes as "the lusory attitude." Athletes can be defined, at least in part, as individuals who take up a lusory attitude vis-à-vis constitutive rules: They choose their constraints or knowingly accept constitutive rules "just so activity made possible by such acceptance can occur."[816]

Yet, a constraint theory of sport is not merely an attempt to define the conditions of the possibility of game playing. It is also an attempt to explain the conditions and actions of game playing that make for skill-maximised game-playing endeavours. For athletes are, on this account, not merely game players but also skilled constraint maximisers. In this respect constraint theory addresses a central element of sport as game playing that Suits leaves largely unexamined. In particular, a constraint theory of sport provides an explanation of what Suits calls, somewhat misleadingly, "rules of skill." [817] From the perspective of the constraint theory of sport, such 'rules', which for Suits are expressed in an injunction such as "keep your eye on the ball," are not really rules at all. Rather, they are those skill-maximising embedded actions that result from choices within constraints and constrainers' constraints and characteristically make for better or worse athletic undertakings.

In "The Elements of Sport" Suits makes clear that his purpose "is to define not well-played games, but games." Constraint theory should thus be viewed as a supplement to rather than a critique of Suits's work. That supplement could be summarised as follows: The function of the lusory attitude (choice of constraints) adopted by athletes is not merely to make game playing possible but also to make for well-played games by enabling choices within those constraints that maximise individual creativity and skill.

A Constraint Theory of Boxing

In the clearing stands a boxer,
And a fighter by his trade
And he carries the reminders of every glove that laid him down
Or cut him till he cried out in his anger and by shame
I am leaving, I am leaving
But the fighter still remains.
Simon and Garfunkel

Remember, the mind is your best muscle. Big arms can move rocks, but big words can move mountains.
Rocky Balboa

Everyone has a plan until they get punched in the mouth.
Mike Tyson

Claims about boxing's history or its role in society provide abolitionists or reformers with important considerations but they don't in the end help to establish the moral status of the sport. An argument that does attempt to establish this is a very simple approach. It claims that boxing is ethical because boxers want to do it. This argument raises questions about much more than sport—the tension between individual preferences and community values.

As rough games go, boxing is one of the oldest and most straightforward. It is "the simplest pageant of all to men fight, rest a minute and fight some more."[818]

The sport of boxing would appear an unlikely candidate for constraint theorising. Indeed, pugilism typically conjures up images of human action far more Hobbesian than Elsterian. In boxing it is not the rationality of constrained maximisation but rather an unbound state of nature that appears to prevail. The boxing ring is frequently alluded to as a kind of observable site of raw egoism in which the virtues of force and fraud and the war of all against all appear in miniaturised form. Boxing, it is often argued, "makes violence central"[819] and entails a mean-spiritedness that is morally unintelligible. The raw struggle for survival leaves boxing vulnerable to social criticism, because boxers survive only by hitting their opponents to the point of eliminating them. But that is not how Italian-Welsh boxer, Joe Calzaghe, saw it. "The reason why I fight is simple: I like to win. I see boxing as an art. Hit and not be hit. Yes, I can take a hard punch to the chin, I've been cut and I've been knocked down but I would never quit. I have the heart of the fighter. I don't crave to be in a war, however, for the sake of being in a war. It is simply about winning."[820]

Sports announcers, popular media, and often boxers themselves do much to provoke this image of boxing as a sport where vicious human actors engage in violent and harmful physical conflict. Ringside commentators and 'experts' commonly describe boxing matches as 'street fights' or 'brawls' and boxers as 'street fighters' or 'brawlers'. The 1974 heavyweight championship fight in Zaire between Muhammad Ali and George Foreman was called the Rumble in the Jungle; an episode of the popular American television series, CSI, opens with a boxing ring as a crime scene in which a professional boxer has been murdered in the course of a bout; and the popular Rocky film series presents the ring death of one of its lead protagonists as nothing less than a murder to be avenged in a future match. To this short list we could easily add Mike Tyson's (and others') wild ring antics or the desire that 'Iron Mike' once expressed at a prefight press conference to eat the children of heavyweight champion Lennox Lewis. Mike Tyson also said "my business is to hurt people." Other sports involve 'elimination' when we play most games, you stand between me and my goal if "I cannot eliminate you." The elimination is mostly symbolic. Momentary elimination is enough in games like football or rugby. Each time I block you your passing I have taken you out of the game. The touched player in baseball illustrates another take on this symbolism. The runner is symbolically 'killed' or rendered inert by a ritualistic contact.

The apparent irrationality and viciousness of boxing and of some boxers is often conjuring in such images and sound bites. But these images and sound bites badly obscure the elaborate matrix of choices and constraints in which boxers act as rational practitioners of their chosen athletic genre. For rumbles and street fights are forms of human action by definition devoid of rationally chosen constraints. Actors engaged in such actions are limited only by the arbitrary motivations of their own emotions, the emotions or threats of others, circumstances, or some combination thereof. In this sense, then, rumbles and street fights are paradigmatic of rationally unbound action. They lack, precisely, the bounds of rational choices designed to maximise creativity and skill that define and make possible athletic endeavours as such.

In boxing, by contrast, one seeks to restrict choice in ways that enable creativity and skill maximisation. The fact that in so many prize-fights individual boxers fail to achieve constrained maximisation is not the result of an absence of constraints. Instead, such failures are primarily the result of the suboptimal soft constraints constitutive of prize fighting. To see why both amateur and professional prize fighting suffer from suboptimal constraints we need only to consider here a few of the many soft constraints of each and the kinds of choices within constraints they enable.

In professional boxing soft constraints are also likely to produce suboptimal athletic endeavours, but for opposite reasons. Longer rounds with longer bouts, round scoring, and the absence of headgear tend to promote a style in which boxers do little other than seek to land a 'big blow' or knockout punch. A telling example of this kind of suboptimization can be seen in the so-called rope-a-dope method invented and popularised by Muhammad Ali during the middle and late years of his professional career. Such a style—in which a boxer leans against the ropes with guard up, enduring round after round of blows in hopes that the opponent will eventually exhaust himself or herself and become easy prey to a knockout punch in later rounds—may, at times, be an effective way to win a professional boxing match. But this amounts to the victory of suboptimal constraints that foster artless inaction, not constrained maximisation.

Muhammad Ali was the master of constraint theory as his rope a dope display so clearly manifested. He exposed himself to punishment in order to weaken his opponent who became a soft target. He then meted out his punishment. That was constraint maximisation par excellence.

As an individual whom I met as a young barrister being one of his legal representatives I was overawed by his charisma and panache. The case involved a dispute with his infamous manager Don King. It was said of Don King that his haystack hair had been grown in order to hide his horns! Whether that was true or not Don King had subjected Ali to inequitable financial constraints which made his return from boxing at that time in his early years disproportionate to his success as a prodigy. I am pleased to say that with assistance of more senior advocates, we were able to remove those constraints so that the fruits of his magnificent, pugilistic talents were enjoyed by him and not unfairly dispersed to others. As I have earlier written Ali was a man of immense charisma both as a man and a professional sportsman, whose ability and personality maximised his ability to overcome all constraints operating on him. He was a genius and hero rolled up into a human being—a Nietzschean overman.

This chapter has drawn on the work of Jon Elster to try to introduce a constraint theory of sport. The overarching thesis of that theory is that competitive sport is primarily an endeavour to maximise skills within the constraints of a chosen athletic genre. Following Elster, I have called such an endeavour an attempt at "constrained maximisation." I then went on to acknowledge that this account shares with Suits's work a conception of constitutive rules and the importance of the 'lusory attitude' for athletic endeavours. It also shows how a constraint theory of sport supplements Suits's discussion of rules of skill and the function of the lusory attitude in important ways. Conceptualising athletes as constraint maximisers not only illuminates what it means to play a game but also sheds light on how skilled excellence in game playing is realised. The function of the lusory attitude (choice of constraints) adopted by athletes is not merely to make game playing possible but also to make for well-played games by enabling choices within those constraints that maximise individual creativity and skill.

Boxing is a case study for a constraint theory of sport. It illustrates the nature of constitutive rules, or soft constraints, and some of the choices enabled within the constraints characteristic of pugilism. Under constraint theoretical analysis, it has revealed that boxing, despite popular stereotypes, does not suffer from an absence of constraints but, rather, from suboptimal constraints in both its amateur and professional versions. From the perspective of constraint theory, it is in sparring—where the constraints of a chosen constrainer play a vital role—that an optimal tightness of bounds and constrained maximisation are most readily realised in the sport of boxing. J.S. Mill famously declared that intervention in personal choice is permissible only where my private actions interfere with your legitimate interests.[821]

A functionalist-inspired critique might direct its attention to the presupposition of voluntary choice on which a constraint theory of sport appears to be based. It would insist that a constraint theory of boxing, and constraint theory more generally, is sociologically naïve. For the notion of voluntary 'choice' on which constraint theory depends assumes a kind of abstract autonomy that is nowhere to be found in the material life of everyday agents, especially those who 'choose' to become boxers. A society that values personal choice can also seek to set limits on risk based even on the nature of freely chosen risks.

A moralist critique of constraint theory might also focus on the issue of voluntary choice, but its criticism would likely move in the opposite direction. That is to say that it would grant that constraint maximisers in general, and boxers in particular, are indeed pursuing their individual goals with relative autonomy. But it would maintain that the ends realised in such a pursuit have no value or are, at least in the case of boxing, legitimated and sanctioned in ways that are morally forbidden in other sports and everyday life. The argument, in other words, would be that intentionally striking others in vicious attempts to harm them is, in fact, morally decisive for criticising what boxers voluntarily choose to do.[822]

One rejoinder to functionalise would be to argue that it is a mistake to overburden what Elster calls the first 'filtering device' of human action. A strong holism of this sport unjustifiably privileges and makes determinate the initial filtering (constraint) mechanism of human action. In so doing, it reifies structural constraints at the expense of individual choice, creativity, and autonomy.[823] It also obfuscates the attitude of the individual constraint maximiser, as well as the nature and function of soft constraints (or constitutive rules) in sports. A constraint theory of sport can readily accommodate the intuition that choosing to pursue constrained maximisation as a boxer, rather than as, say, a golfer or bowler, is never merely the result of disembedded preferences. Yet it can do so without degrading boxers (or golfers or bowlers) to mere bearers of structural forces. Like all other human actors, boxers are embedded agents. This means that they reflexively choose their constraints not from a set of abstractly possible courses of action but, rather, from a feasible subset power of structural constraints but, rather, soft constraints—the constitutive rules of amateur and professional boxing—that make constrained maximisation in the sport so difficult to achieve.

An immediate response to the moralist is somewhat more difficult for constraint theory to formulate. Evaluating the moral merits or demerits of boxing, or any other sport, for that matter, is not something a constraint theory of sport is designed to accomplish. Yet, the constraint theory of boxing developed here does provide some basis for speculation. For example, one could begin to examine more closely the cooperative ethos cultivated in sparring. Part of what the constraint-theoretical analysis of sparring has made explicit is that one spars with a partner, not an opponent. What this suggests is that the mutual self-binding of sparring fosters not viciousness and violence but a unique kind of sociability—a form of reflexive social cooperation not without potential moral value. Further constrain-theoretical discussions of sparring in a boxing gym would need to analyse more fully the sociability of sparring to ascertain whether there is, in fact, something akin to a moral grammar at work in one of boxers' most common but least studied activities.

According to one writer and more generally, "boxing inculcates the value system and behavioural trappings of a 'civilised' society. The irony should not be lost. Those who regularly call for boxing to be banned are often the same people who abhor the rough working-class culture which boxing, at least on individual basis seems best placed to counteract. It could be argued that boxing is one of the few harbingers of bourgeois civilisation to penetrate into the heart of the ghetto."[824] Our shared moral landscape includes continued revision of standards for private and public risk taking. We decide as a society of individuals which occupations are acceptable, which legacies are worth sustaining and which limits to put on consensus risk taking sports are very much a part of and affected by this ongoing process of moral revaluation. Boxing can continue to evolve as society continues to revise its list of risks people should take.

There seems to be increased interest in recreational boxing and a variety of sports and exercise routines like 'cardio-boxing' that are based on the sport. Perhaps most telling of all, women now compete as amateurs and professionals, an opportunity that the male-dominated history of boxing had previously denied them. Is this really the time to prohibit it. For the reasons given above there is only weak moral justification for doing that.

Chapter 35
The Conscious and the Unconscious

I know of no more encouraging fact than the unquestionable ability of man to elevate his life by conscious endeavour.
Henry David Thoreau

The deep well of unconscious cerebration.
Henry James

If the myth is tragic, that is because the hero is conscious.
A. Camus, *The Myth of Sisyphus*

What of imminent will and its designs?
It works unconsciously as heretofore,
Eternal artistries in circumstance.
Thomas Hardy, *The Dynasts*

Genius is the ability to make use of subliminal thought, more than most people.
Frederic Myers

We never do anything well until we cease to think about the manner of doing it.
William Hazlitt

Brain researchers estimate that our unconscious database outnumbers the conscious on an order of ten million to one. This database is the source of your hidden, rational genius. In other words, a part of you is much smarter than you are. The wiser people consult that part.
Michael J. Gelb

D.H. Lawrence said: "The unconscious mind is beyond the pale of reason by definition."

The British psychologist Stuart Sutherland has said:

"Consciousness is a fascinating but elusive phenomenon: it is impossible to specify what it is, what it does, or why it evolved. Nothing worth reading has been written on it."

So perhaps the reader should skip the next few chapters.

Nietzsche saw consciousness as a selfish drive that worked in the service of selfishness and self-cultivation. He warned "The entire surface of consciousness-consciousness-consciousness is a surface—has to be kept clear of the great imperatives. Even the grand words, the grand attitudes must be guarded against."[825]

Bertrand Russell[826] said "Mentality is a matter of degree."[827]

He acknowledged that however, 'mental events' will be certain of the events that occur in heads that have brains. They will not be all events that occur in brains, but only such as cause a reaction of the kind that can be called 'knowledge'. To Russell the adjective 'mental' is one which is not capable of any exact significance. He recognised an important group of events, namely precepts, all of which may be called 'mental'. But considered it arbitrary to say that there are no 'mental' events except precepts. He saw the most essential characteristics of mind being introspection and memory. But even memory in some of its forms is a consequence of the law of conditioned reflexes which is at least as much physiological as psychological, and may characterise living tissue rather than mind.

He considered 'mind' and 'mental' are merely approximate concepts, giving a convenient shorthand for certain approximate causal laws.

In Russell's view, there are two marked characteristics of mind.

First, it is connected with a certain body, secondly, it has the unity of one experience. These two characteristics he says are, one physical, the other psychological.

So every mental event known to us is also part of the history of a living body. He therefore viewed 'mind' as the group of mental events which form part of the history of a certain living body. And he did not confine this to the brain but suggested "that each cell in the body has its own mental life which we regard are ours."[828] The 'unconscious' might therefore be the mental lives of subordinate parts of the body, having occasional mnemic events which we can notice, but in the main separate from the life of which we are 'conscious'. In this way, the mental events connected with one body will be more numerous than the events making up its central 'mind'. This remains a fascinating theory. It puts consciousness in context.

He regarded it as only kind of mnemic effect[829] and not entitled to a special place.

He says that to say that I am 'conscious' of an event is to say that I recollect it, at any rate for a short time after it has happened.

It is to say that a certain event is occurring in me now which is connected by mnemic causation with the event recollected and is of the sort we call a 'cognition' of that event. But events which I do not recollect may have mnemic effects upon me. This is the case not only where we have Freudian suppression, but in all habits which we learnt long ago and have now become 'automatic', such as writing and speaking. Russell said "The emphasis upon consciousness has made a mystery of the 'unconscious', which ought to be in no way surprising."[830]

Russell provides us with a brilliant analogy between physics and psychology.

It was the difference between a postman's knowledge of letters and the knowledge of a recipient of letters. The postman knows the movements of many letters, the recipient knows the contents of a few. Russell says we may regard the light and sound waves that go about the world as letters of which the physicist may know the destination, some few of them are addressed to human beings, and when read give psychological knowledge.

Erwin Schrodinger has said[831] that the distinction between consciousness and unconsciousness is not so sharp. Intermediate degrees between fully conscious and completely unconscious occur.

The key to understanding this is that a succession of events in which we take part with sensations, perceptions and actions gradually drops out of the domain of consciousness when the same string of events repeats itself on a frequent basis. However, it returns to the conscious region, if at such a repetition either the occasion of the environmental conditions met with on its pursuit differ from what they were on previous incidences.

This gradual fading from consciousness is of great importance to many aspects of our lives. It is linked to the process of acquiring practice by repetition. This is a process which Richard Semon[832] has generalised to the concept of Mneme.

On frequent repetition, the responses become ever more reliable as they fade from consciousness. The boy recited his poem, the girl plays her piano sonata "well-nigh in her sleep." Schrodinger says that we follow the habitual path to our workshop, cross the road at the customary places, turn into side-streets etc while our thoughts are occupied with entirely different things. But whenever, says Schrodinger, the situation exhibits a relevant differential—let us say that the road is up at a place where we used to cross it, so that we have to make a detour—this differential and our response to it, intrude into consciousness, from which, however, they soon fade below the threshold, if the differential becomes a constantly repeated feature. Faced with changing alternatives, says Schrodinger, bifurcations develop and may be fixed in the same way. Now in this fashion, he says, differentials, variants of response, bifurcations etc are piled up one upon the other in unsurveyable abundance, but only the most recent ones remain in the domain of consciousness, only those with regard to which the living substance is still in the stage of learning or practising.

"Consciousness one might say, metaphorically, is the tutor who supervises the education of the living substance, but leaves his pupil alone to deal with all those tasks for which he is already sufficiently trained. But I wish to underline three times in red ink that I mean this only as a metaphor. The fact is only this, that new situations and the new responses they prompt are kept in the light of consciousness; old and well-practised ones are no longer so."[833]

The ontogeny not only of the brain but of the whole individual soma is the 'well-memorised' repetition of a string of events that have taken place in much the same fashion a thousand times before.

Its first stages, as we know from our own experience, are unconscious—first in the mother's womb; but even the ensuing weeks and months of life are for the greatest part passed in sleep. The ensuing organic development begins to be accompanied by consciousness only insomuch as there are organs that gradually take an interaction with the environment, adapt their functions to the changes in the situation, are influenced, undergo practice, are in special ways modified by the surroundings. We higher vertebrates possess such an organ mainly in our nervous system. Therefore, says Schrodinger, consciousness is associated with those of its functions that adapt themselves by what we call experience to a changing environment. He says that the nervous system is the place where our species is still engaged in phylogenetic transformation; metaphysically speaking it is the 'vegetation top' of our stem. He summarises his general hypothesis thus:

"Consciousness is associated with the learning of the living substance; its knowing how is unconscious."[834]

When during the celebration of his seventieth birthday, one of his disciples hailed Freud as "the discoverer of the unconscious" he answered "The poets and philosophers before me discovered the unconscious. What I discovered was the scientific method by which the unconscious can be studied."[835]

Arthus Koestler found that "The creative act, in so far as it depends on unconscious resources, presupposes a relaxing of controls and a regression to modes of ideation which are indifferent to the rules of verbal logic, unperturbed by contradiction, untouched by the dogmas and taboos of so-called commonsense. At the decisive stage of discovery the codes of disciplined reasoning are suspended as they are in a dream, the reverie, the manic flight of thought. When the steam of ideation is free to drift, by its own emotional gravity, as it were, in an apparent lawless fashion."[836]

Chapter 36
Libet's Half-Second
(The Time-Delays of Consciousness)

Equipped with his senses, Man explores the universe around him and calls the adventure science.
Edwin Hubble

It is also a good rule not to put over much confidence in the observational results that are put forward until they are confirmed by theory.
Sir Arthur Eddington

Science has explained nothing, the more one knows the more fantastic the world becomes and the profounder the surrounding darkness.
Aldous Huxley

Two experiments have been performed on people and which appear to have rather remarkable implications. They relate to the time that consciousness takes to act and to be enacted. The first of these is concerned with the active role of consciousness, and the second, its passive role.

The first was performed by H.H. Kornhuber and associates in Germany in 1976. A number of people volunteered to have electrical signals recorded as a point on their heads (electroencephalograms, i.e. EEGs) and they were asked to flex the index finger of their right hands suddenly at various times *entirely at their own choosing*. The idea was that the EEG recordings would indicate something of the mental activity that is taking place within the brain, and which is involved in the actual conscious decision to flex the finger.

In order to obtain a significant signal from the EEG traces, it was necessary to average the traces from several different runs, and the resulting signal was not very specific. What was found was that there was a gradual build-up of recorded electric potential for a *full second*, or perhaps up to a second and a half, *before* the finger was actually flexed.

This appeared to indicate that the conscious decision to process takes over a second in order to act. This may be contrasted with the much shorter time that it takes to respond to an external signal if the mode of response has been laid down beforehand. For example, instead of being 'freely willed', the finger flexing might be in response to the flash of a light signal. In that case, a reaction time of about one-fifth of a second is normal, which is about five times faster than the 'willed' action that is tested in Kornhuber's data.[837]

In the second experiment, Benjamin Libet of the University of California, in collaboration with Bertram Feinstein of the Mount Zion Neurological Institute in San Francisco[838] tested subjects who had to have brain surgery for some reason unconnected with the experiment and who consented to having electrodes places at points in the brain, in the somatosensory cortex.

The upshot of Libet's experiment was that, when a stimulus was applied to the skin of these patients, it took about half an second before they were consciously aware of that stimulus, despite the fact that the brain itself would have received the signal of the stimulus in only about a hundredth of a second, and a pre-programmed 'reflex' response to such stimulus could be achieved by the brain in about a tenth of a second. Moreover, despite the delay of half a second before the stimulus reaches awareness, there would be the subjective impression by the patients themselves that no delay had taken place at all in their becoming aware of the stimulus!

The somatosensory cortex is the region of the cerebrum at which sensory signals enter. Thus, the electrical stimulation of a point of the somatosensory cortex, corresponding to some particular point on the skin, would appear to the subject just as though something had actually touched the skin at that corresponding point.

However, it turns out that if this electrical stimulation is too brief—for less than about half a second—then the subject does not become aware of any sensation at all. This is to be contrasted with a direct stimulation of the point on the skin itself, since a momentary touching of the skin can be felt.

Now, suppose that the skin is first touched, and then the point in the somatosensory cortex is electrically stimulated. What does the person feel? If the electrical stimulation is initiated about a quarter of a second after the touching of the skin, then the skin touching is not felt at all.

This is an effect referred to as *backwards masking*. Somehow the stimulation of the cortex serves to prevent the normal skin-touching sensation from being consciously felt. The conscious perception can be prevented ('masked') by a later event, provided that that event occurs within about half a second.

This in itself tells us that the conscious awareness of such a sensation occurs at something like half a second after the actual event producing that sensation! This is known as 'Libet's half-second'.

However, one is not 'aware' of such a long-time delay in one's perceptions.

One way of making sense of this curious finding might be to imagine that the 'time' of all one's 'perceptions' is actually delayed by about half a second from 'actual time'—as though one's internal clock is simply 'wrong' by half a second or so. The time at which one perceives an event to take place would then always be half a second *after* the actual occurrence of the event. This would present a consistent, albeit disturbingly delayed, picture of sense impressions.

From the first of the above experiments, we seem to deduce that a conscious action takes something like a second to a second and a half before it can be effected, while according to the second experiment, consciousness of an external event does not seem to occur until half a second after that event has taken place.

Imagine what happens when one responds to some unanticipated external occurrence. Suppose that the response is something that requires a moment's conscious contemplation.

It would appear, on the basis of Libet's findings, that half a second must elapse before consciousness is called into play, and then, as Kornhuber's data seem to imply, well over a second is needed before one's 'willed' response can take effect.

The whole process, from sensory input to motor output, would seem to require something like two seconds!

The apparent implication of these two experiments taken together is that consciousness cannot even be called into play *at all* in response to an external event, if that response is to take place within a couple of seconds or so!

This has significant indications for the characterisation of sporting abilities of which I will speak later.

If we take these findings at face value, we are driven to the conclusion that we act entirely as 'automatons' when we carry out any action that would take less than a second or two in which to modify a response.

This suggests that consciousness is slow-acting, as compared with other mechanisms of the nervous system.

Consciousness may be, after all, merely a spectator who experiences nothing but an 'action replay' of the whole drama.

Similarly, on the face of it, there would be no time for consciousness to be playing any role at all when, for example, one plays a shot at tennis or cricket.

No doubt the experts at these pursuits would have all the essentials of their responses superbly pre-programmed in cerebella control. But that consciousness should be playing *no* role at all in the decision as to what shot should be played at the time is something that requires careful scrutiny. That is why I believe it has been said you can teach someone how but not when.

No doubt there is a lot in the anticipation of what one's opponent might do, and many pre-programmed responses might be available to each possible action of the opponent, but a *total* absence of conscious involvement at the time is something that it is difficult to accept.[839]

Penrose suggests that we may be going wrong when we apply the usual physical rules for *time* when we consider consciousness.[840]

It is possible that a very different conception may be required when we try to place conscious perceptions into a conventionally time-ordered framework.

The way in which time is treated in modern physics is not essentially different from the way in which *space* is treated and the 'time' of physical descriptions does not really 'flow' at all; we just have a static-looking fixed 'space-time' in which the events of our universe are laid out.

Yet, according to our perceptions, time *does* flow.

Penrose's guess is that there is something illusory here, and the time of our perceptions does not 'really' flow in quite the linear forward-moving way that we perceive it to flow.

The temporal ordering that we 'appear' to perceive is something that we impose upon our perceptions in order to make sense of them in relation to the uniform forward time-progression of an external physical reality.

But even if we accept that consciousness itself has such a curious relation to time—and that it represents, in some sense, contact between the external physical world and something timeless—how can this fit with a physically determined and time-ordered action of the material brain? Are we left with a mere 'spectator' role for consciousness if we are not to monkey with normal progression of physical laws?

Penrose argues for some kind of active role for consciousness, and indeed for a powerful one, with a strong selective advantage.[841]

Not all bodily reactions required intervention by the conscious mind.

Penrose looks at the phenomenon of driving for miles while engrossed in conversation and then discovering that you have utterly no memory of the road, the traffic, your car-driving activities.

It is as if someone else had been driving.

Many theorists[842] have cherished this as a favourite case of "unconscious perception and intelligent action."

But are we really unconscious of all those passing cars, stop lights, bends in the road at the time?

We were paying attention to other things, but surely if we had been probed about what you had just seen at various moments on the drive, we would have had at least some sketchy details to report.

The 'unconscious driving' phenomenon is better seen as a case of rolling consciousness with swift memory loss.

The brain's task is to guide the body it controls through a world of shifting conditions and sudden surprises, so it must gather information from that world and use it swiftly to 'produce future'—to extract anticipations in order to stay one step ahead of disaster[843]. So the brain must represent temporal properties of events in the world, and it must do this efficiently.

The processes that are responsible for executing a task are spatially distributed in a large brain with no central node, and communication between regions of this brain is relatively slow; electrochemical nerve impulses travel thousands of times slower than light (or electronic signals through wires).

So the brain is under significant time pressure.

It must often arrange to modulate its output in the light of its input within a time window that leaves no slack for delays.

On the input side, there are perceptual analysis tasks, such as speech perception, which would be beyond the physical limits of the brain's machinery if it didn't utilise ingenious anticipatory strategies that feed on redundancies in the input.

Normal speech occurs at the rate of four or five syllables per second, but so powerful are the analysis machines we have evolved to 'parse' it, that people can comprehend 'compressed speech'—in which the words are electronically sped up without raising the tone chipmunk-style—at rates of up to thirty syllables per second.

On the output side, many acts must occur so fast, and with such accurate triggering, that the brain has no time to adjust its control signals in the light of feedback; acts such as playing the piano or accurately pitching a rock[844] must be ballistically initiated.

Libet has interpreted his results as raising a serious challenge to materialism: "...a dissociation between the timings of the corresponding 'mental' and 'physical' events would seem to raise serious though not insurmountable difficulties for the...theory of psychoneural identity."[845]

According to Sir John Eccles, a Nobel laureate in medicine for his research in neurophysiology, this challenge cannot be met:

"This antedating procedure does not seem to be explicable by any neurophysiological process. Presumably it is a strategy that has been learnt by the self-conscious mind...the antedating sensory experience is attributable to the ability of the self-conscious mind to make slight temporal adjustments, i.e., to play tricks with time."[846]

Although Libet's experiment has been widely hailed in non-scientific circles as a demonstration of the truth of dualism, few in the cognitive science community share this opinion.

In the first place, Libet's experimental procedures, and his analysis of the results, have been severely criticised.

His experiment has never been replicated, which is reason enough in many quarters to remove his 'results' from consideration. But I disagree. The fact that no one has proved or disproved his theory is not enough. It remains a coherent view.

The sceptical view, then, is that his phenomena simply don't exist. But what if they did?

That is just the sort of question a philosopher would ask, but in this case, there is more than the usual philosophical motivation for asking it.

There is a challenge for materialism of making sense of what Libet calls "a primary phenomenological aspect of our human existence in relation to brain function."[847]

He says:

> "It is important to realise that these subjective referrals and corrections are apparently taking place at the level of the mental 'sphere'; they are not apparent, as such, in the activities at neural levels[848]." It is the first step on a buttered slide back to dualism.
> "The reports by subjects about their different experiences…were not theoretical constructs but empirical observations…The method of introspection may have its limitation, but it can be used appropriately within the framework of natural science, and it is absolutely essential if one is trying to get some experimental data on the mind-brain problem."[849]

Libet claims that his experiments with direct stimulation of the cortex demonstrate "two remarkable temporal factors:"

(1) There is a substantial delay before cerebral activities, initiated by a sensory stimulus, achieve 'neuronal adequacy' for eliciting any resulting conscious sensory experience.

(2) After neuronal adequacy is achieved, the subjective timing of the experience is (automatically) referred backwards in time, utilising a 'timing signal' in the form of the initial response of cerebral cortex to the sensory stimulus.[850]

The 'timing signal' is the first burst of activity to appear in the cortex which occurs only 10 to 20msec after stimulation of the peripheral sense organ.

Libet suggests that the backwards referral is always 'to' the timing signal.

Libet concludes that temporal referral raises problems for materialism (the "theory of psychoneural identity").[851]

Libet concluded that "cerebral initiation of a spontaneous voluntary act begins unconsciously."[852]

This seems to show that consciousness lags behind the brain processes that actually control the body.

It seems to rule out a real (as opposed to illusory) "executive role" for "the conscious self."

He claims that when conscious intentions to act are put into registration with the brain events that actually initiate the acts, there is an offset in the 300–500 msec range.

This is huge—up to half a second—and it does look ominous to anyone committed to the principle that our conscious acts control our bodily motions.

It looks as if we are located in Cartesian Theatres where we are shown, with a half-second time delay, the real decision-making that is going on elsewhere.

Libet has shown that motor action itself, and the sense of agency that comes along with it, depend on neurological events that we do not control, and that happens before our conscious decision.[853]

The results of his experiments indicated that on average, 350 msecs before you are conscious of deciding to move, your brain is already working on the motor processes that will result in the movement.

Before you know it, the readiness potential is already underway, and you are preparing to move.

The brain seemingly decides and then enacts its decisions in a non-conscious fashion, on a subpersonal level, but also inventively tricks us into thinking that we consciously decide matters and that our actions are personal events.

This is just the kind of evidence that some theorists appeal to in order to make the case against free will.

'The initiation of the freely voluntary act appears to begin in the brain unconsciously, well before the person consciously knows he wants to act.

Is there, then, any role for conscious will in the performance of a voluntary act?'[854]

Libet[855] himself answers in the positive: we can still save free will—because there is still approximately 150 msecs of brain activity left after we are conscious of our decision, and before we move. So, he suggests, we have time consciously to veto or modify the movement. One can see this in batsmen adjusting their shot.

The attempt to frame the question of free will in terms of these subpersonal, prenoetic processes is misguided for at least two reasons.

First, free will cannot be squeezed into time-frames of 150–350 msecs; free will is a longer-term phenomenon that depends on consciousness, and in this respect the sense of agency is more than just an accessory.

Second, the notion of free will does not apply primarily to abstract motor processes that make up intentional actions—rather it applies to intentional actions themselves, described at the highest pragmatic level of description.

The temporal framework for the exercise of free will is, at a minimum, the temporal framework that allows for the process to be informed by consciousness.

Once events of conscious deliberation are included in the behavioural feedback loop certain things in the environmental begin to matter to the agent.

Meaning and interpretation comes into the picture.

Conscious interpretation introduces a temporally extended 'looping effect'.[856]

The conscious deliberation of the agent, which involves memory and knowledge has real effects on behaviour.

To the extent that consciousness enters into the ongoing production of action, and contributes to the production of further action, *even if significant aspects of this production take place non-consciously*, our actions are intentional.

Voluntary actions are not about neurons, muscles, body parts, or event movement—all of which play some part in what is happening and for the most part, nonconsciously—but all such processes are carried along by (and are intentional because of) my decision to act—that is, by what is best described on a personal level as my intentional action.

Free choice is not about neurons, muscles, or bodily movement; it's not a mediating executive mental process, which somehow puts the bodily parts into action'[857]; such motor acts are not the 'prototype' of free action.

Free will is about my purposive actions, which are best described in terms of intentions rather than neurons, muscles, reachings, etc.

In this regard, such extra-conscious aspects of body-schematic control are like the vehicle to the volitional content.

What we call free will, however, cannot be conceived as something *purely* subpersonal, or as something instantaneous, an event that takes place in a knife-edge moment located between being undecided and being decided.

If that were the case, it would completely dissipate in the milliseconds between brain events and our conscious awareness.

Free will involves temporally extended feedback or looping effects that are transformed and enhanced by the introduction of deliberative consciousness.

This means that the conscious sense of agency, even if it starts out as an accessory experience generated by the brain, is itself a real force that counts in the formation of our future action.

It contributes to the freedom of action, and bestows responsibility on the agent.

Daniel Dennett[858] has addressed these issues in *Freedom Evolves*. On his view, the processes that constitute free will need not be conscious and need not depend on conscious decision.

The unconscious processes that precede conscious decision in Libet's experiments are part of the system that he calls the person.

Although free will does not depend on consciousness, it requires an extended time-frame.

> "Once you distribute the work done...both space and time in the brain, you have to distribute the moral agency around as well. You are not out of the loop, you *are* the loop. You are that large. You are not an extensionless point. What you do and what you are *incorporates* all these things that happen and is not something separate from them."[859]

In response to the position that Gallagher has outlined, Dennett gives and takes. He writes:

> One commentator on Libet who gets close is Sean *[sic]* Gallagher: 'I think that this problem can be solved as long as we do not think of free will as a momentary act. Once we understand that deliberation and decision are processes that are spread out over time, even, in some cases, very short amounts of time, then there is plenty of room for conscious components that are more than accessories after the fact' (Gallagher 1998). But, then he goes on to say that if the feedback is all unconscious, it will be 'deterministic' but if it is conscious, it won't be. Cartesian thinking dies hard.[860]

Gallagher doesn't disagree with Dennett concerning the role played by nonconscious elements, except that he thinks we are even larger than he thinks— we are not just what happens in our brains.

The 'loop' extends through and is limited by our bodily capabilities, into the surrounding environment, which is social as well as physical, and fed back through our conscious experience into the decisions we make.

The non-conscious embodied processes, including the kind of neurological events described by Libet, and distributed by Dennett, are essential to a free will that is specifically human.

All such relevant processes are structured and regulated by my intentional goals as much as they also limit and enable my action.

When I decide to reach for a drink, all the appropriate physical movements fall into place. These embodied mechanisms thus enable the exercise of free will.

And to the extent that we are not required consciously to deliberate about bodily movement and autonomic processes, our deliberation can be directed at the more meaningful level of intentional action.

What is 'in the loop' is not just the non-conscious processes happening in our brain, but the larger system of body-environment-intersubjectivity.

You are not out of the loop. But the loop isn't just you. It's larger than you. It's you as you interact with the things and with other people in the world.

And it is only in those larger contexts that the issue of free will is at stake.

Chapter 37
The Embodied Mind and
the Cognitive Unconscious

The power of thought—the magic of the mind.
Lord Byron

The mind is a musical instrument with a certain range of tones beyond which in both directions we have infinite silence.
John Tyndall

The two great philosophers of the embodied mind were Maurice Merleau-Ponty and John Dewey. Merleau-Ponty used the word 'flesh' for our primordial experience and sought to focus the attention of philosophy on what he called 'the flesh of the world' as we fell it by living in it. John Dewey saw that our bodily experience is the primal basis for everything we can mean, think, know and communicate.

The mind is inherently embodied, it is mostly unconscious and abstract concepts are largely metaphorical.[861] These are the three major findings of cognitive science. Lakoff and Johnson suggest that these three findings from the science of mind are inconsistent with central parts of western philosophy.

Reason has been taken for two millennia as the defining characteristic of human beings. A radical change in our understanding of reason is therefore a radical change in our understanding of ourselves. According to this view, human rationality is not all what the western philosophical tradition has held it to be. The strands of this argument are as follows.

Reason is not disembodied but arises from the nature of our brains, bodies and bodily experience. It is not the obvious claim that we need a body to reason but that the very structure of reason itself comes from the details of our embodiment.

The same neural and cognitive mechanisms that allow us to perceive and move around also create our conceptual systems and modes of reason.[862]

To understand reason we must understand the details of our visual system, our motor system, and the general mechanism of neural binding.

"Reason is not, in any way, a transcendent feature of the universe or of disembodied mind. Instead, it is shaped crucially by the peculiarities of our human bodies, by the remarkable details of the neural structure of our brains and by the specifics of our everyday functioning in the world."[863]

Reason is evolutionary, in that abstract reason builds on and makes use of perceptual and motor inference present in 'lower animals'. Reason is thus not an essence that separates us from other animals, rather it places us on a continuum with them.

Reason is not completely conscious, but mostly unconscious. Reason is not purely literal, but largely metaphorical and imaginative. Reason is not dispassionate but emotionally engaged.

There is no Cartesian dualistic person, with a mind separate from and independent of the body, sharing exactly the same disembodied transcendental reason with everyone else, and capable of knowing everything about his or her mind simply by self-reflection.

Rather the mind is inherently embodied, reason is shaped by the body, and since most thought is unconscious, the mind cannot be known simply by self-reflection. Empirical study is necessary.

Human beings are not, for the most part, in conscious control of—or even consciously aware of—their reasoning. Although we have a theory of a vast, rapidly and automatically operating cognitive unconscious, we have no direct conscious access to its operation and therefore to most of our thought.

There exists no Fregean person—as posed by analytic philosophy—for whom thought has been extruded from the body.

There is no such thing as a computational person, whose mind is like computer software. The neural structures of our brains produce conceptual systems and linguistic structures that cannot be adequately accounted for by formal systems that only manipulate symbols.

There is no Chomskyan person, for whom language is pure syntax, pure form insulated from and independent of all meaning, context, perception, emotion, memory, attention, action and the dynamic nature of communication.

Moreover, human language is not a totally genetic innovation. Rather, central aspects of language arise evolutionarily from sensory, motor and other neural systems that are present in 'lower animals'.[864]

In asking philosophical questions, we use a reason shaped by the body, a cognitive unconscious to which we have no direct access, and metaphorical thought of which we are largely unaware.

What emerges is a philosophy close to the bone. A philosophical perspective based on our empirical understanding of the embodiment of mind is a philosopher in the flesh, a philosophy that takes account of what we most basically are and can be.[865]

It has discovered, first of all, that most of our thought is unconscious, not in the Freudian sense of being repressed, but in the sense that it operates beneath the level of cognitive awareness, inaccessible to consciousness and operating too quickly to be focussed on.[866]

When we understand all that understands the cognitive unconscious, or understanding of the nature of consciousness is vastly enlarged.

Consciousness goes way beyond mere awareness of something, beyond the mere experience of qualia (the qualitative senses of, for example, pain or colour), beyond the awareness that you are aware, and beyond the multiple takes on immediate experience provided by various centres of the brain.

Consciousness certainly involves all of the above plus the immeasurably vaster constitutive framework provided by the cognitive unconscious, which must be operating for us to be aware of anything at all.[867]

"Conscious thought is the tip of an enormous iceberg" (ibid Lakoff and Johnson p.13).

It is the rule of thumb among cognitive scientists that unconscious thought is 95 per cent of all thought and that may be a serious underestimate. Moreover, the 95 per cent below the surface of conscious awareness shapes and structures all conscious thought. If the cognitive unconscious were not there doing this shaping, there would be no conscious thought.[868]

> "Our unconscious conceptual system functions like a 'hidden hand' that shapes how we conceptualise all aspects of our experience."[869]

Philosophical theories are largely the product of the hidden hand of the cognitive unconscious. It uses metaphor to define our unconscious metaphysics.

Throughout most of our history, philosophy has seen itself as being independent of empirical investigation. It is that aspect of philosophy that is called into question by results in cognitive science. Through the study of the cognitive unconscious, cognitive science has given us a radically new view of how we conceptualise our experience and how we think.

Cognitive science is the empirical study of the mind.[870]

There is no fully autonomous faculty of reason separate from and independent of bodily capacities such as perception and movement. The evidence supports an evolutionary view in which reason uses and grows out of such bodily capacities.

These findings of cognitive science are profoundly disquieting.

Firstly, they tell us that human reason is a form of animal reason, a reason inextricably tied to our bodies and the peculiarities of our brains.

Secondly, these results tell us that our bodies, brains and interactions with our environment provide the mostly unconscious basis for our everyday metaphysics, that is, our sense of what is real.[871]

Cognitive science provides a new and important take on an age-old philosophical problem, the problem of what is real and how we can know it, if we can know it.

Our sense of what is real begins with and depends crucially upon our bodies, especially our sensorimotor apparatus, which enables us to perceive, move and manipulate, and the detailed structures of our brains, which have been shaped by both evolution and experience.[872]

Every living being categorises.

Even the amoeba categorises the things it encounters into food or non-food, what it moves towards or moves away from. The amoeba cannot choose whether to categorise, it just does.

The same is true at every level of the animal world. Animals categorise food, predators, possible mates, members of their species and so on.

How animals categorise depends upon their sensing apparatus and their ability to move themselves and to manipulate objects.

Categorisation is therefore a consequence of how we are embodied. We have evolved to categorise; if we hadn't, we would not have survived.

Categorisation is, for the most part, not a product of conscious reasoning.

We categorise as we do because we have the brains and bodies we have and because we interact in the world the way we do.[873]

To take an example, says Lakoff and Johnson,[874] each human eye has 100 million light sensing cells, but only about 1 million fibres leading to the brain. Each incoming image must therefore be reduced in complexity by a factor or 100. That is, information in each fibre constitutes a 'categorisation' of the information from about 100 cells.

Neural categorisation of this sort exists throughout the brain, up through the highest level of categories we can be aware of.[875]

A small percentage of our categories have been formed by conscious acts of categorisation, but most are formed automatically and unconsciously as a result of functioning in the world.

We have a visual system, with topographic maps and orientation-sensitive cells, that provides structure to our ability to conceptualise spatial relations. Our abilities to move in the ways we do and to track the motion of other things gives motion a major role in our conceptual system.

The fact that we have muscles and use them to apply force in certain ways leads to the structure of our system of causal concepts.

It is not just that we have bodies and that thoughts are somehow embodied. It is that the peculiar nature of our bodies shapes our very possibilities for conceptualisation and categorisation.

"An embodied concept is a neural structure that is actually part of, or makes use of, the sensorimotor system of our brains. Much of conceptual inference, is therefore, sensorimotor inference" (ibid Lakoff and Johnson p.20).

It follows that if concepts are embodied, the locus of reason (conceptual inference) would be the same as the locus of perception and motor control, which are bodily functions.

Lakoff and Johnson asks

"Does reason piggyback on perception and motor control,"[876] their conclusion is that from the perspective of the brain, the locus of all three functions, it would be quite natural if it did.

Lakoff and Johnson provide some interesting thoughts on the meaning of colours.[877]

Our colour concepts, their internal structures and the relationships between them are inextricably linked to our embodiment.[878]

Colours, as we see them, say the red of blood or the blue of the sky are not out there in the blood or the sky. Indeed the sky is not even an object. It has no surface for the colour to be in.

And without a physical surface, the sky does not even have a surface reflection to be detected as colour. The sky is blue because the atmosphere transmits only a certain range of wavelengths of incoming light from the sun and of the wavelengths it does transmit, it scatters some more than others.

The effect is like a coloured lightbulb that only lets certain wavelengths of light through glass. Thus the sky is blue for a very different reason than a painting of the sky is blue. What we perceive as blue does not characterise a single 'thing' in the world, neither 'blueness' nor wavelength reflectance.

Colour concepts are 'interactional', they arise from the interactions of our bodies, our brains, the reflective properties of objects, and electromagnetic radiation.

Philosophically, colour and colour concepts make sense only in something like an embodied realism, a form of interactionism that is neither purely objective nor purely subjective.

Colour is a function of the world and our biology interacting.

Spatial-relations concepts are at the heart of our conceptual system. They are what makes sense of space for us. They characterise what spatial form is and define spatial inference. But they do not exist as entities in the external world. We do not see spatial relations the way we see physical objects.[879]

We use spatial-relations concepts unconsciously, and we impose them via our perceptual and conceptual systems.

We just automatically and unconsciously 'perceive' one entity as in, or across from another entity.

Such perception depends upon an enormous amount of automatic unconscious mental activity on our part.[880]

In Western philosophy there is assumed to be an absolute dichotomy between perception and conception. While perception has always been accepted as bodily in nature, just as movement is, conception—the formation and use of concepts—has traditionally been seen as purely mental and wholly separate from and independent of our abilities to perceive and move.[881]

This picture is false.

Base-level concepts depend on motor movement, gestalt perception and mental imagery, which is carried out in the visual system of the brain.

Colour is anything but purely mental; our colour concepts are intimately shaped not merely by perception as a faculty of mind, but by such physical parts of our bodies as colour cones and neural circuitry.

Spatial-relations concepts are not characterised by some abstract, disembodied mental capacity but rather in terms of bodily orientation.

Motor control has an intrinsic structure. All motor schemes have the same control structure.[882]

Neural structures that can carry out sensorimotor functions in the brain can in principle do both jobs at once—the job of perception, motor control, on the one hand and the job of conceptualising, categorising, and reasoning on the other.[883]

From a biological perspective reason has grown out of sensory and motor systems.[884]

Philosophically, the embodiment of reason is a crucial part of the explanation as to why it is possible for our concepts to fit so well with the way we function in the world. They fit so well because they have evolved from our sensorimotor systems, which have in turn evolved to allow us to function well in our physical environment.

Our embodiment of mind leads us to a philosophy of embodied realism.

Our concepts cannot be a direct reflection of external, objective, mind-free reality because our sensorimotor system plays a crucial role in shaping them.

Chapter 38
Visual Perception, Action in Sport and the Ball of the Century

The appearance of things to the mind is the standard of every action to man.
Epictetus

Among the blind the cross-eyed man is king.
Erasmus

Men, you are all marksmen—don't one of you fire until you see the whites of their eyes.
William Prescott (1726–95)

So much sport can be characterised as visual perception in action. This phenomenon was recently encapsulated by the former legendary darts player Bobby George when he summed up the essence of darts as "look where you throw and throw where you look".

This is yet again another sporting anecdotal truism of metaphysical significance extending beyond the concept of sport.

Darts indeed like golf is a study in the 'focussed intensity' to which I referred to earlier in chapter 6 'The Aesthetics of Sport' and the work of Professor Gumbrecht 'In Praise of Athletic Beauty'.

He referred to athletes and spectators alike "getting lost in focussed intensity".

This is true of most sports, but darts is a prime exemplar of it. No other sport displays such outstanding dexterity and hand eye coordination. That is why it is so pleasing to see it manifest itself at a young age. Witness the success of the sixteen-year-old darts player Luke Littler at the Worlds Darts Championship 2024.

Furthermore, turning to the older generation of darts players like Phil 'The Power' Taylor and the Welshmen Alan Evans and Leighton Rees, these great dart players were a remarkable exponent of it too, but even more remarkably they could exemplify those innate talents having consumed large quantities of alcohol before and during the course of play.

That if I may say so and not tongue in cheek is the magic and mystery of darts and indeed alcohol.

Sport is thus a heightened form of visual perception in action.

Schopenhauer, in Studies in Pessimism 1851, said that every man takes the limits of his own field of vision for the limits of the world. Great sportsmen like great thinkers do not do that but extend their field of vision and indeed their perception to beyond the horizon. That is the hallmark of great thinkers and top-class athletes alike, they see things that ordinary mortals cannot.

Skilled actions in sport have one important thing in common: they exemplify the significant spatio-temporal demands on the top-class athlete in complex and dynamic environments.[885]

These demands highlighted daily in the sports media, have been more precisely quantified in scientific analysis.

In ball games such as cricket, baseball and tennis, projective speeds of between 36 and 45 ms^{-1} have been recorded.[886]

The time window afforded performers in high-level sport are typically measured in thousandths of a second (ms).

It has been demonstrated how cricket batsmen often have only 230 ms to cope with late fluctuations in the flight of a ball approaching at 150 kilometres per hour (kph).[887]

Yet, skilled performers are capable of the most extraordinary precision in matching the spatio-temporal constraints of their sports and activities.

Batting is not primarily a reaction time problem.

The ball is continuously visible during its flight, and not suddenly presented.

Consequently, the problem is one of tracking a moving object, predicting its course and, at some point in its flight, deciding to swing or not. Thus, prediction sub serves the motor response.[888]

Batting is a perceptual problem rather than something analogous to reaction time.

It is doubtful whether a batsman in cricket or baseball for example *can* differentiate between a fast and slow ball *if* the bowler conceals the run-up or wind-up to either ball in such a way that the batsman cannot perceptually distinguish it from other modes of delivery and if the batsman is not aware of the possibility of such an idiosyncrasy on the part of the bowler/pitcher. The whole art of deception in games skills consists in either masking the visual display in such a way that opponents misinterpret the cues or in presenting distracting cues for response.

Bowling is a good example of this. The main attributes of the bowler are speed and/or guile. The fast bowler principally relies on speed alone to beat the batsman. This type of bowling lends itself least to modification and it is therefore more easy to learn to predict the flight path.

The slow bowler however, uses his guile in the manipulation of length, flight and direction in an attempt to confuse the batsman.

The batting situation therefore becomes one in which the batsman has less time to play the fast ball which is more predictable and more time to play the slow ball which is less predictable!

Two further factors in bowling which are worthy of mention are 'spin' and 'cut' (that is deviation after the ball has pitched) and swing (deviation in flight before pitching).

These are additional weapons in the bowler's armoury and on his mastery of them will depend to a large measure, his success.

An example of this was the successful bowler S.F. Barnes—a fast bowler with all the best attributes of a slow bowler. He could deliver away swingers, off-breaks and leg breaks all with sufficient velocity to make any batsman hesitant.

> "A typical over, first two very late outswingers straight enough and well enough up to force the batsman to play off the front foot, then two penetrating off-breaks, the fifth ball a fast leg-break and a leg-break it was rather than a leg cutter—and finally such a delivery as on his great Australian tour bowled Victor Trumper at the height of his powers, the ball swerving from the leg stump on to the off and then breaking back to hit the leg. 'It was the sort of ball', said Charlie Macartney, 'that a man might see when he was tight!'[889]"

More recently, Derek Underwood was a slow bowler with all the attributes of a fast bowler.

The Ball of the Century—Shane Warne to Mike Gatting on 4th June 1993 Old Trafford

The Ball of the Century may not have been the ball that Shane Warne bowled to Mike Gatting in 1993. The ball which SF Barnes bowled to Victor Trumper may have been just as good. But all that shows is that in sport as in life, comparisons are odious. Both would have been amazing deliveries and the perception would have been that they were unique examples of outstanding talent. But that is not how things work in life, in life and sport generally nothing is unique. Most things in life are just further examples of what can happen. I do not believe that historically outstanding talent is about unique performance. It is about exceptional talent manifesting itself at different times. It is true that Shane Warne was a great spin bowler and that he bowled a ball to Mike Gatting which was perceived by him and others as an amazing delivery which could not be played. It was perceived as The Ball of the Century because people wanted it to be. It is true that the ball's combined cut, swerve, spin and change of pace which made it appear impossible to play, but the truth is Mike Gatting could not play it. But I believe others could have but that does not mean others could not and that is the joy of sport, what appears to be impossible can become achievable. And that is why it retains the fascination it does, one never knows. The Ball of the Century is philosophically interesting because it shows that history can repeat itself and frequently does. It is a prime example of Thucydides axiom that "history is philosophy teaching by example," in that sense The Ball of the Century was philosophy. But I know from speaking to Mike Gatting he would put it in slightly more graphic terms!

Batsman do in fact hit balls which they would not have been able to if accepted reaction time measures were taken into account.

In this connection, there *is* experimental laboratory evidence in the form of what has been termed *transit reaction time*. This form of reaction time would not appear to differ—in terms of times recorded—from the measures of simple reaction time.[890]

This statement must, however, be qualified, because under conditions in which the stimulus is present *all* the time and is the focus of attention, expectancy is likely to be at its highest and hence the reaction time will be at its least.[891]

In addition, since the player is not having to wait for the sudden onset of a stimulus it might be expected that he would be able to concentrate on the organisation of the response. This in itself can result in a faster reaction time than if the concentration is on the stimulus.[892]

The task was to press a key, or make some simple movement, at the instant when two objects approaching one another come into exact coincidence.[893]

A player has to initiate his response one reaction time plus one movement time before the ball is in position for hitting or catching. Under such conditions, reaction time is going to be more crucial in fast ball skills than those of a relatively more leisurely pace.

Although studies such as those of Knapp[894], Slater-Hammel & Stumpner[895] and Miller & Shay[896] have drawn attention to the fast simple and choice reaction times of competent games players, the problem is more than that of fast reaction.

As Poulton[897] has pointed out:

"Although reaction time is very much tied up with this kind of skill (fast ball skill) there is no simple relation between them. The man who is good at fast ball games, is not necessarily the man whose reaction time is the shortest when measured. The only certain generalisation is that the man with the long reaction time will not be good at fast ball games."

In a laboratory study of reaction and movement times in cricket-type situations[898], it has been shown that a more or less constant *movement time* in a variety of situations is in the region of 0.3 secs. The movement involved was stepping forwards or backwards in several directions as would be done in an actual playing situation.

Assuming once again a reaction time to a visual stimulus to be in the region of 0.2 secs, it would seem that a time of about 0.5 secs is necessary from the moment a decision is made to initiate a movement, to its completion.

Figure

Speed of Ball	Flight path (60') time
80 m.p.h	0.51 secs
60 m.p.h.	0.68 secs
40 m.p.h.	1.02 secs

The relation of flight path time to the speed of the ball in a cricket situation.

Reference once again to the figure will indicate the limitations of watching the ball all the time in such a situation if the information from the ball is to be used in affecting the stroke. Even for a ball bowled at 10 m.p.h. the stroke will have to be initiated when the ball is approximately 7½' away!

Long jumpers can accurately hit a 20-cm take-off board at the end of a 40-m run-up at speeds of around 10 ms^{-1}).[899]

Research has illustrated the consistent of bat control in national-level table-tennis players by calculating the variability in timing the initiation of an attacking forehand drove. They found that typical values for timing variability in the stroke were between 2.03 and 4.72 ms.[900]

In order to satisfy these task constraints, the sport performer is heavily dependent on the visual system to provide much of the information for perceiving and acting.

Skills might be classified as predominantly perceptual or predominantly habitual. Such sports as cricket, tennis, basketball, and fencing would be *perceptually oriented*, whereas diving, putting the shot, and trampolining would be habitually oriented. In some sports and activities, the prime concern of the performer is the potential changing environment. His reaction cannot be fixed but, rather, depends on the circumstance. Other skills require repetitious practice until the act can be performed as a habit, perceptual need is minimised and the ability to reproduce the same act continuously and consistently is emphasised.

Habitual skills, those requiring a fixed response to a given situation, appear to be associated with stimulus-response theory. The stimulus environment is relatively stable. Desired response can only be achieved through constant practice, and the performer's attention is to the act itself. With successful practice, skill sets in, and the individual may execute the skill as if automatically without any direct concern over the intricacies of the act. Self-paced skills, closed skills, and habitual skills possess many similar characteristics.[901]

The diver's environment consists of a diving board and a pool, with the board a certain height over the pool. This condition is constant, wherever he goes to compete. True, things are never quite the same day to day or place to place, but then, no skill is purely habitual or purely perceptual. Elements of both are necessary for successful skill execution, but emphasis on each is dependent on the nature of the skill. Differences in diving-board structure, pool temperature, and spectators present, among many other factors, will contribute to an altered stimulus environment.

But basically, the diver is an example of an habitual skill. The act is to be repeatedly performed under the condition in which it must ultimately be demonstrated for success to be probable. Any sport where the participant initiates the action rather than having to respond to thrown objects or moving players can be identified as primarily habitual.

Stimulus-response theory appears to have limited value in team sports as well as in many individual and dual sports where perceiving a changing environment is necessary (perceptual skills). Gestalt theory appears to be more appropriate for this situation. However, the initial learning of the basic skills underlying a sport might very well be conditions, using the stimulus response. Common sense would dictate that a skill must be learnt well under stable conditions before the individual will be able to execute it regardless of unpredictable circumstances.

But a fixed response is no use to the performer reacting under varying conditions. The emphasis later would be on the environment, understanding relationships and patterns, utilising gestalt methodology. A cricketer or a tennis player might have beautiful form and execution when hitting against a ball-throwing machine.

A game situation, however, requires much flexibility in response, for the ball now the comes to stroker at varying speeds, with curves and slices, and with indiscriminate bounces. Now he needs to demonstrate his awareness of spatial relationships and an ability to perceive and react according to changing stimuli. In team sports, the same condition exists.

The football player must consider not only his own developed skill, but also his position on the pitch with regard to fellow and opposing players. It is not enough merely to assume a definite response pattern each time a specific situation is present.

The player who mistakenly fixates an offensive or defensive move, who reacts in an inflexible manner to give stimuli will certainly be discovered shortly and advantage will be taken of these conditioned patterns. It is much more desirable to have skills developed to be called into play at any time, regardless of the situation. Externally paced skills, open skills, and perceptual skills contain many similar characteristics.

Perception involves detecting and interpreting changes in various forms of energy flowing through the environment such as light rays, sound waves and neural activation.[902]

The environmental changes which can be perceived from these energy flows over space and time are used to support the goal-directed actions of the athlete.

A cricketer needs to be able to detect and interpret the light information which reflects from the surface of the ball when taking a catch.

When performing this task, he needs precise information to locate the ball in space ('where' information) at a specific point in time ('when' information).

Spatio-temporal information regarding the approaching object must be acquired early so that the approach components of his skeletomuscular system (i.e. muscles of the trunk and joints in the arms and shoulders) may be coordinated in time.

The batsman and catcher also need to be aware of late deviations in flight due to added complications such as spin, swerve or drag effects.

The visual perceptual systems of the cricketer support the balance and postural control necessary as he organises the catching response.[903] They also allow the skilled team games player to recognise patterns in the formation of the opposing players so that the two-handed catching action may be integrated into a tactical sequence of play.

The study of visual perception and action in sport is related to the athlete's need to perceive the spatio-temporal structure of environmental information order to successfully perform actions.[904]

This is not to deny that other forms of sensory information are important, it is just the visual information is the source upon which we rely most.

The word 'information' signifies that it means 'to instil form within'.[905]

Visual perception is the process of picking up environmental information which instils form (of objects, surfaces, events, patterns) with a perceiver.

Since we perceive a three-dimensional world with a two-dimensional projection device (the retina), it follows that geometrical form is that which is instilled in the observer.

Geometrical abstractions are picked up and projected on to the retina to represent the external world. In this sense, the human visual system may be viewed as a 'geometry-analysing engine' It has been argued that visual perception is "the study of mapping from perceptible external objects, through optic information that represents them, to the observer who uses that information."

After the Second Word War cognitive psychologists adopted a 'process-oriented' approach in which the main focus of study became the unobservable and hypothetical mental process, such as perception, attention and memory, which were believed to mediate between the sensory reception of ambiguous stimulus information and movement response.[906]

Cognitions can be defined as any knowledge, opinion or belief about the environment, self, or behaviour potential that an individual might possess.[907]

From the traditional perspective, perception and action are determined by cognitions of the specific context of action: in our case sport settings.

In the study of perception, an important philosophical issue has been the way that the stimulus properties of the environment are sensed and perceived. How do we gain knowledge of our world so that our actions are intentional and successful?

One of the hallmarks of consciousness in biological organisms is intentionality.

For our actions to have intend they have to be purposefully directed towards objects in the environment.

Behaving with intent in a rule-governed, social context like sport requires considerable understanding of the environment. That is, in order to engage in intentional behaviour in sport one needs to interpret precisely what is going on in the world.

The cognitive perspective on the perception-action relationship has been dominated by the Cartesian school of philosophy which emphasises that the reality which we perceive is a kind of mental reconstruction of the environment.

This capacity of the mind to represent the world internally has been compared to the way that a digital computer works. The way that a computer processes information and can represent it within the system in the form of a symbolically coded language has been used as a metaphor for mental processes.

Typically, in cognitive science it has been believed that people attain knowledge composed of symbols which represent external objects in the mind.[908]

Cognition involves manipulating these symbols in an abstract and rule-governed manner according to a syntax. Operating according to these rules is known as computation.

Like a computer, it is argued, the mind 'reads' symbolic representations semantically. We rely on symbolic representations in order to carry out goal-directed activity. This is the essence of cognitivism.

The basis of this computational modelling is rooted in Newell and Simon's (1976) Physical Symbol Hypothesis.[909] They argued that an important characteristic of intelligent behaviour is the capacity to manipulate physical symbols. Their hypothesis was that the symbols used in the encoding language used to represent concepts and programmes in a digital computer "in fact the same symbols that we humans have and use every day in our lives."[910]

In other words, like computers, humans are physical symbol systems. This is how the concept of the representational mind could be physically sustained. The task for cognitive scientists is to specify the physical symbol systems which constitute the human mind or, in other words, 'that constitute systems of powerful and efficient intelligence'.[911]

Traditional theorising on perception and action in humans has emphasised the representation and communication of information in the cognitive system, rather like a hierarchical control system in the engineering sciences.[912]

A major characteristic of control engineering theory is that it posits rules or algorithms for controlling system output. A major assumption is that biological nervous systems operate in the same way as robotic or engineering control systems.[913]

The role of the perceptual systems has been conceptualised as providing the necessary stimulus for the 'release' of a specific programme of action or as contextualising a symbolic movement representation, once initiated.

In this sense, the processes of visual perception may be likened to a series of computations on the raw sensations registered on the retina during sport performance.[914] The basic cognitive science model of perception, that ambiguous sensory cues are compared with information stored in memory before output from the system can occur, has led to the popularisation of the computer metaphor for explaining perception and action in human behaviour.

In planning a motor response, cognitive psychologists have argued that skilled performers use internally represented knowledge to:

- attend to relevant sources of environmental information and to ignore the less relevant cues;
- search the visual field in a systematic and skilful way;

- anticipate events in time-constrained sports before they actually happen; verify the impoverished information which their perceptual systems receive from the environment.

This philosophical approach argued that inferential support was necessary for adequate perception of environmental stimuli because of the ambiguous nature of sensory input.

It has been proposed that, in between sensing and consciously perceiving events or objects in the environment, there was an unconscious, supportive role for the knowledge of the individual.[915]

Cognitive psychology holds that we are unable to make much direct sense of the route feature without the intervention of hypothetical processes which, effectively, reconstruct reality in our heads. Thus, perception is a process of constructing meaning and, by inference, can never be direct.

The task of reconstruction involves important cognitive activities such as remembering (the act of attempting to stereotype, on the basis of long term memory, a stimulus representation) and attending (scrutinising some and ignoring other parts of the environment).

The influence of the philosophical ideas of René Descartes may be seen in the hierarchical nature of most cognitive psychology models of the perception-action relationship.

The machine metaphor has been adopted because the mind is viewed as the organ which controls a dependent physical system (the body).

In early accounts of movement behaviour by information processing theorists, the physical components of the system were seen as merely subservient to the commands issued by the higher levels of the system.[916] Perception may, therefore, be studied separately from action, given this emphasis.

The Cartesian notion of mind-body dualism currently dominates sport-related research on perception and traditional attempts to model motor behaviour.

Recently, there has been a burgeoning interest in models of visual perception which emphasise the capacity of the neurons of the brain to link together in networks or patterns to interpret environmental information.

In these cognitive neuroscientific explanations, "the representations of the world are expressed in terms of activities in neuron-like units, rather than in terms of the construction and storage of abstract strings of symbols."[917]

The relative activity of the neurons as they connect with other neurons provides the basis for visual perception. The patterns of activity in the networks of neurons correspond to precepts. A description of an environmental feature such as the edge of the beam for a gymnast performing a routine would be provided in the brain as a pattern of connections between specific groups of neurons.

It has been pointed out that these connectionist models of visual perception have a number of advantages over traditional information processing models.

These include a better correspondence with existing neurophysiological evidence, the capacity for parallel processing and a greater potential for explaining how learning occurs through pattern formation and associative weighing.

Many theoretical advances in psychology emerged from the rise to ascendancy of the information process paradigm. These included the notion of central intermittency[918], the perceptual moment hypothesis[919] and the foundation of the serial processing idea, the single channel hypothesis.[920]

Some of the most important ideas in this respect were proposed by Poulton.[921]

He argued that two types of predictive information was necessary for successful performance of interceptive actions, such as catching.

First, receptor anticipation information is obtained on the time of arrival of a ball at the catching hand. Receptor anticipation processes are used when performers have a clear view of an object during its approach.

Successful interceptive actions are dependent upon a series of complex differentiations involving 'snapshots' of velocity and distance cues from ball flight which are related to past memories of similar events.

In information processing accounts of timing behaviour, it is traditionally argued that the observer derives time-to-contact (Tc) from a number of physical variables during the relative approach.[922]

These variables include distance, velocity and size information from an object, surface or individual.

Extensive experience in a situation allows the observer to develop an internal algorithm to compute the value of each variable in extrinsically timing an action.

For the computations, it is argued that the observer needs knowledge about the size of an approaching object before perceived information about velocity and distance can be scaled into the algorithm for computing Tc.

The observer indirectly computes Tc by dividing the object's momentary distance from the eye by its current velocity or d/v.[923]

Knowledge about object size is acquired through specific experience in a particular performance setting and is symbolically represented somewhere in the memory component of an information processing system. Thus, the more information stored in memory about the interceptive task, the more likely it is that timing behaviour will be successful.

Because expert performers have access to expansive knowledge bases which are specific to particular sport domains, it has been argued that they only need a limited amount of information from the environment to construct valid perceptions of events.

Consequently, it has been noted that, in the study of catching behaviour, the main questions during the past three decades have 'reflected concern about 'the amount' of information necessary upon which to make decisions rather than the 'nature' of the information per se'.[924]

This was the rationale behind the manipulation of viewing time of a ball (in particular the extent and the location of viewing time along the flightpath) as an experimental variable in many catching studies.[925]

The crucial questions revolved around (i) how much information needed to be present in the information processing system during successful catching performance; and (ii) the exact point in time when that information needed to be accessed by the performer's perceptual systems.

Second, after the performer has interpreted important cues on the velocity and position of the ball, the next prediction concerns when to initiate the movement. Knowledge of the temporal duration of an interceptive action, such as a bat-swing or a reach-to-catch, allows a performer to correct predict when to initiate the movement.

Cognitive theorists argued that extensive practice of a movement allows it to be included in a repertoire of programmed actions in the high-level athlete.[926] It is important to note that the successful selection of the correct movement programme is dependent on skilled perception of ball flight characteristics.

The time constraints of fast ball sports are so restrictive at the highest levels of performance that it is not feasible to readily modify the duration of parts of the movement (e.g. quicken one phase of a biphasic batting action).[927]

This type of variability would increase the programming demands upon the performer.

Rather, the skilled athlete is one who 'buys' time by exploiting the advance signals emitted by the movements of opponents for decision-making and preparation of a response.

Skill in rapid interceptive actions, such as catching and hitting a ball, is based upon the ability to detect and interpret perceptual information through a comparison with an internalised memory structure based on past experiences in similar situations.

Top class players have developed highly sophisticated models of the world which allow them to predict events and to select pre-programmed sequences of movements specifically designed to carry out interceptive tasks.

This explains why skilled athletes never seem to merely react to unexpected events, but appear to operate in the future. They use an 'anticipator mode' of action.[928]

A good example is that of a cricket fielder attempting to intercept a ball hit by a batsman.

An information processing perspective suggests that internal cognitive mechanisms provide the skilled fielder with a sound basis for interpreting very early signals from ball flight, often from advance movements of the batsman in preparing to play a stroke.

Knowledge of the state of the game also provides information on the strategic options for the batsman, since information processing psychology is dominated by the view of the skilled performer as a rational decision-maker.

It is believed that experienced performers form situational probabilities of events to plan actions in advance.

In time-constrained environments, the ability to detect and interpret early cues allows the catcher to quickly prepare the appropriate movements to carry out the interceptive task, because the signals from later ball flight are redundant and carry little additional information.

Furthermore, the ability to programme basic postural and orientational movements in addressing the approaching ball is believed to 'free' the attentional mechanisms of the expert fielder to focus on more sophisticated cues regarding what to do with the ball once it has been intercepted.

For example, while less skilful fielders need to monitor response-produced feedback on the position of the arms and hands in order to get them into the right place at the correct time, expert catchers can use peripheral vision to check whether the ball, once intercepted, needs to be thrown to the wicketkeeper's end or not.

The example above seems to question whether skilful games players need to fixate the ball for the whole of its flight as demanded by the coaching edict: 'Keep your eye on the ball!' Even now the advice of most coaches of ball games is to keep your eye on the ball—regardless of expertise.

But is this appropriate? Rather, at critical moments, experts seem to be able to switch attention between important, alternative sources of environmental information such as the position of opponents and team-mates and the location of surfaces and targets.

The first attempt to adopt a rigour, experimental approach to this question was by Hubbard and Seng.[929] They employed a highly innovative strategy to examine whether professional baseball batters needed to watch the ball for the whole of its flight in order to strike it successfully. They pointed out that top class sport performers often exhibited visual defects as measured on clinical tests.

It seemed that perceptual skill was the basis of batting excellence—a combination of visual ability and extensive experience in sport.

In order to test this assumption, they filmed the batting performance of 29 professional batters during practice.

Through careful positioning of the camera and the use of a large mirror they were able to record the whole duration of 70 pitches from release of the ball to the strike of the bat.

Despite some individual differences, the data seemed to suggest that skilled batters only needed to foveally track the ball up to 2.4–4.5m from the bat.

No further head or eye movements were recorded after this point. Furthermore, the batters seemed to reduce the scope of the motor-control problem by gearing the step before strike to the release of the ball from the pitcher's hand. That is, step duration was regulated by pitch velocity.

Faster pitches induced shorter steps and slower balls warranted longer steps.

What is the significance of this behaviour by skilled batters? This strategy had the effect of allowing the duration of the swing to be kept remarkably constant and independent of ball speed.

From the point of view of the present theoretical discussion on indirect perception, the lack of late eye and head movements could be taken to indicate that extensive practice allowed professional batters to develop an internal representation of the event in order to use early cues from the action of the pitcher and the flight of the ball predictively.

Later stages of ball flight were redundant due to the knowledge of the batters on the characteristics of various types of pitch.

More recent work by Bahill and LaRitz[930] has provided some support for the early findings of Hubbard and Seng (1954) by demonstrating that baseball pitches often reach a level of velocity which exceeds the tracking capabilities of the eye movement system. Yet, baseball batters are capable of the most exquisite timing.

For many psychologists these findings indicated the important role of knowledge founded on past experiences in supporting perception in such time-stressed circumstances.

The approach taken by Hubbard and Send (1954) was at least three decades ahead of its time since it examined the receptor anticipation abilities of skilled athletes and recorded changes in components of batting action *in situ*.

The issue of motor programming in sports contexts was specifically addressed by Tyldesley and Whiting[931] in a study of expert table tennis players.

They argued, like Hubbard and Seng, that inordinate levels of consistency in the output side of the perception-action relationship meant that performers only needed to attend to the input and decision-making components of performance.

Their argument was that 'operational timing', a means of reducing the temporal uncertainty in somewhat predictable environments by practising an action until it is highly consistent in duration, could reduce the processing demands on the performer to that of 'input timing' only.

Input timing refers to the performer's ability to compute velocity and distance information from ball flight characteristics in order to correctly predict the initiation time of an action with a known temporal duration.

They argued that timing could come about through the 'time issue of muscular commands' in the form of 'consistent motor patterning'.[932]

Expert performers, capable of a high level of motor programming due to intensive practice, seem to 'know' the precise duration of a programme, with the result that a degree of freedom is freed up in the perception-action relationship in complex environments.

In other words, for highly skilled athletes, the initial processing demands of 'operational' and 'input' timing have been reduced to the latter.

The result of a high level of task practice is that "the expert will be faced solely with a problem of temporal prediction of when to start a movement sequence which has been planned in its entirety in advance."[933]

An additional benefit is that, as skill develops, the conscious processing requirements typically involved in gaining adequate input timing information is moved to a very early portion of flight, thus freeing up the perceptual mechanisms "to smooth out minor output disturbances" in the ongoing response and for strategic planning.[934]

The term 'attention' is used to refer to three different processes.

First, the construct of attention has been postulated to explain the selectivity of attention (i.e. focussed attention).

Second, it relates to our ability to distribute attention across several concurrent tasks (i.e. divided attention).

Third, it refers to our state of alertness or readiness for action.

Selective attention is viewed as "the preferential detection, identification, and recognition of selected stimulation."[935]

It is the process by which certain information is processed while other information is ignored. An example is the skilled baseball batter's ability to focus only on pertinent aspects of the pitcher's delivery action while disregarding extraneous information.

Selective attention is involved at some level in almost all tasks, since even if the subject only attends to one visual or auditory cue, proprioceptive and interoceptive inputs simultaneously compete for attention.[936]

The second meaning of the term attention relates to the fact that skilled performers can regulate their mental resources or capacity across several concurrent actions.

An example of this is the skilled racing driver who changes gear at a difficult hairpin bend while scanning the upcoming road layout and monitoring the position of opponents in the rear view mirror. This ability to perform two or more tasks concurrently distinguishes between controlled and automatic processing.[937]

Controlled processing is slow, effortful, attention demanding and under conscious control. For example, a golfer uses controlled process in selecting the appropriate line and weight of shot into the green.[938]

In contrast, automatic processing is fast, effortless and not under conscious control (i.e. non-attention demanding).

The skilled golfer is likely to process information automatically while playing a drive shot off the tee.

Clearly, sports require a combination of both automatic and controlled processing. In some situations performers function in a 'reflexive', automatic manner, but in others, they are required to make decisions and process information consciously.[939]

The ability to rapidly and effectively switch between these different modes of processing is viewed as an important characteristic of expertise.[940]

The third usage of the term 'attention' denotes our state of alertness or 'preparedness' for action.

Typically, research in this area has examined the effects of alertness and arousal on sports performance.[941] It is argued that through experience the performer knows the important contextual cues within the display.

Take an attacking situation in football where the ball is played behind the defence into a position on the right wing with the opportunity to cross the ball into the penalty area from the goal-line. The covering defender or 'sweeper', along with the other defensive players, will be recovering back towards goal.

The sweeper would be visually coding information from the display and the most important inputs, selected by the stimulus-analysing mechanisms, would reside temporarily in short-term memory. Simultaneously, the experienced sweeper would be making use of past experiences to establish expectations and probabilities of certain events occurring.

The context in which the event occurs would be important in order to establish pertinence, for a particular class of events.

In the above situation, there is a relatively high probability of the cross being driven hard and low into the 'prime target area' between the goalkeeper and the retreating defenders along the edge of the six-yard box.[942]

Furthermore, the performer is likely to be making use of contextual cues such as the attacker's approach to the ball, preparatory stance, and early part of the kicking technique in order to anticipate ball direction.[943]

In addition, the runs and movements of other players are assumed to be important sources of contextual information.[944]

This combination of information derived from expectations and contextual cues then excites in memory the most pertinent or important event(s) to which attention may be allocated. The interaction of sensory information and past experience allows the performer to select and attend to the most important and informative areas of the display.

A more satisfactory global perspective on attention may be provided by recent development in cognitive science based on parallel distributed processing or connectionist models.[945]

The approach originates from neurophysiology and our understanding of how neurons function within the brain.[946]

A connectionist model involves a neural network of processing units or nodes that are connected by links. These nodes present abstract elements over which meaningful patterns can be defined. Control is distributed at a more local level with each node being aware only of the output from the nodes to which it is connected. Nodes simply receive input from one or more other nodes and, based on this input information, they compute an output value which is transmitted to other connected nodes.

The system is seen as being inherently parallel in that several nodes can carry out their computations at the same time.

It is the pattern of connectivity between nodes which constitutes what the system knows and determines how it will respond. Each node constrains the other and contributes in its own way to the global observable behaviour of the performer.

It has been suggested that knowledge structures are modified through experience by changing the patterns of interconnectivity between nodes. The argument is that with increasing levels of expertise performers develop new connections, lose some of the existing connection and modify the strengths of connections which already exist.[947]

In this way, the system exhibits plasticity in the sense that the pattern of interconnection is not fixed but rather undergoes modifications as a function of experience. This organisation offers a mechanism for learning by suggesting that changes in the activations of elements and the strengths of links occur with practice.

The development of automaticity and enhanced attentional selectivity may therefore result from a strengthening of these connections between nodes (termed 'weights'). A higher weight means that a stronger signal was received along a connection with less resistance. Thus, the system is seen as a more efficient and constantly evolving or adaptive neural network.

As the novice becomes more skilled there is a dramatic improvement in performance.

For example, when learning to parallel turn in skiing, attention may be initially directed towards several aspects of technique, such as the positioning of the skis and ski poles, the forward lean of the trunk, the distribution of body weight on uphill and downhill skis, the angle of the knees and the rotation of the trunk.

It is likely that novice skiers will have to consider each of these individual parts as they perform the skill. Consequently, performance will be rather awkward and not as smooth and efficient as that of skilled skiers.

As novices become more proficient, they are likely to stop thinking about the individual parts of the turn. Rather these components will be grouped together as larger parts so that the entire action becomes much more coordinated and efficient. At this stage, attention may be focussed only on the placement of the downhill ski pole or the imminent terrain of the slope. The rest of the action is carried out effortlessly without consciously thinking about any particular aspect of the technique.

These changes equate to a reduction in the amount of attention which has to be devoted to the technique. This progression enables attentional resources to be allocated to other concurrent activities and to the development of more refined performance strategies.[948]

Moreover, if skilled skiers think too much about what they are doing as they ski down a difficult black run, they may find that there is a deterioration in performance.[949]

From a cognitive perspective, directing conscious attention to the various parts of the skill may disrupt the established motor programme controlling the action. This results in an over reliance on conscious feedback mechanisms or a shift towards a different mode of control during the task leading to what is referred to as 'paralysis by analysis'.

Many researchers have suggested that a possible explanation for the expert's more polished performance could be that as a result of prolonged practice some processing activities cease to make demands on attentional resources.[950]

Following extensive practice, skills can be performed 'automatically' requiring restricted conscious attentional demand. It originally referred to an autonomous or automatic stage of learning.[951]

There has been some disagreement about the best definition of automaticity, but several researches have drawn a distinction between conscious and subconscious processes.[952]

Conscious procedures are referred to as controlled processes. They are regarded as being slow, of limited capacity, requiring attentional resources, and can be used flexibly in changing circumstances. Examples include the tennis player selecting the type of serve to play against her opponent or the golfer selecting the appropriate club to play his approach shot into the green.

Subconscious mechanisms are referred to as automatic processes. They suffer no capacity limitations, are fast, parallel in nature and non-attention demanding, and are difficult to modify once they have been learnt. For example, while executing the tennis serve or golf approach shot, skilled players are likely to be employing subconscious processes. According to this distinction, automaticity results when the performer moves from conscious to subconscious processing.

Another theory attempts to explain why prolonged practice leads to automaticity.[953]

It is suggested that: (a) every time a stimulus is encountered and processed a memory trace of that particular stimulus-response relationship is stored; (b) continued practice with the same stimulus leads to the storage of more and more information about the stimulus and its respective response alternatives; (c) this increase in task-specific knowledge leads to rapid and efficient retrieval of relevant information as soon as the appropriate stimulus is presented; (d) consequently automaticity is based on the efficiency of retrieval of knowledge from memory.

Performance is viewed as being automatic when retrieval is based on 'single-step direct-access' of information from memory. In the absence of practice the task of responding appropriately to a stimulus requires thought and the application of rules. However, prolonged practice leads to an increase in the knowledge base which subsequently permits rapid retrieval of relevant information and fast action. Daniel Kahneman, the Nobel laureate, in Thinking Fast and Slow (2002) draws a distinction between fast and slow thinking. He describes mental life by the metaphor of two agents, called System 1 and System 2, which respectively produce fast and slow thinking. He speaks of the features of the initiative and deliberate thought as if they were traits and dispositions of two characters in your mind. He argues that the intuitive System 1 is more influential than your experience tells you and is the secret author of many of the choices and judgements you make.

Chapter 39
Consciousness, Perception and the Senses in Sport

From the senses originate all trustworthiness, all good conscience, all evidence of truth.
Nietzsche

Oh for a life of sensations rather than thoughts.
Keats

Practise yourself, for heaven's sake, in little things, and then proceed to greater.
Epictetus

The more abstract the truth you wish to teach, the more must you allure the senses to it.
Nietzsche

They broke the mould of solid gold that once made Barry John.
Max Boyce, *The Outside-Half Factory*

A second, a half second in thinking is an expensive luxury with a heavy cost. There is no time to rationalise the predicament. A player responds because it is in the blood, a sixth replay of having been there before.
Barry John

I felt that the moment of a lifetime had come. There was no pain, only a real unity of movement and aim.
Roger Bannister, when breaking the 4-minute mile

Samuel Johnson sagely stated, "Example is always more efficacious than precept."

The athlete's self-awareness in peak moments often differs radically from his everyday sense of self. There is a feeling of detachment. When he broke the four minute mile, Roger Bannister says that halfway through the race "I felt complete detachment."[954]

Dancer Jacques d'Amboise also described a sense of detachment in the supreme moments when he feels in command, that he can do anything with his body. "When you're dancing like that, you seem to be removed. You can enjoy yourself doing it and watch yourself doing it at the same time."[955]

David Hemery, who set a world record at the 1968 Olympics in the 400 metres hurdles, winning by a wider margin than anyone had in 44 years, wrote of the event "Only a couple of times in my life have I felt in such condition that mind and body worked almost as one. This was one of those times. My limbs reacted as my mind was thinking: total control, which resulted in absolute freedom, instead of forcing and working my legs, they responded with the speed and in the motions that were being asked of them."[956]

Kenneth Ravizza, one of the leading researchers into this phenomenon in sport, quotes a number of athletes interviewed about their peak performances, who offer comments such as this by a football player. Everything is right, everything is clicking, nothing is opposing me.[957]

The great soccer player Pelé gives this account of one such experience.

"It was a type of euphoria, I felt I could run all day without tiring, that I could dribble through any of their team or all of them, that I could almost pass through them physically. I felt I could not be hurt. It was a strange feeling and one I had not felt before."[958]

Margharita Laski, who studied the 'triggers' or circumstances, leading up to the experience of ecstasy found that one of the major triggers was movement. She says "the kinds of exercise or movement that seem to be relevant to ecstasy are two: regular rhythmical movement such as walking, jogging along on a horse, riding in a carriage; and swift movement, such as running, flying, galloping."[959]

"An account of bull fighter El Cordobes has described many performances in which he was 'crazy happy'. He was hypnotised by his own success with his animal, unable to think of anything else but that splendid, drunken feeling of power of each movement, each pass of the bull, gave him."[960] One of the most memorable experiences of my life was meeting El Cordobes in a bar of a Spanish hotel and it was quite a night!

John Brodie, former San Francisca 49'er quarterback, has said that "a player's effectiveness is directly related to his ability to be right here, doing that thing, in the moment. All the preparation he might have put into the game—all the game plays, analysis of moves etc—is no good if he can't put it into action when game time comes. He can't be worrying about the past or the future or the crowd or some other extraneous event. He must be able to respond in the here and now."[961] This approach was echoed by Bill Walsh, the coach of the 49ers when he said "The goal of planning is to remove from a tough situation the panic element of 'What the hell are we going to do now?' The less thinking people have to do under adverse circumstances the better. When you're under pressure, the mind plays tricks on you. It's a terrible mistake to let outside forces influence you more than the practical realities of the situation already are."

Arnold Palmer wrote that tournament play golf "involves a tautness of mind but not a tension of the body. It has various manifestations. One is concentration on the shot in hand. The other is the heightened sense of presence and renewal that ensures through an entire round or an entire tournament. There is something spiritual almost spectral about the latter experience. You're involve in the action and vaguely aware of it, but your focus is not on the commotion but on the opportunity ahead. I'd like it to a sense of reverie—not a dreamlike state but the somehow insulated state that a great musician achieves in a great performance. He's aware of where he is and what he's doing, but his mind is on the playing of his instrument with an internal sense of rightness—it is not merely technical—it is not only spiritual; it is something of both, on a different plane and a more remote one."[962]

Michael Murphy said:

"Despite the many long years of instruction, study, practice and training that most athletes put in, they generally do not act consciously when they make outstanding plays. The conscious knowledge of correct and incorrect moves serves as kindling and logs to fire, but in the white heat of the event they are burnt into non-existence, as the reality of the flames takes over—flames originating in a source beyond conscious know-how, melding athlete, experience and play into a single event."[963]

He suggests that we can speak of two stages of conscious and unconscious, neither which could operate without the other. He says because we can talk about it, think about it, read about it, the conscious aspect is familiar to us all; to some it is the only reality.

He acknowledges equal recognition must be given to the lesser known, unknown, unconscious aspect in the state that can be recalled only after the play has been made, when the athlete talks about being 'unconscious', 'out of my mind', 'in the twilight zone', 'out of my gourd', playing 'over my head'. He doesn't know how he made the play—he just did it. It was instinct.

One does not consciously have to plan how to act; instead, one lets the appropriate responses happen of themselves. It's like in singing where moving from one note to another results from letting the musicality take over.

Michael Novak described the phenomenon of unity of self.

"This is one of the great inner secrets of sports. There is a certain point of unity within the self, and between the self and its world, a certain complicity and magnetic mating, a certain harmony, that conscious mind and will cannot direct. Perhaps analysis and the separate mastery of each element are required before the instincts are ready to assume command, but only at first. Command by impact is swifter, subtler, deeper, more accurate, more in touch with reality than command by the conscious mind. The discovery takes one's breath away."[964]

The importance of unconsciousness in sport has been expounded by the German philosopher Eugen Herrigel.[965]

He describes his efforts to learn Zen archery. The Zen master who taught him insisted that "the shot will only go smoothly when it takes the archer himself by surprise. You mustn't open the right hand on purpose."

Former world heavyweight champion Ingemar Johansson found something similar in boxing. He insisted there was something about his right hand—the hand that delivered the knockout punch—that was strange. He told a Life reporter that it worked independently of his conscious mind and was so fast he couldn't even see it. "Without my telling it to, the right goes, and when it hits, there is this good feeling, something just right has been done."[966]

Many athletes have recognised the importance of not acting deliberately during peak moments. They seem to know that conscious thoughts must be held in abeyance.

Catfish Hunter, in describing the perfect game he pitched against the Minnesota Twins in 1968, says "I wasn't worried about a perfect game going into the ninth. It was like a dream. I was going on like I was in a daze. I never thought about it the whole time, if I'd thought about it, I wouldn't have thrown a perfect game—I know I wouldn't."[967]

In the Ultimate Athlete George Leonard writes:

> "Pressing us up against the limits of physical exertion and mental acuity, leading us up to the edge of the precipice separating life from death, sports may open the door to infinite realms of perception and being."[968]

An experience of inner unity is the apex of the sports experience for many athletes. Boxer Randy Neumann says of running to exhaustion "It's an amazing sensation to feel your mind and body become a single force against gravity. This is also the sensation experienced in a rare fight when you pull your whole being together and pit it against an opponent."[969]

Patsy Neal describes the oneness that can be experienced in a moment of competition:

"There seems to be a power present that allows the individual to 'walk on water' or to create miracles in those precious moments of pure ecstasy. He runs and jumps and lives through the pure play process, which is composed of joy and pleasure and exuberance and laughter, even the pain seems completely tolerable in these few precious and rare moments of being, and of knowing that one is just that...a oneness and wholeness."[970]

Athletes also feel a sense of unity with their equipment. Jim Clark, the racing driver, once observed "I don't drive a car, really. The car happens to be under me and I'm controlling it, but it's as much a part of me as I am of it."[971]

John Brodie also once said "Often in the heat and excitement of a game, a player's perception and co-ordination will improve dramatically. At times and with increasing frequency, now, I experience a kind of clarity that I have never seen adequately described in a football story."[972]

Mental alertness and clarity are often associated with sports. In one systematic study, it was found that after vigorous exertion most people are "more alert, they can think more clearly, and more effective mentally."[973]

There is evidence that as a result of this heightened sensitivity, sportsmen often report more strongly etched and vivid perceptions. Not only do their minds feel clear, but they actually do perceive things more fully and vividly. The senses often register things that one never catches in normal consciousness. Athletes often have exceptionally vivid memories of these moments of heightened sensory involvement. It was said of Gerald Davies that being so lightly built he had to develop the instincts of a forest animal. It heightens perception and the adrenalin secreted adds to the surge of acceleration. Likewise of Barry John that he was fragile in build and had the same awareness of physical danger as a forest animal and that helped him to find as much time and as much space as he did.[974]

Jack Nicklaus was said to be able to remember every shot he took in a golf tournament. It appears their experiences were memorable because of the unusual state they were in i.e. state of mind that apparently contributed to their outstanding performances. Murphy asks does their exceptional state of mind determine both the quality of the performance and the vividness of the recollection of it, not the other way around as is usually assumed?

Some athletes report being able to perceive many more details than is customary for them. Barry John was described as "The dragonfly on the anvil of destruction who ran in another dimension of time and space."[975] Maybe Alistair

Maclean got an insight into this when he wrote 'Fear is the key'. It helps unlock the doors of perception. Barry John epitomised the wonder and mystery of sport. After his retirement at the early age of 27 he retained that mystique and presence which I experienced personally on a number of memorable occasions. In real life there was something untouchable about him which mirrored the elusive nature of his playing days. Hs death in February 2024 brought back those vivid and graphic memories of a genius in action and his mesmeric rugby skills.

Barry John and the subconscious

What follows is an account taken from a seminal work on rugby The Fields of Praise: The Official History of the Welsh Rugby Union, (1881–1981). It amounts to a philosophical rugby discussion involving rugby legends Barry John, Carwyn James and Ray McLoughlin. I set it out in extenso because it is so revealing as to the role of the subconscious in rugby and applicable to other sports.

"Now it so happened that 'the sub-conscious' was you might say on Barry's mind that day. The night before, after supper John had fallen into a nice, deep rooted philosophical discussion and disputation with Carwyn and that grand and deep-thinking Irish prop forward Ray McLoughlin, who was defending his insistence…that in sport, taking ball-players skills, balance and fitness for granted, the down-to-earth pragmatism and technical and mechanical excellence were all that were required—and only under those parameters, plus the unknown variables, of luck, bounce and rub of green, some men winners, some losers."

As Barry listened, intrigued Carwyn begged to differ with Ray—sports, he said, was made of all those things the Irishman had listed; but then there could be more, much more—in seemingly straightforward rugby, too, as well as possibly more obviously cerebral sports and pastimes, instinct, intuition, call it what you like, and a player can be nervous in the extreme at the precise moment, or ice cold and calculating, but suddenly, unpractised, an almost "accidental profundity' can invade his mind in a split-atom fraction of a second, and he will do something he had never thought himself capable of had he planned it for a century."[976]

That's what makes one actor, say, or one piece of journalism, or one sporting politician seems streets ahead of the other who has just as much, even more technical and well coached ability—"spiritual, subconscious, transcendental, unknowing—where it came from ruddy instinctive intuition."[977]

Looking back, "I know that try owes a lot to that first feint to drop a goal. To this day I don't know why I didn't go for it—I had enough room to pop it over. But from then on I would just '…intuitively' that not one of opposition around or ahead of me was balanced and sort of 'ready' for me. So I just continued on—

outside one inside the other—all the way to the posts. I know it is funny, but it was all as if I was in a dream, that I had 'placed' the defenders exactly where I wanted them, like poles in the garden to practise swerving. I don't know what you call it—'Transcendental'? Metaphysical? I don't really know the exact definition of those words, but it was just marvellously weird. Like I was down there re-enacting the slow motion replay before the actuality itself had happened. As if I was in a dream state of déja vu, that I was in game, and doing something that had already taken place at another time!"[978]

"A second, a half second in thinking is an expensive luxury with a heavy cost. There is no time to rationalise the predicament. A player responds because it is in the blood, a sixth replay of having been there before."

Talk me through that try, the questioner asks in the tunnel afterwards. "Tell me how it happened" is hardly likely to elicit a lucid exposition of the events the player may have experienced as the main actor at the scene. "For, quite often, in the heat and fury of the contest and the place of the defining moment within it, it is not reason that guided him to do what he did, exceptional and brilliant though the execution may have been, it is intuition."[979]

Roger Kahn in Heroes reports that the baseball player Stan Musial told him that he didn't guess what a pitch would be, he knew. He said "I pick the ball up right away…I see it as soon as it leaves the pitcher's hand. That's when I got to concentrate real hard. If I do, I can tell what the pitch is going to be."

"When can you tell?"

"When it's about halfway to home plate…I can tell by the speed. Every pitcher has a set of speeds. I mean, the curve goes one speed and the slider goes at something else. Well, if I concentrate real good, I can pick up the speed of the ball about the first 30 feet it travels. I know the pitcher and I know his speeds. When I concentrate, halfway in I know what the pitch is gonna be, how the ball is gonna move when it hits the home plate."[980]

Musial felt that concentration was the crucial thing. Murphy thought other factors might be involved. It has been noted that when conscious attention is entered on one thing, this act may allow other, more subtle perceptual abilities to come into play. It is not simply a matter of heightened visual acuity but rather some intuitive factor.

Under special conditions objects appear to be much larger than they are in reality. Cricket and baseball players talk about how large the ball is when they bat, and basketball players see the hoops getting bigger and bigger.

It has been said of Pele that:

"Intuitively at any instant, he seemed to know the position of all the other players on the field, and to sense what each man was going to do next."[981]

In studying brain surgery patients, California Institute of Technology, scientist Dr Colwyn Trevarthon believed he had discovered a second sight system that controls peripheral vision. He hypothesised that this system evolved from the primitive type of visual awareness belonging to birds, reptiles and other animal life, allowing an automatic response to action in the surrounding space. Thus, if there is unexpected movement in that space, Trevarthon conjectured that it "registers first through its second, more primitive system before the classical system became aware of it."[982] (See my chapter on Libet's half second).

There is also evidence that in peak moments in sport, the sportsman's sense of time is altered. Often it is perceived as moving more swiftly, as in the case of a chess player who said "Time passes a hundred times faster. In this sense, it resembles the dream state."[983]

Or time can be compressed into a centred moment in which there is neither before nor after. It has been observed that:

"The sense of time is disorientated. A single play may seem like forever or an innings may seem like only a second. There is no conscious sense of past time or future time. The moment-to-moment passage of time is all that is relevant; in-the-moment perception is all that the player possesses."[984]

The slow motion experience has been described by Jackie Stewart:

"Some days you go out in a race car and everything happens in a big rush. You don't seem to have time to change gears or brake and the corners are all coming up too quickly. You're not synchronised. And thus the most important thing is to synchronise yourself with the elements that you're competing against, the motor car and the track. Your mind must take these elements and completely digest them so as to bring the whole vision into slow motion."

"For instance, as you arrive at the Masta you're doing a hundred and ninety-five m.p.h. The corner can be taken at a hundred and ninety-five m.p.h. At a hundred and ninety-five m.p.h. you should still have a very clear vision, almost in slow motion, of going through that corner—so that you have time to brake, time to line the car up, time to recognise the amount of drift, and then you've hit the apex, given it a bit of a tweak, hit the exit and are out at a hundred and seventy-three m.p.h. Now, the good driver will do this in a calculated way such that as he gets out the other side he'll say 'whew, I did that well.' It wasn't a case of coming out and trying to catch the car and regain control. The driver who's fighting it, who doesn't have a mental picture in advance, will arrive at the corner to find that it's all happening very quickly. He's too heavy on the brake, the car is sliding too much, it's a big, deep breath in and a hope that I get around. Now this man doesn't have it."[985]

Runner Steve Williams says "if you do a 100 right...that 10 seconds seems like 60...Time switches to slow motion."[986]

In the most intense moments of a football game, John Brodie claims that "time seems to slow way down, in an uncanny way, as if everyone was watching in slow motion. It seems as if I had all the time in the world to watch the receivers run their patterns, and yet I know the defensive line is coming at me just as fast as ever. I know perfectly well how hard and fast those guys are coming and yet the whole thing seems like a movie or a dance in slow motion."[987]

The experience of time slowing down has been noticed by spectators, implying that there may be an objective aspect to this experience.

Herbert Saal says of the skill of Mikhail Baryshnikov

> "The most chilling weapon in the arsenal of this complete dancer was his ballon, his ability to ascend in the air and stay there, defying gravity, especially in the double tour en l'air, in which the male dancer revolves two full turns before landing. The Stuttgart Ballet's Richard Cragun can turn three times in a blur of motion. But Baryshnikov did it in slow motion. And it was unbelievable. He blasted off with the hesitation and majesty of a space-ship. He turned once twice—and every thread on his costume was plainly visible as he soared high above the audience like an astronaut looking back at earth."[988]

If slowed perception is a feature of exceptional performance in sport, then most athletes would naturally want to induce this altered perception of time.

A number of athletes especially golfers have recognised the importance of slow motion and have deliberately tried to swing more slowly in the interests of achieving more power e.g. Bobby Jones, Jack Nicklaus.

It is interesting to consider that even though the passage of time has not actually slowed down, nevertheless the fact that it feels as if it has apparently enables the athlete to accomplish more, just as he would if in fact he did have more time, more amplitude.

It has been suggested that altered time perceptions, as well as altered states of consciousness such a hypnogogic imagery and creative reverie, are associated with various brainwave patterns. It has been speculated that time perception is related to brainwaves and that this may explain some superior athletic feats.

"Having seen that time (and/or motion) goes slower the slower the brainwave rhythm, it would not be at all surprising to discover that those with superior skills—great athletes for example—may merely be blessed with basal brainwave firing significantly slower than that of the general population. This may prove to be the critical difference between the 'star' and the 'superstar'. The baseball player firing alpha, for instance, might perceive the ball at no more than half the speed perceived by his teammate firing beta. One firing theta could carefully observe the approach and spin of the ball, examine the stitches, read the label, and have up to four time as much 'time' to regulate the swing of the bat and make his moves."[989]

Roger Bannister in the midst of his record breaking race commented:

"I felt that the moment of a lifetime had come. There was no pain, only a real unity of movement and aim. The world seemed to stand still, or did not exist. The only reality was the next two hundred yards of track under my feet. The tape meant finality—extinction perhaps."[990]

It is sometimes said that athletes have a 'sixth sense'. This seems to boil down to the ability to be in the right place at the right time. It is probably more useful to think of it as a composite of several sensory modalities and abilities such as timing, knowledge, memory and perception. It seems more empathy and intuition than of speed and muscle.

Murphy[991] argues that the emphasis on such factors as empathy and intuition raises the possibility that there may be something more involved and not generally yet recognised. He suggests that in addition to 'luck' or coincidence, and picking up subliminal sensory cues there is an element of extrasensory perception or ESP in sports.

In fact, the 'sixth sense' is an older term for what today is known as ESP or psi ability. ESP includes the ability to obtain information about the contents of someone's mind (telepathy), and the ability to obtain information directly from events themselves (clairvoyance). There is also ESP of the future or precognition. These are all elements of perception and I will treat them as such. Furthermore, it is almost impossible to identify ESP in a sporting context because almost all sports take place in settings where the players are within sensory range of each other. In certain plays that are unusual, although some form of ESP may have occurred, it is more likely that the needed information was available through subtle sensory cues.

"One of the interesting characteristics is the attention paid by hitters to making sure that their trousers are just right around the genitals. It seems doubtful that positioning the genitals should have much to do with hitting a baseball, but in the language of the unconscious the genitals are the symbols of potency and power. It appears in large part to satisfy unconscious not physical needs that a batter must pay close attention to his genitals."[992]

"Many baseball players deliberately try to avoid thinking when they are hitting. This appears to be an attempt to dispel any disturbing fantasies that might creep into the player's mind. The so-called no think school of hitting has a large number of adherents. Yogi Berra has been quoted as saying 'How can I think and hit at the same time.' If the player does not think, he cannot think anything bad that will come into conflict with his conscience. Not thinking therefore serves as a protective mechanism."[993]

Beisser says that:

> "An athlete in a competitive sport must face the goal of winning. To accomplish this he must somehow master or overcome certain unconscious concomitants of winning. The aggression in the game may be tainted with implications of violence and mayhem and the athlete has learnt restrictions of conscience against such murderous thoughts. These thoughts countermand his desire to win so that he unexplainably chokes or falters in the crucial moment. Since these forces opposing victory take place outside the player's awareness, they seem magical. Even though the player logically denies the existence of magic, he finds himself bound by superstition and employing elaborate rituals to defend himself against dire consequences."

British golfer Tony Jacklin, winner of both the US Open and the British Open, admits to having experienced a state of altered consciousness some ten times in his golfing career "It's not like playing golf in a dream or anything like that. Quite the opposite when I'm in this state everything is pure, vividly clear. I'm in a cocoon of concentration. And if I can put myself into that cocoon, I'm invincible."[994] Such a state of 'concentration' might be described in terms of self-hypnosis, but that would reduce and oversimplify the experience.

Jacklin[995] continued "when I'm in this state, the cocoon of concentration, I'm living fully in the present, not moving out of it. I'm absolutely engaged, involved in what I'm doing at that particular moment. That's the important thing. That's the difficult state to arrive at. It comes and it goes, and the pure fact that you go out on the first tee of a tournament and say 'I must concentrate today,' is no good. It won't work. It has to be already here."

Precognition, the ability to know of events before they happen is reputed to go along with episodes of great emotional tension and high physical risk. Motor racing pushes human perception to its utmost limits and beyond. Sterling Moss's chief mechanic, Alf Francis has attested to the fact that on several occasions the Great British racing driver stopped his car just a moment before an axle broke or a wheel bearing lock up. Francis also believed that an elementary telepathic communication sometimes existed between him and Moss when Moss was driving.

Johnny Wilkinson saw the importance of the sub-conscious mind. He observed:

"I'd read somewhere that the subconscious mind's one million times more powerful than the conscious mind. The subconscious mind is where we store everything—perceptions, habits, experience. It forms, in my eyes a lot of who we are and our gut feelings and intuition. Compared to our louder and more recognisable conscious mind, which seems to deal with the surface makers of organising day-to-day life, the subconscious mind is largely silent."[996]

I decided then that I wanted to make good use of that subconscious mind. I want to fill it with pure positivity and inspiration. I want to store in it excellent skills on what I can rely. The way I try to do that today is by living the habit. I want to create, whenever I possibly can to be the change I want to see in the world, as Gandhi said. I try to entertain the most positive thoughts, refraining from the temptation to whinge and moan and cower in the corner. I try not to judge others, but actually to encourage instead. After enough time training these behaviours, the subconscious can look after the rest.

Just thinking and imagining is not enough. You have to practise what you think in order to form solid, effective intuitive responses. I need to see it in action.

"I need to experience the deliberate contact of the ball on my foot and see it travelling through the middle of the post many times a week, to nail down the belief that I am a good kicker."

"The 'game me' is the instinctive, ultra-competitive being—the part of myself which, when fired up, shoots first and then leaves it to be 'everyday me' to ask the necessary questions later. It is more animal than human; in this stage I go by what I see and I do what I have to do."[997]

I reckon we all have a 'game me' inside of us and in my experience it is a creative beast, it is capable of showing you things about yourself that you never knew existed, when deep habits and subconscious learnings are unlocked by the right attitude and the urgent push of pressure and need for action. This creative side tends to surface when there is no time for thought, when there is a small window, only big enough to allow for a split-second scan, a quick once-over, before acting.

"Practising flawlessly in the mind without even venturing anywhere near a field can actually improve my physical skills and begin to close the gap between imagination and reality."[998]

In *Full Throttle: the technology of speed* which was shown on British television in 2000, Alexander Wurz described the nuances of controlling a car at speed and the commentator said that Professor Mike Land of the University of Sussex "believes these images and feeling that a driver has helped him to refine a sophisticated map of the circuit he has created in his mind." A driver, Tomas Scheckter was to cover some laps so that what he was doing could be analysed by Professor Land.[999]

"Scheckter," the commentator said, "glances at the same part of the road as ordinary drivers but it turns out his mental map of the circuit allows him to use the information differently. He doesn't have time to use feed-back information from the side of the track to position the car—he's travelling too fast to react to it—but, because he knows the circuit so well, his brain is able to use the feed-forward information in conjunction with his mental map to predict the tiny reactions he needs to make to position the car on the track."

Perry McCarthy expands the theme.

"A lot of a racing driver's job is to take on information quite coldly. As we go faster and faster and faster we are acting, to a degree, like a micro-processor. You go into a corner and you store what happened to the car out of feeling. It's rather like people who are able to speed-read. They are taking it in intravenously as it were, and that is what we are doing with a car. To articulate our feelings takes an awful lot longer than a lap, and you just know what's happening. You can't say to yourself 'I am approaching this corner, I had a little bit of understeer there last time so maybe if I go a bit deeper on the brakes, turn in a bit harder. I'll do it better.' You're not doing that. It's rather like a flow chart and you say 'right, I'm going there, I'm going there, I'm going there' because of what you felt the last time."

"The feeling is subconscious but you've stored it consciously. It's in you. I can tell you when I come in from a lap, as many, many other drivers can, exactly what the car's been doing at every single corner, under braking, at the turn-in, by the apex, the exist, power on, whatever, I can re-live it."[1000]

All of this is clearly anecdotal but seems empirically to support the theoretical propositions put forward in his book about the role of perception and the unconscious in sport.

The psychologist Mihaly Csikszentmihalyi in his celebrated book FLOW[1001] has done more than anyone else to study this state of effortless attending and the name he proposed for it, flow, has become part of our language. People who experience it as a state of effortless concentration so deep that they lose their sense of time, of themselves, of their problems and their descriptions of that state are so compelling that Csikszentmihalyi has called it 'optimal experience'. Many of the above cited sporting experiences are examples of a sense of flow.

It is a form of heightened creativity. The sportsmen are effectively self-modelling their symbolic perceptions when they realise they have a choice to relax the controls and take the handbrake off. Ernest Rossi in The Psychobiology of Mind—Body Healing, describes such moments as creative breaks: "What we usually experience as our ordinary state of awareness or consciousness is actually habitual patterns of state-dependent memories, associations and behaviour one can conceptualise in 'creative moments' in dreams, artistic and scientific creativity and everyday life as breaks in these habitual patterns. The new experience that occurs during creative moments is regarded as 'the basic unit of original thought and insight as well as personality changes.'[1002] This symbolic modelling seems to induce creative breaks in our habitual patterns which result in the creation of novel ways of thinking, perceiving and being in the world."[1003]

Chapter 40
A View from Everywhere

Between 'everywhere' and 'forever', there is no compromise.
Camus

What you see, yet cannot see over, is as good as infinite.
Thomas Carlyle

Some people think football is a matter of life and death. I don't like that attitude.
I can assure them that it is much more serious than that.
Bill Shankley

Christopher Lasch spoke of sport as 'splendid futility'; Feezell as 'splendid triviality'.

This view of sport rejects any reductive notion, especially any kind of purely instrumental view of sport. If sport is conceived as merely an occasion for winning, the locus of competition, or an instrument for amusement or entertainment, something of the complexity of this form of human activity will have been left behind.[1004] I reject this view. It is far too simplistic. I prefer the sportsman's assessment. As the great Welsh rugby player Gerald Davies pointed out in his Foreword to The First XV "We will conclude that in the larger scheme of things sports is ultimately but trivial goings-on which we gladly embrace to relieve us of today's tedium to make us feel as close as it is possible to feel, that glad confident morning once more. Rugby, whether played as its glorious best or for that matter a simple unadorned victory, uplifts us to make us feel better about ourselves and the world we inhabit."

To emphasise sport as play is not just another form of reductionism.

Play is a concept that allows us to appreciate the multi-layered, complex phenomenon we call 'sport'.

To stress the notion that sport is found in the neighbourhood of play is to uphold that sport is an intrinsic good with its own internal purpose.

In this sense, sport is analogous to Kant's conception of the work of art— "purposiveness without a purpose."

To treat sport as an instrumental good is to fail to take it seriously *as* good in itself.

Christopher Lasch's notion that the problem of sport in contemporary society is not that we take it too seriously from the point of view of the commercial and nationalistic ends for which sport is used.

He says the problem is that we fail to take seriously the intrinsic value of such trivial activity.

To insist that sport is fundamentally trivial is to resist a reductionist view of sport.

That is the only way to take sport 'seriously' qua sport and to see the other ends for which sport is used as secondary.

G.E. Moore once commented that it was neither the world nor science that occasioned his philosophising; rather, the curious and puzzling things that other philosophers *said* about the world and about science caused him to start thinking. Often as in sport, philosophy is more authentic in reaction than initiation.

Some philosophers have been scandalised by his admission, as if such motivation were more appropriate for an undergraduate without a viewpoint of his own than for a mature thinker motivated by a sense of wonder and the idiosyncrasies of his own autonomous reflection.

Sport, Bodily Excellence and Play

The first systematic work on sport by an important sports philosopher, was *Sport: A Philosophic Inquiry,*[1005] by *Paul Weiss.* It is significant because it seemed to show, for some, that genuine philosophical reflection about sport might be possible. In it he offers an early, interesting account of sport. Superficial ideas provoke more profound and philosophical reaction.

He looked first at the Platonic search for the nature of sport.

The philosophy of sport involves a variety of central concerns as evidenced by recent publications which give a sense of the subject matter and scope of such reflection.[1006]

The philosophy of sport involves questions about the nature of sport and its relation to two other important concepts: play and games.

In the preface to his provocative book on sports, Michael Novak responds to the neglect of sport by serious thinkers:

> Considering the importance of sports to humankind—considering the eminence of stadia and gyms and playing fields on university campuses, comparing the size of the sports section to any other in the paper—our intellectual negligence is inexcusable. Only prejudice, or unbelief, can account for it.

"What 'grabs' so many millions? What is the secret power of attraction? How can we care so much."[1007]

In this same spirit, Paul Weiss begins his philosophic inquiry of sport by wondering why so many people are involved in sport.

> Sport does not only interest the young: it interests almost everyone. The fact compels a pause. Why are so many so deeply involved, so caught up emotionally in athletic events? Are they in the grip of some basic drive? Do they only express some accidentally acquired cultural habit of admiration for successful violence? Are they really interested in perfection? Does it perhaps give them a special kind of pleasure?[1008]

Weiss immediately adds that these questions have "philosophic import, dealing as they do—as we shall see—with what is close to the core of man, what he seeks, and what he does."[1009]

For Weiss, a philosophy of sport is embedded in a philosophy of human nature.

A philosophic account of sport answers this guiding question by relating sport to the nature of man himself and the concept of self.

As Weiss says "If a study of sport is to be of philosophic interest, it should show its relation to men's basic concerns. It will then be able to make evidence why sport is pursued almost everywhere."[1010]

A philosophy of sport, in Weiss's sense, would attempt to sift the complex elements involved in athletic experience and find the most basic factor or factors that would account for the fact that sport "interests almost everyone."[1011] It transcends boundaries and makes us to a different world.

What is the 'basic concern' of people that accounts for the widespread interest in sport?

According to Weiss it is "concern for excellence." It is not clear why Weiss thinks this is the most adequate account of why people engage in sport.

Certainly, the pursuit and appreciation of excellence is an importance aspect of human existence, and this is the first premise in Weiss's argument. But it is not its major characteristic. It is much more than that.

"Unlike other beings we men have the ability to appreciate the excellent. We desire to achieve it. We want to share in it."[1012]

If we accept this first premises, how are we led to the notion that sport is primarily understood in terms of people's concern for excellence? That doesn't distinguish from so many other pursuits in life. We all want to achieve excellence in what we do.

Weiss sees sport as an arena in which people pursue the ideal of excellence by means of physical activity. Insofar as we are bodily beings, the pursuit of bodily excellence is a necessary element in a person's quest for self-perfection.

Aristotle called *eudaimonia,* or human flourishing, beyond the perfection of the intellectual and moral aspects of human nature. But that is not unique to the domain and concept of sport.

It is time the young person channels his vitality towards satisfying the desire for excellence, a desire he supposedly shares with the rest of humanity. Young people often want to be great athletes.

"It makes good sense for a young man to want to be a fine athlete; it is not unreasonable for him to suppose that through his body he can attain a perfection otherwise not possible to him."[1013] The reason for wanting to be an athlete may be physical.

The 'athlete' is the category that best describes the participant in sport who "strives to have a fine body and to use it well."[1014]

I do not think this is an adequate account of why young people are attracted to athletic pursuits, and we also have an explanation for why the spectator may be fascinated by sports. The spectator has the ability to recognise and appreciate excellence; therefore, the spectator's fascination with the game is also related to people's attraction to ideality.[1015]

> Few men work at becoming all they can be. Fewer still try to do this by achieving a disciplined mastery of their bodies. But all can, and occasionally some do, see the athlete as an expression of what man as such can be and do, in the special guise of this individual body and in these particular circumstances. In the athlete all can catch a glimpse of what one might be were one also to operate at the limit of bodily capacity.
>
> …By representing us, the athlete makes all of us be vicariously (underlining mine) completed men. We cannot but be pleased by what such a representation man achieves.[1016]

There is little doubt that the athlete's pursuit of excellence is sometimes an important part of sport, but this account of sport is not the complete picture.[1017]

At best, Weiss's theory can stand only as a partial truth and only superficially plausible.

The reason why his view is plausible is that it seems to fit well with our conception of the dedicated athlete but not the philosophical athlete.

As Keith Algozin has suggested "For Weiss the model of the athletic pursuits display of bodily excellence is the Olympic Games. Men and women of all nations have trained their bodies to challenge the resistances of space and time with speed, endurance, strength and accuracy and coordination prescribed by the various particular sports."[1018]

But can the emphasis on the idealised conception of an Olympic athlete capture the reality of the 'universal interest' in sport cited by Weiss.

The mistake here is rather like the mistake made by some philosophers of art who provide a plausible theory of one particular aspect of art but have insurmountable problems when the theory is extended to other types of art.

Weiss is misled in his theory of sport by his suggestion that the crucial factor to be explained by and incorporated into a philosophy of sport is the fact that 'young men' are most involved in sport.

But why should this be taken as the most important clue to a philosophy of sport? It could very well be that even if Weiss's explanation for why young 'men' engage in sport is true, this account would not explain why so many others are interested in sport. The problem is not just Weiss's misleading emphasis on the fact that it is young people who are most absorbed in sport.

Weiss argues that the 'athlete' is the category that best describes the participant in sports, and he wonders why young men want to become athletes, as if the athlete is the embodiment of bodily excellence. He ignores the intellectual and philosophical stimulation achieved through the playing of sport.

The 'athlete' is not just a conception that represents the ideal of the physical being that the person ought to become. It is what he wants to become as a person.

Why do so many people engage in sport? At the cost of offering an answer that seems trivial, banal, or too simplistic, this element of enjoyment or fun, which is an intrinsic part of participating in sports, leads one to consider play as an explanation for the attraction of sport. Sport might well be termed "enjoyment in play."[1019] We find sport in the neighbourhood of play, a particular kind of play. [1020] Play transcends the physical and has philosophical connotations.

Weiss briefly discusses the play of children in relation to a version of the social theory of sport—that the play of children develops them as social beings—but quickly discards the theory as it relates to sport, because "the sport of men and the play of children…are distinct from one another in structure. They have different results."[1021] He was wrong in coming to that trite conclusion.

Weiss does not deal adequately here with play and sport in philosophical terms.

Throughout the central chapters of his work he fails to treat the conceptual dimension of sport that relates to play. He finally discusses play in a later part of the book, only to dismiss play as a category that has little to do with the fundamental aspects of sport and the athlete.[1022] That is Ryle's terminology was a 'category mistake'. His focus purely on 'athleticism' is misconceived.

Neither Huizinga nor Caillois have as a central goal to provide a philosophy of sport. But their discussions of play are enlightening, when viewed as the foundation for a perspective that conceives of sport as play and therefore beyond the concept of sport philosophically and metaphysically.

Huizinga emphasises the freedom of play and insists that "the *fun* of playing resists all analysis, all logical interpretation...It is precisely this fun element that characterises the essence of play."[1023]

Huizinga summarises his notion of play in the following passages:

[play is] a free activity standing quite consciously outside 'ordinary' life as being 'not serious', but at the same time absorbing the player intensely and utterly. It is an activity connected with no material interest, and no profit can be gained by it. It proceeds within its own proper boundaries of time and space according to fixed rules and in an orderly manner. [1024]

...play is a voluntary activity or occupation executed within certain fixed limits of time and place, according to rules freely accepted but absolutely binding, having its aim in itself and accompanied by a feeling of tension, joy and the consciousness that it is 'different' from 'ordinary' life.[1025]

Caillois describes play as free, separate, uncertain, unproductive, and governed by both make-believe and rules.[1026]

Huizinga and Caillois offer these descriptions of play without explicit reference to sports and quite rightly so. They are beyond the concept of sport.

But when we reflect on these descriptions with sport experience in mind, they strikingly reveal the relevant phenomena in a more adequate manner than Weiss's account.

Both Huizinga and Caillois emphasise that play is an activity with which one positively identifies, and in this sense it is free. Weiss explains it this way: "Play is free in the sense that it is carried on by the player only while he desires to engage in it."[1027] Huizinga stresses the freedom of play as it first central characteristic. Play is never necessitated, it is like an 'ornament', he says. It is a gift, added to life, 'superfluous' in the sense that it is not a product of biological, cultural or moral necessity. Closely related to the freedom of play is its separateness.[1028]

Play is 'separate' in the sense that when one plays, there is a movement from the world of ordinary concerns to the immediate involvement in an activity that suspends the ordinary.[1029] Huizinga also speaks here of the 'disinterestedness' of play. When we play, we momentarily transport ourselves from the world of work, moral duty, and human needs—the 'serious' concerns of everyday life—to an alternative world that has an 'as if' quality.

Although play is not 'serious' when contrasted to the world of work and human suffering, it is often wholly absorbing and engaging. It has its own internal 'seriousness' that frees us from our usual everyday worries. Often, but not always, a play world is established, constituted by its own rules and bounded by unique orderings of space and time, internal to the play world.

In this world of play, especially in the competitive playing of games, something is at issue.

Finally, and perhaps most important, the activity is engaged in for its own sake. It is autotelic, intrinsically valued, not instrumentally desired. Play is engaged in for the sake of the intrinsic enjoyment of play itself. One might see the play of sport as a free and immensely enjoyable physical activity engaged in for its own sake. This conception of sport more nearly captures the sense in which it is of 'universal interest'[1030] and reflects the way man behaves in other domains and pursuits.

George Sheehan says the following:

"The intellectuals who look at sport start with the assumption that it must serve something that is not sport. They see its useful functions of discharging surplus energy and providing relaxation, training for fitness and compensation for other deficiencies. What they don't see is that play is a primary category of life which resists all analysis.

Play, then, is a nonrational activity. A supralogical nonrational activity in which the beauty of the human body in motion can reach its zenith. Just as the supralogical feast of Christmas confirms man's unique value and destiny. So the intellectuals are probably as upset with play as the theologians are with Christmas. Men having fun is as mystical and supralogical as the Word made flesh."[1031]

Chapter 41
The Freedom of Play

The love of liberty is the love of others: the love of power is the love of ourselves.
William Hazlitt

Let every nation know whether it wishes us well or ill, that we shall pay any price, bear any burden, meet any hardship, support any friend, oppose any foe, to assure the survival and the success of liberty.
John F. Kennedy

No one can be perfectly free till all are free; no one can be perfectly moral till all are moral; no one can be perfectly happy till all are happy.
Herbert Spencer

Man is born free and everywhere he is in chains.
Rousseau

In every real man, a child is hidden who wants to play.
Nietzsche

Freedom has a thousand charms to show,
That slaves, howe'er contented never know.
William Cowper (1731–1800)

Freedom, one must be very careful with that. In painting as in everything else. Whatever you do, you find yourself once more in chains. Freedom not to do one thing requires that you do another, imperatively. And there you have it, chains.
Picasso

Freedom's just another word for nothing else to do.
Kris Kristofferson

Arguably the greatest desire of mankind is to be free. That is why Freddie Mercury of Queen majestically proclaimed 'I want to break free'. Sport is a manifestation of that desire. It is a derivative of play which is something we do in our free time.

In our English language semantically, the use of the epithet free is used to connote something to be desired, which is positive and beneficial and beyond the norm and creative like free trade, free elections, free media, free thought, free speech, free will and free movement. The hallmark of something being free is at the heart of a civilised democratic society. Sport exemplifies that attribute and characteristic, at least philosophically it should.

The concepts of freedom and liberty are paradoxical and fascinating ones. John Ruskin wrote in *The Seven Lamps of Architecture* (1849) "The treacherous phantom men call Liberty".

Daniel Dennett, who sadly died in April 2024 on the eve of the publication of this book, (see Times obituary April 25th 2024), had a lot to say about free will and consciousness as in Consciousness Explained (1991) and Darwin's Dangerous Idea (1995). As would have been seen from earlier chapters particularly on consciousness I have drawn heavily on Dennett's ideas. Dennett was a paradoxical character himself and although American had been himself heavily influenced by the views of Gilbert Ryle and had taken a DPhil with Ryle when they were together at Hertford College Oxford. He has often been characterised as a difficult character whose appearance it was said was modelled on his admiration of Charles Darwin, his hero. He had a prodigious moustache and a billowing white beard, grown it was teasingly suggested in homage to Darwin. The Times obituary described him as 'physically imposing, self-assured and incapable of suffering fools silently'. He nevertheless had a whimsical side as well as a gift for invective. For example, in 2023 he published a work 'I've been Thinking' described by one British reviewer as a "paradoxically engaging and annoying memoir". He recalled in his last work how one philosopher had said "Dan believes modesty is a virtue to be reserved for special occasions". When asked by the Observer where such intellectual confidence came from, Dennett paused for a moment, smiled a little and answered: "well, being right helps, I guess".

Notwithstanding that reputation, there was a romantic side to him evidenced by the fact that he kept a 42-foot cruiser on the Charles river in Boston named Xanthippe after the wife of Socrates. Readers will forgive my digression but I think it reveals what a free-thinking philosopher he was. This is further illustrated by what he says in the book Freedom Evolves (2003) when he describes poetically the concept of free will. "Free-will is like the air we breathe and it is present almost everywhere we go, but it is not only eternal, it evolves and is still evolving". We can see there the influence of Darwin on his thinking. Dennett had a naturalistic way of seeing things. He saw the brain as a super computer and believed its natural engineering- or neuroanatomy was where everything was located from intelligence and consciousness to sense of self and intentionality.

Readers may have noticed that in about The Author the publishers mention the fact that I am a freeman of the City of London but may not know what that means. I certainly did not until I became one! The medieval term 'freeman' meant someone who was not the property of a feudal Lord but enjoyed certain privileges. In medieval thought freedom and privilege were correlatives. That in the eyes of many may still be the case. Even today the benefits of being a Freeman of the City of London include- the right to drive sheep and cattle over London Bridge, to carry a naked sword in public, and that if the City of London Police finds a freeman drunk and incapable, they will bundle him or her into a taxi and send them home rather than throw them into prison. I am bound to say that I have never availed myself of those privileges nor do I know of anyone who has. But one day perhaps, especially the latter might come in handy. But I jest. Plato believed that myths to justify inequalities of wealth and power were essential to preserve order in society. Are the concepts of freedom and liberty just myths?

One recent commentator Raoul Martinez in a stimulating book Creating Freedom (2017) says "Facing up to the limits on our freedom explodes a number of persistent myths-surrounding individual responsibility, justice, political democracy and the market. Some of these myths persist because they advance the interest of those in power; because they flatter up offering false comfort. The way we think about freedom shapes our view of the present and our vision of the future. It is a lens through which we interpret and evaluate the world, a compass by which we set our course".

I agree with him and that is why I want to explore that concept in the context of play and sport. Martinez also speaks eloquently about the value of truth and to try and follow it beyond the shell that encloses our present understanding, to break through disciplinary boundaries and limits to our imagination. I agree that it is time to re-think the ideal of freedom. Play and Sport could and should be a catalyst for doing that both individually and collectively.

In his book, *On Being Free,* Frithjof Bergmann has presented a radical rethinking of the notion of freedom.[1032]

Bergmann explores a number of different and conflicting views of freedom in the attempt to identify some common structure that might reveal the ultimate presuppositions of the notion of freedom.

He focusses at length on Dostoyevsky's exemplar of freedom, the Undergroundman.

The Undergroundman thinks that a free act is one that 'offends reason' and freedom "demands that rationality be violated."[1033]

A free act is described as an act of "sheer caprice, performed in total independence, in rebellion against every consideration of advantage or of reason."[1034]

Bergmann asks: "What is the experience presupposed by this idea of freedom."[1035]

The answer he gives is:

> "The affirmation of the Undergroundman that one must act contrary to reason to be free presupposes that he experiences his rationality as something other than himself."

Thus, the Undergroundman does not identify even with his own thoughts; the demands of rationality are experienced as coercive forces dictated by order and society.

An act of self-identification is presupposed by his view of freedom, and it is the Other, not his authentic self, that speaks to him in the guise of rationality.

What it means to be free depends upon how the Undergroundman conceives of his 'real' or 'true' self.

Likewise, when other theories of freedom are examined, this same structure is revealed. J.P. Sartre said "Man is condemned to be free because once thrown into this world he is responsible for everything he does."

The Platonic view of freedom, contrary to the Undergroundman's irrationalist position, holds that we are free only when we follow the dictates of reason; we are unfree when our actions are enslaved by and conform to the demands of the passions.[1036]

On this view a free act is one performed according to reason.

Reason is not a coercive force; rationality is experienced as an expression of one's authentic self; and what Bergmann calls 'identification' is presupposed by the theory. Bergmann offers 'a kind of definition' after examining yet a third conception, and the definition is a metatheory, a theory about theories of freedom.

"An act is free if the agent identifies with the elements from which it flows; it is coerced if the agent disassociates himself from the element which generates or prompts the action. This means that identification is logically prior to freedom, and that freedom is not a primary but a derivative notion."[1037]

As the self forms and is formed, we come to realise and experience the way in which some of our actions conform to what we 'really' are.

There is a natural sort of conformity between what we are and what we do, and when we do whatever naturally expresses what we are, there is a sense of liberation.[1038] Rousseau as I have earlier quoted said "Man is born free but everywhere is in chains." Play enables man to unshackle those chains.

Bergmann expresses it this way:

> "Freedom for us is the expression of what we are, of the qualities and characteristics we possess, but in an unpretentious sense: it is the expression of qualities with which we identify."[1039]

Freedom is a 'matching':

> Our outward life has to match our identity or our self if we are to attain freedom. We have to achieve something like geometrical congruence, a mutual fit, a kind of attunedess, like a harmony between two tones. There should be a basic sense of ease, as when two gears spin without friction in a prearranged synchronisation. The usual stress on the difficulty of freedom…should begin to have some slight ring of melodrama and pathos, and just the reverse side should make itself felt: the absence of strain, the collapse of tension, the lightness of freedom, glorious as that of *pure play* [emphasis mine.][1040]

If play is a free activity, and freedom involves identification, then play deeply expresses what I am. Freedom does not mean escape from the world; it means transformation of our entire way of being, our mode of embodiment, within the lived world itself.

Far from being unimportant or frivolous, the free activity of play expresses some aspect of myself that I take to be 'real' or 'authentic'. If I may be forgiven for that reference to the present or the topical and make a comment on the pandemic and lockdown, I am reminded of one of my hero's remarks, Benjamin Franklin, who wrote "They that can give up essential liberty to obtain a little temporary safety deserve neither liberty nor safety."[1041]

Play and the Absurd

Sisyphus was condemned by the gods to the external drudgery of ceaselessly and strenuously pushing a huge stone to the top of the hill, only to see it roll back down, where he must retrieve it and continue his endless toil.

The punishment inflicted upon Sisyphus is not absurd or meaningless simply because he is condemned to lead a life of endless toil.

Here is an image of a being, with a human heart and human desires, confronting a future in which any desire contrary to the desire to push rocks must be externally frustrated.

What is absurd, perhaps, is the incongruity between human purposiveness and necessary frustration.[1042]

Albert Camus interprets the myth. Sisyphus is the image of a being whose whole being "is exerted towards accomplishing nothing."[1043]

This is activity that has no telos; no goods come about, no external justification invests it with meaning.

Richard Taylor also argues that the toil is meaningless because "nothing ever comes of what he is doing, except simply, more of the same…the work is simply pointless."[1044]

Now consider *homo ludens-*. for example, the pick-up basketball player or the ageing tennis player. Is *homo ludens* Sisyphus? Is his play absurd; like the pointless toil of Sisyphus? Admittedly, there is a sense in which nothing comes of the playing.

This is the essential aspect of play that Caillois called 'unproductive'.

As he argues, there may be an exchange of money involved—playing games of chance, for example—but no goods are created by the activity. The labours of Sisyphus result in no stone edifice being constructed; the activity of the player results in nothing tangible being produced.

As Caillois puts it:

> A characteristic of play, in fact, is that it creates no wealth or goods, thus differing from work or art. At the end of the game, all can and must start over again at some point. Nothing has been harvested or manufactured, no masterpiece has been created, no capital has accrued. Play is an occasion of pure waste, see Umberton ECO: waste of time, energy, ingenuity, skill…As for professionals…it is clear that they are not players but workers. When they play, it is at some other game.[1045]

Just as the abilities, interests, and desires of Sisyphus must be eternally wasted in pointless activity, there is a common intuition that the person who spends much of his time playing games is absurdly wasting his "time, energy, ingenuity, skill" in the pursuit of pointless and adolescent tasks.

Caillois' remarks are also suggestive of another point. Sisyphus' labour must inevitably lead to boredom because of the endless repetition of his activity.

Likewise, the game begins, it is momentarily played out, and it ends.

Nothing comes of it, yet it will begin again, over and over, in the playful life of the player who continuously seeks to pursue this pointless repetition. Nothing comes of his play—except more of the same.

Thus it is not prima facie implausible to compare play and the absurd toil of Sisyphus.

However, even in the light of what has just been said about the seemingly pointless, unproductive, and repetitive aspects of play, it would be a mistake to construe the absurdity of play in terms of the image of Sisyphusian labour.

The differences are more striking than the similarities.

First, Sisyphus is condemned and sentenced by the gods. We are to regard his labouring as wholly involuntary

But this stands in stark contrast to the phenomenon of play. As we have seen, play is characterised by freedom.

"Typically, the player freely decides to play, and this very freedom associated with the phenomenon of play suggests a variety of other central aspects. Because the player freely chooses to play, he must identify with his activity. He must see it as very much his own sort of doing, unlike activities that are alienated and seem to be foreign to the one who engages in them."[473]

Play is never constrained or forced, else the activity wouldn't be play.

Moreover, although play is in some respects as unproductive as the toil of Sisyphus, that doesn't mean that play is a meaningless as Sisyphus' toil.

Although nothing comes of play, the free decision to play suggests that one desires to engage in the activity, it is a good.

But it is good not by virtue of any goods that it brings into being; we have already denied that this is the case.

Play is activity that is intrinsically valued.

In fact thinkers such as Robert Osterhoudt have suggested that play is less a description of concrete activities than a way of viewing and living a variety of activities:

> Play is therefore of a different order than concrete activities as such: it is effectively a way of regarding these activities. It is a quality of concrete activity by which the activity(ies) to which it is 'attached' (or in which it inheres) is intrinsically valued, or valued in-and-for-itself, and so voluntarily engaged, of an extraordinary or supra-mundane and disinterested character, and aesthetically ordered.[1046]

Furthermore, there is no 'end' to the toils of Sisyphus. His activity is pointless repetition, unproductive like the play of humans.

While Sisyphus' labour and play may be unproductive in one sense—thus having no 'end'—there is certainly an end associated with much of human play that we do not find in the image of Sisyphus.

When the player in sport chooses to play, he voluntarily commits himself to the rules of the game in which he is playing. The rules describe the proper end(s) of play and also prescribe the only acceptable means by which the end may be brought about.

Within a game there is a telos at the centre of all the activities associated with playing it.

Activities take on a clear meaning by virtue of the rules internal to the game.

A world of transparent meaning comes about. Something is at issue, players freely commit themselves to goals and prescribed behaviour, and something gets settled.

These play activities are purposive insofar as they are oriented towards the ends internal to the play world, and they lead towards a consummation.

There is no real consummation in the activity of Sisyphus; although he momentarily reaches the top of his hill, nothing really gets settled because no 'end' has been at issue.

Thus play is a free activity, intrinsically valued and therefore meaningful, joyous, or happy, ending in a consummation or fulfilment.[1047]

If play is absurd, it must not be absurd by virtue of an exact analogy with the absurdity of Sisyphus' plight, for the analogy is obviously an imperfect one. Absurdity must not exactly be meaninglessness.

Consider Richard Taylor's remarks: "Meaninglessness is essentially endless pointlessness, and meaningfulness is therefore the opposite. Activity, and even long-drawn-out repetitive activity, has a meaning if it has some culmination, some more or less lasting end that can be considered to have been the direction and purpose of the activity."[1048]

Unlike Sisyphus, the player deeply identifies with the activity in which he freely decides to participate. Attitudes, here, are of greatest importance, for suppose that Sisyphus wanted to do what he is, in fact, doomed forever to do by necessity.[1049]

We should think that his activity would then be like the bliss of heaven rather than the torment of hell, for his activities would be in perfect conformity with his desires, and each moment would be the experience of immediate gratification.

What makes Sisyphus the image of an absurd life is the gap between his desires and the reality of his activity; he is burdened by attitudes absurdly inconsistent with his situation. It is difficult to conceive of something 'objectively' absurd, because the absurd seems to be a function of our wills, desires, or interests.[1050]

The player throws himself joyfully into an intrinsically valued activity, and therefore his play seems different from the endless toil of Sisyphus.

Yet this is also a tension between his own attitudes and the reality of playful activity. Unlike Sisyphus, the player regards and lives his activity *as if it* were truly significant and meaningful but, like Sisyphus, he realises that his play really does come up empty.

Nothing comes of it, and he knows that his play isn't *really serious.* We need to examine this tension more closely, but first we must more closely clarify the concept of the absurd.

For Camus, the absurd resides neither within the person alone nor within an irrational world. The absurd arises in the relationship between the person's desire for understanding, unity, and the absolute, and a reality that will not divulge any ultimates.

The absurd for Camus is really a brand of epistemological scepticism, and his early philosophy is the attempt to work out the existential implications of this Archimedean uncertainty.[1051]

He is a failed Cartesian who wants to ground his life only on the basis of certainties, yet must finally draw out the honest consequences of recognising that the only certainty is the absurd:

> "The absurd is born of this confrontation between the human need and the unreasonable silence of the world. This must not be forgotten. This must be clung to because the whole consequence of a life can depend upon it."[1052]

Although Camus' description of the absurd may be limited, insofar as he insists that the key to the necessary absurdity of life rests upon the contingent desire 'to know' in some ultimate sense, his analysis is very suggestive.

The absurd arises in the incongruous clash between our aspirations and the reality we confront.

Thomas Nagel has attempted to clarify the concept of the absurd and show how life as a whole is absurd.

"In ordinary life a situation is absurd when it includes a conspicuous discrepancy between pretension or aspiration and reality."[1053] An absurd situation might be called an instance of 'local' absurdity. Obviously, there are a variety of situations in life that often meet this description, based on the particular contingent desires and aspirations of people.

But Nagel's claim is much stronger, he thinks life is 'globally' absurd:

If there is a philosophical sense of absurdity, however, it must arise from the perception of something universal—some response in which pretension and reality inevitably clash for us all. This condition is supplied, I shall argue, by the collision between the seriousness with which we take our lives and the perpetual possibility of regarding everything about which we are serious as arbitrary, or open to doubt.[1054]

How, according to Nagel, is this seriousness necessarily undermined?

We cannot live human lives without energy and attention, nor without making choices which show that we take some things more seriously than others. Yet we have always available a point of view outside the particular form of our lives, from which the seriousness appears gratuitous. These two inescapable viewpoints collide in us, and that is what makes life absurd. It is absurd because we ignore the doubts that we know cannot be settled, continuing to live with nearly undiminished seriousness in spite of them.[1055]

Perhaps Nagel's analysis is more akin to Camus' discussion of the absurd than first appears, because, although Nagel doesn't simply focus on the mind's desire for understanding, there is still in his elucidation the notion that the absurd is generated by scepticism.[1056]

For Nagel, we do and must live our lives with utmost seriousness, yet we have the power to disengage ourselves in reflection from the immediacy of living and recognise that the very hinges upon which our life turns are themselves arbitrary and nonderivative.

The very contingency of any belief we take as ultimate undermines the seriousness of life.

Reflective detachment provides the permanent possibility of fuelling a doubt that is more real than the abstract Cartesian methodological procedure.

Once again, Nagel argues

But this is precisely what provides universal doubt with its object. We step back to find that the whole system of justification and criticism, which controls our choices and supports our rationality, rests on responses and habits we never question, that we should not know how to defend without circularity, and to which we shall continue to adhere even after they are called into question.[1057]

What is the upshot of this recognition of the absurd?

Just as Camus wanted to draw out the consequences of the absurd in the life of the absurd hero, Nagel also thinks such a recognition has profound implications for our basic attitude towards life.[1058]

This scepticism will certainly not force us to abandon our fundamental hinges or beliefs, but, according to Nagel, "it lends them a peculiar flavour…we return to our familiar convictions with a certain irony and resignation."[1059]

Our seriousness would be mediated by an ironic sense of our own limitations and an unconvinced retreat from dogmatic claims of ultimacy.

An ironic sense of detachment never allows the world to be too much with us.

Such an attitude recognises that Promethean scorn, Sartrean despair, or religious dogmatism are equally invaded by a spirit of seriousness that either denies our finitude, romantically thinks that the situation can be other than it is, or hopes that life can be magically transformed by defiance.

There is a basic absurdity at the centre of play.

Recall that the absurd involves "a conspicuous discrepancy between pretension or aspiration and reality."

The player must at one and the same time embrace the seemingly contradictory attitudes that his play world is a fiction, his commitment to the arbitrary rules of his game is utterly gratuitous; yet he must play *as if it* really mattered, because his decision to play necessitates such commitment.

Without commitment, he isn't really playing, with commitment he must act *as* if his play really mattered, even though it doesn't.

Without pursuit of victory, the play of contests is reduced to mere frolic and the spirit of the play world is lost.

The absurdity of play describes the incongruous collision between the single-minded aspiration of the player and an ontology of play that undermines the seriousness of the pursuit.[1060]

Regardless of whether Nagel is right about life as a whole, the local absurdity involved in game playing is apparent.

There is, consequently, something deeply ironic about the attitude of the player.

He must attempt to balance aspiration and reality in a precarious way.

To become too serious about his play would negate the irony.

The paradigm for such an abuse of the ironic stance of the player is the famous remark, "Winning is not the most important thing: it's the only thing."

Yet equally abusive would be another famous dictum about sport, "It's not whether you win or lose, but how you play the game," as if play were merely the instrument of moral education.

Finally, it would also abuse the irony of the play attitude to emphasise entirely the intrinsic value of play to the exclusion of a serious pursuit of an end—"It's not whether you win or lose, but *that* you play the game."

All of these are only partial truths that overemphasise one side of the dialectic. Play is serious, but unreal.[1061]

You must have it both ways.

Recall Nagel's position that "philosophical scepticism does not cause us to abandon our ordinary beliefs, but it lends them a peculiar flavour…we return to our familiar convictions with a certain irony and resignation."

In this ironic attitude resides the superiority of the attitude of play.

Play is absurd from the standpoint of what we normally take to be serious.[1062]

Yet even this absurdity can be undermined by an altered ontology of play that describes play as what is most important. We can take up either of these incongruous standpoints; play or 'real life' may be taken as absurd.

In life, as well as in play, we must pursue our goals *as if* what we did really mattered, knowing full well that we are also precisely that person who can take a standpoint from which such seriousness is undermined.

Our life remains the same, but, as Nagel says, "We then return to our lives, as we must, but our seriousness is laced with irony."[1063]

But may I conclude by returning to Rousseau's comment in Emile "That man is truly free who desires what he is able to perform and does what he desires." Sportsmen can achieve that freedom.

Chapter 42
Sport and the View from Nowhere

You never really understand a person until you get inside their skin and walk around in it.
Atticus (Harper Lee, To Kill a Mockingbird)

"Isn't football the toy department of life?" American television journalist Mike Wallace asks[1064]. Churchill said he spent his whole life playing with toys.

In a number of essays, many of which have been collected in *Mortal Questions,*[1065] and in *The View From Nowhere,*[1066] Thomas Nagel has analysed a single problem that is one of the most debated issues in philosophy.

Initially, he spoke of this as the problem of 'subjective and objective'[1067]: the conflict or opposition between two very different viewpoints we can take towards ourselves and our experience of the world.

This ignores the reality that we experience life from a particular perspective, (taking with us) the contingencies of particularity as we make our way through the world.

But it is also clear that we can step back from our immediate involvement in life and take a new viewpoint that includes the original, more particular perspective.

The possibility of transcending one's subjective viewpoint and seeing it from a larger perspective appears to be a reality for a complex reflective being.

Nagel speaks of objectivity as the "view from nowhere" and systematically explores this problem in the book by the same name.

His concern is "how to combine the perspective of a particular person inside the world with an objective view of the same world, the person and his viewpoint included" *(VN, 3)*.

Nagel's idiosyncratic approach to the philosophical problems associated with freedom, knowledge, ethics, the metaphysics of mind, and the meaning of life generates both metaphilosophical conclusions and broader notions about basic attitudes towards life and death.[1068] I do not find his approach philosophically erudite or instructive.

In most cases, Nagel neither offers a solution to these perennial philosophical problems nor engages in therapeutic attempts to dissolve our worries. In fact, he shows why our worries are both real and seemingly intractable.

He attempts to show why we find these issues so perplexing and how we might live in the face of such uncertainties. His philosophy is one of confusion—and somewhat disingenuous in some respects.

He says, philosophical understanding resides in proper perplexity:

> "Certain forms of perplexity—for example—about freedom, knowledge, and the meaning of life—seem to me to embody more insight than any supposed solutions to these problems" (VN, 4).

That doesn't really help us a lot.

His approach is helpful in understanding the sources of the forms of perplexity associated with thinking about sports participation but not particularly constructive.

I accept his way of looking at things has important implications for our attitudes towards life in general, not simply our attitudes towards sport.

This is because reflections about sport are related to arguments about what matters in life.

I acknowledge his worries about sport naturally involve worries about the value of these activities, and questions about the value of sports participation lead inevitably to questions about the foundation of our idiosyncratic attachments in life. But he reaches no firm conclusions to the questions.

Why do we care so much about certain apparently trivial activities?

Isn't meaning in life intimately related to the pursuit of worthwhile projects?

One of the sources of philosophical interest in reflecting about sport is found right here: philosophical thinking about sport leads to questions about what in fact matters to us, why it matters, and how we might, in principle, show what should matter to us.[1069]

As Nagel concedes, to gain "a more objective understanding of some aspect of life or the world, we step back from our initial view of it and form a new conception which has that view and its relation to the world as its object…we place ourselves in the world that is to be understood" *(VNA)*.

The distinction between subjective and objective is that the development of objectivity delivers a new way of seeing the world as a whole.

Objectivity has metaphysical implications, since it seems to assume that the way things are is ultimately "a conception of the world which as far as possible is not the view from anywhere within it" *(MQ, 206)*.

The fundamental distinction is a matter of degree, often truth-guided, and offers a view of the world that attempts, in principle to be centreless.

Yet if the situation were so simple, there would be no problem associated with the distinction between subjectivity and objectivity. That is, there would be no problem of combining the two perspectives, since the point of developing an increasingly objective view of life would be to give up appearances in favour of a true, more objective understanding of reality. [1070]

It is always more truthful to negate the subjective in whatever way seems appropriate for a more objective way of understanding.

Nagel suggests, the "internal-external tension pervades human life, but it is particularly prominent in the generation of philosophical problems" (VN.6).

To give just two examples: from the inside of life, I seem to be an agent who performs free actions and for whom autonomy means the ability to do otherwise.

However, when I step back from my actions and see them as events or processes with causal explanations, and I see myself not as some mysterious immaterial nugget but as the locus of various causally produced biological and psychological processes, agency seems to vanish.[1071]

Likewise, from the internal viewpoint, I seem to have privileged access to my own irreducibly subjective mental states. I am directly aware of tastes, sounds, feelings, and so on, and I assume that there's "something it's like" to be a particular human being.[1072]

When I step back, however, and see myself as a part of nature, it's not clear how dualism could be true, in part because it has so little explanatory power when contrasted with various forms of materialism and in part because the interaction between immaterial mental processes and brain processes would be so mysterious.[1073]

In both of these examples, the objective perspective confronts something that resists its reductive or eliminative impulses.

I agree as Nagel remarks, "The trouble occurs when the objective view encounters something, revealed subjectively, that it cannot accommodate. Its claims to comprehensiveness will be threatened. The indigestible lump may either be a fact or a value" *(MQ,* 210).

In the philosophy of mind and the freedom-determinism debate, the objective viewpoint can't accommodate certain 'internal' facts about the self as an agent, whose existence seemingly can't be reduced to a set of complex physical processes.

Any justification of value may be sceptically undermined:

> Objectivity itself leads to the recognition that its own capacities are probably limited, since in us it is a human faculty and we are

conspicuously finite beings. The radical form of this is philosophical scepticism, in which the objective standpoint undermines itself by the same procedures it uses to call into question the prereflective standpoint of ordinary life in perception, desire, and action. Scepticism is radical doubt about the possibility of reaching any kind of knowledge, freedom, or ethical truth, given our containment in the world and the impossibility of creating ourselves from scratch (VN.7).

We are left with a very perplexing situation and this is the problem with Nagel's approach.

Sometimes the objective standpoint offers a truer way of understanding the world.

When we transcend our prejudices and presuppositions, we are able to see the way our particular nature distorts our perspective.[1074]

The success of science constitutes a powerful testament to the belief that objectivity should be our ideal for understanding reality.

But the movement towards objectivity has problems in both directions.

In the direction of subjectivity, it leaves behind some—or much—of what it means to be a particular, highly specific human being, whether it be autonomy, mind, or meaning. It seems to lead to reductionism.

At the other extreme, in the direction of the most 'centreless' form of objectivity, the attempt to recognise our epistemic limitations may inevitably lead to a scepticism or nihilism that undermines both objective knowledge and any values to which we are personally attached.

The problems associated with sports participation are related to the problem of the meaning of life—at least insofar as the problem of the meaning of life arises because we have the capacity to step back from our immediate involvement in life and take a detached viewpoint from which our life seems not to matter much at all. We struggle to get through school, get a good job, raise a family, worry about the bills—for what end?

As Nagel has said: "From far enough outside my birth seems accidental, my life pointless, and my death insignificant, but from inside my never having been born seems nearly unimaginable, my life monstrously important, and my death catastrophic" (VN, 209).

From the outside, from the view from nowhere from beyond the horizon, my life seems to be an accidental and insignificant moment in the entire scheme of things; my attachments as well as accidental as my life, objectively insignificant from the standpoint of a sceptical reflection that denies the ultimate justification of value.[1075]

Now if it is difficult to follow Nagel to the furthest reaches of objectivity from which the problem of the meaning of life arises, it is certainly not difficult to see the way in which the path towards objectivity threatens some of our cherished attachments in life.[1076]

There is often a pervasive spirit of *seriousness* when participants engage in sports. Subjectively they are seriously engaged and committed.

Players are praised for the intensity of their competitive seriousness.

From the internal viewpoint, life often seems to present itself at its subjectively best—or worst—in sports.

Internally, experience is enhanced when structured by the rhythms of games, contests, and seasons.

Sports participation seems to give life some dramatic or narrative shape as the pursuit of athletic success unfolds.

However, subjective attachment to sport is only part of the picture, precisely because we are complex, reflective beings who can step back from our particular involvements and scrutinise them.

From the internal viewpoint, sports participation is serious, sometimes all-consuming, and 'monstrously' important.

But even for people who are not disposed to be very reflective, an objective viewpoint insinuates itself into experience and suggests that sport is relatively unimportant.

How are these two viewpoints to be reconciled?

How can they coexist?

After all, it may be within the same person that the two standpoints clash.

The objective viewpoint claims that sport is not important.

Subjective involvement resists that judgment and wants to return to its pre-reflective attachment unaffected by the objective perspective.[1077]

It is curious that we engage in such unnecessary and seemingly trivial activities? What could be the end or goal of activities that are inherently irrational by ordinary standards?[1078]

For Suits, if the goal of the activity is to bring about a state of affairs in an inefficient manner, the player must have an attitude that recognises this but affirms that the activity itself is still worth doing.

The 'lusory attitude' of the player is "the knowing acceptance of constitutive rules just so the activity made possible by such acceptance can occur."[1079]

In my view, this attitude towards an instrumentally trivial or absurd activity must affirm the intrinsically satisfying or valuable nature of the activity—at least in the genesis of the game or the player's introduction to it.

From the internal standpoint many participants identify strongly with Huizinga's original description of play, stressing the freedom and joy of playing, outside of real-life concerns, with utmost seriousness about and absorption in such an intrinsically valuable but superfluous activity.[1080]

The arguments about the meaning and value of sport participation are similar to the way in which Nagel approaches the problem of the meaning of life.

The difference is simply how far we 'step back' from the immediacy of particular involvement in life and where we find ourselves when we land as a result of the distancing made possible by this reflective metaphor bearing in mind that we can't step into the same river twice. If an objective philosophical reflection finds sport to be trivial and irrational, then, a fortiori, at the furthest reaches of scepticism, whatever values we arguably experience in sport are undermined.[1081]

Nagel argues that a situation is absurd "when it includes a conspicuous discrepancy between pretension or aspiration and reality."

For Nagel, life as such is absurd because it inevitably involves a "collision between the seriousness with which we take our lives and the perpetual possibility of regarding everything about which we are serious as arbitrary, or open to doubt."

From this standpoint, sport is absurd because life is absurd, not simply because our seriousness in sport collides with the recognition that there are many more important things in life than improving one's backhand or winning games. But that doesn't mean sport or life is absurd. It is a non sequitur.

The pressures exerted by the reflective movement towards the "view from nowhere" are real and occasion the fundamental problem, which we will call the "problem of coherent attitudes."

The problem arises because of the claims made from the standpoint of reflective detachment.

> The same person who is subjectively committed to a personal life in all its rich detail finds himself in another aspect simultaneously detached; this detachment undermines his commitment without destroying it—leaving him divided. And the objective self, noticing that it is personally identical with the object of this detachment, comes to feel trapped in this particular life—detached but unable to disengage, and dragged along by a subjective seriousness it can't even attempt to get rid of it. (*VN,* 210).

The problem of coherent attitudes is the problem of how to respond to such worries, which Nagel calls the "discomfort of objective detachment." Objective transcendence brings with it some inevitable degree of disengagement or dissociation from the original subjective engagement, and the problem is how the objective self rejoins the subjective self.[1082]

For the reflective sports participant, the problem is significant. "Some of us feel a constant under-tow of absurdity in the projects and ambitions that give our lives their forward drive. These jarring displacements of the external view are inseparable from the full development of consciousness" (FtV211).

From the internal standpoint, seriousness is uncontested.

The pursuit of excellence, the exuberance of play, and the satisfaction of victory give meaning to the life of the sports participant in the context of life generally.

Yet once the self expands to include the external point of view with its concomitant judgments, the objective self returns, reconstituting the activity.

If a person now finds, from an external point of view, that a significant part of his life is objectively insignificant, how is he to deal with this?

Sport is meaningful and significant. Sport is trivial and absurd.

As Nagel says: "The objective self is a vital part of us, and to ignore its quasi-independent operation is to be cut off from oneself as much as if one were to abandon one's subjective individuality. There is no escape from alienation or conflict of one kind or another" (VN, 221).

The appropriate attitude for the reflective sports participant is irony.

This is a somewhat unusual way of speaking of irony. It refers neither to ironic uses of language nor to the irony of unexpected events.

Rather, irony refers to "an attitude of detached awareness of incongruity."[1083]

Irony is a way to regard sports participation, including the pursuit of athletic excellence and the desire for victory, *as if* it really matters, while at the same time recognising that it is relatively trivial in the larger scheme of things.[1084]

Irony is a unifying attitude that is positive, not negative; it is an awareness of the paradoxical nature of sport as competitive play, serious nonseriousness or nonserious seriousness.

Irony is an attitude that embraces the basic incongruity of our intrinsic devotion to triviality, our celebration of absurdity every time we compete intensely and play games seriously.

Joel Feinberg nicely describes irony as a 'cosmic attitude' appropriate to the human situation in general.

His description also fits the athletic ironist's attitude towards sport.

Irony is

> a state of mind halfway between seriousness and playfulness. It may even seem to the person involved that he is both very serious and playful at the same time. The tension between these opposed elements pulling in their opposite ways creates at least temporarily a kind of mental equilibrium…One appreciates the perceived incongruity much as one

does in humour, where the sudden unexpected perception of incongruity produces laughter. Here the appreciation is more deliberate and intellectual.[1085]

Humility is an attitude that promotes harmony and integration between the internal and the external.

It is a consequence of the objective perspective on sport (and life). In general, humility works against our tendency to inflate the significance of our personal projects and successes.

In the sports world, it works against the athlete's natural reaction to our culture's glorification of athletic success.

For the sports participant, it realistically deflates the seeming global pretensions of our merely local athletic concerns.

It is a way to come to grips with our capacity to step back from our participation in sports while at the same time sustaining our commitment.

As Nagel says about this attitude:

> Humility falls between nihilistic detachment and blind self-importance. It doesn't require reflection on the cosmic arbitrariness of the sense of taste every time you eat a hamburger. But we can try to avoid the familiar excesses of envy, vanity, conceit, competitiveness, and pride—including pride in our culture, in our nation, and in the achievements of humanity as a species. *VN, 222).*

Competition is essential to sport, since it provides the possibility for achieving the internal goods made possible by participating in such a practice.[1086]

On the other hand, since sport involves participation in activities whose external significance is originally minimal, and whose value is, for the most part, intrinsic to the activity, sport inevitably is associated with play.

If sport is objectively insignificant and participation is absurd, it is more appropriate to participate without gravity or solemnity, to reinforce the possibilities of play with an attitude expressed in a wry smile rather than a grimace of discomfort.

In a broad sense, the development of moral sensibilities and moral character is the attempt to mould a self whose actions and dispositions can be endorsed by a more objective viewpoint.

In Nagel's words, "Morality is a form of objective reengagement. It permits the objective assertion of subjective values to the extent that this is compatible with the corresponding claims of others" *(VN, 222).*

If, from some relatively objective viewpoint, we agree that sport is, in itself, relatively trivial and insignificant, there is certainly also a standpoint outside of the immediacy of internal attachment that affirms the importance of the moral qualities made possible by sports participation.[1087]

In sport, we must confront the possibility of isolation and failure.

Sport is an arena in which courage and responsibility may be developed. The sportsman in the Arena is also the Man in the Arena.

Since sport is rule-governed and embodies standards of achievement for those who attempt to become good, players may exhibit an understanding of justice as they respond to their own competitive situation in relation to others.

Since they must compete with others who challenge them to become better, they may develop respect for opponents, without whom the achievement of excellence would be impossible.[1088]

The Man in the Arena thrives on the struggle and the challenge that life represents.

Because sport requires officials to enforce rules, sport offers the opportunity to respect the guardians of order within the game.

Since sports develop historically, participants may develop a respect for tradition and past excellence that expresses a certain broadness of appreciation and opens the self to the rich possibilities of historical consciousness.[1089]

In short, sport is an arena of the kind that Roosevelt described within which it is possible to develop and display the excellence of good moral character, and the development of good character can be endorsed by an objective viewpoint.

As Nagel says about various forms of morality, they involve to some degree "a position far enough outside yourself and other people to reduce the importance of the difference between yourself and other people, yet not so far outside that all human values vanish in a nihilistic blackout" *(VN,* 222).

This position should not be misunderstood.

It's not as if the only justification of sport participation is the old character building argument. From the internal viewpoint, sport is significant because of the personal satisfactions and values involved, whatever they are.

From a more objective standpoint, sport is insignificant and the collision between our seriousness and its objective insignificance makes our participation absurd.[1090]

From the most extreme objective vantage, the view from nowhere delivers a judgment of global insignificance on our life. Yet one way to bring a degree of harmony to the opposition between subjective and objective is to see the way in which "morality as objective reengagement" reduces the opposition.

This means that developing attitudes associated with sportsmanship is an appropriate way to respond to the paradoxical nature of sport.[1091]

Chapter 43
Courage

Two things stand like stone,
Kindness in another's trouble,
Courage in your own.
Adam Lindsey Gordon (1833–1870)

Nothing splendid has ever been achieved except by those who dared to believe
that something inside of them was superior to circumstance.
Bruce Barton

The bravest are the tenderest,
The loving are the daring.
Bayard Taylor (1825–1878)

Even the bravest of us rarely has the courage for what he really knows.
Nietzsche

Anxiety is fear of oneself.
William Steke

Unbounded courage and compassion joined,
Tempering each other in the victor's mind,
Alternately proclaim him good and great,
And make the hero and the man complete.
Joseph Addison (1672–1719), The Campaign, line 219

The great epochs of our life are at the points when we gain courage to rebaptise
our badness as the best in us.
Nietzsche

Not always actions show the man; we find who does a kindness is not therefore
kind
Who combats bravely is not therefore brave,

417

He dreads a death-bed like the meanest slave.
Alexander Pope

Let me assert my firm belief that the only thing we have to fear is fear itself.
Franklin Roosevelt

Most people, even the present generation, when asked about the personification of courage, would nominate Winston Churchill and his statesmanship during the Second World War. It is not surprising therefore, that in his fascinating book 'Great Contemporaries' published in 1937, he characterised courage as "rightly esteemed because it is the quality which guarantees all others."

Courage is related to conflict. Soldiers and sportsmen are characterised as heroes when they take on a superior opponent and win. They like David become legends for overcoming Goliath.

Sport, like war, is often cited as man's ability to display courage under the pressure of the moment. Those domains are not the only ones where that can be said. In the legal world too I have often thought that the forensic difference in ability between advocates turns not so much on intelligence, fluency or knowledge of the law but on the ability to fight one's cause, to never bend under pressure, to display courage. So what is courage, often characterised as the most universally admired virtues. As André Comte-Sponville has perceptively pointed out, its prestige seems not to depend on the society, the period or even, for the most part, the individual.[1092] I have dealt with this topic in some detail in my chapter on Sport and War but I now look more conceptually at the subject.

Courage is the stuff of heroes John Kennedy asked about his heroic behaviour said "It was involuntary they sank my boat."

It is a universal concept. Courage can take many forms. Each civilisation has its fears and its corresponding forms of courage. But it has a common character—the capacity to overcome fear. Courage is the virtue of heroes and throughout the history of sport there are so many who have been elevated to that status for their courage on the field of play. From the Greeks onwards, as we have seen, the athlete; the sportsman has been admired for his courage and made heroes. This is a major attraction of sport, namely as the song says "to search for one's heroes." Creativity is an adjunct of courage. As Henri Matisse said "Creativity takes courage." This is true of sport as well as art. It took courage for Michelangelo to create the Sistine Chapel. It takes courage to score a goal at Wembley, a try at Twickenham, a century at Lords. Psychological courage is needed to stand up and perform in the arena.

The coach of the 1971 British Lions tour was Carwyn James, who perhaps most memorably pinned down what made JPR Williams possibly the greatest full back in rugby union history "like a forest animal, he was blessed with the sixth sense for the presence of danger. It was an element he sought and loved. Fearless, and compromising. The competitor of competitors".

It is not just a question of admiration. Beauty is also admired though it is not a virtue.

"Courage is not a virtue" says Voltaire "but a quality shared by black guards and great men alike." An excellence, then, that is neither moral nor immoral in itself. Sport, it is said, is the pursuit of excellence. Where does courage fit in. Intelligence and strength too, are admired and they can serve good as well as evil as is exemplified by warfare. But it is arguable that courage is a more telling excellence and sport is a manifestation of that excellence. Courage in and of itself retains its ethical aura even it has been said, "in a bastard."[1093] But that doesn't prove it is always a virtue. To say of someone "He is cruel and cowardly," the two adjectives add up, if I say "He is cruel and courageous," they subtract from each other.

What we respect about the display of courage in sport or in war or any other form of human activity is its culmination in self-sacrifice. It is the acceptance of incurring of a risk without selfish motivation. If it is not always a form of altruism it is a species of disinterestedness, detachment or a distancing from self.

That is possibly why we find courage morally worthy of respect especially in sport where it is on public display and a spectacle. If as Sir Anthony Hopkins said, "Virtue is not photogenic;" courage certainly is and explains why so many people watch sport so avidly.

Courage is primarily a low sensitivity to fear. This could be because either it is minimally felt or because it is easily or even pleasurably withstood. In my experience, the greatest of sportsmen have that characteristic.

It could be described as the courage of daredevils, of the coolheaded, of those who love a fight—the courage of 'tough guys'![1094]

On that level courage may have nothing virtuous about it. Courage constantly commands respect. Man is fascinated about courage whether it is purely physical or strictly self-serving courage. It has been called 'a premoral respect'.[1095] It testifies to greater self-possession, dignity, and freedom. Pope in 'The Odyssey of Homer' saw the nobility of it when he commented: "And what he greatly thought he nobly dared." Marcel Duchamp said with insight "Art is not about itself but the attention we bring to it." The same is true of sport.

Courage commands respect. But the fascination with courage can be dangerous. Morally courage proves nothing. But it can be explained by the fact that the courageous person manifests an inclination to wrench himself away from the sway of instincts and fears. It can be seen as mastering oneself or one's fears. It is a mastery that, though not always moral, is the necessary—though not sufficient condition of all morality. The Greeks and Romans saw courage as a mark of virility (the word *andreia*, which means courage in Greek, and the word *virtus* in Latin comes from *aner* and *vir* respectively; root words that denote man, not in the general sense but man as opposed to woman). It is sometimes said of a daring act that 'it took balls' which tended to show that physiology, at least in a figurative sense, counts for more than morality. It focussed on physical or martial courage. In my view, this is far too limiting an emphasis. Women can and do demonstrate this kind of courage as well, but from the moral standpoint. Such demonstrations prove nothing. This pathological courage as Kant would call it or passionate courage, to use Descartes' term, is useful but primarily to the person who feels it which, in my view, is why it is devoid of any properly moral value of its own. By contrast, courage as a virtue always presupposes some term of selflessness, altruism or generosity. Virtuous courage does not rule out a certain insensibility to fear or even a certain relish for it. But it does not presuppose for it. But it does not presuppose them as Comte-Sponville has pointed out.[1096]

This kind of courage is not absence of fear but the capacity to overcome it by a stronger and more generous will. It is no longer just physiology: it is fortitude, moral strength in the face of danger (see my Chapter on Self Overcoming and Nietzsche for a more comprehensive consideration of this theme generally). It is fortitude, moral strength in the face of danger. It is no longer a passion: it is a virtue; one that is the precondition of all the others. It is no longer the courage of the tough: it is the courage of the gentle, and of heroes. One of my great sporting heroes was John Charles, the Welsh footballer, who was known as the 'gentle giant'. I now understand the significance of this. There is also a connection between courage and truth.

The distinction between courage and bravery is also a deeply philosophical one and more than a semantic one and relevant to the conscious and unconscious themes developed elsewhere in this book. I will return to this theme later in this chapter.

Courage and Truth

Plato tried to reduce courage to knowledge (in the Laches and the Protagoras) or to opinion (in the Republic): courage, he says, "is the science of things that are or not to be feared," or more modestly, the "constant safeguard of an upright and legitimately accredited opinion on the things that are or are not to be feared."

What both these formulations ignore is that courage presupposes fear and that the burden of courage is simply to standing up to it.[1097] In the words of the Tom Petty song 'Never back down'.

Fear is both a necessary and sufficient element in courage and it doesn't matter whether fear is justified or illegitimate, reasonable or unreasonable. Don Quixote shows courage in tilting at windmills, whereas science for all the reassurance it offers, has never made any one courageous. It has been said no other virtue is more resistant to intellectualism.[1098]

Courage is not knowledge but a decision, not an opinion but an act.[1099]

Aristotle had already said it. Nor indeed was Socrates right in asserting that courage was knowledge.[1100] That is why courage needs more than reason alone.

"Reasoning tells us what to do and if it must be done, but it doesn't tell us that it must be done; much less does it do what it says."[1101]

Reason is reason. But not so much with will or with courage, which is simply will at its most determined, and in the face of danger or suffering, at its most necessary. Reason is in every instance, universal and anonymous, whereby courage is always singular and personal.

Such courage is called lucidity. It is the courage of truth, though there is no truth that by itself can make us lucid. All truth is eternal, while courage has meaning only in finite, temporal terms, in duration. This is the kind of truth we see on display in many sporting moments. The courage to persist and endure, to live and to die, to hold out, fight, resist, persevere. Spinoza calls tenacity the "desire by which each one strives, solely from the dictate of reason, to preserve his being."[1102]

But courage resides in the desire, not in reason; in the striving, not in the dictate. To be courageous to persevere in our being.

Like in any other virtue, courage exists only in the present (see my chapter on Epicurean and Stoic Philosophy, Chapter 3).

The fact that we once had courage does not prove that we will have it again or even that we have it now.

In short, courage doesn't have to do only with the future, with fear, with imminent harm; it has to do with the present and is always much more a matter of will than of hope.

Courage and Bravery

"Brave men act" writes Aristotle, "for honour's sake which can be translated as 'for the beauty of the courageous action' or 'for the love of good'."[1103]

Life and especially sport teaches us that we need courage to withstand despair and also that despair can sometimes give us courage. When there is nothing more to hope for, when there is nothing more to fear, then against all hope, courage is at hand to face the present fight, the present suffering, the present action. That is why, Rabelais explains "according to true military discipline one must never reduce his enemy to the point of despair for such necessity multiplies his strength and increases his courage."[1104]

The distinction between these concepts as I suggested earlier may be that it exemplifies the difference between the conscious and unconscious. But there is more to it than that. Courage is a part of a man's character but bravery relates to how a man reacts to an event. A brave man does what he does when he reacts to an event. He may not be reacting to a fear but behaves in the way he does because that is what he is—a brave man doesn't think of the consequences of his actions but does what he does because of what he is and what he wants to do and what he wants to be. He is reacting to the voice within himself which says "Do it." And he does. This mentality is so prevalent in outstanding sportsmen and I suspect soldiers too.

Courage is not the strongest force. As I point out later chance or luck plays a vital role and is often determinative of the outcome of the contest.

It is true as Virgil said in the Aeneid that fortune favours the brave. At least it can, but not always because for every man there are things he can and things he cannot endure; whether or not he will encounter before he dies the thing that will break him depends as much on chance as courage. For a further exposition of this theme (see my last chapter on luck).

The realisation of that is why I suspect, Shakespeare concluded in Henry IV "Oh God that one might read the book of fate."

Courage is clearly related to genius and talent. It leads to the discovery of new worlds and a reality, being and the horizon. André Gide in his journal perceptively observed, adopting an earlier statement by Christopher Columbus, "Man cannot discover new oceans unless he has courage to lose sight of the shore."

Chapter 44
The Roots of Virtue

See how the world its veterans rewards
A youth of frolics, an old age of cards.
Pope

As Montaigne identifies in his Essays, "Virtue shuns ease as a companion. It demands a rough and thorny path."

Freud introduced us to the study of "the much furrowed ground from which our virtues proudly spring." The relationship between virtue and vice is self-evident. Montaigne also noted of himself that "I find that the best virtue I have in it some tincture of vice." In the preceding chapter on courage I refer to Churchill's emphasis on it as being "the quality which guarantees all others."

Nietzsche cautioned that "he who fights with monsters should be careful lest he thereby becomes a monster. And if thou gaze into the abyss, the abyss will also gaze into thee". Dostoyevsky wrote "If God did not exist, everything would be permitted."

What are the genetic roots and the evolutionary rationale of certain basic qualities that have since ancient times been called 'virtues'? Virtue is an evolutionary concept. As Milton observed well ahead of his time "Virtue could see to do what virtue would." The development of psychoanalytic thought has focussed on 'ego-strength'. It has suggested that human virtue should be reconsidered as an 'inherent strength', 'an active quality'.[1105]

These can be defined as the unwritten rules of groups of humans, backed up by feelings of right and wrong. They are ideas in society about what is right and what is wrong. This translates to rules about what to do and what not to do, including how to interact with other people in the group. Moral codes have a long evolutionary history in humans. Morality binds people into groups. It gives us tribalism, it gives us genocide, war and politics and it also gives heroism, altruism and sainthood.[1106] Morality is virtue in action and sport has its own morality.

Sport is fertile territory to explore the question 'by virtue of' what qualities, can man claim to be, or to be able to become, humanly strong.

We value what we call virtue. But there is a dilemma which Darwinian biology and Freudian psychology share. They have focussed on what is considered man's 'lower nature': the descent and evolution of the genus man from a prehuman state of animality; the emergence of civilised man from degrees of savagery and barbarism; and the evolution of individual man from the stages of infantility. Darwin's theory has been generalised as a "tooth and claw struggle for survival in which the crown of creation would go to what T.H. Huxley called the 'gladiatorial' type of man (see more on this in my chapter on Sport and War)."

The relationship between the modern sportsman and the gladiators of old is more than just of historic interest. The war like spirit of the gladiator is arguably reflected in the intrinsic aggressive attitude of the competitive sportsman.

Freud's theory of inner conflict has been over-simplified. It clings to the earliest formulation of this conflict and conceives of it as anyone's inner tooth and claw struggle between ravenous instincts (the impersonal Id) and cruel conscience/the moralistic 'Super-Ego'.

Thus the moral alterations seemingly implicit in Darwin's and Freud's discoveries were over-dramatized.

The double myth of an inner and outer struggle has made it difficult to come to terms with the question of man's moral strength.

Freud created the concept of the Ego. Ego is an age-old term which stood for the unity of body and soul and in philosophy in general for the permanency of conscious experience. Psychoanalysis has not concerned itself with matters of soul and has assigned to consciousness a limited role in mental life by demonstrating that man's thoughts and acts are co-determined by unconscious motives which prove him to be both worse and better than he thinks he is. But this also means that his motives as well as his feelings, thoughts and acts, often 'hang together' much better than he could (or should) be conscious (see my earlier chapters in the role of the unconscious in sport and generally). The ego then is analogous to what it was in philosophy in earlier usage; a selective, integrating coherent and persistent agency central to personality formation.[1107] The distinct roles of passivity and activity in man's thinking are particularly prevalent in man's behaviour and especially as we have seen in sport.

Where do man's virtues then stand in the process of evolution? It has been said that what we call virtues are "the specifically human qualities of strength which are implicit in man's evolution."[1108]

The Greeks and the Romans had their own words for virtue. Their ancient words suggested expertness while everyday words have countless connotations: to optimists they make virtues sound easy accomplishments, and to pessimists, like idealistic preferences.

One writer[1109] has described the basic virtues as hope, will, purpose and skill as the rudiments of virtue developed in childhood, of fidelity as an adolescent virtue; and of love, care and wisdom as the central virtues of adulthood.

The argument is that the virtues, step by step, become the inner strength of the human life cycle which has evolved in each individual and his generation giving renewed support to the institutions and traditions of society including sport, which in turn are to safeguard the process of virtue-formation in all successive generations. I will briefly look at the suggested virtues and show how closely related and interdependent they are.

Hope

Walk on, walk on with hope in your heart and you'll never walk alone.
Football anthem Liverpool AFC

The dream of a waking man.
Aristotle

He that lives upon hope will die fasting.
Benjamin Franklin

Helen Keller in Optimism published in 1903, an American writer and social activist well ahead of her time perceptively saw the close relationship between hope, optimism and achievement not limited to sport when she said "optimism is the faith that leads to achievement; nothing can be done without hope".

Musonius Rufus in his lectures said "the human being is born with an inclination towards virtue." I agree and would argue that the notion of original sin has weighed down human kind for centuries. In reality, we are made up to help each other and to be good to one another, to me it is an important evolutionary concept. We wouldn't have survived as species otherwise.

Napoleon Bonaparte said "leaders deal in hope" by which he meant that political leaders use hope as a political tool to encourage those who throughout their existence have lived in the hope of a better life. Hope like fear is something which relates to a future state of affairs. In the same way that people fear what is to come people aspire to the hope of something better in their lives. Hope is a fundamental and intrinsic aspect of man's desires, it can be characterised as good and fear as bad. In truth, they are both capable of leading people astray in that decisions made out of fear and or hope are often misguided and make people behave not according to reason but out of whim and irrationally.

Hope is generally regarded as good, fear is generally regarded as bad. To a Stoic they are both the same, both are projections into the future about things we do not control. Both are the enemy of the present moment that you are actually in and are contrary to the Epicurean philosophy of living in the moment. Both mean you are living a life in opposition to Nietzsche's amor fati and they are both related to each other. As Seneca said in Moral Letters:

"Hecato says 'cease to hope and you will cease to fear'…The primary case of both of these ills is that instead of adapting ourselves to present circumstances we end our thoughts too far ahead."

But not withstanding all this hope springs eternal.

Hope, it is said, is both the earliest and the most indispensable virtue in the state of becoming alive.[1110] Poetically it has both been described as the "most hopeless thing of all"[1111] and "The entity of those that are not yet."[1112]

Some have called it the deepest quality namely confidence. Hope must remain, even where confidence is wounded and trust impaired. But there is something in the anatomy of hope which suggests that it is the most dreamlike of ego qualities. This thought is ascribed to Aristotle by Diogenes Laertius who when asked went hope is answered 'The dream of waking man'.

An exclusive condition of hopefulness would be a paradise in nature, a Utopian social reality and a heaven in the beyond. It is an intrinsic ingredient of the concept of sport. And hope leads inexorably into conflicts between self-will and the will of others which at heart is what sport represents. "Hope springs eternal in the human breast," says Pope and "travels tough, nor quits when we die."

Will

The will when declared, is an act of sovereignty and constitutes law.
Rousseau

That to live by one man's will will be the cause of all men's misery.
Richard Hooker

"Conscience," said Heidegger, "is the will to be guilty."

The rudiments of Will are acquired, in analogy to all basic qualities, as the ego unifies experiences on fronts seemingly remote from each other; awareness and attention manipulation, verbalisation and locomotion. As Nietzsche pointed out in Beyond Good and Evil "The will to overcome an emotion is ultimately only the will of another, of several other emotions."

It is most dependent for its verification on the charity of fate. (See later my chapter on luck.) Maybe that is what Pope had in mind in his Essays on Man when he suggested that "in faith and hope the world will disagree. But all mankind's concern is charity."

The training of eliminative sphincters can become, in sport and in other domains the centre of the struggle over inner and outer control. A sense of defeat can lead to deep shame and a compulsive doubt whether one ever really willed what one did or really did what one willed.

If will is built securely in the early development of the ego it survives, as hope does: for the maturing individual gradually incorporates a knowledge of what is expectable and what can be expected of him. Often defeated he nevertheless learns to accept the existential paradox of making decisions which he knows 'deep down' will be predetermined by events, because making decisions is part of the evaluative quality inherent in being alive; will or ego strength depends, above all, on the sense of having done one's active part in the chain of the inevitable. There is no better exemplar of this proposition than sport.

Purpose

This is very much linked to the concept of play which I have dealt with at length in this book. Man must develop in 'mere' phantasy and play and sport the rudiments of purpose, a temporal perspective giving direction and focus to concerted striving. Play is to the child what thinking, planning and blueprinting are to the adult a trial universe in which the conditions are simplified and methods exploratory, so that past failures can be thought through, expectations tested. In the toy world the child plays out the past, often in disguised form, in the manner of dreams and he begins to master the future, by anticipating it in countless variations of repetitive themes. In taking the various role, images of his elders into his sphere of make-believe he can find out what it feels like to be like them before fate forces him to become like them. In that sense sport like play is a make believe dream world beyond the experience of reality in the fun world.

It may be that the evolutionary function of infantile play and the adult sport affords an intermediate reality in which the sense of purposefulness can disengage itself from the fixation on the past by giving it a mythological order and quality.

Purposefulness attaches itself to reality in sport and life generally.

Skill

Skill is simply the facility in doing something.

I have looked at this concept at length earlier in the context of sport and more generally as the hallmark of excellence in achievement. I now consider briefly its evolutionary and philosophical roots as a fundamental virtue. George Eliot observed "Tis God gives skill but not without men's hands: He could not make Antonion Stradivari's violins without Antonio." The rudiments of skill add method to hope, will and purpose.[1113] All cultures have their logic and their truth which can be learnt by exercise, usage and ritual.

Sport is a paradigm example of the embodiment of skill and provides man with a concrete and abstract demonstration of the workings of reality in this world and beyond. It provides man with cooperative and competitive encounters, it is a self-verification of lasting importance in the use of his physical and mental coordination. It gives man a sense of competency without which there can be no strong ego. It provides him with the hope that he can match an ever-increasing section of manageable reality with his growing capacities.[1114]

Fidelity

William Barrett said "Faith for Protestantism is the irrational and numinous centre of religion."[1115] This echoes Martin Luther's famous curse "the Whore is reason."

This is a concept which relates to "the sense of identity." Again a topic explored earlier in this book. It has been argued[1116] that there is a particular ego-quality which emerges with it and from it, fidelity. This word combines a number of truths or alternatively adhere: high accuracy and veracity in the rendering of reality; the sentiment of truth, as in sincerity and conviction; the quality of genuineness as in authenticity; the trait of loyalty, of 'being true'; fairness to the rules of the game; and all that is implied in devotion; as freely given but binding vow. Cultures, societies, religions, and certainly sport offer the nourishment of some truth in rites and rituals of confirmation as a member of a totem, a clan, or a faith, a nation or a class, which henceforth is to be his super family; in modern times we also find powerful ideologies which claim and receive loyalty. Loyal and legal are kindred words. He who can be loyal can bind himself legally.

Love

Love knoweth no laws.
John Lyly

Love—an object intercourse between tyrants and slaves.
Goldsmith

Heaven has no rage like love to hatred turned. Nor hell fury like a woman scorned.
William Congreve

If philosophy is, as the etymology of the word signifies, then Socrates was a genuine philosopher, a lover of wisdom even though he did not claim to know about this love.
William Barrett

Love is an ambivalent virtue. In his Essay on Criticism, Pope exclaimed "what dire offence from amorous causes springs." This is a mutual search for a shared identity for the mutual verification through an experience of finding oneself, as one loses oneself, in another. Many forms of love can be shown at work in the formation of the various virtues. It is the selflessness of joined devotion, which anchors love in a mutual commitment. Intimate love is the guardian of that elusive and all-pervasive power. It guarantees individuals identity in joint intimacy. All of this is necessary for the human equivalent of those rituals by which birds select each other for mating and nesting. Nietzsche observed "One loves ultimately one's desires, not the thing desired."[1117]

That in man various kinds of 'love', rather than interactive certainty must animate this affiliation and associations, is one reason for his clannish adherence to styles and ideas which he will defend. That results in the righteous defence of the endangered identity (religious, racial or ideological) for which there seems to be no parallel in the animal world. Freud observed that mature geniality alone guarantees that combination of intellectual, sexual mutuality and considerate love, which anchors man in reality.

In life there is a long hope for kindredness beyond blood-bonds. Sport is yet again an example of this virtue in man which results in the task of caring for others and shared ideas. Fellowship is a form of love and pervades the world of sport in both the players' and spectators' love and respect for the game and the spirit of competition.

Care

Why ae we fond of toil and care? Why choose the rankling thorn to wear?
J.M Usteri

Man has evolved as a caring animal. This form of caring has been characterised as generativity in the form of selfless caring and a need to take care of whatever one generates and leaves to the next generation. The climate change movement is a topical example of this.

Man's creation of all caring gods was not only an expression of his need for being taken care of but also a projection of a human agency. This agency is a further example of the concept of beyond the concept of sport, the subject matter of this book. This agency has to be strong enough to guide man's propensity for freely causing events and creating conditions which prove to be beyond him. It is vital that man must learn not only to develop but to understand and restrain his capacity for unlimited invention and wisdom in sport and other domains.

I have a duty of care to my fellow man, my neighbour. And the concept of a duty of care is at the heart of English law. That is a fundamental principle of our legal system.

Wisdom

Go teach Eternal Wisdom how to rule then drop into thyself and be a fool.
Pope
*The wisdom of mankind creeps slowly subject to every doubt that can retard
Or fling it back upon an earlier time.*
Richard Horne

The wise through excess wisdom is made a fool.
Ralph Waldo Emerson

Bertrand Russell said, "to conquer fear is the beginning of wisdom."

I am not so sure I agree with Pope when he stated "in youth and beauty wisdom is but rare."

I also hope in my own domain that Samuel Johnson was right in his view that "The law is the last result of human wisdom acting upon the human experience."

Wisdom again is an evolutionary concept. People who have wisdom can see beyond the horizon. It is most often attributed to those who have great experience of life and have learnt from their experiences the lessons of life and what the future holds. They can see beyond the narrow frontiers of daily life on earth and can look at things in broader perspective. But it is not limited to the old and not beautiful. Wisdom is timeless and genderless! However, I agree with Cervantes that "Time ripens all things. No man is born wise."

It is the ability to maintain the wholeness of experience even as the body faculties gradually fall apart but often hopefully before that process takes place.

Wise people can envisage problems in their entirety and conceptually.

This is what integrity means. It is a living example of the 'closure' of a style of life. Only such integrity can balance the despair of a limited life coming to a conscious conclusion but continuing 'beyond'.

Evolution has made man wiser. It can also elongate man's life beyond the period of procreative power and in so doing has given the old a once new role of imparting wisdom and experience in the young in all domains of human endeavour including sport. Witness the way retired sportsmen remain in the game to impart their knowledge and advice to the new generation. Old players remain in the game to pass on their wisdom and the new players enthusiastically lap it up! Wisdom is a difficult concept. T.H. Huxley observed "logical consequences are the scarecrows of fools and the beacons of wise men."[1118]

Conclusion

The aforementioned look at virtues shows that they are deeply rooted in evolution and are invariably unconscious processes of which we are not aware. Sport provides rich material for the investigation and acknowledgement of the importance of these largely unconscious processes. I have dealt extensively with these matters earlier in this book. This chapter has attempted to look beyond and behind the role of the conscious and unconscious in sporting activities.

Virtues and vices are ever present in sport. They are on constant display. They show the individuality and diversity of man in action. Sport shows man's reaction to the diversity of conditions. We can ascribe a long range (beyond) meaning to the conformist, idiosyncratic and deviant alike.

Sport is an exercise in 'adaptation'. Lord Byron in Childe Harolde looked at this when he spoke of man "living wisdom each studious year." It shows that in his history man has realised this blueprint only in fragments.

Sport increases our philosophical and psychological understanding of man by the recognition of the body's wisdom, of the power of the unconscious and the ego's functions and limitations.

This increased margin of consciousness and beyond consciousness is itself a major step in evolution. It enables man to see beyond the horizon, to visualise new moral alternatives in terms of virtues and vices. Sport is a striving for perfection and excellence both abundant and adaptive. It mediates more realistically between his inner and outer world than do the fatal compromises resulting from moral absolutes. Sport is the great leveller which exposes virtues and vices alike. That may be why Albert Camus concluded that he had learnt ethics from sport.

Chapter 45
Sport, Character and Virtue

Falsehood and delusion are allowed in no case what so ever, but as in the exercise of all the virtues, there is an economy of truth. It is a sort of temperance, by which a man speaks truth with measure that he may speak it the longer.
Edmund Burke

Human good turns out to be the active exercise of the soul in conformity with excellence or virtue.
Aristotle

You base football player.
Shakespeare, *King Lear*

Our religion is made so as to wipe out vices, it covers them up, nourishes them, incites them.
Montaigne.

It's hard to see where the MCC ends and the Church of England begins.
J.P. Priestley

A man's character is his guardian divinity.
Heraclitus

That man lives badly who does not know how to die well.
Seneca

The only thing necessary for the triumph of evil is for good men to do nothing.
Edmund Burke

Be more concerned with your character than your reputation, because your character is what you really are, while your reputation is merely what others think you are.
John Wooden

The best effect of fine persons is felt after we have left their presence.
Ralph Waldo Emerson

And though hard be the task, keep a stiff upper lip.
Phoebe Cary (1824–1871)

That which an age considers evil is usually an unreasonable charge what was formally considered good—the atavism is of an individual old ideal.
Nietzsche

De La Rochefoucauld in his moral maxims said: *Our virtues are most frequently but vices disguised.*

Morality is the herd-instinct in the individual.
Nietzsche

Clever men are good but not the best.
Thomas Carlyle

Virtue is her own reward.
Dryden

The greatest offence against virtue is to speak ill of it.
William Hazlett

Veracity is the heart of morality.
T.H. Huxley

Sweet are the slumbers of a virtuous man.
Addison

Hypocrisy is a tribute which vice pays to virtue.
De La Rochefoucauld

Well-timed silence hath more eloquence than speech.
Martin Tupper

Dost thou think that because you are virtuous, there shall be no more cakes and ale.
Twelfth Night, Shakespeare

I relish no pleasure unless I can share it.
Montaigne

Know then this truth—enough for man to know virtue is happiness below.
Pope

I've studied men from my topsy-turvy close, and I reckon, rather true, some are fine fellows; some, right scurvy, most, a dash between the two.
Juggling Jerry, George Meredith

Virtue is like precious odours—most fragrant when they are increased or crushed.
Francis Bacon

We should look long and carefully at ourselves before we pass judgement on others.
Molière

The end of moral philosophy is not knowledge but action.
Aristotle

The web of our life is of a mingled yarn, good and ill together: our virtues would be proud if our faults whipped them not; and our crimes would despair if they were not cherished by our own virtues. All is well that ends well.
Shakespeare

Moral virtue may, test itself in pleasure and pain.
Aristotle

The wrong way always seems the more reasonable.
George Moore

The preponderance of pain over pleasure is the cause of our fictitious morality and religion.
Frederick Nietzsche

All I knew most surely about morality and obligations I owe to football.
Albert Camus

No man can climb out beyond the limitations of known character.
Robespierre

Young men have more virtue than old men, they have more generous sentiments.
Samuel Johnson

Believe, believe, believe!
Sir Anthony Hopkins

Virtue is not photogenic.
Sir Anthony Hopkins

Virtue could see to do what virtue would.
Milton

Thomas Huxley in his wonderful work *Materialism and Idealism* pondered if some great power would agree to make one always think what is true and do what is right, on condition of being turned into a sort of clock and wound up every morning before I got out of bed. I should instantly close with the offer. Belief in oneself results inputting virtues into action. André Gide wrote "The belief that becomes truth for me is that which allows the best use of my strength, the best means of putting my virtues into act..." It is the essence of life, sport and the Man in the Arena.

There are a number of questions related to sport and character building that are philosophically interesting and relevant. Often in sport as in life we do what we feel we have to do. As Quintilian observed "We give to necessity the praise of virtue."[1119]

Francis Bacon got it right when he observed "Prosperity doth best discover vice, but adversity discover virtue"—"On Adversity."

Certainly, the question whether sport does, in fact, cause the development of good character is in part an empirical one. As Marcus Aurelius identified "the happiness and unhappiness of the rational, social animal depends not on what he feels but on what he does; just as his virtue consists not in feeling but in doing."[1120] And Pope saw virtue as predominant. "Know then his truth enough for man to know virtue alone is happiness below."

But there are a host of other issues that can be addressed philosophically.

What is it to have character? Jonathan Swift controversially stated that "a nice man is a man of nasty ideas."[1121] But I believe as Shakespeare said in Measure for Measure "virtue is bold and goodness never fearful."

Good character seems to involve the question of virtue, for virtues are taken to be beneficial and praiseworthy traits of character.

What is a virtue such that participation in sports might develop it?

Juvenal in his satires said "Nobility is the one virtue."[1122] There is a certain nobility about great sportsmen.

If sport did develop character or virtue, how would it do this?

Are there moral dangers involved in sports participation?

Nietzsche said "The preponderance of pain over pleasure is the cause of fictitious morality and religion." This was a twisted gloss on Aristotle's view that "moral virtue manifests itself in pleasure and pain." Aristotle's view was that sport manifests even moral virtue though in Nietzsche's opinion it was a fictitious morality.

Are there natural threats to building character in sport?

Which virtues might we expect to be developed by playing sports?

Are there virtues specific to sports participation?

It has been said that a man of pleasure is a man of pain. Is that the hallmark of the man of sport?[1123]

Are there other important virtues that we would not expect to be developed by sport?

I would like tentatively to answer some of these questions by looking at some work in moral philosophy on the virtues?[1124]

In *After Virtue* Alasdair MacIntyre offers a conceptual account of the virtues that is particularly powerful as a way to get at some of the above questions.

He argues that the natural home of the virtues resides in sustaining what he calls 'practices', which, as we will see, include playing a given sport.

He says of his theory that "this kind of conceptual account has strong empirical implications; it provides an explanatory scheme which can be tested in particular cases."[1125]

Hence, his theoretical account might help us make sense of the conflicting evidence found in various empirical studies concerned with sport and character.

In *After Virtue* MacIntyre attempts to write a philosophical history that will help us understand the problematic characteristics of our current moral discourse.

For MacIntyre our current moral debates are characterised by disagreement because the rivals in these debates move validly from incommensurable moral premises.

The problem is that "we possess no rational way of weighing the claims of one against another"[1126] because we possess these premises as mere fragments, dislodged from the historical context in which they originally and specifically made sense.

We experience moral language in an odd and paradoxical manner because we take moral reasons as impersonal criteria for action, yet the incommensurability of conflicting principles suggests that our ultimate moral principles are arbitrary, our moral positions nothing but expressions of private attitudes and feelings.[1127] As Alexander Pope in his 'Essay on Man' observed "Not always actions. We find who does a kindness is therefore kind."

Only Nietzsche seemed to diagnose accurately the moral ills of modernism.

We are left, in MacIntyre's view, with either an acceptance of Nietzsche's view that rational justification of morality fails and morality is "a set of rationalisations which conceal the fundamentally non-rational phenomena of will,"[1128] or a return to some kind of premodern [or postmodern?] Aristotelian view of morality.

To understand the dilemma of modernism and especially to appreciate the Aristotelian option, MacIntyre's narrative reconstructs the development of the notion of virtue in the ancient world and later.

By the time the reader reaches the unifying chapter on the nature of the virtues he has encountered rich discussions of the virtues as lived and written about in Homeric society, in the Athens of Sophocles and Plato, and later, Aristotle, and in the world of New Testament writers and medieval thinkers.

The problem is the dazzling diversity of these accounts of virtues. "They offer us different and incompatible lists of virtues; and they have different and incompatible theories of the virtues."[1129]

Homer emphasises physical excellence embodied in the warrior.

Aristotle emphasised *phronesis* [practical wisdom] and the virtues of an Athenian citizen.

Faith, hope and love are emphasised in the Christian tradition.

Homer emphasises qualities of character necessary for fulfilling specific social roles.

Aristotle stresses the role of virtue in relation to the telos of human nature.

Christian virtues are also understood teleologically, but the good life for the Christian is in substantial ways quite different from the good life for the Athenian.

Is there a "core conception of a virtue" that can be derived from these different notions? MacIntyre's strategy is to understand virtue in relation to what he calls a 'practice'.

Consider his central theses on the nature of the virtues:

By a practice, I am going to mean any coherent and complex form of socially established cooperative human activity through which goods internal to that form of activity are realised in the course of trying to achieve those standards of excellence which are appropriate to, and partially definitive of, that form of activity, with the result that human powers to achieve excellence, and human conceptions of the ends and goods involved, are systematically extended.[1130]

MacIntyre specifically mentions football in his discussion.[1131]

Crucial to his analysis is the distinction between internal and external goods.

Suppose I teach my child to play football and he becomes an excellent young player. He might receive public recognition, a college scholarship, and even a professional contract.

That is, he might achieve a variety of *external* goods—fame, education, money—from playing football.

The list of possible internal goods associated with the practice of football appears to be quite long, including all of those skills necessary for becoming an excellent player or coach.

A practice involves standards of excellence and obedience to rules as well as achievement of goods.[1132]

> External goods are…characteristically objects of competition in which there must be losers as well as winners. Internal goods are indeed the outcome of competition to excel, but…their achievement is a good for the whole community who participate in the practice.[1133]

Now, how are we to connect these insights to the notion of virtue?

> A virtue is an acquired human quality the possession and exercise of which tends to enable us to achieve those goods which are internal to practices and the lack of which effectively prevents us from achieving any such goods.[1134]

Whenever we engage in a practice we necessarily involve ourselves in a community of shared expectations, goals, and standards.[1135] To quote Blaise Pascal: "There are only two kinds of men: the righteous, who think they are sinners; and the sinners who think they are righteous."

As MacIntyre says, we subordinate ourselves to others with whom we share the experience of the practice. But we never know in which of these categories we belong; if we did know, we would already be in the other! [1136]

Certain things immediately follow from this.

We must recognise both the authority of the tradition that has developed the standards informing a practice and authorities who are knowledgeable about the tradition and its standards.

Some will be better than others in acquiring the abilities that enable a participant to achieve the goods internal to a practice. Those who are better *deserve* their lot, alters consciousness in the direction of unselfishness, objectivity, and realism is to be connected with virtue.[1137]

For Iris Murdoch, the locus of moral development resides in experiences of 'unselfing', in which the self pierces the veil of vanity and self-absorption and begins to see things as they are. She says:

"The self, the place where we live, is a place of illusion. Goodness is connected with the attempt to see the unself...to pierce the veil of selfish consciousness and join the world as it really is."

In her 1970 masterpiece, The Sovereignty of Good and reflecting the theme of this book she recognised to appreciate "infinitely many kinds of beautiful lives" is to step outside the self "beyond its particular conceptions of beauty— which included in her view moral beauty—and walking beside it with humble, non-judgemental curiosity about the myriad other selves afoot on their own paths, propelled by their own ideals of the Good." This was essentially to go "beyond the horizon."

While not claiming anything exclusive about the realm of the aesthetic, she emphasises the moral power of experiences of natural beauty and art, in which the 'brooding self' gives itself over to the reality of the world and forgets its obsessions.

Good art confronts us with a more objective and less jaded sense of things.

"Art transcends selfish and obsessive limitations of personality and can enlarge the sensibility of its consumer. It is a kind of goodness by proxy. Most of all it exhibits to us the connection in *human* beings, of clear realistic vision with compassion."[1138]

Murdoch does not claim that art is preeminent in its role as educator of our moral vision.

Recalling Plato's mistrust of the artist and his emphasis on mathematics as a *techne* (craft), she suggests that the virtues appear in many different kinds of human activities besides the arts. In sciences, crafts and intellectual disciplines we see displayed "such concepts as justice, accuracy, truthfulness, realism, humility, courage as the ability to sustain clear vision, love as attachment or even passion without sentiment of self."[1139]

When we engage in a practice we encounter others, both in an immediate and a historical sense.[1140]

Hence the virtues mediate, in a positive sense, our human relationships internal to the practice. But something else happens in the historical life of practices:

Practices must not be confused with institutions. Chess, physics and medicine are practices, chess clubs, laboratories, universities, and hospitals are institutions. Institutions are characteristically and necessarily concerned with...external goods. They are involved in acquiring money and other material goods; they are structured in terms of power and status, and they distribute money, power, and status rewards.[1141]

MacIntyre's distinction between practices and institutions, between internal and external goods, provides an enlightening way to view this much-talked about and much criticised phenomenon. Institutions are required to sustain practices, but the institutionalisation of a practice carries clear moral dangers.

When the primary focus of a practice is on competition for contingent external goods provided by institutional structures, a natural tension develops between the virtues required to attain internal goods, and qualities of character, vices, that may be helpful in the pursuit of fame, wealth, and power.[1142] But genius and saintliness confer no special rights. Mozart had to pay for his bread like any other person and St Assisi would have neither more nor fewer rights than anyone else.

The ethos of contemporary sport, related to the orthodoxy of character building, emphasises winning—sometimes at all costs.[1143]

Now, paradoxically, we are in a position to see why the emphasis on winning, in the context of the institutional pressures associated with practices can undermine the moral message that is central to the orthodox view. That is especially true of sport.

For if winning is the key to achieving external goods, and the pursuit of external goods is overemphasised, then one may as well cheat to get the goods, that is, to win.

But to cheat is to admit that the overruling reason to engage in a practice is to achieve *external* goods.

It is to ignore the internal goods of the practice, the standards of excellence generated by the tradition, and the virtues required for pursuing and perhaps achieving such excellences.

To cheat, as MacIntyre says, "so far bars us from achieving the standards of excellence of goods internal to the practice that it renders the practice pointless except as a device for achieving external goods."[1144]

Such a view is very much what Christopher Lasch has in mind when he speaks of the 'trivialisation of athletics'.[1145]

In the tradition of Huizinga, Callois, and other play theorists, he sees sport as a form of play precisely because it involves a game, an arbitrary construction of rules for no reason other than making the activity in question possible.[1146]

"The degradation of sport, then, consists not in its being taken too seriously but in its trivialisation. Games derive their power from the investment of seemingly trivial activity with serious intent."[1147]

In MacIntyre's language, the point of sport as a practice is seriously to engage in the pursuit of relatively trivial internal goods.[1148]

We don't take sport seriously when we allow the wholesale invasion of external goods as its primary raison d'etre.

To do that is to ignore the internal goods of the practice.

Again, paradoxically it appears that the more appropriate way to defend sport's moral possibilities is to emphasise the 'worthlessness' of sport; not to see it as a path to wealth and fame but as an occasion for momentary, enjoyable commitment and the achievement of a particular kind of excellence.

Surely the most important sense of winning and competing courageously is to see winning as an internal good of significant yet trivial proportions.[1149]

Perceiving sport as 'splendid triviality' is perceiving it without illusion, in Murdoch's sense.

We can still stress the value of winning as important because it is an essential part of the practice of a sport.

But to overemphasise the value of winning is to diminish the numerous other internal goods of sports, and it is to point in a direction in which it becomes more difficult, not less, to be a good person.

Anthony Quinton in his essay 'Character and Culture' is interested in resuscitating the concept of character both in our cultural life and as an object of philosophical interest.

What is character? Quinton claims that "it is in essence resolution, determination, a matter of pursuing purposes without being distracted by passing impulse."[1150]

He believes this view is consistent with Plato's emphasis on prudence, courage and moderation, since these virtues "are all dispositions to resist the immediate solicitations of impulse."[1151]

For example, as Aristotle also understood, courage is a response to our natural feelings of fear, while prudence is "settled resistance to whim."[1152]

Quinton sums up his notion of character or important qualities of character.

> They are, generally speaking, ways of deferring gratification, of protecting the achievement of some valued object in the future from being undermined by the pull of lesser objects near at hand.[1153]

It is the disposition of habit of controlling one's immediate, impulsive desires, so that we do not let them issue in action until we have considered the bearing of that action on the achievement of other, remoter objects of desire. Understood this way, character is much the same thing as self-control or strength of will.[1154]

Our moral sense comprises a loose collection of intuitive rules of thumb and emotional responses that may have developed as a result of a number of functions which conceptually have little to do with ethics. These spontaneous reactions combined with our rationality, may explain how man decides on what is right or wrong.

"Intuitions about action, intentionality and control feature heavily on our perception of what constitutes an immoral act. The emotional reactions of empathy and disgust also significantly influence our judgments as to who deserve moral protection and those who do not. But one view is that this is a dangerous idea and that intuitions are a poor guide to moral truth and can badly lead us astray in making moral decisions."[1155]

Does sport build character? The best that can be said for this view is that sport can help build a part of character. But the notion of 'character' is sufficiently rich and complex and the social, economic, and psychological pressures are so great that we should expect a very mixed moral result from sports participation.[1156]

The corrupting power of the institutionalisation of sport and thus the emphasis on the pursuit of external goods produces an atmosphere in which internal goods are often underplayed or compromised.

Winning, as the avenue to external goods, becomes the dominant ethos, and moral shortcuts naturally evolve.

With the competition for external social goods comes the diminution of generosity of benevolence as key virtues specific to sports participation.

Thus the character building view is a fragment.

It ignores these difficulties and seems to focus only on instrumental virtues that help us achieve the goals we might have in a competitive, business-oriented social world.[1157]

We are served best by attending to Wittgenstein's admonition to beware of our simplifying "craving for generality," respect particularities, attend to the phenomena, and see how various human experience is.[1158]

In the same vein, Carlyle's assessment of virtue and character was that "We are firm believers in the maxim that for all right judgment of any man or thing it is wise but, nay, essential to see his good qualities before pronouncing on his bad."

This is as true of sport as in life. Carlyle's prose was mirrored poetically in the prayer of Eli Jenkins in Dylan Thomas' Under Milk Wood as follows:

"We are not wholly bad or good
Who live our lives under milkwood
And thou I know will't be the first
To see our best side, not our worst."

But I think Robert Louis Stevenson somewhat pessimistically but correctly was of the view that "Our frailties are invincible, our virtues barren; the battle goes sore against us to the going down of the sun."[1159]

Chapter 46
Fear, Herd Instinct, the Lockdown and the Lack of Wisdom

Fear of danger is a thousand times more terrifying than danger itself.
Daniel Defore (Robinson Crusoe)

A free man is one who lives under the guidance of reason, who is not led by fear, but who directly desires that which is good.
Spinoza

In life's last scene what prodigies surprise,
Fears of the brave, and follies of the wise.
Samuel Johnson

It is much safer to be feared than loved when, of the two either must be dispensed with.
Machiavelli (The Prince)

I referred earlier to the concept of complementarity. Herd instinct and fear are examples of complementarity. They are ambivalent philosophically and metaphysically. Let's take herd instinct Nietzsche wrote 'morality is the herd instinct in the individual'. I would add fear is the herd instinct in morality. Does this mean they are virtues and vices? Our virtues can be vices and vice versa. They are correlatives. Alistair Mclean said 'fear is the key' and so in my view is herd instinct. In Nietzschean terms, they can be forces beyond good and evil. Let's start with the good and to use a sporting example. It was said of Gerald Davies, the legendary Welsh wing three quarter that:

> "being so lightly built, Davies had to develop the instincts of the forest animal. He knew his body was frail, so he had to depend on speed and alertness and quickness of thought and foot work. Fear is an important element in the make-up of such a player, just as it is in the forest animal. It heightens perception and the adrenalin secreted adds to the surge of acceleration."[1160]

Herd instinct requires further consideration. In light of recent events, I have dealt with the ambivalent nature of fear and herd instinct conceptually and will now focus on those recent events.

Collective fear stimulates herd instinct. It generates impulses of cruelty and unreasonableness. So was it in the French Revolution when dread of foreign armies produced the reign of terror. Neither a man nor a crowd nor a nation can be trusted to act proportionately, wisely or reasonably under the influence of a great fear. There is no better example of this in recent times than the British reaction to the Covid virus. The imposition of a total lockdown and the countries' compliant reaction to it was made under the influence of great fear and even worse great panic. It was and not simply with the benefit of hindsight a disproportionate, irrational and misconceived reaction. It caused enormous harm to the well-being of this country and not just economic harm. It was in essence a decision vitiated by fear. The effect of fear is pervasive and damaging. It is for this reason that poltroons are more prone to cruelty than brave men and are also more prone to superstition.

It is a curious fact of life that men and women who are brave in all respects including facing death and who will have the courage to die gallantly will not have the courage to say or even to think, that the cause for which he or she has asked to die is an unworthy one. Obloquy is to most men and women more painful than death; that is the reason why in terms of collective excitement, so few men and women venture to dissent from the prevailing opinion and as a result do not act reasonably.

Coco Chanel declared "the most courageous act is still to think for yourself—aloud."

Clearly, Coco had been reading Aristotle who said "you will never do anything in this world without courage."

It takes courage to express views that differ from public opinion or current creeds whether it be scientific views or religious beliefs. Plato said the wise shall lead and rule and the ignorant shall follow. Unfortunately that is invariably not the case, sometimes our leaders are not only wise but are fools. All our leaders historically have, like Disraeli wanted to be on the side of the angels but it is not only our leaders but individuals themselves who in the words of Dryden in Absalom and Achitophel are "stiff in opinion always in the wrong."

Obviously, I am not arguing that public opinion is necessarily wrong, I do not adopt Thomas Carlyle's view that "popular opinion is the greatest lie in the world." Public opinion is not necessarily wrong but it often is. A more balanced view is to be found again in the earlier cited work of Dryden who proclaimed: "the most may err as greatly as the few."

Eccentrics, bigots and cranks are not necessarily wrong, all opinions should be considered on their merits. However fear, panic, superstition are not conducive to good decision making. Turning to the domain of sport in my experience and knowledge panic does not result in sound judgement in the field of play. Anecdotally I was told on good authority when the Welsh rugby team was under the captaincy of one Welsh captain were in difficult situations they were told unequivocally "don't fucking panic!"

That usually worked and is sound advice in all domains.

I would like to make a further observation on the philosophical significance of the lockdown. The discovery that man can be scientifically manipulated and that the government can turn large masses this way or that as they choose, is one of the major causes of our misfortunes. Communities and nations are being moulded by modern methods of propaganda under the guise of being "led by the science." Science is now the new opium of the masses, being used to brainwash and manipulate public opinion. I accept that this may be being done by well-meaning politicians and that as a result of fear and panic I also accept that governments everywhere cannot be completely virtuous and always wise. That would be an Utopian ideal. But all governments should agree to inculcate no dogmas. That should be the people's legitimate expectation of their government, not to mislead them. I do not want my views to be misinterpreted and I am not saying that unusual eccentric or rogue ideas should not be considered but they should be considered as I have said on their merits.

Fear is a beyond concept and the fear of death especially so. It is reasonable to fear dying because the evidence is that it can be an excruciating and painful experience but the fear of death is a different matter. No one has experienced death. The experience of being dead is unknown. Death is beyond life.

The fear of death is illogical. Why should we fear death when it is most likely only to be a return to a state of existence that preceded our birth?

Fear is an essential component of sport, the fear of failure and one's demise. It is a component part of what I have earlier characterised as a meditation on death, non-existence and sporting resurrection.

Although we may not fear death there is no evidence that we should look forward to it. There is nothing in our pre-existent state that excites our longing for the prospect of a posthumous existence. We are satisfied to have begun life when we did. We do not regret not having set out on our journey sooner and feel that we have had quite enough to battle our way through since. We do not consider the 6000 years of the world before we were born as so much time lost to us. We are perfectly indifferent about the matter. The same should be true of death.

We do not grieve that we did not live earlier so why should we fear death when the world around us continues to exist.

It is true that the pang of parting to leave some cherished purpose unfulfilled is repugnant to us.

The love of life is an habitual attachment not an abstract principle. Simply to 'be' does not content man's natural desire; we long to be in a certain time, place and circumstance.

The mountaineer will not leave his rock nor are we willing to give up our present mode of life with all its advantages and disadvantages for any other life that could be substituted for it.

It is the end of life we fear not the state of death. No young man ever thinks he shall die. He may believe that others will or ascent to the principle that 'all men are mortal', but he is far enough from bringing it home to himself individually.

Youth, buoyant activity, animal and sporting spirits, hold absolute antipathy with old age as well as with death.

André Maurois in The Art of Living (1940) said "growing old is no more than a bad habit which a busy man has no time to form".

On a personal note I agree with what he said and try to live my life accordingly. I want to remain busy until the day I die and then in Dylan Thomas terms "rage, rage against the dying of the light".

William Hazlitt a 'great writer and thinker', had a lot to say about this topic and many others too. As he put it "we do not want to rip up old grievances, nor to renew our youth like the phoenix, nor to live our lives twice over, once is enough. As the tree falls so let it lie. Shut up the book and close the account once and for all. Hazlitt- Selected Essays- on The Fear of Death (1822).

John Donne said:

"No man is an island,
Entire of itself,
Every man is a piece of the continent,
A part of the main.
If a clod be washed away by the sea,
Europe is the less.
As well as if a promontory were.
As well as if a manor of thy friend's
Or of thine own were:
Any man's death diminishes me,
Because I am involved in mankind,

And therefore never send to know for whom the bell tolls;
It tolls for thee."

But that does not mean my death diminishes me. On the contrary my death may enhance the memory of me.

To end on a lighter note I would like to draw on an anecdote contained in Bertrand Russell's wonderful work 'Unpopular Essays' in which he describes how on one occasion a man came to ask him to recommend some of his books as he was interested in philosophy; Russell did that, however the man returned the next day and said he had found only one statement he could understand and that one seemed to him false. Russell asked him what it was and he said that it was the statement that Julius Caesar was dead. When Russell asked why he did not agree the man drew himself up and said "because I am Julius Caesar." Russell amusingly commented that this example may suffice to show that you cannot make sure of being right by being eccentric!

Bernard Shaw when making notes on his play Caesar and Cleopatra said of Julius Caesar: "He's a man of great common sense and good taste meaning thereby without originality or moral courage." Maybe that was a different Julius Caesar from the one that introduced himself to Bertrand Russell!

Chapter 47
Respect for the Game

That is shall hold companionship in peace with honour, as in war—Coriolanus.
Shakespeare

The truth of the matter is we always know the right thing to do. The hard part is doing it.
Norman Schwarzkopf

You have too much respect upon the world: they lose it that do buy it with much care.
Shakespeare, *The Merchant of Venice*

If you obey all the rules you miss all the fun.
Katherine Hepburn

Many speak of sportsmanship as "respect for the game," as devotion or commitment that transcends particular triumphs and failures. Doubtless most competitors do love the competition independently of the winning (or love the winning in large part because of competition) but it is not clear how this love extends to 'the game' itself. It is even murkier why this love should extend beyond the player's playing lifetime why should a competitor care about some abstraction or constellation of rules of play and principles of fair play.

The phrase at first seems too vague to be very helpful in understanding what it means to be a good sport or to function as a basis for making ethical judgments about conduct in sport.

This sort of criticism has been given its quintessential expression in the fictional world of *the* baseball movie, *Bull Durham*. Crash Davis, journeyman catcher and tutor of the talented but erratic young pitcher, Ebby Calvin 'Nuke' La-Loosh, is forced to explain why he dislikes his young, undisciplined pupil. Because, Crash explains, "You don't respect yourself, which is your problem, but you don't respect the game—and that's my problem." During a conversation with a college track coach, he condemned the behaviour of certain athletes as an "insult to the sport," showing a lack of respect for the activities to which he was devoted.

Respect for the game often functions in the sports world as an important or even fundamental action-guiding, ethical principle: one ought to respect the game, or one ought to respect the type of activity characteristic of the sport in which one is involved.[1161]

Sports participants ought to act in such a way that their activity expresses, embodies, or is consistent with an attitude appropriate for engaging in sport.

The assumption is that action expresses attitudes, and a certain over-arching attitude, 'respect', ought to be embodied in sports.

The suggestion that respect for the game, far from being merely a vague, sentimental, and nostalgic aspect of sports talk, is actually the description of a significant attitude that may describe the unity of the good sport's perspective.

The virtue of sportsmanship should be viewed in Aristotelian terms, as a mean excessive seriousness and excessive playfulness and the disposition to act consistently in light of this mean.

But respect for the game is more suggestive and fecund. It is possible to analyse this notion, to say something about the constituent aspects that are suggested by its uses in the sports world and that arise because of a perspective understanding of the nature of sport.

Respect for the game also calls attention to the attitude or perspective of the good sport in a more explicit manner than the concept of sportsmanship.

Even in its everyday usage in the sports world, 'respect' suggests an attitude of esteeming or valuing some phenomenon in the sense that one must attend to something outside one's self, larger than the self, a reality that demands that one takes account of its nature, value, or interests.

It is common to perceive the ethical viewpoint as a perspective from which some larger set of interests or demands must be considered beyond the selfish or narrow interests of an individual.

Central to an understanding of sport are the notions of play, game, and practice, in MacIntyre's sense.

Each of these notions leads to a narrowing of one's reflective focus, towards the relevant sport-related phenomena that are set apart from other worldly phenomena in important ways.[1162]

This narrowing of one's reflective focus leads to a description and conceptualising of sport in terms of various interpretive responses to the notions of the internal and external.[1163]

To think of sport in terms of play turns one's attention inward, towards a world of play with its intrinsic rewards and internal attractions.

The freedom of play suggests that we are liberated as we identify with the activity and we value it for its own sake.

It is constituted by rules, and in most cases it is neither incorrect nor misleading to speak of such activities as games.

When we esteem or value a sport, our attention is fixed on the activities internal to the game, where space, time, and distinctive meanings are defined in terms of a play world.

New goals, intrinsic to such a world, become significant.

We appreciate sport as the playing of a certain kind of game, whose very existence arises because of the rules that define it and whose rules are constructed merely to bring such an activity into being.

A game is a contest with another, an opponent, without whom the contest would not occur.

In sport, I am opposed; I enter a world in which others act against me or my goals in quite explicit ways.

This opposition, however, takes place in the context of a practice.

Participation in a sport is involvement in a practice, a "socially established cooperative human activity."

Participation in a practice involves the attempt to realise internal goods.

Such goods are defined in terms of standards of excellence characteristic of the practice.

The attempt to achieve goods internal to a practice involves the desire to realise a certain kind of excellence, which is possible only if one understands and submits to the authority of such standards and is obedient to the rules or guidelines that define the practice.[1164]

The internal goods of a practice are available only to practitioners; such goods represent the possibility of extending human powers with regard to the means and ends that constitute the practice.

Unlike the external goods available to those outside of the practice, (goods subject to worldly competition), the internal goods of a practice, while occasioned by the competition to achieve excellence, are good for the entire practice community.

They represent goods for the practice itself, not simply for the individual who improves, transforms, or revolutionises the practice. In this sense, internal goods are shared goods.

Thinking of sport in terms of the concept of a practice (as well as the concept of play) changes the ethical tone and atmosphere of sports participation.

Because sport is essentially competitive, it is natural for some to think that the desire to win trumps all other sport-related goals, and the opponent must be treated as the enemy in a zero-sum game.

Alienation, estrangement, separation, conflict—no wonder that some find competition to be necessarily problematic and even harmful for human relationships and psychological well-being.[1165]

However, the exclusive focus on competition and winning leaves out practice-related concerns that shift our ethical focus in a significant way.

A practice is a cooperative human activity, in which a community of practitioners have shared concerns, expectations, and goals.

In sport it is the game that binds the community together, and the game is a larger reality that generates the standards of excellence over against which the individual defines his own achievements and his sport-related identity.

It is the game that demands the attention of the players attempting to achieve a particular kind of goodness, and it is the game that benefits when achievements are understood as shared goods.

Thus it appears that there is a structure of involvement in sports, with various moments that arises as a participant becomes increasingly devoted to sport.

It is within this structure that respect for the game can be understood and such an attitude seems naturally to arise among members of the practice community who become serious about the possibilities related to becoming a good player.

In general, respect for the game, as an ethical principle, often functions as the basis for a judgment about the relative importance of the sport in question and the relative unimportance of one individual player or participant.[1166]

Given a certain understanding of the sport, with its history and traditions, a natural continuation of such an interpretation is humility and appreciation for one's place in the tradition.[1167]

Respect for Rules

Since games are defined by the rules without which these activities would not exist, to respect the game is to respect the rules that define the activity and make possible the attempt to become good—and win.[1168]

Respect for the game requires a stringent respect for fair play, or the intention to play according to the implicit agreements embodied in the rules.

By violating the agreements that make the sport possible and by attempting to gain an unfair advantage over an opponent, the cheater trivialises the sport in question by reducing it simply to a means by which other goods are achieved.

The cheater reduces sport to an instrument for achieving external goods, rather than the locus for the possibility of achieving an intrinsically satisfying set of internal goods occasioned by agreements with others.

Personal Honour

Maybe all these arguments can be subsumed under the concept of personal honour. There is a view that sportsmanship reflects one important outcropping of the massive bedrock of honour that runs throughout human life in all times and places. Honour is found not simply in love and war (for a consideration of this topic see the later chapter War and Sport) nor in hospitality and feuds, nor in social class and religion; most emphatically honour is not confined to distant 'tribal', 'warrior', or 'traditional' societies. Honour lives also in the playing of sport and games. Honour in sport is sportsmanship.[1169]

But honour is a subjective contextual and relative concept. It is all in the eyes of the beholder and the circumstances. In the eyes of Julius Caesar, Brutus was a honourable man. And maybe he was. Honour is not the divining rod of what is right and wrong in sport or in any other domain.

All I can conceptually conclude is that as with thieves there is honour amongst sportsmen. It is not definable and cannot be categorised but when I see it I can recognise it. That is one of the great pleasures of sport.

Chapter 48
Respect for the Spirit of Competition, the Spirit of Play and Traditions and Customs, Achievement and Excellence

Government and cooperation are in all things the laws of life; anarchy and competition the laws of death.
John Ruskin

Is there no respect of place, persons, nor time in you.
Twelfth Night, Shakespeare

The Survival of the Fittest.
Herbert Spencer

"Man is a political animal," said Aristotle; Clausewitz, the great philosopher of war, a child of Aristotle went further to say that a political animal is a war making animal. Essentially man is a competitive animal but there is no inconsistency in Aristotle's and Clausewitz's characterisation and my own. The essential hallmark of politics, war and sport is competition. Politics is about the competition between politicians of differing political ideas, war is about the competition between different states for supremacy and sport about competition between sportsmen of varying sporting talents. Clausewitz, a veteran of the Napoleonic wars who used his years of retirement to compose what was destined to become the most famous book on war simply called On War, actually wrote that war is the continuation "of political intercourse with the intermixing of other means."

Warfare, he says, is almost as old as man himself and reaches into the most secret places of the human heart, a place where self dissolves rational purpose where pride reigns where emotion is paramount, where instinct is king. My view which neither Aristotle or Clausewitz dared confront is that man is a thinking animal in whom the intellect and instinct direct the urge to hunt and to have the ability to kill. Sport intrinsically reflects those tendencies (see Chapter 27 earlier on Sport and War).

Sport is an essentially competitive activity and is an illustration of Alexander Pope's observation in The Rape of the Lock; "What mighty contests rise from trivial things."[1170]

In sport, something is at issue and opponents come together to seek resolution within the play world of the practice community constituted by the rules.

Players assume that their opponents are motivated by the same goals and are equally serious about the pursuit of excellence and the desire for victory.

Hence competition means esteeming or valuing the seriousness characteristic of those who participate in the sport.

This should lead to the grateful recognition that the values internal to the sport are not possible unless worthy opponents are available. When a player is praised for his respect for the game, part of what might be meant involves the sense that a player is obliged to attempt to play as hard as he can, with intensity and devotion, to not give up under difficult circumstances, to be the best opponent he can be in terms of standards of good play.[1171] A good example of this is to be found in the obituary of David Duckham—The Times, January 11th 2023 where it is recorded that he was gracious to all opponents and in particular he said of Gerald Davies, whom he marked so many times that "You know exactly what Gerald is going to do. He is going to run at you at pace, then step off his right foot. Trouble is, you know you can't do anything about it."

In this sense, an appreciation of the competitive nature of sport may lead the player to consider the perspective of his opponent, who is also seriously engaged and equally susceptible to disappointment and discouragement.

I agree with Mathew Syed that sporting rivalry is not an add-on to sport, it is sport. He has recently observed those invented games are about winning and losing but they are also the complex relationship between competitor, spectator, icon and fan. As he insightfully comments "When we watch, we lose ourselves in the human aspect of the struggle projecting our hopes and neuroses onto our heroes and seeing ourselves in reflection. It is why, in a sense we become part of the action too."[1172] Syed's view is that our preference for a sportsperson is often a value statement about yourselves. Chris Evert appealed to Middle America. Martina Navratilova to the sexual counter culture. Martin Hagler was beloved of the blue-collar worker, the nine-to-five of American industrial life. Sugar Ray Leonard with his fast feet and megawatt smile, was the icon of the aspirational middle class, who perceived a Hollywood ending in almost every one of his fights, particularly those that climaxed with showboating. But the greatest was the greatest Muhammad Ali. He was a super hero of the 20th century. He was a man beyond the horizon—He "floated like a butterfly." But "stung like a bee." "His only fault was that he didn't know how good he was." When I met him professionally as a barrister not as a boxer I was enthralled. I have never met anyone who combined the spirit of mind with the presence of body. The only other person who got close was the greatest bullfighter of all time—El Cordobes who I have previously mentioned. Both those men were of the stuff that legends are made of—the heroes of this world and beyond. They were the greatest competitors and magnificent human beings.

Sport is replete with rivalries. Syed's pick with which I would not disagree is

(1)   Roger Federer, Rafael Nadal, Novak Djokovic.
(2)   Muhammad Ali, Joe Frazier, George Foreman.
(3)   Chris Evert v. Martina Navratilova.
(4)   Seb Coe v. Steve Ovett.
(5)   Ayrton Senna v. Alain Prost.
(6)   Sugar Ray Leonard, Thomas Hearns, Roberto Duran, Marvin Hagler.
(7)   Monica Seles v. Steffi Graf.[1173]

The list can be extended to cover historic international and national rivalry.

1.   England v. Germany—Football.
2.   USA v. Russia—Ice Hockey.
3.   England v. Wales—Rugby Union.
4.   Brazil v. Argentina—Football.
5.   `England v. Australia—Cricket.
6.   India v. Pakistan—Cricket.
7.   British Lions v. New Zealand—Rugby.
8.   The Ryder Cup—Golf, USA and Europe.
9.   Oxford and Cambridge—Cricket and Rowing.
10.   Eton v. Harrow—Cricket.

Rivalry at a personal and international level is at the heart of the focussed intensity which sport creates in player and spectator alike. I have already referred to this in my earlier chapter on the aesthetics of sport (see Chapter 6).

Rivalry has a downside. Intense rivalry can spill over to aggression and hatred as in the case of local rivalries particularly in the case, it seems of association football but not significantly in the case of rugby football where ironically the greater aggression and physicality on the field of play is not generally reflected—violence off the field as in soccer—witness the intense rivalries in local derbies which takes on the character of clanlike and ganglike behaviour. Rivalries between cities, such as Manchester City and United,

Glasgow Rangers and Celtic, Arsenal and Tottenham, Barcelona v. Real Madrid (El Classico). These have the hallmarks of civil wars and reflect the relationship I have earlier identified between sport and war yet may still in Syed's terms "Tell us about who we are." Between players rivalry can result in admiration. David Duckham said that Mike Gibson of Ireland was the most 'complete player' he ever played against and that Gerald Davies of Wales was the "greatest three-quarter of his generation in the world."[1174]

Competition involves respecting an opponent who is the condition for the possibility of achieving the goods made possible by participation in a sport.

In an ambitious title, the book 'How Soccer Explains the World'[1175] Franklin Foer, (an American writer as the title evidences—Soccer not Football) provides a graphic example of hostile and historic rivalry not being eradicated by globalisation. Red Star Belgrade, the most successful team in Serbia having a following of unruly fans capable of terrific violence who "occupy a place of honour, and more than that." It had become a bastion of nationalism in Yugoslav papers—and for that rather—across the world war had been a metaphor for sports. Teams would battle and attack: they had impenetrable defences and strikers who fired volleys.

As a matter of philosophical generality sporting hostile rivalries have remained as potent as ever. Sport is not a force for change in removing political ideological and other prejudices. For example, the concept of a religious war lives on. Witness that most intense ruling between Catholicism and Protestation crystallising in the Reformation. The rivalry believes Celtic and Rangers fans is the epitome of a historic religious enmity which globalisation has not tempered and which football has perpetuated. Sport is not, in that instance, a force for good and tolerance but division and hatred.

Respect for the Spirit of Play

It is hardly surprising to insist that respect for the game means respect for a playful activity, a reminder that a game is first of all an activity whose value is internal to the activity and relatively trivial in relation to other human activities and values.

Sport is competitive play, serious nonseriousness, splendid triviality, joyful uselessness.

Respect for the Game's Traditions and Customs

One of the most interesting but difficult aspects of respecting the agreements that are relevant for fair play arises because a sport is more than simply a collection of explicit rules.

The actual playing of a sport also involves a prescriptive atmosphere in which rules are interpreted in various ways and expectations arise concerning what is appropriate conduct in relation to both written and unwritten rules.[1176]

Reference to the customs and traditions of a sport must sometimes be related to other principles, or to other aspects of respect for the game or sportsmanship to evaluate the appropriateness of the behaviour in question. The furore that arose out of the MCC's dropping of the Varsity match and the Eton v. Harrow game is an indication of how tradition is so important to the concept of sport.

There is no algorithm to help us distinguish those customs that ought to be preserved from those that should be rejected, but there is also no mystery concerning the kinds of reasons to which one might appeal in ethical reflection.[1177]

Sledging may have become a part of the game in many sports like cricket and baseball but that doesn't mean such behaviour is justified. These actions fail to respect opponents, the very nature of athletic competition, as well as basic norms of civility and decency.

Respect for Achievement and Excellence

Since practices have standards of excellence, the achievement of internal goods is possible, and the player of a sport must recognise the relevant standards in order to become good.

To respect the game is to recognise and appreciate the possibilities for achievement made possible by participating in the practice.

Such recognition is the basis for accurate judgments about people's own abilities and may fuel the desire for improvement, as well as a humility that is joined with realistic expectations for achieving the internal goods of the practice.

Respect for excellence is also crucial for relations with opponents, because reactions to winning and losing are related to judgments about what constitutes being good in the sport.

Respect for the game requires accepting responsibility for defeat because the opponent played better and graciously acknowledging that the *game* determined the outcome, not external factors generated by excuse-mongering.[1178]

Our sports heroes are those who have played the game the way it ought to be played, in terms of the ideals of excellence that have developed and the moral qualities that are required for such achievements.

George Will has a useful comment:

> It requires a certain largeness of spirit to give generous appreciation to large achievements. A society with a crabbed spirit and a cynical urge to discount and devalue will find that one day, when it needs to draw upon

the reservoirs of excellence, the reservoirs have run dry. A society in which the capacity for warm appreciation of excellence atrophies will find that its capacity for excellence diminishes.[1179]

One of the most successful coach in professional sports was Phil Jackson, whose teams won numerous NBA championships.

His book, *Sacred Hoops,* is a fascinating work. It gives a profound insight into professional sport as evidenced by the following example:

> It was Super Bowl Sunday. When we arrived at the hotel I told the players to get some pizzas and beers after practice and watch the Super Bowl in their hotel rooms. "You guys need to get together and remember what you're doing this for," I said. "You're *not* doing it for the money. It may seem that way, but that's just an external reward. You're doing it for the internal rewards. You're doing it for each other and the love of the game."[1180]

Jackson says that at one point in his coaching career, "I realised I needed to become detached emotionally and put the game in the proper perspective."[1181] That perspective I suggest, lies on the horizon and beyond.

Chapter 49
Self-Overcoming and Nietzsche

I may justly say, with the hooked nosefellow of Rome "I came, saw and overcame."
Shakespeare, *Henry IV*

"Sir, you have wrestled well, and overthrown more than your enemies."
Shakespeare, *As You Like It*

No path of flowers leads to glory.
D.L.A. Fontaine

Genius and heroism are madness.
Nietzsche

Flinch not, neither give up nor despair.
Marcus Aurelius

Under peaceful conditions the militant man attacks himself.
Nietzsche

I teach you the overman. [1182]
Man is something that shall be overcome.
What have you done to overcome him?[1183]
Nietzsche

The strongest man in the world is the man who stands alone.
Henrik Ibsen

Push on—keep moving.
Thomas Morton

A man can be destroyed but not defeated.
Ernest Hemingway

Determination is the wake-up call to the human will.
Anthony Robbins

Grammarian, orator, geometrician, painter, gymnastic teacher, physician, fortune-teller, rope-dancer, conjuror—he knew everything.
Juvenal, Satires iii 76

The art of living is more like wrestling than dancing in so far as it stands ready against the accidental and the unforeseen, and is not apt to fall.
Marcus Aurelius, Meditations

Nietzsche is a fascinating philosopher. He was a man of his times but remains an enigmatic and paradoxical character whose genius remains with us. If there was a physical monument to his existence and importance, which there should be, I am sure this woke world would want to pull it down. However, all that reflects is his cerebral significance and relevance to our world and its concepts. As Elias Canetti observed "most philosophers have too narrow a conception of the variability of human customs and potential". Nietzsche was not he was the epitome of breadth, variety and contradiction. I find his writings stimulating and provocative. Walking with Nietzsche takes you beyond the horizon to the high mountains of thought. His ideas are metaphysically sportive in origin and destination and a challenge to overcome.

The concept of overcoming is at the heart of life and sport. In that basic way sport replicates life. The sportsman has to overcome the challenges that sport presents and to do so within the limits of his potential and the restrictions of his domain. Nietzschean philosophy is at its metaphysical core.

Throughout Thus Spoke Zarathustra, Friedrich Nietzsche offers an account of human value and meaning as "self-overcoming."[1184]

Michael Monahan has compendiously and perceptively illustrated this. (See the Journal of the Philosophy of Sport, 2007) I am grateful to Mr Monahan for his research and analysis.

Monahan points out rather than offering a clear and precise model or a set of universal principles, Nietzsche stresses growth, change, self-criticism, and self-improvement.

Indeed, much of his critique of morality (and philosophy) is targeted at what he sees as its insistence on fixed and universal values, principles, and goals.

The value of human endeavour, for him, lies not in strict adherence to some fixed standard or set of standards but rather in the constant transcendence of those standards. The study of sport is, in many ways, very much in keeping with this aspect of Nietzsche's ethos and, thus, serves as an excellent vehicle for exploring and exemplifying Nietzschean self-overcoming. At the same time,

461

taking seriously Nietzsche's account of self-overcoming can help illuminate sport and life as an ongoing *practice* of self-overcoming. I will draw primarily (though not exclusively) from *Thus Spoke Zarathustra*.[1185]

Overcoming, and especially self-overcoming, is one of the central themes of *Thus Spoke Zarathustra*. They are fundamental themes of sport, the sportsman and the Man in the Arena.

The meaning of the *Übermensch*, we are told, is that humanity is something that must be overcome.[1186] Put more pithily, it is overcoming that puts the *uber* in *Übermensch*. Life, as well, is "that which must always overcome itself."[1187]

Overcoming conjures images of struggle (both physical and more 'spiritual') and this is surely crucial to understanding the concept and its application to sport.

But it also has associations with success in general and resiliency in the face of hardship and, as directed towards oneself, connotes introspection (but not pointless navel-gazing), self-knowledge, and self-critique.

All of these themes find their place both in *Zarathustra* the text and in the eponymous character himself.

Overcoming conjures images of conflict and struggle as does sport. That is why I have earlier described Muhammad Ali as the epitome of the overman. Other great sporting legends such as Pele, Maradona are in that category.

Nietzsche's account of self-overcoming has a healthy dose of struggle with oneself and with others.

It entails struggle with oneself insofar as one seeks to transcend one's limitations (physical and mental) and move towards ever more sophisticated, expressive, beautiful and potent modes of action and expression. Such 'internal' conflicts with fear, insecurity, ignorance, and hubris are all part of the ongoing process of self-overcoming for Nietzsche.

Ultimately, it is in struggling with and overcoming ourselves in this way that we live up to our full potential as human beings. Elster's constraint theory discussed earlier, is a derivative philosophy especially as it relates to boxing and the martial arts and other sports such as mountaineering, sailing, skiing and other challenging outdoor pursuits. These sports either involve overcoming the self or the environment in which they take place.

At the same time, struggle with others can often hone and facilitate this internal process, provided that those with whom we struggle are in a position to offer a genuine challenge and are not mere 'pushovers'.

Thus, we are urged by Zarathustra to seek enemies with whom to struggle—enemies of whom we can be proud, such that the successes of our enemies are our own.[1188]

We should not seek out those who can be easily overcome but, rather, those whose talents and abilities will offer the kind of challenge necessary to truly test our own capacities.

It should be stressed that it is the challenge itself that is the point here, not merely the annihilation of enemies.

While discussing our pride in our enemies, Nietzsche draws a distinction between 'hating' Hasen) and 'despising' (*Verachten*) and urges us to eschew the latter attitude for the former.

What is crucial in this distinction is that the original German makes clear the sense in which one can harbour respect and regard for what one hates, but to despise something is to see it as unworthy of regard or respect.

This is an admittedly awkward distinction to make in English (and relies on what is effectively a Nietzschean term of art in German), and so one must bear in mind the sense in which hatred requires time, effort, and attention.

One must take the object of one's hatred seriously—hatred cannot entail indifference or disregard.

Hatred in this context must be understood as divorced from despising in this sense and also as purged or resentment.

Indeed, for Nietzsche, the line between genuine friends and our enemies is indistinct at best.

Even our friends, he tells us, should present us with struggles:

"In a friend one should have one's best enemy. You should be closest to him with your heart when you resist him."[1189]

True friends and, thus, true expressions of love, therefore, are not blindly accepting and complacent but rather push the beloved to continuous self-improvement. As Oscar Wilde cryptically put it enemies stab you in the back but friends stab you in the front.

Self-overcoming is indeed a kind of struggle with one's enemies, one's friends and with oneself, though, all of these struggles are intimately bound together in practice.

Furthermore, struggles with oneself and with others are likewise bound together in practice for struggling with others I better, and thus overcome, myself, and in struggling with myself, I make myself a more worthy and valuable adversary.

All of this struggle and conflict is not an end in itself.

Ultimately, the point of all this agonism (both with oneself and others) is not destruction for its own sake but rather the creation of something higher[1190]—the transcendence of given standards of value and the creation of something new out of the ashes of the old.

According to Nietzsche, humanity, both as individuals and as a whole, is "a bridge and no end," and so we must reflect on and overcome ourselves so as to redeem the present and the past through the creation of a new future[1191].

We are most human when we are pushing the boundaries of human expression (aesthetic, moral, political, sporting).

We must avoid preserving ourselves[1192] and become instead "procreators and cultivators and sowers of the future."[1193] Complacency, stasis, self-satisfaction, and the ossification of values are all symptoms of decadence.

Self-overcoming (and the kinds of struggle that entails) is thus first and foremost the means whereby we avoid decadence and manifest the highest forms of human flourishing.

In terms of morality and aesthetics, the highest form of value is not particular value as such (good will, eudemonia, pleasure, and the absence of pain) but rather the creation of new values.

We are urged to smash the old 'tablets of values' and inscribe new ones, all the while bearing in mind that these new tablets are themselves only provisional and must eventually be overcome, as well.[1194]

In her book *Beyond Nihilism: Nietzsche Without Masks,* Ofelia Schutte emphasises Nietzsche's valorisation of 'becoming' over and against 'being'.[1195]

Overcoming, creating, esteeming, growing, changing, maturing, struggling—all the recurring themes in Nietzsche's corpus emphasise processes of becoming rather than states of being.

Nietzsche never once refers to a particular individual as 'an overman'.

The overman is a concept; it is something towards which we strive—"a bridge and not an end."[1196] It is yet again a 'beyond' concept.

Self-overcoming, then, is not something that one accomplishes. It is not an end state to be achieved. At no point can one say "Now I have overcome myself, my work is done here." As with my own discipline law (and now philosophy) in sport there is always more to be achieved.

Instead, self-overcoming should be understood as a never-ending process of self-appraisal, self-critique, and self-improvement.

The moment one gives up on this process or decides one has finished is the moment, for Nietzsche, when one ceases to be fully human in the most important sense—one has at this point abandoned the change and movement of becoming for the complacency, decadence, and ossification of being.

Self-Overcoming and Sport

By way of establishing definitions and a common framework, I am limiting my definition of Sport to include only the styles and systems that can roughly be called 'traditional' in that they emphasise internal training, respect for the (often ancient) roots of the techniques, and style or 'art' as much as, if not more than, practicality and efficiency. A properly Nietzschean genealogy of any sport would reveal roots in these more practical pursuits, but in the contemporary context most of what we study in the martial arts is less than truly practical in the more mundane sense. The styles and techniques that comprise the martial art developed, even within the Buddhist traditions, out of the very 'practical' goal of protecting the peoples of what is now the Korean peninsula from foreign conquerors, but the feudal Korean sword techniques, or the empty hand form (*Hyung*) are not what one could call practical martial techniques in an age of conceal and carry.[1197] Over time, the martial arts have become more than a mere means to (an admittedly important) end, they have truly become an art. Martial arts function in the age of handguns in a way analogous to calligraphy in an age of e-mail.[1198]

But this lack of instrumental practicality, this lack of efficiency, is exactly where Nietzsche would begin to locate the value of the sport (as an aesthetic expression and as a practice of self-overcoming). It is in the indifference to practical efficiency that one creates room to turn the martial arts into an avenue for human creativity and self-expression. This is because efficiency must always be subservient to the demands of specific goals within specific technological contexts—the means are always subordinate to the ends.

What is special about sport, however, from the Nietzschean perspective, is that it must always be a practice and never a product. Like one of Nietzsche's own favourite metaphors, dance[1199],[1200] the 'art' of the sport is only ever rendered through movement and is never a static 'artefact'. One's own body becomes the medium for a kind of artistic expression that is always transient and ephemeral. It is kinetic rather than static, a process of becoming rather than a manifestation of being. Of particular importance is the fact that it is never perfected or completed. Each iteration of a form of a technique is a kind of practice, not only in the sense that it is a kind of bodily movement but also in the sense that it is never in itself a perfect terminus of that form or technique. The sportsman never expects to truly perfect his or her art; it is always and can only ever be an ongoing striving for something beyond which we have previously experienced.

It is in its character as a process of striving that sport can be best understood as a kind of Nietzschean self-overcoming. For Nietzsche, the true worth of humanity is proved through resistance to and overcoming of stasis, rigidity, and the ossification of beliefs, values, and 'truths'. This he refers to metaphorically as 'the spirit of gravity'—as a kind of inertia that we must constantly resist in order to manifest our highest potential.[1201] Life itself, Nietzsche claims, is self-overcoming—it is power (will to power) expressed through the decay and destruction of the old and the creation of the new. When he first commits to training, the sportsman effectively sets out to undertake a rigorous regime of breaking down and ridding herself of old habits, both physical and mental. He works towards the shaping of new ones with an eye towards an ideal that is never completely grasped by the practitioner (as one improves, one is able to perceive with more subtlety and depth, reveal ever more layers of imperfection) and cannot ever be fully achieved.

One might say that training is simply repeating the same movement or set of movements over and over, but this is simply not the case. As one becomes more and more familiar and comfortable with the movements through repetition, one actually attains a better grasp of their function. This occurs on two levels. First, epistemically, as one practices the technique or form, one comes to understand better how and why it functions and how to improve it. All of us who have trained seriously for some time can recall moments when, after the nth repetition of a form, some new 'breakthrough' is made, and we come to see and understand the form in a new way, which in turn opens up new foci for future training. Second, physically, one's body becomes increasingly better at executing the movements—a joint-locking technique, for example, can actually change from being at first a series of steps, grabs, and twists into one smooth movement. Most important, there is a reciprocal relationship between these two levels. As one's physical competence increases, new insights are revealed on the more 'epistemic' level, which in turn enhances physical performance.

But the perfection of these techniques or forms is never achieved. We could always use more practice, more training. There is always room for improvement. To entertain the idea that one has perfected some aspects of the sport is to miss one of its most basic, though clichéd, insights: The practice of martial arts is about the journey itself, not the destination. Or, as Nietzsche might prefer, it is about becoming, not being. Thus, the sportsman has dedicated himself to a never-ending process of self-improvement and self-critique. He has himself become a medium for an artistic process of self-expression. Through his training he is, in essence, recreating herself in an explicit and self-conscious way. This is, in large part, when sport serves as such an excellent example of Nietzschean self-overcoming.

There is also an important epistemic aspect of sport in relation to Nietzschean self-overcoming, in the sense of knowing the material, as well as in the method by which one learns the material. First, consider the way in which one learns and studies. To be sure, a good instructor can make a huge difference in the quality of one's training, but in the end, the value of the instruction is contingent on the self-discipline, patience, and motivation of the student. Instructors can demonstrate, correct, cajole, chastise, and otherwise aid the student in learning and improving the material, but he cannot simply hand over a collection of propositions to be memorised and repeated.[1202] The student must on his own take up and practice what he has been taught, and it is only through this practice that we may say that he has truly learnt the technique of form in question. Indeed, the best instructors are not always the ones with the most trophies or the most effective techniques or the prettiest forms, but rather the ones who are best able to motivate and guide their students through what is ultimately an exercise in self-discipline. Sports as self-overcoming, is, in this way, an internally directed effort, but insofar as the goal is physical/practice movement, it cannot be understood as completely solipsistic, or indeed individualist.

Nietzsche can be very helpful in addressing this problem, which is a matter of our understanding of the conditions of knowledge as such, especially within a specifically kinetic context. Nietzsche's epistemic position is subject to several interpretations, much of which revolve around whether his 'perspectivism'[1203] amounts to relativism and, if so, of what kind. Regardless of the ultimate merit of such interpretations, however, it is at least clear that Nietzsche did not understand *himself* to be a relativist in this sense.[1204] He clearly thought that some ways of thinking and valuing were better than others. His epistemology allows for qualitative distinctions among knowledge claims based, in part, on certain kinds of epistemic virtues (some of which are agent centred, some of which are methodological) and a rather robustly naturalised empiricism. Thus, it is not the case that all knowledge claims are of equal merit, but this has far more to do with the way in which we go about offering and supporting our knowledge claims than it does with their content on its own. That is, the standard as Nietzsche describes it[1205] is focussed more on process than product.

At the end of the day, Nietzsche can be understood as a fallibilist—he holds that all knowledge claims, no matter how good they may be, are always, as a matter of principle, provisional. There is, once again, quite a wide array of difference in the exact interpretation of Nietzsche's epistemology, but there is general agreement on the admittedly basic account I am offering here [1206]; [1207]; [1208]; [1209]. Thus, while the interpretation of Nietzsche remains at best an inexact science, there is strong textual support for the general claim that knowledge, for Nietzsche, is never a *fait accompli* but rather an ongoing process. It is, once again, a manifestation of becoming, rather than a state of being.

So, returning to the question of 'knowing' a particular sport the problem lies in thinking of the answer to the question as an all-or-nothing proposition. As a kind of feat, such knowledge is impossible. If at any point I think that I know a weapon in the sense that I have learnt all there is to know about it and that my use of it is perfected, that can only be understood as either a deliberate self-deception or a manifestation of hubris. There are two important elements of taking this approach seriously. First is the fact that the continued validity of my knowledge claim is contingent on my continued practice in the sport—my knowledge lies, in part, in my continued commitment to improvement. If I have stopped practising, my knowledge claims is immediately suspect.[1210] Second, this understanding of knowledge clearly admits of degrees. The knowledge, therefore, lies in the overcoming (as an ongoing process) of ignorance and incompetence. I know the sport if I am engaged in the process of improving my skills with it. The moment I begin to think of the knowledge as a manifestation of being instead of becoming, I have lost the real meaning of it.

Thinking of sport as a manifestation of self-overcoming does not commit one to an understanding of the practice of the sport as solipsistic or radically individualistic. It is importantly self-directed, but there is also a strong sense in which this is a fundamentally social endeavour. I need instructors and exemplars in order to get the most out of my training. There is certainly an element of training that is individualistic. I need to take the time to go over, repeat, rehearse—in short, to train—on my own. But if I do not open myself up to the scrutiny and critique of others, if I do not have real people with whom to practice, if I do not have real (and 'worthy') partners against whom to test myself, my training becomes far less effective and will ultimately stagnate. Indeed, opponents of this sort are fundamental to my own efforts towards self-improvement. By way of illustration, we find that *Zarathustra* is constantly going away to his mountain to be with himself, then coming back down to share and test what he has learnt, and then returning to the mountain, only to come back down again later. The point again is to reject all-or-nothing dualisms. I cannot accomplish much completely on my own, but I cannot hide behind or lose myself among others, either (like the person whose training is driven solely by the need to beat all opponents). Ultimately, both of these manoeuvres can be understood as way of hiding from oneself. The people who spend all of their training time hiding alone in their backyards cannot test themselves against others and learn from their observations and insights, and those who seek only to best their opponents are focussed solely on the weaknesses and strengths of others (their opponents) and come to grasp their own capacities only in light of those others' strengths and weaknesses. Self-overcoming, therefore, is about not becoming complacent, either in solitude or in company.

Sport can be fruitfully understood as a kind of manifestation of Nietzschean self-overcoming. It serves as a helpful example of this Nietzschean concept, and thinking about the sport through the lens of self-overcoming illuminates important aspects of the methodology and ethos of sports training and practice and life. It is about testing oneself, working through failure, and always, to use an important Nietzschean concept, *"becoming what we are."*[1211] One cannot *be* a sportsman in the static sense of having some property or possessing some status; one is a sportsman only to the extent that one is always striving to *become* a better sportsman and to experience what lies beyond the horizon. The same applies to becoming a more complete person.

However, I need to conclude this chapter by pointing out that there are significant limitations in extending Nietzschean philosophy to the concept of sport and beyond. He was Professor of Greek at Basel University in Switzerland. He therefore would have had an in-depth understanding of Greek sport and its concepts. The philosophy of Nietzsche was the rejection of moral values and the "slave morality" of Christianity. His ideal was the 'overman' or 'Superman 'who would impose his will on those deemed to be too weak and worthless to be anything but slaves.

Until the 20th century, his beliefs remained largely ignored or opposed by Conservatives and Socialists alike, but support for modern Totalitarianism has often been claimed, wrongly in my view, as stemming from Nietzschean philosophy. The true sportsman is not an overman. Unlike the overman the true spirit of sport is to respect one's opponents and admire them for their talent and ability. It is true that the word 'overcome' can be equated to beat, be victorious, master, overpower and to prevail. But sport is much more complex than that. In this context it is significant to note that Nietzsche in his mature years suffered a complete mental breakdown in 1889 from overwork and loneliness. This was not withstanding his philosophy of overcoming and self-overcoming. It seems therefore that ultimately, he had difficulties in achieving that himself. But I am not surprised by that. As I explain in my chapter 51 (Sledging) many sportsmen likewise seem to have failed to overcome their own frailties in overcoming others. Karl Marx described religion as the "opium of the masses" and in my opinion there may be a tendency on the part of the sportsman to use sport as an opiate to disguise their own weaknesses. There can be no doubt that sport and indeed work can be used as an escape from reality and amount to the antithesis of 'self-overcoming' in Nietzschean terms.

Chapter 50
Sporting Metaphors,
Competition and Capitalism

The essence of metaphor is understanding and experiencing one kind of thing in terms of another.
George Lakoff and Mark Johnson

All perception of truth is the detection of an analogy.
Henry David Thoreau

An iron curtain has descended across the continent.
Winston Churchill

It is a poor sport that's not worth the candle.
George Herbert (1593–1632)

Aware of every one against everyone.
Thomas Hobbes (Leviathan)

You give 100% in the first half of the game, and if that isn't enough in the second half you give what's left.
Yogi Berra

This is a well-documented topic. I feel like Sir Isaac Newton did when he metaphorically explained "If I have seen further it is by standing on the shoulders of giants". Great sportsmen must feel that too. I therefore write this chapter with a degree of trepidation as the concept of sport and beyond lends itself readily to metaphor and the spirit of competition. So here goes!

Sport is a metaphor for life. This book itself is a metaphor about the Man in the Arena and beyond the horizon. It focusses on the metaphorical bifurcation of mind and body and looks at the reality beyond the concept of sport. It takes us to a new dimension of what is real yet far away and beyond the norm. Some concepts cannot be adequately defined but only recognised when we identify them. Metaphor, simile and allegory are in that category as they take many forms and are contextual. Philosophy, law and sport have a common thread. It all depends on the circumstances. Abstract concepts have practical connotations and manifestations. Sport is a paradigm example of that proposition.

Sport is a tool for understanding aspects of life beyond the concept of sport. That is why its philosophical significance has such import in understanding the nature of reality.

Wittgenstein said the meaning of words depends on their application. Metaphor is a classic example of that proposition.

Metaphor is the figurative use of terms without indication of their figurative nature in contexts to which they are not literally applicable. The power of metaphor cannot be understated. I have already referred to the philosophy of José Ortega Y Gasset, a great Spanish philosopher in my earlier chapters. He wrote a great deal on the significance of sport. He used metaphor and analogy on a regular basis to expound his views on the relationship between Sport and the Human Condition, Sport and the Good Life, Sport and the State. He said of metaphor: "The metaphor is perhaps one of man's most fruitful potentialities. It's efficacy verges on magic and it seems a tool for creation which God forgot inside one of His creatures when He made him."

Whenever we explain or communicate a concept by likening it to something else, we are using a metaphor. The two things may bear little resemblance to each other, but our familiarity with one allows us to gain the understanding of the other.

Metaphors are symbols and as such they can create emotional intensity even more quickly and completely than the traditional words we use. Metaphors can transform us instantly. All great teachers—Buddha, Mohammed, Confucius, Lao Tzu have used metaphors to convey their meaning to the common man. Regardless of religious beliefs, most would agree that Jesus Christ was a remarkable teacher whose message of love has endured not only because of what he said but also the way in which he said it. He didn't go to the fishermen and tell them he wanted them to recruit Christians; they would have no reference for recruiting. So he told them he wanted them to become 'fishers of men'.

Metaphors can empower us by expanding and enriching our experience of life. This is especially so in terms of our appreciation and understanding of the concept of sport.

Philosophically the metaphor 'Life is a game' is probably the most used sporting metaphor. Donald Trump has converted this in many of his speeches to 'Life is a test'. In sporting language you either win first place or you lose, there's no in between. This may be interpreted to mean it's going to be tough you'd better be prepared. For some people life, like sport, is a competition. That might be fun but could also mean that there are other people you have to beat, that there could only be one winner.

For some people, life is a game. How might that colour your perceptions? Is it in Thoreau's terms, the 'perception of truth'? Life might be fun—what a concept. It might be somewhat competitive. It might be a chance for you to play and enjoy yourself a lot more. Some people say "If it's a game then there are going to be losers." Other people ask "Will it take a lot of skill?" It all depends on what belief you attach to the word game; but wit that metaphor "you have a set of filters that is going to affect the way you think and the way you feel."[1212]

That is the law which again and again throws bourgeois production out of its old course and which compels capacity to intensify the productive forces of labour, because it has intensified them, it, the law which gives capital no rest and continually whispers in its ear: "Go on! Go on! (Karl Marx, *Wage Labour and Capital*)."[1213] I am indebted throughout this chapter to Ann E. Cudd of the Department of Philosophy at the University of Kansas for her seminal and illuminating article Sporting Metaphors Competition and the Ethos of Capitalism (The Journal of the Philosophy of Sport, 2007).

That article admirably collates and embellishes much of the learning found in the authoritative writings of Lakoff and Johnson, A Kohn, Karl Pribram and T J Roberts, to which I refer in this chapter and elsewhere in the book and whose contributions are identified and attributed to in the End Notes. I congratulate Ann E. Cudd, currently President of Portland State University, on her perceptive analysis of this fascinating topic at the heart of sport and philosophy and philosophy and sport.

Metaphors themselves are comprised of a number of interrelating components to which are called symbols. Metaphor is right at the bottom of being alive. It is now the whole fabric of mental connections held together. Neuroscientist Karl Pribram wrote "Analogical reasoning sets in motion a self-reflective process by which metaphorically speaking brains come to understand themselves."[1214]

The English language is replete with metaphors that use sport to describe daily life as a kind of game. Many of our sports metaphors date back to an earlier time when the most popular games were games of chance, poker and horse racing. In the latter part of the 19th century, the team sports of baseball and football (to be followed by basketball and hockey and later still by lacrosse and football) began their ascendancy to the top of people's imagination. At the same time, capitalism was becoming the dominant economic system in the western world. As capitalism became more industrial, team sports gained in popularity[1215]. These sports are face to face and hard hitting, emphasising quick, strategic decision making; athletic moves; and team play. Throughout the phases of 19th- and 20th-century capitalism, sporting metaphors were used to reflect how those characteristics were to be found in all our activities ranging from personal to managerial to fiduciary[1216].

This connection between sports and capitalism is reflected in and emphasised by our metaphorical language connecting sports and work. No metaphor is more powerful than competition and the idea of the competitive market as a winner-takes-all, no-holds-barred dogfight. There are metaphors that illuminate the competitive aspects of capitalism and its focus on winning as well as metaphors that emphasise cooperation and ways that capitalism improves the lives not only of the winners but also of all who choose to play the game by its rules. Although sports metaphors invoked to describe capitalist competition may appear to cast an unflattering light on both capitalism and sport, on a deeper analysis those metaphors appeal to many of us because they reveal a closer resemblance to the Latin root of the word 'competition' and its cooperative, pareto-improving implications. Just as healthy competition in sports requires cooperation, healthy capitalism is also ultimately, a cooperative endeavour. Metaphors imported from and expanded through our experiences of sport reveal many, while concealing other, aspects of capitalism.

1. Metaphors and Culture

Although we tend to think of metaphor as primarily linguistic embellishment, two influential scholars of metaphor, George Lakoff and Mark Johnson, ascribe to metaphor a dual and ubiquitous role in thought and language. Metaphor, they argue, helps us categorise and understand our perceptions by constructing connections between distinct concepts. "The essence of metaphor is understanding and experiencing one thing in terms of another."[1217] When two concepts are connected by a metaphor, we use the more familiar concept to understand the less and use the scripts for action in one realm as a guide for action in the other. Thus, Lakoff and Johnson argue that metaphors reveal much about the way we think and act and are not just window dressing for our language. Indeed, they see metaphor as systematising our very conceptual systems.

Metaphor structures human thought and guides our thinking. We use metaphor to think everyday thoughts. Our conceptual system hangs together in part by means of the relations of the metaphors we think with and use linguistically. We heap metaphors upon metaphors, as a quick glance at this sentence reveals: It contains a construction metaphor (heap), a spatial metaphor (upon), a time metaphor (quick), and two vision metaphors (glance, reveals). Although we are familiar with the injunction not to mix metaphors, our linguistic expressions would be nearly empty without such promiscuous mixing. Some of these metaphors are more obviously metaphors than others. Those that are less obvious or appear to be literal expressions are simply more fundamental to our conceptual scheme. On Lakoff and Johnson's analysis, we can speak literally and only the most mundane, practical, and immediate experiences; all other talk is metaphorical. Andrew Orteny points out three remarkable properties of metaphors: In possibility, vividness of compactness. Because metaphors embody that "something, vague, unknown or hidden, they give form to the inexpressible." Because they make use of everyday concrete things to illustrate intangible, complex and relational aspects of life, they are vivid and memorable.[1218]

Metaphor both reflects and constructs cultural foci. Lakoff and Johnson argue, "The most fundamental values in a culture will be coherent with the metaphorical structure of the most fundamental concepts in the culture."[1219] They propose that spatial relations form the most fundamental metaphors in our culture: in-out, up-down, near-far. They claim that one can trace the causal chain of metaphorical relations through a logical hierarchy of implications. For example, up is typically metaphorical for good, down for bad. Up is also metaphorical for more and down for less. If we examine a claim like "inflation is up," we can see that the "up is more" metaphorical relation takes precedence over the "up is good" relation. Thus, the metaphorical relation "up is more" is more fundamental to our conceptual system than "up is good."[1220] Carl Jung noted that there is always something more to a symbol than meets the eye and no matter how much a symbol (or metaphor) is described, its full meaning remains elusive.

Terence Roberts highlights another aspect of metaphor that about the development of capitalism through sporting metaphors.[1221] Through a discussion of Richard Rorty's theory of metaphor, Roberts reveals how metaphors "alter logical space," enlarging the space of what is considered possible. Roberts illustrates this through a discussion of the development of different bowling techniques in cricket. For Roberts, sporting technique is a language-without-words game, and new techniques or moves in a sporting practice are like new metaphors in language. They extend the language by making possible new thought combinations or analogies, but they also are open to contestation and critique. Similarly, new techniques for bowling create new combinations of batting and defensive techniques and challenge ideas about what was considered good or even legal bowling. Some techniques work for a while and then are defeated (perhaps only to reappear in subsequent generations) by new batting techniques. Roberts's discussion of cricket and metaphor was aimed at showing how sporting practice is a kind of language game, but the important point is to see that metaphor is a way of expanding our thinking and revealing new aspects of a rule-governed practice, as well as challenging the justification of the rules.

Metaphors focus our attention on some aspects of concepts and divert our attention from others. For example, the metaphor of game, as in "life is a game," focuses our attention on how life reflects the following sorts of aspects of games: Games have definite beginnings and ends; there are rules; games often bring pleasure to the players but sometimes pain, as well; there are winners and losers; and so on. This example also reveals how metaphors can be false or misleading, though. While many games have only one winner and many losers, life does not seem to be like that at all. Metaphors also disguise aspects of concepts, such as how the game-life metaphor diverts our attention from the fact that life does not have well-defined constitutive rules and is open ended in terms of the allowable strategies that players can pursue.

Because metaphors are coherent with cultural values and reflect and construct our cultural obsessions while diverting our attention from other values and meanings, metaphor is therefore intimately connected with the construction and expression of ideology. By 'ideology' is meant a system of beliefs, mostly or at least partly political, that has implications for action. While liberal political theorists are content to talk about ideology in this rather benign way, radicals point out the obfuscating features of ideology[1222], such as Marx's famous discussion of commodity fetishism.[1223] Lakoff and Johnson's theory of the role and function of metaphor lends itself to both liberal and radical theories of ideology in that it explains how ideas hang together through metaphorical connections and how those connections can distort, disguise, or divert our attention from other aspects of those ideas. Thus, metaphors can powerfully support ideological movements by highlighting their positive features and disguising the negative ones.

Capitalism constitutes the ruling ideology of our age. There are metaphorical relations that connect our conceptual scheme and support such pride of place to capitalism. The metaphors imported from sport through which we understand capitalism and formulate our politics reveal and conceal much about our culture that lies beneath our immediate, conscious experience of it. Metaphors embody and define the intangible and abstract but this process limits and constrains perceptions and actions to those which make sense within the logic of the metaphor. Metaphors are therefore both descriptive and prescriptive.

2. Sporting Metaphors

Most theories of sport define it as activity that meets at least the following three criteria: Persons voluntarily engage in the activity, the activity is rule governed, and the activity poses a competitive challenge to the persons engaged in it. Perhaps the most influential definition of games, which was proposed by Bernard Suits and as discussed earlier posed the following definition[1224]. A game involves (a) the prelusory goal, a state of affairs specifiable independently of the rules of the game, that the players are trying to attain; (b) the (lusory) means for attaining the goal permitted by the constitutive rules; and (c) the lusory attitude on the part of the players that they accept the constitutive rules. The crucial insight in this definition of 'game' is that a game involves voluntarily using only the means allowed by the rules to reach the goal, and these means are characteristically not the most effective means for doing so. A sport for Suits is a game that also requires physical activity. Thus, some theories emphasise that sport must be physically challenging[1225] in order to rule out such games a poker or chess, but I would rather cast a wider net for my purposes here. I am not trying to set out a novel definition or understanding of sport, but rather I am trying to examine capitalism and its ethos through sports metaphors that are commonly used to discuss the business of capitalism.

By capitalism is meant an economic system whose core, defining feature is that it allows private ownership of the means of production, that is, of capital inputs to production. In such a system, under very minimal assumptions of differences in preferences and/or initial distribution of capital inputs, markets will develop, including markets for labour. The definition of capitalism entails that these markets are to be free of undue government intervention so that we may enjoy freedom of movement and enjoy the products of work and trade.

The most important aspects of the concept of sport for its analogy to capitalism revolve around the rule-governed nature and competitively challenging nature of the activity. Sport is governed by two kinds of rules: constitutive rules that define what moves are permitted and how the game is scored and rules of decency and fair play. Metaphorical relations that come out of the constitutive rule-governed nature of sport include 'foul', 'fair', 'in the ballpark', 'extra innings', 'from the word go', 'tackle', 'score', 'no holds barred', 'down for the count' and 'level playing field'[1226]. The injunction to 'play by the rules' is a standby of business ethics. The primary metaphor from the rules of decency and fair play that surround sport is that of the 'good sport'.

Sports pose a competitive challenge to their players in several ways. They often pose physical challenges of skill, athleticism, stamina, or endurance. Sports always pose mental challenges by requiring quick and effective decisions, the ability to assess opponents (and teammates) and react to their strategic decisions, and emotional strength in the form of confidence, determination, flexibility, and persistence. Each of these aspects of challenge in sports engenders metaphorical relations that form a part of our conceptual scheme of capitalism.

Metaphorical relations that begin in the physical challenges of sport may seem to poorly fit the challenges of economic life in capitalism. But many such relations are metaphorical for the determination and persistence that are rewarded by success in business. "No pain, no gain" might be used by managers to justify a decision to streamline a company, despite the complaints by workers and public officials that the company might endure (not to mention the physical suffering of the workers themselves, who can only cynically be described as gaining from their layoff). A long session at work is a 'marathon', and someone who works hard will 'go the extra mile'. One who is deft is 'on the ball'. A person with determination might 'make a comeback' after suffering a 'setback'.

Decision making in games such as poker, chess, and gambling holds many similarities with the decisions made by capitalists and is a rich source of metaphors. Game theory, an important theoretical model of capitalist interaction, exploits this analogy explicitly. Many metaphors that are useful in understanding capitalism, then, have to do with taking risks. Entrepreneurs are said to 'take a shot' at developing new products, even when the 'stakes are high' and more timid persons would 'bet on it'. The stock market's most secure capital are its "blue chip stocks." When there is little information on which to make a prediction, however, "all bets are off." Managers must be good at making decisions and taking responsibility for the results. They might "toss an idea around" with their associates, but ultimately they must "call the signals" and "make a move" or just "go for it." If they find that they are outdone by a competing firm because they are "out of their league," then the 'buck stops' with the manager.

What about the voluntariness criterion of sport—are there important metaphorical relations engendered by this aspect of the concept that form a part of our concept of capitalism? Some metaphors that might fit this bill would be 'free agent', 'for the love of the game', 'call the signals', 'freestyle', or perhaps 'go for it'. But metaphors that reveal coercion or force are just as common here. Consider 'backed into a corner', 'pinned', 'cut one's losses', 'tackled', 'in over one's head' or 'fall guy'. On Jan Boxill's view of sport, sport must be voluntarily engaged in if it is to have positive moral value for us. The same can be said of capitalism. Yet capitalism has been criticised as coercive. Indeed, Boxill contrasts sport with work in capitalism to illuminate how sport, unlike work, is free, unalienated activity. The metaphorical ambivalence, is reflected in our ambivalent feelings about capitalism and its social costs and benefits.

Sport and capitalism are analogous in some ways and disanalogous in others. Sport has been characterised already as an activity that is voluntary, rule-governed, and competitively challenging. Capitalist interaction is likewise rule-governed and voluntary, if it is legal. The economic models of capitalist interactions are termed 'competition', as in 'perfect competition', or 'monopolistic competition'. In broad terms, sport and capitalism both describe systems that structure large portions of most of our daily lives. There are important voluntary and non-voluntary aspects to each of them. Sport can be avoided but at the cost of not being able to speak the lingua franca of contemporary popular discourse. Capitalism can be avoided at a cost, as well, although the cost may be one's ability to survive. Whole subcultures live among us that avoid sports or capitalism, such as academics and the Amish, but neither is completely free of what they shun; they live on the periphery, not over the edge. For members of this culture, opting in to either sport or capitalism is far easier than opting out.

Some argue that a crucial disanalogy between sport and capitalism is that the former is not a serious, life-and-death matter, whereas the latter is. Francine Hardaway criticises sports metaphors to describe aspects of contemporary culture as 'doublespeak' that makes us think that the serious business of winning and losing in life is as negligible as if it were just a game[1227]. Hardaway seems here to define as serious only pursuits that make life meaningful. Clearly sport can do this for many of us, as Suits points out in his discussion of the seriousness of sport.[1228] Leisure activities have become the intrinsically valuable activities in terms of which we feel justified in working hard. Suits also notes that we have become pluralists about what counts as valuable leisure activities. Sport is as serious as art or religion, two other systems of meaning in our lives, each of which is at least a serious rival to capitalism as one of the main sources of meaning and value for us. Sports practitioners can be extremely dedicated to their games, and we often admire them for that dedication. Sport can be a life-or-death matter, though it typically is not, even though competitive sport often injures its participants. So sport is as serious as "winning and losing in life." If art can be life, then sport can be life, as well.

3. Competition: The Central Metaphor

Competition describes a situation that determines a winner (and therefore the nonwinners, or losers) under commonly known criteria for winning (losing), and usually awards some prize or recognition. Sports are competitive by definition. They determine their winners by their constitutive rules; they are what Alfie Kohn calls structurally competitive because the whole point of their structure is to determine a single winner, or at least when to give up trying to select one and declare a tie. Competitive sports paradigmatically pit players against each other, although solo sports pose a competitive challenge by setting difficult criteria for success that relatively few can reach. In a sport competition if the players are trying to play at all they are necessarily trying to win and not to lose. Since there can be only one winner and the others must be losers, there is a necessarily zero-sum aspect to sports. Competition in sports tends to breed a psychology of intensity and self-perfection bordering on narcissism and egotism. Thus, the often quoted Vince Lombardi statement that "winning isn't everything; it's the

478

only thing." But if it is the only thing that matters, then morality or decency, to say nothing of beauty, does not matter. Maximising participating and contributing to education do not matter, either. In such an atmosphere, it makes sense to try to get away with breaking or hedging the rules whenever it gives an advantage.

Capitalism is metaphorically, not literally, competitive. There is, after all, no literal winner or loser in capitalism. Competition and capitalism are thus distinct concepts, but their connection is deep and enduring, and so we normally fail to note the metaphorical nature of the connection. Neoclassical economic theory, the theory of capitalist economies, uses the terms 'competition' in a special sense. A competitive market is an idealised model of a capitalist market in which it is assumed that there are no barriers to entry or exit, the agents involved take no interest in the others' interests and pursue the maximum satisfaction of their own interest, subject to their budget constraints; and there are no transaction costs. Each seller in a competitive market can have no effect on prices; each faces a horizontal demand curve and so makes no decision on setting prices. Sellers compete to stay in the market, that is, to keep their costs low enough to be able to cover all their costs (including their own entrepreneurial labour costs). In the theory, winning is just a matter of staying in the game. Monopolistic competition turns out not to be an oxymoron for the neoclassical economist but rather another term of art, in which there are assumed to be many similar goods but none exactly alike (e.g., different brands of cereals), and each seller faces a downward-sloping demand curve. The metaphor of competition in economic theory, then, aptly describes the attempt to maximise success, which is the satisfaction of the agent's interests given her budget constraints. That is, each can choose a price at which to sell their good within a range of options, only one of which is optimal. In the sense of seeking the maximisation of self-interest, the theoretical agents of capitalism are definitional egoists, but not necessary egotists.

The culture of competition in capitalism, like that in contemporary sport, does, however, breed a psychology of intensity, greed, and egotism. The popularity of Donald Trump and his reality TV show *The Apprentice* testified to this fact and may explain his subsequent election to be President of the United States. He became a cultural icon for his egotistical braggadocio. The contestants for the role of Trump's apprentice seemed willing to go to great lengths to impress him with their willingness to climb over the bodies of their competitors in quest of his affirmation. Capitalism pits not only sellers against sellers but also sellers against buyers. Sellers want to get the highest price from the buyer, who wants to pay the least for the good. There are perfectly legal and decent ways to get the highest price or to pay the lowest one. But moral and legal violations of laws meant to ensure fair competition in capitalism are also not rare. Thus, competition can inspire bad behaviour in the question to win at any cost.

Is competition to be positively evaluated, then? Kohn lodges three main objections to structural competition, which he also describes as mutually exclusive goal attainment[1229]. First, he claims that (ironically) it is inefficient. If people work together they can join their energies in whatever enterprise and attain a better outcome. If two businesses share knowledge, they can produce better and cheaper products; if two sporting rivals train together, they can encourage each other to higher levels of achievement. Second, competition causes psychological damage to the individual in the form of lower self-esteem (though, presumably, not in the winner) and performance anxiety. Third, competition causes social damage in that it damages relations between persons who must see each other as rivals.[1230] A fourth, related, objection is raised by Torbjörn Tännsjö, who argues that our culture of competition is fascistic in that it encourages us to disrespect the weak, the infirm, and the disabled[1231].

Kohn's objections assume that it is possible to eliminate competition. This is not always the case. In a running race some or all might improve their time, but there is one who is fastest. If only one person can have a certain job, then there is necessarily a winner. Robert Nozick argued that with any variable skill there will always be a relative component of success or achievement[1232]. So we cannot assume with Kohn that it is possible to eliminate competition in sport or in capitalism, or even in another kind of play or another kind of economic system that might replace them. That said, if Kohn's arguments stand, then competition might still be viewed as a necessary evil. The psychological and social effects to which Kohn refers ring true. Competition does raise anxiety and lower self-esteem in (those who see themselves as) frequent losers. It does cause rivals to be diffident and hostile to one another, at times even inciting violence. Although it is true that these are not necessary components of competition, given our psychology, we can expect them.

Competition could be valuable either intrinsically or instrumentally. In sports, a case can be made for the intrinsic value of competition, since the very definition of sport requires competition. Thus, if sport is valuable in itself, competition is, as well. The intrinsic value of sport consists in the fact that it is an unalienated activity, offering the opportunity for meaningful self-expression and community with others[1233]. By unalienated is meant that the activity is freely chosen and would be freely chosen even without any external reward for the activity. By this test, sport is an unalienated activity, but work is not. That is, most of us would not work, but almost all of us who play sports would continue to do so without its external rewards. Of course there are exceptions: the child who is forced by his parents to play a sport he does not like, the woman who watches football with her partner only for the opportunity to be with him, the professional athlete for whom the game has become too much like work to enjoy, or the man who plays golf in order to cultivate high-paying clients. In these cases sport is alienating and not intrinsically valuable. But these are the exceptions, where the external reward (or fear of punishment) is the only reason that these persons participate in their sport. Most of us find some value in the sport that is intrinsically motivating enough for us to continue playing it, if we play it at all. It is characteristic of unalienated activity that we can find ourselves lost in the activity, focusing on its internal goals and strategies, unaware of the world external to it. Athletes will say that they play the game for such moments of 'flow', which are intrinsically motivating.[1234] Now one might object that although this shows how sport is intrinsically valuable, it does not follow that competition is intrinsically valuable. Rather, some aspects of sports are intrinsically valuable, such as the fact that it involves play and games, but not competition. However, competition makes it possible for most persons to focus enough to have those intense experiences of flow that are so valuable. Thus, there is at least one aspect of sport that is intrinsically valuable and that thrives on the competitive nature of sport.

Some defenders of capitalism will claim that it has intrinsic value because it is tantamount to a type of freedom. People are free in capitalism to contract to buy or sell products or services in the market in which others are free to accept or reject or offer competing contracts. Self-ownership is thus a hallmark of capitalism. This contrasts sharply with earlier forms of economic organisation in which many persons did not own their own labour power, which could be commanded by others. This shows only that self-ownership, which is intrinsically valuable, typically results in competition. It does not show that competition is intrinsically valuable.

Likewise, sport is best described as cooperative competition. Drew Hyland's discussion of the place of competition in the dialectic of sport is helpful here[1235]. He argues that sport does not have the single goal of winning but rather a set of *teloi* that make sense of the dialectic of athletic competition, which consists in the attempt of each competitor to try to overcome or 'negate' the other, in a single game or over a series of them. This negation can lead to alienation and the sense that the whole point is to beat the other team, perhaps even violently, or it can lead to a more benign form of negation, which "can even be an occasion for friendship, for striving or questioning together."[1236] The *teloi* of competition in sport, he argues, must include *teloi* such as friendship and mutual excellence, however. Alienated competition must be seen as a defective mode because seeking to win by all necessary means is the negation—the endpoint—of competition itself. The dialectic cannot be continued. Similarly, competition in capitalism cannot have as its end the goal of draining the other competitors in the market of all their wealth, as that would be the termination of market interaction.

Although there can be only one winner in many games, there must be competitors who adopt a lusory attitude, that is, agree to play by the rules and compete at their best, in order for there to be a meaningful sporting context. There must be athletes and officials who agree to the rules that will govern the sport. There must be cooperation (even while there will also be competition) among team members in team sports. In individual sports, competitors often critique each other's performances and help coach them to better performances. Simon's concept of sport as a mutual question for excellence[1237] helps explain why cooperation is as important as competition for sport. Without the cooperation of a community of players, coaches, and supporters, a sport cannot thrive and, thus, neither can the athlete who participates in that sport. Although athletes and sports commentators may occasionally forget this, they cannot successfully continue to participate without some willingness to cooperate for the good of the competition. 'Competition' comes from the Latin word *compete*, which means "to strive after something in company or together[1238]." These roots are clear when we understand competitive sports as a mutual question for excellence.

The competition metaphor tends to distract us from this cooperative aspect of both sports and capitalism in many cases. For example, in discussions of international trade, the ability of capitalistic competition to raise the well-being of all parties is often confused with the idea that there can be only one winner and one loser. Likewise, in sport the goals of maximising participation or the educational value of sport are sometimes seen as a choice between sacrificing or being sacrificed to the goal of competition, rather than as potential compatible outcomes of competitive athletes.

The metaphorical concept of competition in capitalism has an ambiguous meaning and value for our culture. While it conjures an image of the egotistical, striving Donald Trump and workers trying to undercut each other in a race to the bottom, on reflection, it also connotes a cooperative quest for excellence that lifts up the overall level of welfare in society and counts on the cooperative interaction of all members of the community. Competition combines with other sporting metaphors to make this ambivalence of capitalism go even deeper into the discourse of our culture.

4. Sporting Metaphors of Cooperation and Competition
Level Playing Field

An important sports metaphor that extends and refines the metaphorical concept of competitiveness is the level playing field. This metaphor conjures up the image of a flat field in which no team is forced to play uphill and where round balls roll evenly, without surprising bounces. In sport, having a literal or metaphorical level playing field is important so that an effective challenge can be mounted by roughly equal players who are making their best efforts to win within the rules of the game. This, after all, is the point of sport—to test one's own and one's opponents' skills by attempting to meet each other's challenge within the rules. This is the meaning of Simon's claim that sport is a mutual quest for excellence or Hyland's understanding of sport as mutual striving.[1239] A level playing field is necessary not only for a good match but also for a fair one. If the playing field is not level, then the challenge is greater on one side and lesser on the other. It may lead to a blowout, which is a situation in which the competitive challenge no longer exists, and is unlikely to be either 'mutual' or a successful 'quest for excellence'. The metaphor of the level playing field thus connotes both fairness and the requirements for a good, successful satisfying competition.

A level playing field is important in capitalism also to ensure competition as a way to maintain the balance between firms, consumers, and workers that I discussed previously as an important component in a successful capitalist economy. Because competition is the central metaphor of capitalism, the metaphorical relation of the level playing field as preserving and enhancing competition is mirrored in our capitalist discourse. For example, in debates over laws governing antitrust or trade subsidies parties often argue that such laws will make or disrupt a level playing field.

Unlike the metaphor of competition, the metaphor of the level playing field itself carries little ambivalence for us. Fairness is an unmitigated good; a level playing field is a requirement of justice. However, like competition, the metaphor of the level playing field is clouded by ambiguity in its meaning. Achieving or recognising a level playing field is politically loaded, reflecting the political ambiguity of what it means to enhance or inhibit competition, let alone ensure justice. A good example of this is debates over international trade policies (i.e., 'fair trade') where some parties will argue that subsidies or quotas are needed to level the playing field for companies since they face higher taxes or environmental standards, while others will point to their labour costs as the obstacle to a level playing field. On the other side are countries whose firms produce at lower costs because their workers do not have the luxury to forego work and income in order to demand better environmental standards or higher wages. They complain that there is not a level playing field when other countries' policies effectively prohibit consumers from buying goods made by poorer workers, thus further impoverishing them. What counts as fair thus depends on how one describes the conditions on which the competition is based.

Playing by the Rules

An important metaphor for understanding fair competition in capitalism is the concept of rules and what it is to "play by the rules." Rules imply guides for behaviour that are commonly known, or at least assumed by most of those playing the sport to be commonly known, and enforced to a greater or lesser degree so that if one is in clear violation of the rules one can expect negative consequences. In sport we can distinguish two kinds of rules: constitutive rules and rules of decency and fair play. The constitutive rules of a game are explicit and formal and describe the aim, the allowable moves, and the penalties for violating them. Rules of decency and fair play describe what moves, strategies, and behaviours are informally allowed. Since sport is a practical activity, even the constitutive rules have to be interpreted and tend to change organically over time to fit external and internal circumstances.

In capitalism, we can distinguish similar kinds of action-guiding principles, namely, laws that are the constitutive rules of capitalism, ethics, and the rules of decency and fair play. Laws define property and property rights and thus are definitive of the economic system itself. If persons can be property, then we have slavery. Likewise the law of property can make a system feudal or capitalist or socialist by assigning certain sets of rights and obligations to persons based on their historical relations to material and other wealth-conferring or—creating objects. Property rights define theft, which is a particular salient way of failing to play by the rules. In our contemporary culture, we talk about people who play by the rules and manage to make a living or fail to because of some kind of hard luck. Those who steal or cheat are said to be not playing by the rules and, therefore, in need of legal or social sanction. Law also defines fair and unfair competitive practices and prevents monopoly power, collusion, or insider trading from thwarting competition. Although there are many ethical prescriptions for individuals in all societies, the ones that could be said to be the ethical prescriptions specific to capitalism are the ones that concern the behaviour of firms, businesspersons, and managers more than workers, who are constrained mainly by the laws regarding theft of one kind or another. Businesses are expected to show a certain amount of generosity or philanthropy in order to be said to be playing by the rules.

In either the legal or ethical sense, then, "playing by the rules" concerns the maintenance of the competitive environment as a mutually beneficial, or a cooperatively competitive, one. Thus, the metaphorical use of the positively normative phrase playing by the rules supports competition in capitalism in a way that, on inspection, reveals the important cooperative element of competition.

Teamwork

The metaphor of the team and teamwork is explicitly cooperative. Typically it conveys the notion of cooperation when used to describe workers or managers in a firm. Yet the sense of cooperation is ambivalent in this metaphor, as well. A sports team is not a purely cooperative situation but rather more of a cooperative competition. Each team member has an interest in working well with the team to defeat an opponent, but each also wants to play individually well enough to maintain her position in the starting line-up. While those on the bench would like the team to win, and so will work to improve the play of the other team members in practice, each would also like to outshine the others in order to earn a starting position. As the metaphor is imported into capitalist discourse, it can also carry with it the implicit understanding that while teams work together, teammates are not only altruistically motivated to maximise the performance of the team, but are also interested in their individual standing on the team. One is often encouraged to be a team player when one is being asked to sacrifice individual interests for the good of the team.

In economic theory companies which might be considered the quintessential economic teams, are sometimes modelled as having a single set of desires or preferences, as represented by a single utility function, in which the utility of the company is positively proportional to financial wealth (i.e., money). But just as team members in sport have their own desires to succeed as individuals, individual employees' utilities are opposed to each other in that each one wishes to gain individual wealth (and perhaps also status), even if it costs the company as a whole some financial wealth. Economic theorists have more recently noted this competitive situation within, as well as between, capitalist firms[1240].

Slam Dunk; Step Up (to the Plate)

This brings us to two sporting metaphors, each of which emphasises the individual achievements and roles of team players. A slam dunk in basketball is an especially violent dunk. The basketball players performs a slam dunk in order to intimidate the other team, a team-oriented motivation, and impress them and the fans with the individual dunker's athleticism and raw power, an individualistic motivation. A slam dunk is also nearly a sure thing—one almost cannot miss the shot when the ball is stuffed into the basket. The metaphorical use of this term particularly conveys either or both aspects of the concept. A product that is a "can't miss" success is a slam dunk. A person who makes a particularly good business presentation is said to have made a slam dunk, meaning that it was a display of individual virtuosity in business acumen and that it will certainly succeed.

Taking credit for one's individual actions and performances implies that one also takes responsibility for one's actions. The metaphor of "stepping up" or to "step up to the plate" reveals this dual aspect of individual responsibility. "Stepping up to the plate" literally describes a baseball player as he comes up to bat and steps into the batter's box at home plate. At this point he is the one offensive player on whom the immediate future of the game rests. There is no other player at that moment who can affect the game like the batter will, particularly if he should hit a home run. Metaphorically, an individual who steps up to the plate (or simply steps up) is taking the responsibility for attempting to either secure her organisation on the same successful course or to effect some change in course that will help it succeed where it had been in some sense failing or to take responsibility for not doing so. Only an individual can take moral responsibility, and being willing to do so indicates courage, just as it requires courage to stand in the box as a pitcher throws a baseball in one's general direction at lethal speed.

Beyond the language of sport

There is so much sporting ancestry that lies behind and beyond much of what we read, write and say. The sayings and metaphors from sport form a rich component part of our English language and many have philosophical and conceptual importance and significance.

Consider the ideas of:

When the ball is in your court, pulling no punches, it is time to play your hand, to carry all before you, to win at a canter, moving the goal posts, hitting below the belt, carry weight, in the swim, kick into touch, being snookered.

The terminology of cricket, horse racing, hunting, boxing, card games and all sports is found everywhere. With my love of cricket which others have spoken of, I cannot conclude this chapter without referring to two of my favourites from that domain- 'stumped' and on a 'sticky wicket'. In my professional life especially, I have tried to avoid being' stumped' and to deal with playing on a 'sticky wicket'.

Stumped

In addition to being bowled and caught, players batting in games of cricket can be dismissed by being stumped. This requires agility and quick thinking on the part of the wicket-keeper, stationed behind the wicket. If the player batting misses the ball and is beyond the crease when he or she stands to face the bowling, the wicket keeper can break the wicket with the ball and claim a 'stumping'. In the context of this book, I draw the readers attention to being 'beyond the crease', emphasising its wider and metaphorical significance of being 'outwitted', 'defeated' and 'at a loss'.

As a batsman I hated being stumped which was in my view the worst way of being out. Professionally too I strive to avoid 'being outwitted' defeated and 'at a loss'. But it can happen albeit, thankfully rarely. However, I will leave others to be the judge of that and in professional practice the Judges usually are!

Playing on a sticky wicket

Given that the ball is always bowled so that it bounces on the wicket before reaching the batsman, the condition of the wicket is of vital importance in deciding how to play the ball. A sticky wicket is one of the most difficult to play on. Rain-soaked but fast-drying it can make the ball cut sharply to one side and rise abruptly. Batting on a 'sticky wicket' requires concentration and skill to avoid mistiming the stroke and thereby presenting a catch to the fielding side or allowing the ball to hit the wicket. Playing in Wales I often played on a 'sticky wicket'. Professionally too one not infrequently appears in court on a 'sticky wicket' where care and vigilance are needed in dealing with an awkward and unpredictable situation. The ability metaphorically to deal with a 'sticky wicket' is the hallmark of a good advocate.

Chapter 51
Sledging

First they ignore you, then they laugh at you, then they fight you, then you win.
Mahatma Gandhi

Talk is cheap.
Sir Thomas More (c.1600)

Talk is cheap, it don't cost nothing but breath.
K.C. Strahan (1929), Footprints

Words are but words and payes not what men owe.
Chapman v. Shirley (1639) Ball V.I.

In an earlier chapter I have considered the relationship between war and sport. There is no doubt that sledging is a form of psychological warfare. I have dealt earlier in chapter 29 with this matter in my discussion of Sun-Tzu's seminal work *The Art of War*.

In America it is called trash talking. Sledging is an Australian word in origin, used in connection with cricket but in other sports too. It is a beyond concept. If you asked a cricket devotee which team historically had displayed the most prowess in cricket they would probably say Australia. It is therefore no coincidence that Australia has produced some of the best or perhaps the worst sledgers. Think of Rodney Marsh, Dennis Lillee, Glenn McGrath, Shane Warne, Steve Waugh and his brother Mark to name but a few. Sledging is a metaphorical concept which extends beyond the boundaries of cricket and sport generally. It has been described as "a comment or series of comments generally negative, taunting or aggressive in tone made by one or more competitors in the hope that the said comment will affect the mental state of another competitor".

In this chapter I further explain its philosophical and psychological significance and in modern parlance a species of sporting gas lighting.

Sledging is often condemned as unsportsmanlike behaviour in the philosophy of sport.[1241] This condemnation results from the conception of sports as tests of purely athletic skills and from perceived incompatibility with the notion of sportsmanship. This idealisation of sports is actually a very narrow conception and hardly applicable to many modern sports. A broader understanding of what abilities sports test shows that sledging may well be an acceptable practice in certain circumstances. [1242]

Sledging is a revealing topic. Sport is often more about what is going on in your head rather than your body. Certainly, all sports require the display of courage. The late JPR Williams to whom I have already referred was a superb manifestation of that virtue. Perhaps even more than courage the ability to perform under great pressure is the hallmark of great sportsmen and women. Keith Miller the truly legendary cricket all-rounder had been a fighter pilot in the Australian air force during the Second World War. After he had retired he was once asked "Keith when you played cricket did you ever suffer from stress?" To which he enigmatically replied "No, stress is a Messerschmitt up your arse". In other words everything in life is relative and Einstein was right. Stress is a distraction to otherwise focussed performers. It interferes with and disturbs concentration. That is the purpose of sledging. As I will make clear later it is not my intention to set out a host of examples of sledging. All sportsmen will be familiar with many. However, I cannot resist from referring to my favourite example which I admire more for the reaction rather than the original sledge. It is as follows James Ormond when playing for England arrived at the crease to be told by Mark Waugh "what are you doing out here? There is no way you are good enough to play for England". But Ormond dramatically came back with the burn saying "maybe not, but at least I am the best player in my own family" This was in reference to his brother and Captain of the Australian team Steve. Playing sport can be extremely stressful. There are many examples of talented players giving up playing because they could not handle the stress. Marcus Trescothick the England batsman is a name that immediately comes to mind.

I also want to touch upon the concept of post-traumatic stress. Many sportsmen, after their playing days are over, find it difficult to cope with life. There are many examples especially in cricket. There are a plethora of examples given by David Frith in his revealing work 'By their Own Hand' (1990) of former cricketers who have taken their own lives after their playing days are over. They have lost their purpose in life and the heightened experience of playing sport at the top level. And I can understand why. They experience stress and devoid of stress and excitement they have to settle for the ordinary everyday experience of ordinary life without that experience. They no longer see a worthwhile purpose in life. They have lost what they once experienced as a special form of human existence. What is there left to look forward to, what is the point of it all? Peter Roebuck a talented cricketer with a good mind jumped out of a window to his death in a hotel in Cape Town on the 12th November 2011. Why did he do it?

Perhaps he had lost hope as to experiencing anything more than he had previously experienced whilst playing cricket. Like Alexander of Macedonia and Eric Bristow there were no more worlds to conquer. Roebuck in 1990 had written the foreword to David Frith's book. That is a sad and tragic irony. In Roebuck's entertaining book It Never Rains (1994) he wrote poignantly the following:

"Some people have predicted a gloomy end for this writer. One former colleague said so to my face in September 1986. It will not be so. The art is to find other things which matter just as much which stretches just as far". Sadly, he did not do so. In Boris Pasternak's An Essay in Autobiography (1959) a supposedly exemplary definition of the implications of suicide are given "but a man who decides to commit suicide puts a full-stop to his being, he turns his back on his past, he declares himself a bankrupt and his memories to be unreal. They can no longer help or save him, he has put himself beyond their reach. The continuity of his inner life is broken, his personality is at an end". I do not agree with that proposition it is far too pessimistic about man and the choices he has in life. As Mae West said "you only live once but if you do it right once is enough". I a far from convinced that one should be uncomfortable about Peter Roebuck's suicide and all those sportsmen who have taken their lives. It may be that the true philosophy of the suicide taker 'is this as good as it gets? It doesn't get any better than this'. To conclude from the fact of suicide that the taker of one's life is as a result of unhappiness, depression and or anxiety is a logical non-sequitur. Suicide is the exercise of the most fundamental human right-freedom to take one's own life. We did not ask to be born. We had no choice. We could not select our parents. That is the lottery of life to which I will refer in chapter 52 on Luck and Sport.

It is quite feasible that Peter Roebuck and all those cricketers and indeed all sportsmen who have sacrificed their lives did so because that was what they felt comfortable with doing, having lived a meaningful and pleasurable life.

I have written earlier (see the Prologue) about Socrates' opinion that all philosophy is a meditation on death. In my view as I have said before, sport itself is a symbolic meditation on death. I can add here that it may in some circumstances be an actual meditation. So many sportsmen (not only cricketers) give up on life after their playing days are over. The reasons for that are manifold. Albert Camus claimed suicide is prepared within the silence of the heart, as is a work of art. Sport is a species of art form. It is significant that sport not just cricket is responsible for exposing man's inner frailties which in later life can result in disillusionment which become as Frith described 'a lethal weapon turned in upon oneself'.

One would not expect to find a chapter in a philosophical work about sledging. It seems to be too basic a form of human behaviour-slagging people off. But, philosophically, I find it fascinating. Why do we like to slag people off? It is not because we do not like them it is because we want to put them off. We want to distract them so that they do not perform as well as they can. The concept is not limited to sport. It is a universal human trait. In my world of advocacy and the courts it is employed on a regular basis. How does one weaken a strong opponent? The answer is to distract and disturb them, to make them lose their focus, to disrupt their flow. Often the best opponents are the easiest to distract. They have strong egos and believe they are beyond the norm. In life as in sport talented people are especially vulnerable to criticism and distraction. It does not have to be deliberate distraction. Appearing in a play entitled Redemption, American actor John Barrymore was infuriated by the audience's coughing. When the noise resumed during the next act, Barrymore tugged a huge fish from inside his clothes and hurled it across the footlights. "Chew on that you walruses", He bayed, "while the rest of us get on with the libretto". He had been distracted and had lost his focus.

Sledging should be looked upon as a psychological tool of great import. It is a form of anticipatory retaliation for being so competent as to what you do. It is not a cruel device but a highly complimentary acknowledgement of talent and ability. There are some extremely amusing but rather simplistic anecdotes about great sportsmen being sledged by equally great opponents which I will not repeat here as most sportsmen will be familiar with many of them. Carwyn James infamously said to the Welsh team "get your retaliation in first". Likewise, with sledging the tactic is to get your sledging in as soon as you can and await the response and generally the more hostile the reaction the more successful the sledging. Psychological warfare is at the heart of all sporting contests and is a most effective form of offence and/or defence, the more hostile the reaction the more effective is the sledging.

Traditionally sports are understood to be tests of athletic skills constituted and bound by sets of rules defining appropriate actions and penalties for infractions. The aim of a contest is to see which party is better at the athletic skills involved in the sport. Clearly, actual competitions do not always pan out this way. Bad calls by the referee may affect the outcome, weather affects play, injuries may cripple one side. In an ideal match, however, the better athlete or team, in terms of athletic abilities relevant to the sport at hand, will win. This is what competition is supposed to determine.

The problem with this account of sport and competition is that it is lacking the capacity to deal with many modern sports, particularly those played professionally around the world. What it misses, which has been pointed out by Kretchmar, is that there are skills of a purely competitive nature, which we cannot and should not preclude from sports, at least at certain levels of play.[1243]

These skills are cultivated solely to improve one's chances of winning a competition and cannot be described as purely athletic skills. In fact, many of them are purely mental skills.[1244] I want to mention two of these skills to illustrate the fact that they cannot be extricated from our modern sports without getting much of what is so fascinating about them.

Most sports, or at least most team sports, are underdetermined. This under-determination means that there is considerable flexibility in the constitutive rule set for a side to decide how to go about achieving the competitive goals of the sport in question. Thus, in many sports this means that wildly different strategies, tactics, formations, and so on are available to the sides involved in the competition. The use of a specific strategy or formation can hardly be called an athletic skill and, yet is often central to the determination of the victor in a given contest. Consider the case of American football, wherein head coaches get as much praise as the players in many cases. In this instance it is because the coaches are the ones designing the plays a team uses, and the choice of plays is central to the game. American football would be exceedingly dull if it weren't for the strategy and play calling involved. This is not meant to take away from the importance of executing the plays properly, but if they are not the right plays (whether in terms of situation or the opponent's counter), even if they are executed to perfection, the team in question is not going to win.[1245] The importance of team strategy and plays is obviously not limited to American football, but it is probably the clearest example of a sport in which superior tactics can beat superior athletes on a regular basis,[1246] and we do not consider this a flaw in the competition when it happens. This point is hinted at by one commentator when he says "it would be incorrect to say that Jones is playing good tennis if he is hitting crisp ground shots when intelligent play calls for charging the net."[1247] The competitive test is not a test merely of athleticism but also of the intelligent application of athletic and competitive skills to the rules and situations that arise.

What these examples illustrate is the centrality of competitive skills, not just athletic ones, to the notion and value(s) of sports. Without the recognition of these skills, our understanding and evaluation of sports miss crucial elements of what is happening in the competition. So how then is an individual match to be considered a test solely of athletic skills? Quite simply, we need to reconsider what it is that is happening in any specific instance of competition.[1248] If we take a view of "competition as a mutual question of excellence,"[1249] then the range of skills being tested if broadened significantly. The opponents enter into "an implicit social contract under which both competitors accept the obligation to provide a challenge for opponents according to the rules of the sport."[1250]

Given the varieties of sports, levels they are played at, and cultural importance given to any specific one, it is probably not possible to say that all sporting contests are mutual quests for excellence. Because of this difficulty, the remainder of this discussion is limited to the upper echelons of professional sports.[1251] What is being tested at the highest level of sport ideally is which side is more skilful at achieving the goal of the sport in question. This may not sound terribly controversial, but by incorporating nonathletic skills into the realm of what is being tested by competition, the view of sports becomes better able to cope with and evaluate nonathletic skills. By looking at the more holistic concept of better competitor rather than better athlete we give ourselves a much more robust picture of what is going on in sports. The reason this holistic viewpoint is not often advocated is that it places too much emphasis on winning. If what a sport is testing is which side is the better competitor, then, so long as nothing contrary to the rules takes place, the better competitor on the day will win.[1252] It looks too much like a win-at-all-costs view of sports. One would not advocate that this view of sport be applied to something like Little League baseball or village green cricket but at the professional level it is not clear what else the competitors would be doing. Individuals may be playing for sheer enjoyment, but professional sports are not. The reason these professional leagues are set up (in sporting terms, not economic) is to find out which dedicated competitors are going to win.[1253] To put it simply, regardless of any individual's reason for participating or watching, winning matters. This is not to suggest that winning is the only thing that matters, but it is a significant indicator of success.

The two crucial skills are strategy and determination. They often interact, and one can design strategies specifically to undermine the determination of one's opponent. Under less holistic concepts of sport this would be something to be avoided. If the aim is to determine which side is better at the given set of athletic skills, then undermining the determination of one's opponent to perform at their best is antithetical to the purpose of competing. However, we quite often praise as brilliant strategy, and I would say rightly so, plays designed to do just that.[1254] This is for the simple reason that these sports are tests not simply of the athletic skills involved but also of the ability to compete.

In American football the long pass immediately after a start or restart (i.e., halftime, time-out, etc.) is often used to take the wind out of the sails of the opponent. When it works the opposing side is often stunned and takes a bit of time to regroup. This is simply good strategy, which is why it is attempted with some regularity. Similarly, in baseball, pitching inside is often used to scare the batter off the plate so that subsequent pitches on the outside corner will not be as easily hit. Or in football, the hard tackle can be used to dissuade the attacking player from being quite so eager to run forward with the ball. This does not mean an illegal tackle but rather a well-timed legal tackle that the defender knows is likely to result in some pretty heavy bodily contact.[1255] These are all very common ploys used to undermine the determination of the opposing side.

These examples show that attacking the mental game or competitive skills of an opponent is often done. From that point of view it is simply a question of whether it *should* be done. Clearly, if one is playing a friendly game of tennis where the outcome does not matter, it would be odd to attempt to psych out one's opponent.[1256] On the other hand, if one is playing in a true competition where the aim is to see who is the better competitor, a play designed to undermine the opponent's determination should not necessarily be considered out of line.[1257] The reason for this is the holistic understanding of competition involved here. If a competitor is allowed to use something to her advantage, then her opponent should be allowed to attempt to counter that advantage or skill, as long as what the opponent does is within the rules. Similarly, part of the quest for excellence is going to be testing all facets of an opponent and oneself, including how well they stay focussed on the task at hand.

There is one objective to the use of the holistic conception of better competitor as the evaluative standard that needs to be considered. An individual or team may be a great competitor without being skilful at some of the central skills of the sport in question.[1258] In these cases, the competitor wins by not engaging in the competition in that way. Is this a flaw in the competition? One of the central skills involved in most sports is the ability to dictate what skills are at play in any given contest. We see this often in the expression "let's play our game" or the idea of controlling the tempo of a game. The challenge involve in competition is to counter the skills of the opponents, and one way defences do this is to deny them the opportunity to use their strengths. If a team generally plays a slow and methodical offence, quick pressure may be the perfect way to counter their style of play. Or, if a team is heavily reliant on three-pointers, guarding tightly on the three-point line may be called for. The objection seems to implicitly rely on the idea that controlling the manner in which the game is played somehow negates the challenge presented rather than being an appropriate response to it.

The holistic understanding of competition means that there may be skills involved in a sport that have no purpose except testing the skills involved in how the opponent presents a challenge, such as awareness.[1259] These are the type of actions and skills generally referred to by the term 'gamesmanship'.[1260] Gamesmanship is usually considered to be a questionable approach in the philosophy of sport, with the notable exception of Howe's essay 'Gamesmanship'.[1261] It is taken to be antithetical to the notion of sportsmanship and the essence of the sporting contest. Howe notes, however, that it can also be used to bring out the best is one's opponent. A well-timed "bring it" can dramatically alter the contest in terms of focusing one's opponent and signifying respect for her and, as such, should be allowed and possibly even encouraged.[1262] In this case gamesmanship is to be allowed because it signifies respect for the opponent and encourages her to compete at her best. For Howe, gamesmanship can be divided into two classes—weak and strong.[1263] The stronger forms are to be discouraged and actually suggest a failure on the part of the gamer to engage properly in the contest as mutual challenge. Instead, they seem to be a rejection of it in favour of viewing winning as the sole indicator of success. The weaker forms are strategies that not only challenge the recipient but also make the gamer a better player.

These types of gamesmanship can be a competitive skill; actions falling under its designation are intended to affect the outcome of play, and as long as they are within the rules of play, they should be allowed.

One such skill is sledging. A skilful sledger can manipulate her opponent's level of play, bringing out the best in or demoralising the opponent. There should not be anything objectionable (at least in most cases) about encouraging one's opponent to play at her best through verbal means, whether stated as simple encouragement or as a challenge. The case of demoralising one's opponent, though, does seem to be questionable. However, the interesting feature of sledging, and why it should be considered an acceptable or weak form of gamesmanship, is that the intent behind it does not directly correlate to its effects. A verbal challenge intended to get the opponent to perform better may instead make them realise that they are not playing as well as they should, and some competitors only go into a greater slump. Similarly, many players use attempts to demoralise them as fuel to perform even better. Sledging should be understood as a test of the opponent's mental commitment to the contest at hand. As a skill to be used it should be engaged in in order to make sure that one is getting the challenge one deserves. This does not mean that it cannot be used to negatively affect the opponent; rather, such instances should be understood as a failure on the part of the opponent to meet the challenge presented. This may sound particularly harsh and unforgiving, for one would not want to lose to an opponent who lacks the ability to stay focussed or raise their game because this aspect of the competition was not challenged. When considered from the holistic view of competition, it should be obvious that so long as the sledge is within the rules of the game, so there is no sledging while a golfer is in their backswing, it is just one more option in the arsenal of competitive skills designed to make sure the challenge presented is all that it can be.

If you even dream of beating me you'd better wake up and apologise.
Muhammad Ali

The central issue when considering sledging, is whether it is compatible with sportsmanship. In terms of being a competitive skill within the bounds of this idea of sport, sledging is largely a question of timing. In cricket, for instance, taunting an opponent after getting out does nothing to affect play and would just seem to be rude. The player being taunted is already removed from play. They can do nothing to counter it and it does not serve as a competitive skill. On the other hand, sledging while the player is at the wicket can be a very handy strategy for breaking their concentration, undermining their determination, or distracting them from the game play. Concentration, determination and strategy are all clear examples of competitive skills that are open to being challenge in the course of competition. If sledging is a method of defending against a determined opponent or even just testing their ability to fully challenge the opposition by breaking down that determination, then a skilful competitor may wish to develop that skill. They may, however, choose to counter in any number of ways. All I am suggesting here is that appropriately timed sledging should be considered acceptable as a means of addressing the competitive skills of an opponent. It is not outside the bounds of play by virtue of being antithetical to the notion of the competition.

Sportsmanship can be understood in a variety of ways, all of which require sledging to be defended as within the bounds of sporting behaviour. Briefly, sportsmanship can be understood as fair play, honour, or ethics applied to the sporting realm.[1264]

Sportsmanship as fair play is the idea that competitors should behave in a manner consistent with the spirit and norms of the game. Loland and McNamee explain the fair-play ideal: "If voluntarily engaged in sporting games, keep the ethos of the game if the ethos is just and if it includes a proper appreciation of the internal goods and the attitude of playing to win."[1265]

The example they note on the question of fair play is the norm in football; when a player is injured the side with the ball should kick it out of bounds. On restart, the opposing side then throws the ball back to the other team. This way neither side gains an advantage resulting from the injury. The challenge for sledging posed by this notion of fair play is that it is unfair to the side receiving the taunts. In response to this it seems to suggest that sledging is an extremely powerful weapon unavailable to both sides or outside the bounds of the sport. Unless fairness is taken to mean all competitions should be close ones, efficacy is not a reason for precluding sledging just as it is not a reason to preclude a wicked spin bowler. Obviously in most cases it is going to be an available option to both sides in a competition, so claiming unfairness because of unequal availability is a nonsense. The final part of the challenge, that it is outside the bounds of the sport, is certainly applicable in some cases. Sledging is not outside the bounds of sport generally, but this does not mean that it is always within them. So, the ethos of a sport may preclude sledging, but this is by no means a conceptual necessity or even desirable in all cases.[1266]

Sportsmanship as honour raises a slightly different challenge for the sledger. This view of sportsmanship is that it "is part or all of that group's [the collective of competitors in question] honour code of competition."[1267] On this account, the only problem for the sledger is if the code of honour involved precludes it. Although this is certainly the case at many levels of play or all sports, it is by no means obviously precluded from many levels of professional sports. In fact, it would seem to be within the bounds of many of them.[1268] Furthermore, if the central goal of these levels of sports is to present a complete challenge to win, then it could be considered dishonourable not to use every legal skill available to strive towards victory.

The final notion of sportsmanship is that it is the player who behaves ethically in pursuit of victory. Initially this challenges the sledger because taunting someone is at least prima facie unethical. It is simply wrong to taunt people no matter what the situation. This actually points to the problem with this understanding of sportsmanship generally. Sportsmanship cannot be the simple application of ethics to sport, because quite often sports are contradictory to certain ethical edicts. It is wrong to punch people; therefore, boxers can never be sportsmen. There goes a sport historically renowned at the realm of great sportsmen. It is probably wrong to throw things at an individual's head. Well, there go cricket and baseball. Tackling people is mean…no more footy, rugby, gridiron.

It is disrespectful to taunt an opponent, particularly given the content of many instances of sledging. What needs to be noted is that content and intent are often opposed and that it is the intent behind sledging that should be considered when evaluating whether or not it is disrespectful. Given the account of competition, if the opponent is not a threat that needs to be countered by sledging, then its use is not called for and cannot be considered skilful. If I am playing Frisbee against a team that I know simply does not have the physical skills to compete against my team, then it would be disrespectful to say 'Scoreboard' after they celebrate finally scoring a goal. But there is no denying that sledging can be disrespectful but it is not necessarily disrespectful. Quite often the use of sledging is motivated by respect for the opponents' ability, it can be a legitimate test of the nature of the challenge the opponent presents. When opponents are evenly matched in terms of athletic skill, what differentiates between the two are the competitive skills. The use of sledging in that case is going to be a recognition that it is necessary to is test the opponent's mental determination. Thus, when used, sledging is a token of respect for or recognition of the opponent's ability, not an ethically impermissible treatment of her as a mere obstacle to be overcome.

Sledging is not an inherently unsportsmanlike behaviour, except under conceptions of sportsmanship that are untenable for modern professional sports. Any lingering unease about sledging is likely to be a result of unease over the questionable motives and content of many instances of sledging, and this is where the focus of the debate needs to shift. What are we prepared to accept as legitimate verbal interaction? Is an honest test of an opponent's commitment to providing a full-fledged challenge of our own abilities, accepting the often antagonist wording involved, an action we really want to preclude from sports? If we answer this affirmatively, there need to be significant reasons, given by the nature of the competition involved, and not the mere unreflective dismissal that seems to have been the norm.

Chapter 52
Luck and Sport

For those whom God to ruin has design'd
He fits for fate, and first destroys their mind.
John Dryden

Man is not the creature of circumstances; circumstances are the creatures of
men.
Benjamin Disraeli

Remember that you are an Englishman and have consequently the first prize in
the lottery of life.
Cecil Rhodes

A consistent man believes in destiny, a capricious man in chance.
Benjamin Disraeli

It is in your moments of decision that your destiny is shaped.
Anthony Robbins

Fortune favours the brave.
Virgil Aeneid

Men must pursue things which are just in the present and leave the future to
divine Providence.
Francis Bacon

Oh God that one might read the book of fate.
Shakespeare, King Henry IV

The human heart has a tiresome tendency to label as fate only what crushes it.
Kafka

I don't like it but I guess things happen that way.
Johnny Cash

Oh, I am fortune's fool.
Romeo in *Romeo and Juliet* by Shakespeare

The sad vicissitudes of things.
Laurence Sterne

The harder I work the more luck I have.
Thomas Jefferson

The more I practise the luckier I get.
Gary Player

Shallow men believe in luck.
Ralph Waldo Emerson

Luck favours the prepared mind.
Chinese proverb

Diligence is the mother of good luck.
Benjamin Franklin

Show me a good loser and I'll show you a loser.
George C. Scott, The Hustler

You've got to know when to hold 'em
Know when to fold 'em
Know when to walk away
And know when to run
You never count your money
When you're sittin' at the table
There'll be time enough for countin'
When the dealin's done.
The Gambler (Kenny Rogers)

No philosophical work about the concept of sport and the Man in the Arena would be complete without the examination of the concept of luck. It is an existential concept. In Creating Freedom (2017) a work to which I have already referred, Raoul Martinez writes "the abilities and capacities we possess can also be chalked up to good fortune. Whether we have the brain of Isaac Newton or the speed of Usain Bolt is really a matter of chance". With respect I think there is much more to it than that. If I had to put it in algebraic form it would be

$x + y = z$

x focus, talent, discipline, practice, perseverance, resilience and endurance.

y luck, genetics circumstance and opportunity.

z genius, excellence, creativity, achievement in specific domains.

A E Housman put it poignantly when he wrote in his Final Poems in 1922:

Little is the luck I've had,
And oh, 'tis comfort small,
To think that many another lad,
Has had no luck at all.

I am bound to say that these lines from the author of *A Shropshire Lad* metaphorically move me to tears every time I read them.

In a sense the sportsman is a personification of Kingsley Amis' Lucky Jim. He or she is blessed with talent and sometimes touched by genius. They can do what others cannot do. They can ride their luck and get away with it.

Douglas Jerrold in *Meeting Trouble Halfway* in 1859 said, "That some people are so fond of ill luck that they run half way to meet it." I am inclined to think that is as true now as it was then. La Rochefoucauld said "although men flatter themselves with their great actions, they are not so often the result of a great design as of chance."[1269] Richie Benaud's aphorism was that captaincy was "90 per cent luck, 10 per cent skill."[1270] But I suspect this view just reflected his approach to cricket: "Keep it simple." A bit too simple for me!

Nietzsche profoundly got to the essence. "My formula for greatness in a human being is amor fati: that one wants nothing to be other than it is, not in the future, not in the past, not in all eternity. Not merely to endure which happens of necessity, still less to dissemble it—all idealism is untruthfulness in the face of necessity—but to love it."[1271]

More recently, Nassim Taleb in the Black Swan (2007)[1272] referred to the notion of a narrative fallacy to describe how flawed stories of the past shape our views of the world and explanation of the future. They assign a larger role to talent than to luck. But I believe that most if not all successful people would attribute their success as much to luck as talent. Being in the right place at the right time is luck. The ability to make the most of the opportunity is what talent truly represents.

The view that luck spoils the sports contest, or at least tends to undermine its point, follows from what might be called the Skill Thesis. According to the Skill Thesis, competitive sports contests are tests of the competitors' skills designed to determine which opponent is more skilful in the sport being played. [1273] If a contest is decided by luck, it has not determined which of the participants is most skilful and so the game is spoiled or, at the very least, has not determined which of the competitors has best met the test of competition. It has been said "unlucky losers...provide another category of failed athletic contests."[1274] Some say that "competition in sports is supposed to be a test of the athletic ability of persons." Other remarks also suggest such a view.[1275] Does luck spoil the sports contest? If so, as someone has asked, should the possible influence of luck on sports, good and bad, be minimised?[1276] I totally disagree with this view. Luck is what makes sport so exciting and unpredictable. Conceptually too it is fascinating. The vicissitudes of life is at its essence and represents reality.

Concerns about the role of luck in sport, reflected larger concerns about how luck should influence our moral evaluations—concerns raised famously by Thomas Nagel in his paper 'Moral Luck'.[1277] If A and B both drive while intoxicated, and A hits a pedestrian but B, who is just as intoxicated as A, does not injure anyone simply through pure luck, is B any less blameworthy than A? If we view even our character and our skills as unearned consequences of what Rawls has called 'the natural lottery', it remains unclear how much of our behaviour is truly under our control and how much is attributable to the luck of the draw. Indeed, if one pushes the natural-lottery argument hard enough, it suggests that desert is not a fundamental moral notion and perhaps not applicable to the world of sport at all. If so, the athlete or team that plays best cannot deserve to win in any important sense because the notion of desert has, at most, limited applicability to sport, and perhaps to all other areas of life, as well. So while the Skill Thesis may suggest that luck spoils the game, some versions of the lottery argument suggest that because of the luck of the initial draw of talents, skills, and abilities, overall outcomes in sports ultimately are more the result of luck than we might think. If so, the significance of claims of athletic desert and merit is greatly reduced, assuming they are not expelled from the playing field altogether. Daniel Kahneman in Thinking Fast and Slow (2011) has said that "luck plays a large role in every story of success; it is almost always easy to identify in every story or success a small change in the story which would have changed a remarkable achievement into a mediocre outcome."

The view, advanced by Sigmund Loland[1278] is that luck may not always undermine the good sports contest in the way suggested previously, and in some of its manifestations, may even be compatible with the Skill Thesis.[1279] It all depends on the kind of luck involved. Indeed, in some cases, it may be plausible to say the athlete in question deserved to be lucky. Fate is our destiny. We can do nothing about it. We become what we become and probably what we deserve. We can do nothing about it and so what. That is life. Take it on the chin and bear it. It is your true desert. It is all part of the tragedy of human existence. We are all born to die and be no more. Sport, as I said in the Prologue is a metaphorical symbol of death and self-sacrifice which lies beyond the horizon.

Luck does not spoil a good sports contest and does not always show that the best athletes have failed to win. As Loland puts it, "Where luck follows skill, there is no serious threat to…meritocratic distribution of advantage."[1280] Rather, the abilities of skilled athletes created the potential for luck to work in their favour, and so on the stronger version of this thesis, they did deserve to be lucky. At the very least, they met the challenge of the contest better than their opponents. In such cases, luck does not spoil the game but can even enhance it and make us appreciate the skills of the successful players even more than otherwise.

Of particular interest, is scepticism about the role of luck in sport arising from the kinds of concerns raised by John Rawls in *A Theory of Justice*. These arise from what has been called the natural lottery, that is, the initial distribution of generic endowments and favourable environmental circumstances into which the individual is born. Since these factors were not under the person's control, he or she can claim no credit for them or for the benefits that arise largely as a result of their influence. More strictly, as George Sher has formulated the argument, if one individual has a competitive advantage over another due to differences in initial circumstances that were under the control of neither, the favoured individual can claim no credit for the successes that flow from that undeserved head start.[1281] Witness currently the debate about transgender sportspersons.

Many of us would reply that in athletics it is not our initial innate talent that determines athletic success but what we do with it—how hard we work to develop it, our choice of strategies, our commitment to training, and the like. Rawls responds, however, that "the assertion that a man deserves the superior character that enables him to make the effort to cultivate his abilities is equally problematic: for his character depends in large part upon fortunate family, social and genetic circumstances for which he can claim no credit."[1282]

It is unclear, however, how such remarks of Rawls are to be taken, I doubt that they express an overall commitment to some form of hard determination or incompatibilism. Rawls's focus in *A Theory of Justice* is on assessing the fairness or unfairness of structural inequalities in the basic structure of societies, not on defending a metaphysical approach to free will. If so, the natural-lottery argument is best understood as an appeal to our considered judgment that in a system where initial advantages are unfairly distributed, the claims of the successful that they deserve to be on top should at best be taken with a large grain of salt. Perhaps even in such an unfair system, someone can deserve praise for individual moral choices—for example, for helping someone in trouble—but large-scale economic success and failure averaged over a large population are often or generally the result of unearned initial circumstances for which we can take no credit.

David Carr recently has criticised attributions of luck in competitive sport on the grounds that outcomes in athletics reflects and frequently are best explained by unearned initial distribution of talents and abilities for which the competitors can take no credit. Carr's article is interesting and important, not only because of its analytic acuity but also because it draws out the apparent implications for luck from some apparently simple and plausible assumptions.

Carr himself denies that he is a hard determinist or that he rejects all applications of the concept of personal responsibility. Rather, Carr acknowledges that some qualities, let us call them moral qualities, are under the athlete's control. He specifically mentions courage and self-control and presumably would allow other qualities such as dedication, commitment, integrity, and the capacity to maintain composure and focus in the face of difficulty, as well. However, Carr suggests that not only innate athletic ability but also such qualities as the ability to see and take advantage of strategic opportunities (strategic intelligence) or the ability to benefit from repetitive training, drills and exercises reflect initial endowments and so are undeserved. As Carr says of strategic intelligence, "there is no reason—assuming such intelligence to be the usual mixture of training and native wit—to give him or her any special credit for it."[1283]

The point is reinforced by George Sher's observation that in any complex activity or practice, there are multiple combinations of talents, skills, and capacities that can contribute to success.[1284] In athletics, one player may have more innate ability than others, but his competitors may have different combinations of characteristics, such as the capacity to exert greater effort or the patience and courage to develop new techniques even if the learning process may involve temporary setbacks. These qualities may allow them to overcome their initial disadvantage in native talent. Accordingly, if we acknowledge that some of the factors contributing to athletic success are under the athlete's control and that these include higher order capacities grouped under the umbrella concept of strategic capabilities, athletic performance is not simply the result of the natural lottery and is often the sort of activity for which participants can claim at least partial credit (or take at least partial blame).

Of particular interest in this context is the point made by sociologist Dan Chambliss, in his award winning book, *Champions: The Making of Olympic Swimmers*. In his book, Chambliss describes his experiences as an observer at the Mission Viejo swim club during the selection of swimmers for the American team in the 1984 Olympic Games.[1285] During his study, Chambliss attended virtually all team practices and meetings and had full access to athletes and coaches. His special interest was to advance a theory as to why some athletes made the Olympic team and others did not.

In a chapter titled 'The Mundanity of Excellence', Chambliss argued that differences in natural ability do relatively little to explain who made the cut and who did not. Indeed, past a certain surprisingly minimal point, adding time to workouts had little if any effect. Rather, he found that such factors as not cutting corners in practice (for example, using perfect form in making turns rather than being sloppy), changing techniques, and favourable attitudes towards practice (i.e. regarding practice not as a chore but as one of the centres of the athlete's social life) all played major roles.

Chambliss's work suggests that the kinds of factors most clearly affected by the natural lottery or the luck of the draw, such as innate ability and innate character traits, play only a partial role in affecting athletic outcome. Other factors including psychological and moral ones that are likely to be under the athlete's control, such as the willingness to treat each practice as if it were in actual contest, play a significant role, as well.[1286]

A conceptual framework that allows for claims of luck based on our responsibility for some of our actions may have a moral rather than a metaphysical base and so be defensible regardless of the truth of determination. That is, the distinction between what is and is not under our control may be justified by what might be called its moral function rather than its metaphysical basis or appropriateness.

An analogy with the law of torts may be useful here. Torts is that area of law where it is determined who should bear the costs of an injury. For example, should it be the injured person (as when I hit my thumb with a hammer) or another agent (as when you carelessly swing your hammer, not noticing my thumb is in the way).

One approach to determining who should bear the cost is to find the real cause of the injury. If the real or true cause is my action, I am responsible and should bear the cost, but if the real cause is your action, you are responsible and should bear the cost. A problem with this approach is that it may not be a simple matter of fact which contributing factor is the real cause. Were you to blame for not checking that my thumb was out of the way before you swung your hammer, or should I have taken more care in keeping my thumb to myself once I saw you were carrying a hammer? The issue seems to be a moral one, not one of simple causation.[1287]

Thus, on a second approach, we can think of torts as the attempt to distribute the costs of injury fairly. On this view, it is more akin to a moral than a metaphysical enterprise. Thus, if a golfer is struck by another player's errant shot, who should bear the costs of the injury? Courts have ruled that as long as players take due care to warn others of mishit shots, usually by calling out a warning, they normally are not responsible for injuries caused by their wild slices or hooks. Rather, there are certain risks inherent in golf, and golfers are regarded as voluntarily assuming them when entering the course. Right or wrong, this seems like a moral argument about assuming responsibility for one's actions. On this view, the law of torts is moral, not metaphysical.

A similar approach can be applied to desert claims in sports. One traditional line of argument that reflects this approach attempts to reconcile determinism with responsibility and personal desert by relying on utilitarianism. For example, rewarding merit (and in some cases penalising substandard behaviour) may create incentives for acting in desirable ways and so improve the general welfare. However, even many of those favourably disposed to some forms of utilitarianism find such an approach unsatisfactory, since it at most seems to show that it is useful to treat people as responsible agents, not that it also is fair or just to do so.

A second nonutilitarian approach may be more successful in providing an ethical basis for desert claims. This approach ties the practice of acknowledging our deserts to the Rawlsian primary goal of self-respect. Thus, it is noteworthy that in *A Theory of Justice*, Rawls rests his argument for the principle of equal opportunity, which in his theory takes priority over other aspects of his difference principle, to self-respect. For example, he argues that:

> If some places were not open on a basis fair to all, those kept out would be right in feeling unjustly treated even though they benefited from the greater efforts of those who were allowed to hold them…They would be justified in their complaint… because they were debarred from experiencing the realisation of self which comes from a skilful and devoted exercise of social duties.[1288]

However, if a realisation of self comes from "a skilful and devoted exercise of social duties," isn't that because we take pride in our performance or at least view it as something for which we are responsible? It is not viewed merely as something that happens to us but something we do. One suggestion is that taking this responsibility perspective is justifiable not primarily, or at least not only, because it reflects debatable metaphysical presuppositions but rather because it promotes self-respect and assigns due weight to what might be regarded as the central characteristics of personhood.

The major argument made is that desert-based claims of merit should be recognised in sport because the practice of doing so is crucial to such matters as self-respect, fulfilment, and mutual acknowledgement of our status as persons. As Claudia Card suggests, "the recognition of one's deserts is...important to one's self-respect, even to one's personal integrity and identity, just as recognition of oneself as politically the equal of one's peers (also) is important to one's self-respect."[1289]

But is this to misunderstand Rawls's suggestion about self-fulfilment? Perhaps Rawls's point is not that we should take pride in our achievements, which may be largely due to the luck of the draw in the natural lottery, but that carrying out socially useful tasks and exercising skills is something from which humans gain respect and that they find fulfilling. That to me is the fundamental essence of sport and the true motivation of the sports player and athlete. Maybe Lady Luck should be the patron saint of sport and its heroes 'the lucky man'.

The Epilogue

Sir Laurence Olivier in writing the foreword to Hamlet The Film and the Play,[1290] said:

> "I am no writer. One of the dramatic critics—A Scottish one at that— remarked to me the other day, after perusing a so-called essay of mine in some theatre magazine: 'if you don't stop trying to write, I warn you I shall try to act Macbeth, and I'll make you come and see it!' I must admit I saw the fellow's point—even though he was a dramatic critic! The cobbler should stick to his last, the player to his part, and the film director to his film script."

I have not stuck to my last, so please forgive me.

In the foreword to this book, Simon Barnes, in his inimitable style said "Sport is part of a whole person's pleasure in life: it needs neither apology nor dressing up." I hope I have done neither. I have tried to convey in a lucid way the philosophical essence of sport and how its relationship with philosophy is a very real one in that it takes one beyond the horizon of our existence. Thinking and playing are correlatives. And philosophy and sport reveal so much about who we are, the nature of our identity and our significance in the scheme of things and the intrinsic and extrinsic nature of life and the world in which we live.

I hope what I have written is clear even if you disagree with it. In saying that I am mindful of the practice of the Roman judges who, when they could not understand a lawsuit, put two letters on the suit: N.L., non liquet; this is not clear.

In my legal world, it is not necessarily about being right but about being clear.

Schopenhauer said "The profundity of the idea of looking upon the real world as a spectre is the mark of philosophical capacity."

Sport is a spectre, in the words of Wordsworth "a lovely apparition sent to be one moment's ornament." The living in the now and in the moment. But it is much more than that as I believe this book has shown.

It is a symbolic and metaphorical representation of death which as Jean Cocteau saw is "the very condition by which one rises to the truth." It represents 'la difficulté d'être'.

As the Turk said in Voltaire's Candide "our work keeps at bay the three great evils: boredom, vice and necessity."

Sport takes us to Kipling's 'another country' beyond the horizon—excitement, virtue and the unnecessary.

Through sport the human mind derives the knowledge of being. It provides an insight into what preceded our existence, into what is to follow, into what supports the present and in what manner we have received life.

It helps us understand how our brain is capable of ideas and memory. It shows in what manner our limbs obey every motion of the will.

"Sport is a 'wolf to man' even in our civilised society. The primitive lives on in us and embellishes the richness and profundity of our being."

There is no philosophical distinction between the physical and mental. The only real distinction is between the physical and the metaphysical. Plotinus was right in urging contemplation of eternal things but he was wrong in thinking of this as enough to constitute the good life. Philosophy or contemplation, if it is to be valuable, must be married to practice; it must inspire action.

Hence the relationship between philosophy and sport which has inspired this book.

Sport does give us a different insight into the nature of reality and in many ways in a more profound way from science. It is metaphysics—beyond physics.

Very few of us work at the cutting edge of theoretical physics discovering for ourselves the way the world really works. The closest most of us will get is the pop-science shelf and these as has been pointed out have been dominated by the opinions of the Italian physicist, Carlo Rovelli in 'Helgoland' who describes a world made not of particles but of the relations between them. That sort of approach leaves most of us cold and no wiser!

In a different approach to the classical idea of the physical creation—atoms and the void—Frank Wilczek in Fundamentals Ten Keys to Reality, replaces that vision of spacetime, self-propagating fields and properties![1291]

Wilczek's ten keys to reality aren't to do with the 19 physical constants that were considered by Martin Rees, the astronomer royal, in his 1990's pop-science heyday.

It seems the focus now is on the spirit of things. When Wilczek describes the behaviour of an atom, gone are the usual Bohr-ish mechanics in which electrons leap from one nuclear orbit to another. Instead, we get the music of the spheres, a poetic understanding of fields and not a fragment of matter in sight.

The reality of the world is and never will be explained by the physicist. Philosophical speculations are more likely to get a fuller grasp of it. This is what Wilczek calls the 'philosopause'. Physics has come to a stop.[1292]

Wilczek's view is that physicists can pursue a theory of everything they like but they won't find it because if they did find it, they wouldn't understand it. His conclusion is that "Really, this should not come as fresh news. Humans themselves know many things that are not available to human consciousness, such as how to process visual information at incredible speeds or, how to make their bodies stay upright, walk and run."[1293]

At the outset of this book, I quoted Max Born who said "I am now convinced that theoretical physics is actual philosophy."

Maybe the true proposition is that philosophy gives one a more profound insight into the nature of reality than physics. Each chapter in this book attempts to do that by way of metaphysical analysis of the component parts of that reality beyond the concept of sport. It is to use Wilczek's terminology 'Philosopause' The terminology is new but the idea is not. Albert Camus observed as I quoted earlier, that everything he knew about morality and obligations he learnt from football. Sport teaches us not just about moral philosophy but empirical metaphysic generally. It takes us beyond the horizon and the cutting edge of theoretical physics. It gives us a comprehensible and perceptive understanding of man and his world. And as Plato concluded: "Science is nothing but perception." By way of postscript, I hope that none of these chapters is marked 'NL' in the Roman way!

In conclusion I would like to extract a piece of dialogue referred to in the "London Letter" by Francis Cowper in the New York Law Journal (28th Aug 1961).

It is between a Judge and F.E. Smith, 1st Earl of Birkenhead, a great wit, orator, advocate, parliamentarian and a close friend of Winston Churchill.

Judge: I have read your case, Mr Smith, and I am no wiser now than when I started.

Smith: Possibly not, My Lord, but far better informed.

I sincerely hope that the reader having read this book, is not only better informed but much wiser now than when they started reading, about the relationship between Sport and Philosophy and Philosophy and Sport.

Friedrich Nietzsche wrote in the Maxims and Arrows section of *Twilight of the Idols* (1888) 'What does not kill me makes me stronger'.

Having written this book and survived I feel stronger. I can see more clearly now beyond the concept of sport and beyond the horizon.

End Notes

[1] The speech has its place in sports history. Before 1995 the World Cup, Nelson Mandela gave a copy of the passage to Francis Pienar, Captain of the South African rugby team, who defeated the favoured All Blacks of New Zealand.

[2] The Mind of God—Science and the Search for Ultimate Meaning—Paul Davies, Simon & Schuster 1992.

[3] Grant Barley, Philosophy Nov Issue 153 December 2022—January 2023.

[4] Camus, The Myth of Sisyphus.

[5] Friedrich Schlegel—Lucinda and the Fragments (Minneapolis MN 1972, p.253).

[6] Radical Sacrifice, Terry Eagleton—Yale University Press 2018).

[7] Quoted in Raymond Williams, Modern Tragedy (London, 1966, p.116).

[8] Ibid Eagleton p.35.

[9] Ibid Eagleton, p.35.

[10] Man and his symbols, p.3—Jung.

[11] An introduction to the principles of Morals and Legislation, University of London, 1970.

[12] Preface page VII. The Philosophy of Sport An Overview, R.G. Osterhoudt published 1991.

[13] Ibid.

[14] Do not all charms fail. At the mere touch of cold philosophy.
There was an awful rainbow once in heaven. We know here wool, her texture; she is given in the dull catalogue of common things.
Philosophy will clip an angel's wings.
Conquer all mysteries by rule and line.
Empty the haunted and groomed mind—unweave a rainbow—Keats-Lamia (1820).

[15] Lamia part 1, 227-238.

[16] Introduction p6 Skin in the game Nassim Nicholas Taleb Allen Lane 2018.For more on his concept in the context of sport—see Ibid.

[17] Huizinga, J. *Homo Ludens: A Study of the Play Element in Culture,* Boston: Beacon Press, 1955

[18] Sports, Philosophy, and the Quest for Knowledge, Heather L Reid. Journal of Philosophy of Sport, 2009, 36, 40-49

[19] Aristotle. "Metaphysics." In *The Complete Works of Aristotle, 2 vols*. J Barnes (ed.). New Jersey: Princeton U.P., 1984

[20] Pindar. Olympian Odes. Pythian Odes, Trans. William H. Race. Cambridge, MA: Harvard University

[21] Little, W., Fowler, H., and Coulson, J. *The Oxford Universal Dictionary. Oxford:* Oxford U.P., 1955

[22] Laertius, Diogenes, *Lives of Eminent Philosophers, vol I translated by R.D. Hicks, Cambridge*

[23] Benjamin Lowe. The Beauty of Sport. A Cross Disciplinary Inquiry. Prentice-Hall Inc 1977, p.171

[24] Hellas: A Short History of Ancient Greece, New York, Pantheon Books 1948

[25] (pp.183-186)

[26] (see Robert F. Thuma, The Grace of Man, Pittsburgh pa: The Myers and Shinkte Company, Printers, 1897)

[27] Fairs, ibid p.7

[28] Fairs, (When was the Golden Age of the Body?" Journal of the Canadian Association of Health, Physical Education and Recreation 37 no.1 (1970) 11-24)

[29] Ibid Fairs p.23.

[30] Toynbee "Some Notes on the Paintings of Contemporary Games" The London Magazine (1961), 57-60

[31] (Foreword the first XV Parthian 2011 Gerald Davies).

[32] Bertrand Russell said that Aristotle is Plato "diluted by common sense". In science in order to make progress often you have to defy what historian Daniel Boorstin referred to as "the tyranny of common sense". Aristotle wrote "what everyone believes is true" cf. Plato.

[33] 'Game theory', quite apart from Ortega's playful efforts at it from the 1920s forward, developed extensively from the late 1940s to peak in the 1970s, but it continues to be exploited for various fields and purposes into the present. Aside from what I have read to that effect in their works, "play theory" among postmodernists—or among the "modernists" of the 1920s—seems to be almost wholly for theatre and sports, but I see something akin to Ortega's idea and praxis in Karl Sigmund, *Games of Life: Explorations in Ecology, Evolution and Behaviour* (New York: Oxford University Press, 1993) and, regarding the "unsayable" in Ortega's linguistics: Sanford Budick and W. Iser, eds., *Languages of the Unsayable: The Play of Negativity in Literature* (New York: Columbia University Press, 1989)

[34] OC, 7:344-45; WP, 114-15

[35] OC, 5:377, 7:350

[36] OC, 7:322-23

[37] OC, 7:323; WP, 83

[38] OC, 7:422

[39] OC, 7.429; WP 238

[40] OC, 8:296; IPL 309

[41] IPL, 320

[42] IPL, 322

[43] The Concept of Mind by Gilbert Ryle (First published by Hutchison 1949, from Introduction to Daniel Dennett).

[44] "The Lived Body as Phenomenological Datum" by Calvin O. Shrag, in Philosophic Inquiry in Sport, edited by Morgan & Meier. Champaign III, Human Kinetics Publishers 1988 p.110.

[45] Meier Klaus, "Embodiment in Sport and Meaning" in Morgan & Meier ibid p.97.

[46] Marcel Gabriel, Metaphysical Journal, Chicago, Henry Regenery Co. 1952 p.333.

[47] Philosophy of Sport, Drew Hyland Paragon Press 1990, p.96.

[48] Ibid page 97.

[49] Ibid p.99.

[50] The First XV—Foreword Parthian 2011.

[51] Daniel A. Dombrowski, Contemporary Athletics and Ancient Greek Ideals 2009. University of Chicago Press.

[52] (see his Introduction to Philosophy of History). The opposite is true of Greek thinking (p.29).

[53] Miller, Stephen. 2004a *Ancient Greek Athletics*. New Haven: Yale University Press. Martinkova, Irena. 2001 "Kalokagathia: How to Understand Harmony of a Human Being". *Nikephoros* 14:21-28

[54] Ibid Dombrowski, p.4

[55] ibid Dombrowski, p.5.

[56] Ibid Dombrowski

[57] Diogenes Laertius 1950 *Lives of Eminent Philosophers,* Trans. R.D. Hicks. Cambridge: Harvard University Press, 1.3.4. Spivey, Nigel 2004. *The Ancient Olympics: A History.* Oxford: Oxford University Press, p.32

[58] Aristotle. 1984. *The Complete Works of Aristotle,* Ed. Jonathan Barnes, 2 vols. Princeton: Princeton University Press, 2.2387

[59] Ibid Dombrowski, p.15

[60] Miller, Stephen, 2004a. *Ancient Greek Athletics*. New Haven: Yale University Press p.46

[61] Harris H.A.1964. *Greek Athletes and Athletics*. London: Hutchinson pp. 76-77. Finley M.I. and H.W. Pleket. 1976 *The Olympic Games: The First thousand Years*. New York: Viking Press, p.5)

[62] Miller, Stephen, 2004a. *Ancient Greek Athletics. New Haven:* Yale University Press. Chap 9. Harris H.A. 1964 *Greek Athletes and Athletics*; London: Hutchinson, chap.5. Finley, M.I. and H.W. Pleket 1976. *The Olympic Games: The First Thousand Years*. New York: Viking Press, chap. 6

[63] Ibid Dombrowski, p.14.

[64] Miller, Stephen, 2004a. *Ancient Greek Athletics*. New Haven: Yale University Press p.177

[65] Miller, Stephen, 2004a. *Ancient Greek Athletics*. New Haven: Yale University Press pp.176-179 Harris, H>A. 1964. *Greek Athletes and Athletics*. London: Hutchinson, Chap.8. Finley M.I., and H.W. Pleket. 1976. *The Olympic Games: The First Thousand Years*. New York: Viking Press, chap. 7y)

[66] Ibid Dombrowski p.24

[67] Weiss, Paul. 1995 a "Reply to Daniel A Dombrowski," In *The Philosophy of Paul Weiss*. ed. Lewis Hahn. Chicago: Open Court. p.656

[68] Weiss Paul. 1995 a "Reply to Daniel A. Dombrowski" In *The Philosophy of Paul Weiss, ed. Lewis Hahn.* Chicago: Open Court. P660

[69] Keenan Francis 1973, "The Athletic Contest as a 'Tragic' Form of Art". In *The Philosophy of Sport,* ed. Robert Osterhoudt. Springfield IL: Charles Thomas; Holowchak, Andrew. 2002a "Moral Liberalism and the Atrophy of Sport." In *Philosophy of Sport* ed. Andrew Holowchak. Upper Saddle River, NJ: Prentice-Hill; also Best David 2002. "The Aesthetic in Sport." In *Philosophy of Sport*, ed. Andrew Holowchak. Upper Saddle River, NJ: Prentice-Hall Cordner, Christopher. 2002 "Differences between Sport and Art." In *Philosophy of Sport,* ed. Andrew Holowchak. Upper Saddle River NJ. Prentice-Hall.

[70] Ibid Dombrowski p.39

[71] Weiss Paul, 1969. Sport: *A Philosophic Inquiry*. Carbondale: Southern Illinois University Press, pp. 35-36.81

[72] Huizinga Johan. 1955 *Homo Ludens: A study of the Play-Element in Culture*. Trans R.F.C. Hull Boston: Beacon Press pp.18-19

[73] Huizinga Johan 1955. *Homo Ludens: A Study of the Play-Element in Culture*. Trans R.F.C. Boston: Beacon Press pp.87, 130, 143,145

[74] Huizinga, Johan 1955. *Homo Ludens: A Study of the Play-Element in Culture*. Trans. R.F.C. Hull Boston: Beacon Press pp 159-160

[75] Huizinga Johan. 1955 *Homo Ludens: A Study of the Play-Element in Culture.* Trans. R.F.C. Hull Boston: Beacon Press, foreword

[76] Huizinga Johan 1955. *Homo Ludens: A Study of the Play-Element in Culture.* Trans. R F.C. Hull Boston: Beacon Press pp. 1-2

[77] Huizinga Johan 1955. *Homo Ludens: A Study of the Play-Element in Culture.* Trans R.F.C. Hull. Boston: Beacon Press pp 5-6.

[78] Huizinga Johan 1955 *Homo Ludens: A Study of the Play-Element in Culture.* Trans R.F.C. Hull. Boston: Beacon Press, pp.7-10

[79] Suits Bernard. 1967 "What Is a Game?" *Philosophy of Science* 34: 148-156

[80] Huizinga Johan 1955. *Homo Ludens: A Study of the Play-Element in Culture.* Trans. R.F.C. Hull. Boston: Beacon Press, pp.10-11

[81] Huizinga Johan 1955. *Homo Ludens: A Study of the Play-Element in Culture.* Trans R.F.C. Hull. Boston: Beacon press p.19. Guardini, Romano. 1997. *The Spirit of the Liturgy.* Trans. Ada Lane. New York: Crossroad pp.61-71

[82] Huizinga Johan 1955. *Homo Ludens: A Study of the Play-Element in Culture.* Trans. R.F.C. Hull Boston: Beacon Press p.21

[83] Huizinga Johan 1955. *Homo Ludens: A Study of the Play-Element in Culture.* Trans R.F.C. Hull. Boston: Beacon Press pp.22-27. Rahner Hugo 1965. *Man at Play.* New York; Herder and Herder. Also see Thomas Aquinas's [1972] *Summa Theologiae* Ilasllae q.168 a.2

[84] Huizinga Johan 1955

[85] Riezler Kurt 1941 "Play and Seriousness" *Journal of Philosophy* 38: p515

[86] Hyland Drew, 1984. *The Question of Play.* Lanham MD: University Press of America 1990

[87] Huizinga, Johan. 1995. *Homo Ludens: A Study of the Play-Element in Culture.* Trans. R.F.C. Hull Boston: Beacon Press p.45

[88] Huizinga Johan 1955. *Homo Ludens: A Study of the Play-Element in Culture.* Trans. R.F.C. Hull. Boston: Beacon Press, pp.46-51

[89] Huizinga, Johan. 1955 *Homo Ludens: A Study of the Play-Element in Culture. Trans. R.F.C. Hull Boston: Beacon Press, chap.4*

[90] Parry, Jim. 2002. "Violence and Aggression in Contemporary Sport". In *Philosophy of Sport,* ed. Andrew Holowchak. Upper Saddle River, NJ: Prentice-Hall.

[91] Huizinga Johan. 1955 *Homo Ludens: A Study of the Play-Element in Culture.* Trans. R.F.C. Hull Boston: Beacon Press, pp.102-104

[92] Barthes Roland 2007. *What is Sport?* Trans. Richard Howard. New Haven: Yale University Press, pp.9, 25, 30, 37, 47, 63.

[93] ibid Dombrowski, p.88

[94] Huizinga, Johan. 1955.*Homo Ludens: A Study of the Play-Element in Culture.* Trans. R.F.C. Hull Boston: Beacon Press, pp.146-149

[95] Feezell, Randolph, 2004a. *Sport, Play, and Ethical Reflection.* Chicago: University of Illinois Press, pp.83-84

[96] Feezell, Randolph, 2004a. *Sport, Play and Ethical Reflection.* Chicago: University of Illinois Press, p.85

[97] Ibid Dombrowski, p.105

[98] Feezell, Randolph, 2004a. *Sport, Play and Ethical Reflection.* Chicago: University of Illinois Press, p.4, emphasis added

[99] Feezell, Randolph 2004a. *Sport, Play and Ethical Reflection.* Chicago: University of Illinois Press, p.5.

[100] Feezell, Randolph, 2004a. *Sport, Play and Ethical Reflection.* Chicago: University of Illinois, Press p.48

[101] Ibid Dombrowski p.114

[102] Feezell, Randolph, 2004a. *Sport, Play and Ethical Reflection.* Chicago: University of Illinois Press, pp.57,77

[103] Feezell Randolph 2004a. *Sport, Play and Ethical Reflection.* Chicago: University of Illinois Press p.68

[104] C.F. Feezell, Randolph, 2004a. *Sport, Play, and Ethical Reflection.* Chicago: University of Illinois Press p.57, 70, 77

[105] Ibid Dombrowski p.119

[106] Nagel, Thomas 1986 *The View from Nowhere,* New York: Oxford University Press pp. 217-218; Feezell Randolph 2004a. *Sport, Play and Ethical Reflection.* Chicago: University of Illinois Press, p.72

[107] Feezell Randolph, 2004a. *Sport, Play and Ethical Reflection.* Chicago: University of Illinois Press p.72

[108] Ibid Feezell p.73

[109] Ibid Feezell p.74 cf.152

[110] Ibid Feezell, p.74; Feinberg Joel. 1984 "Absurd Self-Fulfilment." In *Philosophy and the Human Condition,* ed. Tom L. Beauchamp William T. Blackstone and Joel Feinberg. Englewood Cliffs, NJ: Prentice-Hall

[111] Ibid Feezell, pp.74-75; Nagel, Thomas. 1986. "*The View from Nowhere* New York: Cambridge University Press, p.222

[112] Corlett, John. 2002. "Virtue Lost: Courage in Sport" In *Philosophy of Sport,* ed. Andrew Holowchak Upper Saddle River. NJ: Prentice-Hall

[113] Ibid Feezell, p.76; Feezell, Randolph and Craig Clifford. 1997. *Coaching for Character.* Champaign IL; Human Kinetics

[114] Ibid Dombrowski p.121

[115] Johann Wolfgang von Goethe, *Faust*, 9381.

[116] The Philosophical Athlete by Heather L. Reid published 2002 by Carolina Academic Press, Durham, NC.

[117] Ibid, 4685.

[118] Ibid, 6487-500.

[119] Ibid, 9370.

[120] Ibid, 9419.

[121] Ibid, 9377-84.

[122] Ibid, 9411-18.

[123] Johann Wolfgang von Goethe, Letter to Zelter of October 19, 1829, in *Goethes Briefe*, ed. K.R. Mandelkow, Munich 1967, vol.4, p.346.

[124] "Book of the Cup-Bearer" ("Das Schenkenbuch"), in *Goethes Werke*, Hamburger Ausgabe (hereafter *HA*), vol.2, p.94.

[125] O. Spengler, *Der Untergang des Abendlandes*, Munich 1923, vol. I, p.11.

[126] J. Hintikka, *Time and Necessity*, Oxford, 1973, p.86.

[127] S. Morenz, *Die Zauberflöte*, Münster, 1952, p.89.

[128] Ecce Homo iv 283.

[129] Cicero, *De finibus*, I, 18, 59.

[130] Ibid, I, 18, 60.

[131] Seneca, *Letters to Lucilius*, 15, 9.

[132] Epicurus, fr. 240, p.567 Arrighetti = Stobaeus vol. III, 17, 22 Hense.

[133] Cicero, *De finibus,* I, 19, 63.

[134] Epicurus, *Ratae Sententiae*, 19, p.127 Arrighetti.

[135] Seneca, *Letters to Lucilius*, 74, 27.

[136] Aristotle, *Nichomachean Ethics*, 10, 3, 1174a17ff; cf. H.-J. Krämer, *Platonismus und hellenistische Philosophie*, Berlin/New York 1971, pp.188ff.

[137] J.M. Guyau, *La morale d'Épicure*, Paris 1927, pp.112ff.

[138] Horace, *Odes*, 2, 16, 25f.

[139] Epicurus, *Gnomologicum Vaticanum* §14, p.143 Arrighetti.

[140] Horace, *Odes*, I, 11, 7.

[141] Horace, *Letter*, I, 4, 13.

[142] Cf. M. Gigante, *Richerche Filodemee*, Naples 1983, pp.181, 215-16.

[143] Lucretius, *On the Nature of Things*, 1033-6.

[144] Epicurus, *Gnomologicum Vaticanum*, §33, p.146 Arrighetti.

[145] Epicurus, *Letter to Menoeceus,* §§124-5, p.108 Arrighetti.

[146] Horace, *Odes*, 3, 29, 42.

[147] Without worries (*sine sollicitudine*), because he knows that on that day he has received all that it was possible to have, and that there is nothing left to be desired.

[148] Seneca, *Letters to Lucilius*, 12, 9.

[149] Lucretius, *On the Nature of Things*, 3, 16-17.

[150] The most accessible text of this saying has been preserved by Clement of Alexandria, *Stomata*, V, 14, 138, 2. Cf. the commentary of A. Le Boulluec in *Clèment d'Alexandrie, Stromates, V,* Paris 1981, p.369.

[151] L. Robin, *Lucrèce, De la Nature, Commentaire des livres III-IV*, Paris 1926, repr. 1962, p.151.

[152] Lucretius, *On the Nature of Things*, III, 947-9.

[153] Marcus Aurelius, *Meditations*, 9, 6.

[154] Ibid, 7, 29, 3;p 3, 12, 1.

[155] The principal text may be found in *SVF* 2, 509 [=Arius Didymus fr. 26 Diels, in Stobaeus vol. I, p.105, 5ff. Wachsmuth]; for a commentary cf. P. Hadot, "Zur Vorgeschichte des Begriffs Existenz," *Archiv für Begriffsgeschichte* 13 (1969), pp.118-19.

[156] Marcus Aurelius, *Meditations*, 6, 32, 3.

[157] Ibid, 12, 1, 1-2.

[158] Ibid, 12, 3, 3-4.

[159] [*Circumcidenda*. Literally, to cut a circle off the bark of a tree, thereby pruning or even killing it—Trans.]

[160] Seneca, *Letters to Lucilius*, 78, 14.

[161] Seneca, *On Benefits*, 7, 2, 4-5.

[162] Cicero, *De finibus*, 3, 14, 45.

[163] Plutarch, *On Common Conceptions*, 8, 1062a.

[164] Marcus Aurelius, *Meditations*, 5, 2.

[165] Ibid, 2, 5, 2; 7, 69.

[166] Seneca, *Letters to Lucilius*, 101, 10.

[167] Marcus Aurelius, *Meditations*, 2, 14, 3.

[168] Seneca, *On Benefits*, 7, 3, 3.

[169] Plutarch, *On Common Conceptions*, 37, 1078e.

[170] Marcus Aurelius, *Meditations*, 6, 37.

[171] Ibid, 10, 5.

[172] Ibid, 10.21.

[173] Seneca, *Letters to Lucilius*, 66, 6.

[174] Conversation with J.D. Falk, in F. von Biedermann, ed., *Goethes Gespräche,* Leipzig 1910, vol.4, p.469.

[175] "An Grafen Paar," I *Goethes Sämtliche Werke* (Cottaxhe Jubiläumsausgabe), Stuttgart 1902, vol.3, p.13.

[176] Johann Wolfgang von Goethe, *Egmont*, Act 2.

[177] Johann Wolfgang von Goethe, "Lebensregel," in *Sprüche,* 97ff, in *Goethes Werke*, *HA*, vol. I, p 319.

[178] Goethe, Letter to Zelter of October 19, 1829, in *Goethes Briefe*, vol.4, p.347.

[179] Goethe, Letter to Sickler of April 28, 1812, in ibid, vol.3, p.184.

[180] Johann Wolfgang von Goethe, *Vermächtnis*, in *Goethes Werke*, HA, vol. I, p.370.

[181] J.P. Eckermann, *Gespräche mit Goethe,* Wiesbaden 1955 (conversation of November 3, 1823), p.61.

[182] *Maximen und Reflexionen*, 314 Hecker = *Goethes Werke*, *HA*, 752.

[183] Goethe, *Faust*, 12104.

[184] "Eins und Alles," in *Goethes Werke*, *HA*, vol. I, p.368.

[185] "Selige Sehnsucht," in *Diwan*, in *Goethes Werke*, *HA*, vol.2, p.18.

[186] Letter to August von Bernstorff of April 17, 1823, in *Goethes Briefe*, vol.4, p.63.

[187] *Book of Suleika*, in the *East-West Divan*, in *Goesthe Werke*, *HA*, vol.2, p.70.

[188] "Testament," *loc. cit.*

[189] "Why am I so wise"—Ecce Homo.

[190] Gadamer. Truth and Method. Second Revised Edition trans. Revisions Weinscheimer and Marshall. New York Continuum 1995, pp.102-112.

[191] ibid Gadamer p.103.

[192] ibid Gadamer, p.103

[193] ibid Gadamer p.105.

[194] ibid Gadamer, p.105.

[195] Friedrich Schlegal, "Gespräch über die Poesie," *Friedrich Schlegels Jugend-schriften*, ed. J. Minor (1882), II, 364. [In the new critical edition of Schlegel, ed. E. Behler, see part I, vol.2, ed. Hans Eichner, pp.284-351, and p.324 for this citation.]

[196] ibid Gadamer p.106.

[197] ibid Gadamer p.107.

[198] ibid Gadamer p.108.

[199] ibid Gadamer p.109.

[200] ibid Gadamer p.111.

[201] ibid Gadamer p.112.

[202] ibid Gadamer p.112

[203] Frank Keating, The Great Number Tens (1993).

204 (see George Orwell 'The Sporting Spirit, Tribune, London December 1945)

205 The Times Obituary, January 13 2023.

206 An English Tradition by Jonathan Duke-Evans 2023 OUP.

207 (Times 1st October 2009).

208 (Times 15th October 2009).

209 (The Times, October 9th, 2009).

210 (The Times, October 12th 2009).

211 Short Books, 2006.

212 Jay Cookley's "Ethics, Deviance and Sports," A Critical Look at Crucial Issues" in Alan Tomlinson and Scott Fleming eds, Ethics Sport and Leisure Crises and Critiques, Chelsea School Research Centre Edition, Vol.1, (Aachen, Germany: Meyer and Mayer Veriag, 1995 pp. 13-20).

213 Robert L Simon Internationalism and Internal Values in Sport Journal of the Philosophy of Sport of v 27 (2000 1-16). J.S. Russell "Are Rules All an Umpire Has to Work With" Journal of Philosophy of Sport v.26 (1999) 27-49.

214 Journal of Philosophy of Sport 1998 XXV P1-22

215 (ibid 11).

216 See the writings of William J Morgan.

217 Cambridge, MA Harvard University Press, 1986.

218 (J.S. Russell ibid p.31).

219 (The Concept of Law, Oxford University Press, 1961).

220 (Hart ibid p.126).

221 (see Dworkin ibid p.1341).

222 (J.S. Russell ibid).

223 (see Rorty R "The Banality of Pragmatism and the Poetry of Justice" South Californian Law Review 63: 1811-1819, 1990

224 Journal of the Philosophy of Sport, 2002, XXIX, 182-186 ©2002 by the International Association for the Philosophy of Sport. **The "Hand of God"?: Essays in the Philosophy of Sport**, by Claudio M. Tamburrini, Published 2000 by Acta Universitatis Gothoburgensis, Box 222, SE-405 30, Sweden (167 pp.) *Reviewed by Heather L. Reid reid@alpha.morningside.edu, who is with the Department of Philosophy at Morningside College, Sioux City, IA 51106.*

225 Ibid 16.

226 Ibid p.19.

227 Ibid p.28.

228 Ibid p.36.

229 Ibid p.62.

230 Ibid p.121.

[231] Ibid p.122.

[232] Ibid p.151.

[233] Ibid p.154.

[234] New Scientist Essential Guide No.1 The Nature of Reality 2020.

[235] Body in Mind p. xix.

[236] Parmelin Hélène Intimate Secrets of a Studio at Notre Dame de Vie, New York, Abrams 1966.

[237] "The Aesthetic in Sport"—Philosophic Inquiry into Sport, Morgan W.J. and Meir K.V. (Eds) Champaign, In Human Kinetics, pp.377-389.

[238] Ralph Venning (1620-1673).

[239] By Hans Ulrich Gumbrecht. Published 2006 by Belknap Press at Harvard Press, Cambridge, MA.
Reviewed by Ted Leland, who is with the Department of University Advancement, University of the Pacific.
Journal of the Philosophy of Sport, 2007, 34, 100-101 © 2007 International Association for the Philosophy of Sport.

[240] Paul Davies, The Mind of God 1992, p.175, Simon & Schuster Ltd.

[241] Alan Richards, From Carwyn a memoir (1984).

[242] Patsy Neal in Sport and Identity Dorrance Philadelphia 1972 p.16

[243] Connotations of Movement in Sport and Dance by Eleanor Metheny, Dubuque, Iowa: Wm Brown Co, 1965

[244] Ibid page 20

[245] (ibid Neal p.21)

[246] Richard Rothschild, Reality and Illusion. New York: Harcourt, Brale and Company 1934, p.48

[247] (Ibid Neal p,50)

[248] Richard Rothschild Reality and Illusion. New York Harcourt Brace 1934 p.77

[249] Kretchmar, Scott R and Harper, Williams A "Must we have a Rational Answer to the Question, Why does Man play?" Journal of Health, Physical Education, Revelation 40: 57-59 March 1969).

[250] (Slusher ibid p.15)

[251] Ibid Neal p.63

[252] Neal ibid p.64

[253] Zen in der Kunst des Bogershießens. Constanced 1948, Munich 1951

[254] Zazengi, translated by T.Hirata Ein Lebenin Zen. In: U. v. Mangoldt: Höhlen, Klöster, Ashrams, Weilheim 1962

[255] It is to be noted by the way that the Sanskrit word for self, atman, originally meant breath, and is etymologically identical with the German word for breath. Atem.

[256] (Tendai-) Shôshikan (Shujûshikanzazenhôyô). Taishô Issaikyô No.1915

[257] Ibid Herrigelm, p.45

[258] This concentration of consciousness is the literal meaning of the word "Zen" itself, "Zen" being the translation of the Sanskrit word, 'dhyâna', which is usually rendered as "meditation" but which is closer in meaning to "recollection".

[259] Ibid p.47

[260] Ibid Herrigel pp.47-48

[261] From Zen Golf by Dr Joseph Parent Collins Willow

[262] Andrew Holowchak "Games as Pastimes in Suits' Utopia: Meaningful Living and the 'Metaphysics of Leisure" Journal of the Philosophy of Sport 2007, 34, 88-96

[263] Suits, B *The Grasshopper: Games, Life and Utopia* Orchard Park, NY: Broadview Press, 2005 pp 27-28

[264] Suits, B. *The Grasshopper: Games, Life and Utopia.* Orchard Park, NY: Broadview Press 2005, pp.28-29

[265] Suits, B. *The Grasshopper: Games, Life and Utopia.* Orchard Park, NY: Broadview Press, 2005. P.149

[266] Suits B. *The Grasshopper: Games, Life and Utopia.* Orchard Park, NY: Broadview Press 2005 pp.54-55

[267] Suits B. *The Grasshopper: Games, Life and Utopia* Orchard Park NY Broadview Press 2005 p.44

[268] This seems to be a rather naïve view of the psychogenesis of artistic creations.

[269] Suits B. *The Grasshopper: Games, life and Utopia.* Orchard Park NY: Broadview Press 2005

[270] Suits B. *The Grasshopper: Games, Life and Utopia.* Orchard Park, NY: Broadview Press 2005

[271] Suits B *The Grasshopper: Games, Life and Utopia.* Orchard Park NY Broadview Press 2005

[272] Suits B *The Grasshopper: Games, Life and Utopia.* Orchard Park NY Broadview Press 2005

[273] Suits B. *The Grasshopper: Games, Life and Utopia.* Orchard Park, NY: Broadview Press 2005

[274] Suits B. *The Grasshopper: Games, Life and Utopia.* Orchard Park NY: Broadview Press 2005.

[275] Suits B. *The Grasshopper: Games, Life and Utopia.* Orchard Park NY: Broadview Press 2005

[276] Suits B *The Grasshopper: Games, Life and Utopia.* Orchard Park NY: Broadview Press 2005

[277] Grant Bartley, Philosophy, November-December 2022—January 2023 Issue 153.

[278] The Times, Jan 4th 2023, List of Great Singers, Saube Dion.

[279] Culture and Value (1931, in Culture and Value 1980).

[280] Why I am so wise. (Neitzsche written in 1888 but not published until 1908).

[281] See August 1994—American Psychologist 726-745. Expert Performance, Its Structure and Acquisition. Ericsson and Charness.

[282] Newell and Simon (1972). *Human problem solving.* Englewood Cliffs, NJ: Prentice-Hall.

[283] Gardner (1983, 1993a, 1993b). *Frames of Mind: The theory of multiple intelligences.* New York: Basic Books. *Creating minds*: New York; Basic Books. *Multiple Intelligences: The theory in practice.* New York: Basic Books.

[284] (Murray, 19869b). Poetic genius and its classic origins. In P. Murray (Ed.,) *Genius: The history of an idea* (pp.9-31). Oxford, England: Basil Blackwell.

[285] (Murray, 1989b, p.11). Poetic genius and its classic origins. In P. Murray (Ed.,), *Genius: The history of an idea* (pp.9-31). Oxford, England: Basil Blackwell.

[286] (quoted in Murray, 1989b, p.12). Poetic genius and its classic origins. In P. Murray (Ed.,), *Genius: The history of an idea* (pp.9-31). Oxford, England: Basil Blackwell.

[287] (Bull, 1987). *A translation of Giorgio Vasari's* Lives of the artists (2 vols.). New York: Viking Penguin.

[288] (Barolsky, 1991). *Why Mona Lisa smiles and other tales by Vasari.* University Park: Pennsylvania State University Press.

[289] (Bull, 1987, Vol.2, p. xxvi). *A translation of Giorgio Vasari's* Lives of the artists (2 vols.). New York: Viking Penguin.

[290] (Barolsky, 1991). *Why Mona Lisa smiles and other tales by Vasari.* University Park: Pennsylvania State University Press.

[291] (Boase, 1979).

[292] (quoted in Boase, 1979, pp.251-252).

[293] (Barolsky, 1992). *Giotto's father and the family of Vasari's Lives.* University Park: Pennsylvania State University Press.

[294] (quoted with original italics in Murray, 1989b, p.28).

[295] (Bate, 1989). Shakespeare and original genius. In P. Murray (Ed.), *Genius: The history of an idea* (pp.76-97). Oxford, England: Basil Blackwell.

[296] Ericsson, Krampe, & Heizmann, 1993. Can we create gifted people? In CIBA Foundation Symposium 178, *The origins and development of high ability* (pp.222-249). Chichester, England: Wiley.

[297] (Gardner, 1973). *The arts and human development.* New York: Wiley.

[298] Ibid Gardner p.189.

[299] (Gardner, 1973, p.189). *The arts and human development.* New York: Wiley.

[300] (p.233).

[301] (p.385).

[302] (p.386).

[303] (p.386).

[304] (pp.385-386).

[305] (1993a, 1993b).

[306] (pp.40-41.

[307] Takeuchi & Hulse, 1993. Absolute pitch. *Psychological Bulletin, 113*, 345-361.

[308] (p.355).

[309] (p.356.)

[310] Takeuchi & Hulse, 1993. Absolute pitch. *Psychological Bulletin, 113*, 345-361.

[311] Pariser, 1987. The juvenile drawings of Klee, Toulouse-Lautrec and Picasso. *Visual Arts Research, 13, 53-67.*

[312] (p.123).

[313] (p.122).

[314] (Ho, 1989). *Yani: The brush of innocence.* New York: Hudson Hills.

[315] (1980, 1986).

[316] (Barlow, 1952. *Mental prodigies.* New York: Greenwood Press; Feldman, 1986. *Nature's gambit: Child prodigies and the development of human potential.* New York: Basic Books).

[317] (p.239).

[318] (p.239).

[319] (Bloom, 1985). Generalisations about talent development. In B.S. Blook (Ed.), *Developing talent in young people* (pp.507-549). New York: Ballantine Books.

[320] (see Ericsson, Krampe, & Tesch-Römer, 1993. The role of deliberate practice in the acquisition of expert performance. *Psychological Review, 100.* 363-406

and Howe, 1990. *The origins of exceptional abilities*. Oxford, England: Basil Blackwell).

[321] (Forbes, 1992). *The Polgar sisters: Training or genius?* New York: Henry Holt.

[322] (Ericsson & Faivre, 1988. What's exceptional about exceptional abilities? In I.K. Obler & D. Fein (Eds), *The exceptional brain: Neuropsychology of talent and special abilities* (pp.436-473). New York: Guilford Press; Feldman, 1986. *Nature's gambit: Child prodigies and the development of human potential*. New York: Basic Books).

[323] (Ericsson & Faivre, 1988. What's exceptional about exceptional abilities? In I.K. Obler & D. Fein (Eds), *The exceptional brain: Neuropsychology of talent and special abilities* (pp.436-473). New York: Guilford Press; Howe, 1990. *The origins of exceptional abilities*. Oxford, England: Basil Blackwell; Treffert, 1989. *Extraordinary people: Understanding "idiot savants"*. New York: Harper & Row.).

[324] (For a review see Ericsson & Faivre, 1988)

[325] (p.279).

[326] (p.297).

[327] (p.546).

[328] (p.609).

[329] (p.609).

[330] (Malina & Bouchard, 1991). *Growth, maturity and physical activity*. Champaign, IL: Human Kinetics.

[331] (Ericsson, Krampe, & Tesch-Römer, 1993). The role of deliberate practice in the acquisition of expert performance. *Psychological Review, 100.* 363-406.

[332] (For reviews see Regnier, Salmela, & Russell, 1993, and Starkes & Deakin, 1985)

[333] (Chase & Simon, 1973. The mind's eye in chess. In W.G. Chase (Ed.), *Visual information processing* (pp.215-281). New York; Academic Press; see Ericsson & Smith, 1991a, for a review. Prospects and limits of the empirical study of expertise: An introduction. In K.A. Ericsson & J. Smith (Eds.), *Towards a general theory of expertise: Prospects and limits* (p.1-39). Cambridge, England: Cambridge University Press).

[334] (Doll & Mayr, 1987). Intelligenz und Schachleistung—Eine Untersuchung an Schachexperten [Intelligence and achievement in chess: A study of chess masters]. *Psychologische Betträge, 29, 270-289.*

[335] (Ericsson, Krampe, & Tesch-Römer, 1993). The role of deliberate practice in the acquisition of expert performance. *Psychological Review, 100.* 363-406.

[336] Hulin, Henry, and Noon (1990)

[337] (see Ericsson, Krampe, & Tesch-Römer, 1993, for a review). The role of deliberate practice in the acquisition of expert performance. *Psychological Review, 100.* 363-406.

[338] (see Ericsson, Krampe, & Tesch-Römer, 1993, for a review). The role of deliberate practice in the acquisition of expert performance. *Psychological Review, 100.* 363-406.

[339] (quoted in Galton, 1908). *Memories of my life.* London: Methuen.

[340] (p.290).

[341] (p.291).

[342] Ericsson, Krampe, and Tesch-Römer (1993)

[343] (Elo, 1986). *The rating of chess players, past and present* (2nd ed.). New York: Arco.

[344] (Elo, 1986). *The rating of chess players, past and present* (2nd ed.). New York: Arco.

[345] (Camerer & Johnson, 1991). The process-performance paradox in expert judgment: How can the experts know so much and predict so badly? In K.A. Ericsson & J. Smith (Eds.,), *Towards a general theory of expertise: Prospects and limits* (pp.195-217). Cambridge, England: Cambridge University Press.

[346] (Chi et al., 1988; *The nature of* expertise. Hillsdale, NJ: Erlbaum. Ericsson & Smith, 1991b; *Towards a general theory of expertise: Prospects and limits.* Cambridge, England: Cambridge University Press).

[347] (Salthouse, 1991a). Expertise as the circumvention of human processing limitations. In K.A. Ericsson & J. Smith (Eds.,) *Towards a general theory of expertise: Prospects and limits* (pp.286-300). Cambridge, England: Cambridge University Press.

[348] (Levin & Addis, 1979). *The eye-voice span.* Cambridge, MA: MIT Press.

[349] (Sloboda, 1985). *The musical mind: The cognitive psychology of music.* Oxford, England: Oxford University Press.

[350] (Biederman & Shiffrar, 1987). Sexing day-old chicks: A case study and expert systems analysis of a difficult perceptual-learning task. *Journal of Experimental Psychology: Learning, Memory and Cognition,* 13, 640-645.

[351] (Howe, 1990). *The origins of exceptional abilities.* Oxford, England: Basil Blackwell.

[352] (Ericsson, Krampe & Tesch-Römer, 1993). The role of deliberate practice in the acquisition of expert performance. *Psychological Review, 100,* 363-406.

[353] (Feldman, 1980). *Beyond universals in cognitive development.* Norwood, NJ: Ablex.

[354] (Platt, 1966). General introduction. In J.E. Meade & A.S. Parkes (Eds.,), *Genetic and environmental factors in human ability* (pp. ix-xi), Edinburgh, Scotland: Oliver & Boyd.

[355] (p.23).

[356] (Csikszentmihalyi, 1990).

[357] Absalom and Achitophel, Part I, line 27.

[358] Philosophy Now, Issue 153, December 2022/January 2023).

[359] The Times, December 30th 2022.

[360] Motor Learning and Human Performance 1975 Robert N. Singer, Second Edition Macmillan Publishing Co.(New York)

[361] See E.T. Layton (Technology as Knowledge. Technology and Culture 15, 31-41

[362] (Ingold ibid p.434)

[363] Motor Learning and Human Performance, Robert N. Singer.

[364] (Ingold ibid p.444)

[365] Ibid p.459

[366] (ibid p.459)

[367] P.464 ibid

[368] (see Lave J (1990) The Culture of Acquisition and the Practice of Understanding in Cultural Psychology: Essays on Comparative Human Development: ed. J.W. Stigler, R.A. Schweder and G. Herdt pp.309-327. Cambridge University Press

[369] (Lave ibid p.323)

[370] The Physiology of Motor Learning Cerebral Palsy Review by Hellebrandt July-August 1958 pp.9-14

[371] Conrad R 'Timing' OP Vol/29 No.3 1995 ppl173-181

[372] FC Bartlett "The Measurement of Human Skill" B.M.J. No.4510, June 14th 1947 pp.835-38

[373] Anne J and Kay "Skilled Performance" O.P. Vol.30 1956 pp.112-17

[374] Fundamentals of Skill by A.T. Welford. Methuen & Co. Ltd. 1968

[375] Paillard J. (1960) 'The patterning of skilled movements', in *Handbook of Physiology—Neurophysiology* III (ed. J. Field H.W. Magoun and V.E. Hall). Washington: American Physiological Society pp.1679-1708

[376] Advanced by Anders Ericsson.

[377] Polanyi and Prosch, H. Meaning. Chicago. University of Chicago Press, 1975

[378] Ibid p.34

[379] ibid p.31

[380] Ibid Moe pp.173, 175

[381] Ibid p.38

[382] ibid pp. 42-43

[383] Ibid p.97

[384] Ibid p.96

[385] Ibid p.96

[386] Ibid p.60

[387] Twain M. "Taming the Bicycle". In the Complete Works of Mark Twain: What is Man? New York. Harper 1917, pp 295-296

[388] Ibid pp 287-288

[389] Ibid Twain pp.294, 296

[390] Howard P. Riding High: Shadow Cycling the Tour de France: London: Mainstream Publishing, 2004

[391] The Mastery of Movement, Rudolf Laban 2nd ed revised by Lisa Ullman, McDonald and Evans 1960.

[392] Ibid Laban p. v. Preface to the First Edition.

[393] Ibid Laban p.10.

[394] Ibid Laban p.17.

[395] Ibid Laban p.17.

[396] Ibid Laban p.17.

[397] Ibid Laban p.18.

[398] Ibid Laban p.18.

[399] Ibid Laban p.18.

[400] Ibid Laban p.18. Gliding is essentially a sustained and direct movement with gentle touch. In gliding, man and his deity are enveloped in the experience of the infinity of time and cessation of the drag of weight, but they are actively concerned with the directional clarity of their movements. Many dances of the aborigines of Africa, Asia, Polynesia and America show this feature of gliding in their dance rituals; and the picture and statues of their gods are represented as figures making gliding gestures. Gods floating over the water show in ritual, or pictorial representation, a yielding attitude towards the motion factors of time, weight and space. Floating is a sustained gentle and flexible movement, mirroring a state of mind of a similar content.

[401] Ibid Laban p.19.

[402] Ibid Laban p.19.

[403] Ibid Laban p.21.

[404] Ibid Laban p.22.

[405] Ibid Laban p.23.

[406] Ibid Laban p.23.

[407] Ibid Laban p.24.

[408] Ibid Laban p.25.

[409] Ibid Laban p.27.

[410] Ibid Laban p.37.

[411] Ibid Laban p.37.

[412] Ibid Laban p.88.

[413] Ibid Laban p.87.

[414] The quick and the dead—when reaction beats intention proceedings of the Royal Society—Andrew E. Welchmanet a (rspb royal society publishing org February 3 2010).

[415] Kurata K and Tanji 1985—J. Neurophysio (53 pp.142-152).

[416] Waszaket al 2005 Ex p Brain Res 162 pp.346-356 Oblix Haggard 2004 Ex p. Brain Res 156 pp.518-523.

[417] Ibid Kurata & Tanji 1985; Romo & Schultz 1987; Ex p. Brain Res 67 pp.656-662.

[418] Ibid Obhid Haggard.

[419] Harper Collins, 1996, p.170.

[420] (New York Review of Books, November 22, 1990, pp.44-50).

[421] Moshe Feldendrais, Awareness Through Movement (New York: Harper & Row, 1977, p.46).

[422] I am greatly indebted throughout this chapter and elsewhere to Professor Shaun Gallagher's influential work "How the Body shapes the Mind" 2005 Clarendon Press, Oxford. In this book he contributes to the formulation of that common vocabulary and develops a conceptual framework that will avoid both the overly reductionistic approaches that explain everything in terms of bottom-up neuronal mechanisms, and the inflationist approaches that explain everything in terms of Cartesian, top-down cognitive states.

[423] For example, Sherrington C. 1953 *Man on His Nature*, 2nd ed. New York: Doubleday often uses the term in this way. Since we have no awareness of neural events which 'register the tension at thousands of points they sample in the muscles, tendons, and ligaments of [a] limb', he maintains that 'I perceive no trace of all this 'proprioceptive activity]' (p.248). Sherrington, however, goes on to consider a proprioceptive awareness of movement. More recently, Fourneret, P., and Jeannerod, M. 1998. Limited conscious monitoring of motor performance in normal subjects. *Neuropsychologia*, 36: 1133-40 conducted experiments that showed subjects to be completely unaware of proprioceptive signals generated by their own movements.

[424] Thus O'Shaughnessy B. 1995 Proprioception and the body image. In J. Bermúdez, A. Marcel, and N. Eilan (eds.), *The Body and the Self* (pp.175-203). Cambridge, Mass: MIT speaks of proprioceptive awareness. Sheets-Johnstone describes proprioception as follows: 'Proprioception refers generally to a sense of movement and position. It thus includes an awareness of movement and position through tactility as well as kinaesthesia, that is, through surface as well as internal events, including also a sense of gravitational orientation through vestibular sensory organs. Kinaesthesia refers specifically to a sense of movement through muscular effort'. Sheets-Johnstone, M. 1990. *The Roots of Thinking*. Philadelphia: Temple University Press.

[425] This is a distinction clearly made in Bermúdez, J.L., Marcel A., and Eilan N. 1995 *The Body and the Self* Cambridge, Mass: MIT and Bermúdez J.L. 1998 *The Paradox of Self-Consciousness*. Cambridge, Mass: MIT.

[426] Recently, however, this subject has been addressed by two notable studies— Hurley S. 1999. *Consciousness in Action*. Cambridge, Mass: Harvard University Press and Sheets-Johnstone M. 1999a. *The Primacy of Movement*. Amsterdam: John Benjamins.

[427] The very onset of consciousness may be directly related to prenatal movement and accompanying proprioception. Still, I do not mean to endorse a motor theory of perception—a theory that would claim that mental states are produced by movement or muscle discharges related to interaction with objects. I want to treat movement as a *constraint on* rather than a *cause* of perception. For a review of motor theories of cognition, see Scheerer E. 1984. Motor theories of cognitive structure: A historical review. In W. Prinz and A.F. Sanders (eds), *Cognition and Motor Processes* (pp.77-97). Berlin: Springer.

[428] Martinez-Conde, S., Macknik, S.L., and Hubel, D.H. 2000. Microsaccadic eye movements and firing of single cells in the striate cortex of macaque monkeys. *Nature Neuroscience*, 3(3): 251-8.

[429] Jeannerod, M. 1994. The representing brain: Neural correlates of motor intention and imagery. *Behavioural and Brain Sciences*, 17: 187-245.

[430] Gallese, V. 1998. Mirror neurons: From grasping to language. Paper presented at Tucson III Conference: Towards a Science of Consciousness (Tucson 1998).

[431] Rizzolatti, G., Fogassi, L., and Gallese, V. 2000. Cortical mechanisms subserving object grasping and action recognition: A new view on the cortical motor functions. In M.S. Gazzaniga (ed.), *The New Cognitive Neurosciences* (pp.549-52). Cambridge, Mass: MIT.

[432] Campos, J.J., Bertenthal, B.I., and Kermoian, R. 1992. Early experience and emotional development: The emergence of wariness of heights. *Psychological Science*, 3: 61-4.

[433] Thelan, E. 1995. Time-scale dynamics and the development of an embodied cognition. In R.F. Port and T. van Gelder (eds.), *Mind as Motion: Explorations in the Dynamics of Cognition* (pp.69-100). Cambridge, Mass: MIT.

[434] This is not to deny that the developmental relation between cognition and movement goes both ways. Thus, development of motor competence is essentially linked to the maturation of intrinsic and primitive recognitional abilities in early infancy to more explicit object recognition based on increased sensitivity to object properties (Bermúdez 1998; Russell 1996).

[435] Bushnell and Boudreau 1993.

[436] Lockman J.J., and Thelan, E. 1993. Developmental biodynamics: Brain, body, behaviour connections. *Child Development*, 64: 953-9.

[437] Panksepp, J. 1998. The periconscious substrates of consciousness: Affective states and the evolutionary origins of the self. *Journal of Consciousness Studies*, 5(5/6): 566-82.

[438] Kinsbourne, M. 1975. The mechanisms of hemispheric control of the lateral gradient of attention. In P.M.A. Rabbitt and S. Dornic (eds), *Attention and Performance*. London: Academic Press.

[439] Lempert H., and Kinsbourne, M. 1982. Effect of laterality of orientation on verbal memory. *Neuropsychologia*, 20: 211-14.

[440] Grubb, J.D. and Reed, C.L. 2002. Trunk orientation includes neglect-like lateral biases in covert attention. *Psychological Science*, 13: 554-7.

[441] Di Pellegrino, G., Fadiga, L., Fogassi, L., Gallese, V., and Rizzolatti G. 1992. Understanding motor events: A neurophysiological study. *Experimental Brain Research*, 91: 176-80.

[442] Gallese V., Fadiga, L., Fogassi L., and Rizzolatti, G. 1996. Action recognition in the premotor cortex. *Brain*, 1991: 593-609.

[443] Rizzolatti, G., Fadiga, L., Gallese, V., and Fogassi L., 1996. Premotor cortex and the recognition of motor actions. *Cognitive Brain Research*, 3: 131-41.

[444] Renaud Barbaras 1999. *Le désir et la distance.* Paris: J. Vrin. 2000 Perception and movement: The end of the metaphysical approach. In F. Evans and L. Lawlor (eds)., *Chiasms: Merleau-Ponty's Notion of Flesh.* Albany: State University of New York Press suggests a close connection between movement and perception and finds within this insight the beginnings of a radically non-traditional, non-metaphysical way of thinking of the self. I do not pursue this route here. It would entail giving up a certain concept experience that I fine useful for bridging the

phenomenological tradition and discussions in cognitive science. I would note, however, that the connection between movement and perception is also founded in metaphysical accounts of the self. For example, Fichte's concept of 'intellectual intuition' is not unrelated to these issues. He writes: 'I cannot take a step, move hand or foot without an intellectual intuition of my self-consciousness in these acts; only so do I know that *I* do it, only so do I distinguish my action, and myself therein, from the object of action before me. Whoever ascribes an activity to himself, appeals to this intuition' (1794: 38).

[445] Darwin, Charles, 1987. *Charles Darwin's Notebooks, 1836-1844*, Paul H. Barrett, Peter J. Gautrey, Sandra Herbert, David Kohn, Sydney Smith (eds.), Ithaca: Cornell University Press, p.564.

[446] *(De Anima 403b 21-23).*

[447] Husserl, Edmund. 1977. *Phenomenological Psychology,* trans. John Scanlon. The Hague: Martinus Nijhoff, p.101.

[448] Husserl, Edmund, 1989. *Ideas Pertaining to a Pure Phenomenology and to a Phenomenological Philosophy: Book 2 (Ideas II)* trans. R. Rojcewicz and A. Schuwer. Boston: Kluwer Academic Publishers, p.102.

[449] The Primacy of Movement 1999, John Benjamins Publishing Company.

[450] Ibid Sheets-Johnstone Introduction xvii.

[451] Kunn, Steven L., 1995. *Mousterian Lithic Technology: An Ecological Perspective*, Princeton: Princeton University Press, p.5.

[452] Lieberman, Philip, 1983. "On the Nature and Evolution of the Biological Bases of Language". In Eric de Grolier (ed), *Glossogenetics*, New York: Harwood Academic Publishing, 91-114.

[453] 1972 "Primate Vocalisations and Human Linguistic Ability". In Sherwood L. Washburn and Phyllis J. Dolhinow (eds), *Perspectives on Human Evolution, 2.* New York: Holt, Rinehart and Winston, 421-37.

[454] Laitman, Jeffrey T. 1983, "The Evolution of the Hominid Upper Respiratory System and Implications for the Origins of Speech." In Eric de Grolier (ed), *Glossogenetics.* New York: Harwood Academic Publishing, 63-90.

[455] Hockett, Charles, 1960. "The Origin of Speech". *Scientific American* 203: 89-96.

[456] Bennett, Jonathan, 1971. *Rationality.* London: Routledge & Kegan Paul.

[457] Wilson, E.O. 1972. "Animal communication". *Scientific American* 227: 52-60.

[458] Dennett, Daniel C. 1991. *Consciousness Explained.* Boston: Little, Brown and Company.

[459] Ibid Sheets-Johnstone, p.4.

[460] Sheets-Johnstone, Maxime. 1992b. "Taking Evolution Seriously". *American Philosophical Quarterly* 29(4): 343-52. 1996a. "Taking Evolution Seriously: A Matter of Primate Intelligence". *Etica & Animali* 8: 115-130.

[461] Ibid Sheets-Johnstone p.132.

[462] Claesges (1964) enumerates six moments of kinaesthetic consciousness: time, space, horizon, world (which he says subsumes the previous three moments), body, and self. He identifies these moments without reference to the originariness of movement—its "beforehand giveness" in primal animation—thus without reference to the origination ground of our sense-making. His broad equation of kinaesthetic consciousness with these moments and his concern to show that *the world* is pregiven (thus, by his definition, kinaesthetic consciousness is pregiven) contain no reference to movement itself.

[463] See, for example, ibid Husserl 1989-351: "But the animal and, in the first place, human beings can also be regarded as reality or nature, and we can here distinguish again between the animal as intuitive unity and the animal as unity of modes of behaviour;" and ibid Husserl 1989: 142: "Without the soul, it [the psychic subject] is unable to stand alone; and yet again, it is a unity which in a certain sense encompasses the soul and which is at the same time so prominent that it dominates the general way of speaking about human and animal subjects."

[464] Carruthers, Peter. 1989. "Brute Experience". *The Journal of Philosophy* 86(5): 258-69.

[465] Fink Eugen, 1995. *Sixth Cartesian Meditation*, trans. Ronald Bruzina. Bloomington: Indiana University Press.

[466] Husserl, Edmund. 1983. *Ideas Pertaining to a Pure Phenomenology and to a Phenomenological Philosophy, Book I (Ideas I)*, trans. F. Kersten. The Hague: Martinus Nijhoff, p.127.

[467] Husserl, Edmund. 1989. *Ideas Pertaining to a Pure Phenomenology and to a Phenomenological Philosophy: book 2 (Ideas II)*, trans. R. Rojcewica and A. Schuwer. Boston: Kluwer Academic Publishers, p.185.

[468] Husserl, Edmund, 1989. *Ideas Pertaining to a Pure Phenomenology and to a Phenomenological Philosophy: Book 2 (Ideas II)*, trans. R. Rojcewicz and A. Schuwer. Boston: Kluwer Academic Publishers.

[469] What I want tangentially to suggest is that the ontological ground Ronald Bruzina has uncovered in the context of his elucidation of Fink's emendations to Husserl has its epistemological corollary in the primary of movement: the ontological temporal is coterminous with the epistemological kinetic: the latter too is the grounding ground that defies constitutional explication insofar as it is

already there prior to and refusing constitution, but is at the same integrated with it.

[470] Husserl, Edmund. 1989. *Ideas Pertaining to a Pure Phenomenology and to a Phenomenological Philosophy; Book 2 (Ideas II)*, trans. R. Rojcewicz and A. Schuwer. Boston: Kluwer Academic Publishers, p.273.

Landgrebe, Ludwig. 1977. "Phenomenology as Transcendental Theory of History," trans. J. Huertas-Jourda and R. Feige. In Peter McCormick and Frederick Elliston (eds), *Husserl: Expositions and Appraisals*. Notre Dame, Ind: Notre Dame University Press, 101-13.

[471] Sheets-Johnstone, Maxine, 1990. *The Roots of Thinking*. Philadelphia: Temple University Press, p.29.

[472] C.f. Sokolowski (1972: 76): "[A]ll these motions of joints [i.e. the kinesthesis] [are] preceded by the activity of motion. And [they are preceded] first of all [by] being awake. There is no basic consciousness without being awake and being awake is one of the basic data. We have to thank our bodies for it." The experience of motion, of course, is not *preceded* by the "activity of motion" but coincident with it. Sokolowski's point, however, about being first of all *awake* is suggestive of being first of all *alive*.

[473] See, for example, Spitz (1983), in particular, the essays, "Life and the Dialogue" and "The Evolution of Dialogue".

[474] This empirical fact is strongly if indirectly supportive of Husserl's analysis of empathy (1973: Fifth Meditation) in its suggestion that an attentiveness to the movement and actions of what is living is central to our existence.

[475] This textual model is well exemplified by Derrida (1976 and in other writings of his as well). For a critique of this textual model, see Sheets-Johnstone 1994. Chapter 4: "Corporeal Archetypes and Postmodern Theory."

[476] Husserl, Edmund. 1989. *Ideas Pertaining to a Pure Phenomenology and to a Phenomenological Philosophy: Book 2 (Ideas II)*, trans. R. Rojcewicz and A. Schuwer. Boston: Kluwer Academic Publishers, p.346.

[477] Obviously, this description of infant experience is not phenomenological. There is nevertheless good—even excellent—reason, to think that, like phenomenologically derived insights, it is true to the truths of experience.

[478] By "distinctive" ways of moving, I do not necessarily mean thoroughly unique. Species-overlapping patterns—including bipedality (higher primates other than humans are bipedal but not *consistently* bipedal)—are apparent in many everyday human movement behaviours.

[479] Husserl, Edmund. 1980. *Ideas Pertaining to a Pure Phenomenology and to a Phenomenological Philosophy: Book 3 (Ideas III)*, trans. Ted E. Klein and William E. Pohl. The Hague: Martinus Nijhoff, p.137.

[480] Ibid Sheets-Johnstone p.138.

[481] Merleau-Ponty, Maurice, 1962. *Phenomenology of Perception*, trans. Colin Smith. New York: Routledge & Kegan Paul, p.198.

[482] Ibid Sheets-Johnsone, p.138.

[483] Landgrebe, Ludwig. 1977. "Phenomenology as Transcendental Theory of History," trans. J. Huertas-Juorda and R. Feige. In Peter McCormick and Frederick Elliston (eds), *Husserl: Expositions and Appraisals*. Notre Dame, Ind: Notre Dame University Press, 101-13.

[484] Gibson, James J. 1979. *The Ecological Approach to Visual Perception.* Boston: Houghton Mifflin Company.

[485] Husserl, Edmund. 1989. *Ideas Pertaining to a Pure Phenomenology and to a Phenomenological Philosophy: Book 2 (Ideas II)*, trans. R. Rojcewicz and A Schuwer. Boston: Kluwer Academic Publishers.

[486] Sheets-Johnstone. I have changed the original descriptive term "areal quality" (Sheets-Johnstone 1966) to "amplitudinal quality". Sheets-Johnstone, Maxine. 1996. *The Phenomenology of Dance*. Madison: University of Wisconsin Press. (Second editions: London: Dance Books Ltd. 1979; New York; Arno Press 1980).

[487] Ibid Sheets-Johnstone p.143.

[488] Ibid Sheets-Johnstone p.161.

[489] For a more detailed clarification of a *transcendental clue*, see pages 244-45. Husserl, Edmund. 1973a. *Cartesian Meditations*, trans. Dorion Cairns. The Hague: Martinus Nijhoff.

[490] Selected Letters of Dylan Thomas (ed. Fitzgibbon) p.48.

[491] Ibid Sheets-Johnstone p.490.

[492] Ibid Sheets-Johnstone p.491.

[493] Ibid Sheets-Johnstone, p.491.

[494] Merleau-Ponty, Maurice. 1962. *Phenomenology of Perception*, trans. Colin Smith. New York: Routledge & Kegan Paul (182).

[495] Merleau-Ponty, Maurice. 1962. *Phenomenology of Perception*, trans. Colin Smith. New York: Routledge & Kegan Paul (196).

[496] Ibid Sheets-Johnstone p.493

[497] Wittgenstein, Ludwig, 1963. *Philosophical Investigations*, trans. G.E.M. Anscombe. Oxford: Basil Blackwell p.107.

[498] Merleau-Ponty, Maurice. 1962. *Phenomenology of Perception*, trans. Colin Smith. New York: Routledge & Kegan Paul (181-82).

[499] Ibid Sheets-Johnstone p.494.

[500] Merleau-Ponty, Maurice. 1964e. "Eye and Mind," trans. Carleton Dallery. In James M. Edie (ed), *The Primacy of Perception*. Evanston, IL: Northwestern University Press, 159-90 (178).

[501] Ibid Sheets-Johnstone p.494.

[502] Keeton, William T. and James L. Gould. 1986. *Biological Science*, 4th ed. New York: W.W. Norton & Co. p.554. An egregious and lamentable effort should be pointed out in Keeton's and Gould's text: In their introduction, they state that "To early 'mechanistic' philosophers like Aristotle and Descartes, life was wholly explicable in terms of the natural laws of chemistry and physics". A reading of *De Anima* should be required reading for all biologists, along with *The History of Animals, Parts of Animals, Movement of Animals, Progression of Animals and Generation of Animals* and also some excellent commentary texts, especially what is considered "the bible" with respect to Aristotle's biology: *Philosophical Issues in Aristotle's Biology*, edited by Allan Gotthelf and James G. Lennox.

[503] Ibid Sheets-Johnstone p.510.

[504] Ibid Sheets-Johnstone p.510.

[505] The Sense of Space by David Morris, State University of New York 2004, pp.96-100.

[506] Infinite Jest by David Foster Wallace.

[507] Matter and memory by Henri Bergson 1991, New York, Zone Books.

[508] Ibid The Sense of Space, p.97 by David Morris.

[509] "How Brains make up their Minds" by Walter J. Freeman, Weldenfield and Nicolson, 1999, p.18.

[510] Freeman ibid pages 18-19.

[511] Dreyfus, H.L. "The Primacy of Phenomenology over Logical Analysis." *Philosophical Topics*, 27(2), 1999, p.3.

[512] Breivik. G, 'Skilful Coping in Everyday Life and in Sport: A Critical Examination of the views of Heidegger and Dreyfus' in *Journal of the Philosophy of Sport*, Volume XXXIV, issue 2, 2007, pp.116-17.

[513] Heidegger, M. *Being and Time*, San Francisco, Harper, 1962.

[514] Breivik, 'Skilful Coping in Everyday Life and in Sport: A Critical Examination of the views of Heidegger and Dreyfus' p.117.

[515] Heidegger, *Being and Time*, p.97.

[516] Heidegger, *Being and Time*, p.97.

[517] Heidegger, *Being and Time*, p.98.

[518] Heidegger, *Being and Time*, p.98.

[519] Breivik, 'Skilful Coping in Everyday Life and in Sport: A Critical Examination of the views of Heidegger and Dreyfus' p.118-9.

[520] Breivik, 'Skilful Coping in Everyday Life and in Sport: A Critical Examination of the views of Heidegger and Dreyfus' p.120-21.

[521] Breivik, 'Skilful Coping in Everyday Life and in Sport: A Critical Examination of the views of Heidegger and Dreyfus' p.121-22.

[522] Heidegger, M. *The Basic Problems of Phenomenology*, Revised Edition, Bloomington and Indianapolis, Indiana University Press, 1984, p.163.

[523] Dreyfus, H.L. "The Primacy of Phenomenology over Logistic Analysis," *Philosophical Topics*, 27(2), 1999, p.66 citing Richard Mitchel in *Mountain Experience. The Psychology and Sociology of Adventure,* Chicago, University of Chicago Press, 1983.

[524] Breivik, 'Skilful Coping in Everyday Life and in Sport: A Critical Examination of the views of Heidegger and Dreyfus' p.122.

[525] Breivik, 'Skilful Coping in Everyday Life and in Sport: A Critical Examination of the views of Heidegger and Dreyfus' p.123.

[526] Heidegger, *Being and Time*, p.98.

[527] Heidegger, M. *History of the Concept of Time*, Bloomington and Indianapolis, Indiana University Press, 1984, p.163.

[528] Dreyfus, *Being-in-the-World. A Commentary on Heidegger's Being and Time*, p.67.

[529] Breivik, 'Skilful Coping in Everyday Life and in Sport: A Critical Examination of the views of Heidegger and Dreyfus' p.126-27.

[530] Breivik, 'Skilful Coping in Everyday Life and in Sport: A Critical Examination of the views of Heidegger and Dreyfus' p.127.

[531] Heinrich, B. *Why We Run. A Natural History*, New York, Harper Collins, 2001, p.231.

[532] Heinrich, B. *Why We Run. A Natural History*, New York, Harper Collins, 2001, p.248.

[533] Breivik, 'Skilful Coping in Everyday Life and in Sport: A Critical Examination of the views of Heidegger and Dreyfus' p.130.

[534] Breivik, 'Skilful Coping in Everyday Life and in Sport: A Critical Examination of the views of Heidegger and Dreyfus' p.130.

[535] Breivik, 'Skilful Coping in Everyday Life and in Sport: A Critical Examination of the views of Heidegger and Dreyfus' p.131.

[536] Breivik, 'Skilful Coping in Everyday Life and in Sport: A Critical Examination of the views of Heidegger and Dreyfus' p.132.

[537] Polanyi and Prosch Meaning, Chicago, University of Chicago Press 1975.

[538] Polanyi's "From-To" Knowing and His Contribution to the Phenomenology of Skilled Motor Behaviour. Journal of the Philosophy of Sport, 2009, 36, 76-87 © Human Kineties, Inc. by Peter Hopsicker.

[539] Ibid Moe p.177.

[540] Ibid Moe p.168.

[541] Ibid Moe p.177.

[542] Ibid Moe p.171.

[543] Ibid Moe p.173

[544] Moe VF "Understanding the Background Conditions of Skilled Movement in Sport: A Study of Searle's Background Capacities" Sport, Ethics and Philosophy 1131, 2007, 299-324 at p.318.

[545] Ibid Moe p.174.

[546] Ibid Moe p.175

[547] Ibid Moe p.171.

[548] Ibid Moe p.173.

[549] Ibid Moe p.175-176.

[550] Ibid Moe p.177-178.

[551] Ibid Moe 2007, p.301.

[552] Ibid Moe 2007, p.318.

[553] Ibid Breivik, p.116.

[554] Ibid Breivik, p.116.

[555] Ibid Breivik, p.131-132.

[556] (Seth Vannatta in A Phenomenology of Sport: Playing and Passive Synthesis Journal of the Philosophy of Sport, 2008, 63-72 at page 67).

[557] (See Husserl, Edmund. Analyses Concerning Passive and Active Synthesis: Lectures in Transcendental Logic. Trans by Anthony Steinbock, Dordrecht: Kluwer Academic 2001 at pages 39-43).

[558] See Husserl ibid pp.41-43.

[559] Husserl ibid p. xii.

[560] Husserl ibid p.631.

[561] Ibid p.121.

[562] Ibid p.164.

[563] Ibid p.184.

[564] Ibid p.185.

[565] Ibid p. xi.

[566] Vannata ibid p.71.

[567] The Human Hand. A Psychological study by Géza Révész 1958, Routledge & Kegan Paul Ltd p.1.

[568] Ibid Révész p.12.

[569] Ibid Révész p.14.

[570] Ibid Révész p.15.

[571] Ibid Révész p.16.

[572] Ibid Révész p.17.

[573] Ibid Révész p.19.

[574] D. Katz Der Autbau der Tastwelt, Leipzig, 1925.

[575] Ibid Révész p.23.

[576] Ibid Révész p.46.

[577] Ibid Révész p.62.

[578] Aristotle Metaphysics, VII, Oxford Clarendon Press. Trust by W.D. Ross.

[579] Ibid Révész p.62.

[580] The Hand, Frank R. Wilson, Vintage Books 1998, p.289.

[581] (The quotation is taken from J.Z. Young: Doubt and Certainty in Science: A Biological Reflection on the Brain, Oxford 1960).

[582] Ibid Wilson p.307.

[583] Wilson ibid p.7.

[584] Ibid Wilson p.9.

[585] Ibid Wilson p.10.

[586] See Hands by John Napier (Revised by Russell H. Tuttle, Princeton University Press, Princeton New Jersey 1993).

[587] (See S. Culin, Games of the North American Indians) (New York, Dover 1975).

[588] (ibid Bell p.188).

[589] An unpublished dissertation for MIT quoted at p.105 ibid Wilson.

[590] The Hand—A Philosophical Inquiry into Human Being by Raymond Tallis, Edinburgh University Press 2003, pp.60-61.

[591] (John Napier, The Roots of Mankind (London: George Allen and Unwin, 1971, p.176)).

[592] Ibid Tallis page 126.

[593] (F. Mechsner, D. Kerzek, G. Knoblich and W. Prinz: Perceptual basis of bimanual co-ordination, nature 414, 2001, pp.69-73).

[594] See later section of visual perception.

[595] See 'The unitary hypothesis: a common neural circuitry for novel manipulations, language, plan-ahead and throwing' in Gibson and Ingold Tools, Language and Cognition in Human Evolution.

[596] Ibid p.13.

[597] (Tallis says he owes this information to a chance reading by Georgina Ferry about the work of the experimental psychologist Peter McLeod 'The Cricketer's Eyes', Oxford Today, Trinity Issue 2001, p.30).

[598] (The relationship between agency and mechanism, between reason—informed behaviour and its automatic components is visited at many points in this trilogy. The most sustained examination of the issues is in Tallis Vol.2: I am; A Philosophic Inquiry into First Person Being).

[599] A Rewritten Account by David Sudnow, The MIT Press, Cambridge Massachusetts 2001.

[600] (Foreword ix).

[601] Ibid Dreyfus Foreword XI.

[602] Ibid Foreword Note 2, p. xii.

[603] Samuel Tode's Body and World (MIT Press 2001).

[604] Ibid Preface XIX.

[605] Ibid Preface xix.

[606] Preface ibid xix.

[607] (Preface page 2).

[608] Preface 2.

[609] Footnote 13. See pages 88-97 ibid The Ways of the Hand.

[610] Ibid p.91.

[611] Sudnow ibid pages 97-97.

[612] Ortega y Gasset, J. Historical Reason. New York: W.W. Norton 1984 [1940] p.96

[613] Ortega y Gasset J. Historical Reason. New York W.W. Norton 1984 [1940]

[614] Ibid.p.24

[615] Ortega y Gasset, J. Historical Reason. New York: W.W. Norton 1984 [1940] (emphasis in original) p.96

[616] Ortega y Gasset, J. Mediations on Hunting, Belgrade MT: Wilderness Press 1995 [1942] p.36

[617] Ortega y Gasset J. Historical reason New York: W.W. Norton 1984 [1940] p97

[618] Ortega y Gasset J. Historical Reason New York: W. W. Norton 1984 [1940] p.97

[619] Ortega y Gasset J Historical Reason New York: W. W. Norton 1984 [1940] p97

[620] Suits B "The Elements of Sport". In Philosophic Inquiry Into Sport, Morgan W.J. and Meier, K.V. (Eds) Champaign II: Human Kinetics, 1995 [1973], pp8-15

[621] Suits, B. The Grasshopper: Games, Life and Utopia. Edinburgh: Scottish Academic Press, 1978

[622] Suits, B The Grasshopper: Games Life and Utopia. Edinburgh: Scottish Academic Press 1978

[623] Wittgenstein L Philosophical Investigations. Oxford: Blackwell 1958

[624] Suits N The Grasshopper: Games, Life and Utopia, Edinburgh: Scottish Academic Press 1978, p.178

[625] Throwing Madonna—Essays on the Brain 1983 McGraw-Hill (Book Company).

[626] Pagel, W., Medieval and Renaissance Contributions to Knowledge of the Brain and its Functions (F.N.L. Peyter, Ed., Charles C. Thomas, Springfield III. 1958.

[627] Meynert T., The Brain of Mammals. In Manual of Human and Comparative Histology, Vol.II (S. Stricker, Ed/New Sydenham Society, London 1872: 367-537).

[628] Broca, P., Sur la faculté du language articulé. Bull. Soc Anth (Paris 1865 6: 337-393.

[629] Frank Benson. The Dual Brain Hemispheric Specialisation in Humans, The Guildford Press, London 1985, p.4.

[630] Ibid Benson p.4.

[631] Ibid Benson p.5.

[632] Ibid Benson p.5.

[633] Sperry, R.W., Some effects of disconnecting the cerebral hemisphere. Science 1982, 217: 1223-1226.

[634] Quoted by Eran Zeidel in ibid Benson p.8 from an interview with Roger Sperry. Omni 1983 5:68-75, 98-100.

[635] Ibid Zaidel p.7.

[636] Public Lecture presented at the Smithsonian Institute December 1977, in the Frank Nelson Doubleday Lecture Series on "The Human Mind". See Consciousness, Personal Identity, and the Divided Brain ibid Benson p.11-26.
Surgical division of the brain divides nearly all direct connections mediating cross-talk between the left and right hemispheres. This includes those fibre systems that normally interconnect left and right halves of the cortical field for

vision. As a result the visual perception of objects in each hemisphere becomes restricted to half the normal field of view, cut off sharply at the vertical midline and centre of gaze. The left hemisphere sees things in the right half of the visual field, using either one or both eyes, things to the left are perceived by the right hemisphere. Interconnections are severed also between the cerebral representations that the right and left arms are legs, including both the primary sensory projections and also the main motor controls for skilled movement. Hence things felt with the right hand are perceived mainly in the left hemisphere, which governs related motor adjustments of the same hand. Conversely motor co-ordination and factual perception for the left hand are mediated predominantly by the right hemisphere.

[637] Sperry ibid p.12.

[638] Sperry ibid p.15.

[639] Sperry ibid p.15.

[640] Sperry, R.W., Zaidel, E. and Zaider, D. "Self-recognition and social awareness in the deconnected hemisphere" Neuropsychologia 1979, 17: 153-166.

[641] Sperry ibid p.17.

[642] Sperry p.20.

[643] Puccetti, Brain bisection and personal identity.
Bogen, J.E. BrJ Phil Sci 1973 23: 396-414. The other side of the brain 11. An appositional mind. Bull, Los Angeles, Neurol Soc 1969, 34: 135-162.

[644] Ibid Sperry, p.23.

[645] Armstrong, D.M., A Materialist Theory of the Mind. Routledge & Kegan Paul, London 1968.

[646] Joseph, R., The Right Brain and the Unconscious, Plenum Press, 1992, p.116.

[647] Ibid Joseph p.33.

[648] Ibid Joseph p.33.

[649] D. Benson and M. Barton, "Disturbances in Constructional Ability," *Cortex 6* (1970), 19-46; A. Benton, "visuoperceptive, visuospatial and visuoconstructive disorders". *Clinical Neuropsychology*, K.M. Heilman and E. Valenstein, Eds., (pp.186-232) (Oxford: Oxford University Press, 1979); F.W. Black and B.A. Bernard, "Constructional Apraxia as a Function of Lesion Locus and Size in Patients with Focal Brain Damage," *Cortex 20* (1984), 111-120; R. Calvanio, P.N. Petrone, and D.N. Levine, "Left Visual Spatial Neglect is Both Environment-Centred and Body-Centred," *Neurology 37* (1987), 1179-1183; E. DeRenzi, *Dsorder of Space Exploration and Cognition* (New York: Wiley, 1982); H. Hecaen and M.L. Albert, *Human Neuropsychology* (New York: Wiley,

1978); Y. Kim, L. Morrow, D. Passafiume, and E. Boller, "Visuoperceptual and visuomotor Abilities and Locus of Lesion," *Neuropsychologia 2* (1984), 177-185.

[650] Ibid Joseph p.34.

[651] A.G. Dodds, "Hemispheric differences in Tactuo-Spatial Processing" Neuropsychologia 16 (1978), 247-254. J. Hom and R. Reitan, "Effects of Lateralised Cerebral Damage on Contralateral and Ipsilateral Sensorimotor Performance," Journal of Clinical Neuropsychology 3 (1982) 47-53.

[652] Joseph ibid p.48.

[653] D. Falk "Brain Lateralisation in Primates and its Evolution. In Hominids "Yearbook of Physical Anthropology 30 (1983) 107-125. D. Falk "Hominid Paleoneurology" Annual Review of Anthropology 30 (1987b) 107-125.

[654] Joseph 1982.

[655] Ibid Joseph p.50.

[656] P.F. MacNeilage, M.G. Studdart-Kennedy and B. Lindblom "Functional Precursors to Language and Lateralisation" American Journal of Physiology 246 (1984) 187-201.

[657] Ibid Falk 1987a, b.

[658] Mlot C "Unmasking the emotional unconscious" Science 1998, 2801 [5366] p.1006.

[659] Zeman A. "Consciousness" Brain 2001, 124(7), pp.1263-89.

[660] Ian McGilchrist, The Master and His Emissary. The Divided Brain and the Making of the Western World 2009, Yale University Press.

[661] Pockett, S. "On subject back-referral and how long it takes to become conscious of a stimulus: a reinterpretation of Libet's data, consciousness and cognition 2002 11(2), p.144-161.

[662] Ibid McGilchrist, p.187, other commentators have expressed similar sentiments. Libet's results seem to be in tension with our commonsense picture only because they suggest positing volitions that initially are not conscious. Rosenthal D.M. "The Timing of Conscious States" Consciousness and Cognition, 2002, 11(2) pp.215-220 at p.219. ibid Jaynes 1976 and Lakoff and Johnson (1999) (p.13) make the same point about the l9ow levels of conscious activity needed for most mental life, however sophisticated.

[663] Ibid Galin, Joseph.

[664] Ibid McGilchrist p.188.

[665] Vaihinger H, The Philosophy of 'As If', trans. C.K. Ogden, 2nd ed., Routledge & Kegan Paul, London, 1935, p.7 [Die Philosophiedes AIS Ob 911].

[666] McNeill D., Hand and Mind: What Gestures Reveal About Thought, University of Chicago Press, Chicago 1992.

[667] Ibid McGilchrist, p.189.

[668] McGilchrist p.191.

[669] McGilchrist p.191.

[670] Ibid Joseph p.68.

[671] Ibid Joseph p.68.

[672] Ibid Joseph p.69.

[673] Ibid Joseph p.70.

[674] Ibid Joseph p.70.

[675] Ibid Joseph p.117.

[676] Ibid Joseph p.115.

[677] Ibid Joseph p.118.

[678] The structures and cell assemblies that make up the limbic system include the hypothalamus, the amygdala, the hippocampus and the septal nuclei. The hypothalamus represents the emotional core of our being. It is also intimately involved in all aspects of sexual behaviour. The amygdala discerns the emotional significance of all aspects of experience. It also receives visual and auditory perceptual information. The hippocampus is involved in all aspects of memory and learning. The hippocampus in the right brain is concerned with visual, emotional, tactile and non-verbal memories; in the left brain, it stores verbal and mathematical memories. The septal nucleus is involved in the capacity to form emotional and social bonds with others.

[679] Ibid Joseph p.119.

[680] Ibid McGilchrist p.12.

[681] Ibid McGilchrist p.12.

[682] David McNeill, Hand and Mind What Gestures Reveal About Thought. The University of Chicago Press, 1992, Paperback Edition 1995.

[683] Merleau-Ponty, Phenomenology of Perception, 1962, p.181 London: Routledge & Kegan Paul.

[684] Kendon A. 1972. Some relationships between body motion and speech. In A. Siegman and B. Pope (eds), Studies in Dyadic Communication, 177-210. New York Pergamon Press.

[685] Ibid McNeill p.35.

[686] Ibid McNeill p.25.

[687] Ibid McNeill p.26.

[688] A prototypical gesture passes through three phases—see Kendon Gesticulation and Speech: Two aspects of the process of utterance. In M.R. Key

(ed.), the relation between verbal and non-verbal communication 207-227. The Hague: Moyton. There is the first preparation for the gesture: the hand rises from its resting place and moves to the front away from the speaker. Then there is the stroke, the main part of the gesture: the hand moves backward from the preparation phase and ends up near the shoulder. Finally, there is the retraction, the return of the hand to quiescence: the hand falls back to the rest position. The preparation and retraction phases are optional, but the stroke is essential. If there is a preparation phase it regularly anticipates by a brief interval the co-expressive linguistic segment. See A. Kendon 1980 and A. Kendon 1972. Some relationships between body motion and speech. In A. Siegman and B. Pope (eds), Studies in dyadic communication, 177-210, New York: Pergamon Press.

[689] Saussure F. de (1916) 1959 Course in general linguistics (W. Baskin trans.). Reprint, New York: Philosophical Library.

[690] These terms refer to the relationship of parts to wholes in gestures, in language, parts (words) are combined to create a whole (a sentence); the direction thus is from part to whole. In gestures, in contrast, the direction is from whole to part. The whole determines the meanings of the parts (thus it is "global"). In language movement, the relationship of words to meaning is analytic. Distinct meanings are attached to distinct words. In gestures, however, one gesture can combine many meanings (it is "synthetic").

[691] McNeill D. 1985b "So you think gestures are non-verbal?" Psychological Review 92: 350-371.

[692] Ibid McNeill p.23-24.

[693] Ibid McNeill p.31.

[694] Ibid McNeill p.245.

[695] Ibid McNeill p.246.

[696] Ibid McNeill p.246.

[697] Ibid McNeill p.246.

[698] Ibid McNeill p.246.

[699] Kevin Tuite (at ibid McNeill p.247). Argument marking in the gestural "verb". MS, Department of Psychology, University of Chicago. Towards a production-based model of gesture. Somiotica.

[700] Ibid McNeill p.248.

[701] Ibid McNeill, p.331.

[702] Kimura D. 1976 "The neural basis of language qua gesture." In H. Whitaker and H.A. Whitaker (eds.) Studies in neurolinguistics, N of 2, 145-156, New York: Academic Press.

[703] Ibid McNeill, p.332.

[704] Trevarthen 1987b. Split-brain and the mind. In R.L. Gregory (ed) The Oxford Companion to the Mind 740-747. Oxford: Oxford University Press.

[705] Gazzaniga M.S. 1970 The Bisected Brain, New York: Appleton-Century-Crofts.

[706] Ibid McNeill p.344.

[707] Ibid McNeill p.344.

[708] "Speak the speech. I pray you as I pronounced it to you, trippingly on the tongue; but if you mouth it, as many of your players do, I had as life the town-crier spoke my lines. Nor do not saw the air too much with your hand, thus; but use all gently; for in the very torrent of your passion you must acquire and beget a temperance that may give it smoothness..."

"It offends me to the soul to hear a robustious periwig-pated fellow tear a passion to tatters, to split the ears of the groundlings, who, for the most part, are capable of nothing but inexplicable dumb-shows and noise; I would have such a fellow whipped; it out herods. Herod; pray you, avoid it."

"Be not too tame, neither, but let your own discretion be your tutor; swift the action to the word, the word to the action; with this special observance, that you o'er step not the modesty of nature;"

For anything so overdone is from the purpose of playing, whose end, is to hold 'twere the mirror up to nature; to show virtue her own feature, scorn her own image, and the very age and body of the time, his form and pressure."

"Now this overdone, though it make the unskilful laugh, cannot but make the judicious grieve; the censure of which one must in your allowance outweigh a whole theatre of others oh, there be players that I have seen play, and heard others praise and that highly, not, to speak it profanely."

[709] Professor Rod Walters.

[710] "Plum in Your Mouth," Andrew Taylor, Harper Collins 2006.

[711] (Alun Richards First XV ibid p61).

[712] John Keegan—A History of Warfare, Hutchinson London 1993, p.139.

[713] Ibid p.155.

[714] Ibid p.156.

[715] Competitive Sport's Imitation of War: Imaging the Completeness of Virtue. Journal of the Philosophy of Sport, 2002, XXIX, 16-37 © 2002 by the International Association for the Philosophy of Sport by Norman Fischer. N. Fischer nfischer@cau.edu is with the Department of Religion and Philosophy at Clark Atlanta University, Atlanta, GA 30314.

[716] (22: p.123).

[717] (2: 148b10-15).

[718] (6: p.111).

[719] (22: p.223).

[720] (19: 403d8-404b7).

[721] (1: 1117b-1-7).

[722] (1: 1117b8).

[723] (1: 1117b16).

[724] (1: 1117a32-35).

[725] (16).

[726] (15).

[727] (24: pp.180-183).

[728] Butcher, R., and Schneider, A. "Fair Play As Respect for the Game." *Journal of the Philosophy of Sport* 25: 1-22, 1998.

[729] Corlett, J. "Virtue Lost: Courage in Sport." *Journal of the Philosophy of Sport* 23: 45-57, 1996.

[730] (12, pp.19-20).

[731] (1:1177b1-26).

[732] (3: 1258a18-b7).

[733] (24: pp.179-183).

[734] (1: p.225; 14, pp.75-78).

[735] Dikkers, S. (Ed.) *Our Dumb Century*. New York: Three Rivers Press, 1999.

[736] (24: pp.39-41).

[737] (15).

[738] (3: 1257b10-58a13).

[739] Hegal, G.W.F. *Hegel's Philosophy of Mind* (Being Part Three of the *Encyclopaedia of the Philosophical Sciences*, 1830; Wallace, W., Trans.). Oxford, UK: Clarendon Press, 1971.

[740] Hyland, D. "And That Is the Best Part of Us..." *Journal of the Philosophy of Sport* 5:36-49, 1978.

[741] Hyland, D. "Competition and Friendship". *Journal of the Philosophy of Sport* 4:27-38, 1977.

[742] (12: pp.3-8).

[743] Hyland, D. "Opponents, Contestants and Competitors: The Dialectic of Sport". *Journal of the Philosophy of Sport* 11:63-70, 1984.

[744] Kojeve, A. *Introduction to the Reading of Hegal* (Nichols, J. Trans.). Ithaca, NY: Cornell University Press, 1969.

[745] (3: 1257b10-58a13).

[746] (1: 1117b16-20).

[747] Methany, E. "The Symbolic Power of Sport". In: *Sport and the Body.* Philadelphia: Lea and Febiger, 1979, pp.213-227.

[748] Nietzsche, F. "Homer's Contest". In *The Portable Nietzsche* (Kaufmann, W., Trans.). New York: Penguin, 1959, pp.32-39.

[749] (1: 1095b27-28).

[750] (22: 225-226).

[751] (8: par. 562).

[752] Kahneman—Thinking Fast and Slow, London, Penguin, 2012, pp.24-25).

[753] Why We Fight—Mike Martin, 2018, C. Hurst & Co (publishers Ltd)

[754] Ibid Mike Martin, p.3.

[755] Ibid Martin, p.4.

[756] Create Space, Independent Publishing Platform, 2017, p.26.

[757] David Brown, The Times, November 25th 2022.

[758] Times, 25th November 2022.

[759] Darwin's Athletes: How sport has damaged Black America and preserved the Myth of Race by John M. Hoberman Houghton Mifflin, 1997. See Review Essay—Journal of Philosophy of Sport 1999 XXVI—pp.105-112—John Valentine.

[760] Journal of the Philosophy of Sport, 1999, XXVI, 105-112 © 1999 by the Philosophic Society for the Study of Sport. Review Essay: Darwin's Athletes: How Sport Has Damaged Black America and Preserved the Myth of Race by John M. Hoberman (Houghton Mifflin, 1997), John Valentine, Shawnee State University.

[761] My references throughout are to Mariner paperback cited Mariner page number (whatever that is) so for example the endnote for page 3 would be mariner ibid 3.

[762] Ibid Hoberman-Mariner p.9.

[763] Ibid p.9.

[764] Ibid p.17.

[765] Ibid Hoberman-Mariner p.6.

[766] Ibid p.9.

[767] Ibid p.53.

[768] Ibid p.46.

[769] Ibid p.3.

[770] Ibid p.46.

[771] Ibid p.20.

[772] Ibid pp.9-10.

[773] Ibid p.99.

774 Ibid p.240.

775 Ibid p.230.

776 Origin of Species Chicago: William Benton 1952 1: p.71.

777 Ibid p.60.

778 Ibid xxvi.

779 Ortega y Gasset J "Man the Technician" In Towards a Philosophy of History, Ortega Y Gasset J (Ed.) Urbana and Chicago: University of Illinois Press 2002 [1941]

780 David Inglis "Meditations on Sport: On the Trail of Ortega y Gasset's Philosophy of Sportive Existence" Journal of the Philosophy of Sport 2004 XXXI, pp78-96 at p.81

781 Ibid pp.129-139

782 Ibid p.132

783 Ortega y Gasset, J Historical Reason New York W.W. Norton 1984 [1940] p.24

784 Wittgenstein L Philosophical Investigations. Oxford: Blackwell 1958

785 Winch P The idea of a Social Science and Its Relation to Philosophy. London Routledge 1958

786 Ortega y Gasset J "The Sportive Origin of the State" In *Towards a Philosophy of History*, Ortega y Gasset J. (Ed.). Urbana and Chicago: University of Illinois Press 2002 [1941], p.17

787 (ibid p.19)

788 Ortega y Gasset J. "The Sportive Origin of the State". In *Towards a Philosophy of History, Ortega y Gasset J. (Ed.).* Urbana and Chicago: University of Illinois Press, 2002 [1941] p.18

789 Marx, K *Economic and Philosophic Manuscripts of 1844.* Moscow: Progress 1981 [1844]

790 Ortega Y Gasset J. "The Sportive Origin of the State" In *Towards a Philosophy of History,* Ortega y Gasset J.
(Ed.). Urbana and Chicago: University of Illinois Press 2002 [1941] p.31

791 Ortega y Gasset J. "The Sportive Origin of the State" In *Towards a Philosophy of History Ortega y Gasset J.* (Ed). Urbana and Chicago: University of Illinois Press 2002 [1941] p.32

792 Ortega y Gasset J. "The Sportive Origin of the State". In *Towards a Philosophy of History, Ortega y Gasset J. (*Ed). Urbana and Chicago: University of Illinois Press 2002 [1941] p.25

[793] Ortega y Gasset J. "The Sportive Origin of the State". In *Towards a Philosophy of History, Ortega y Gasset J.* (Ed). Urbana and Chicago: University of Illinois Press 2002 [1941] p.28

[794] Ortega y Gasset J. "The Sportive Origin of the State". In *Towards a Philosophy of History, Ortega y Gasset J.* (Ed). Urbana and Chicago: University of Illinois Press 2002 [1941]

[795] Ibid p.32

[796] Ibid Inglis p.87

[797] From Sport as Strategic Action: A Habermasian Perspective: Sport, Ethics and Philosophy Vol.1, No.1 April 2007 page 33 by Andrew Edgar, Jursen Habermas was a German philosopher and social theorist.

[798] (page 36 ibid)

[799] (Wittgenstein 1978 Philosophical Investigations Chpt 7)

[800] (see T.W. Adorno 1984 Aesthetic theory London; Routledge & Kegan Paul)

[801] Boxing: The Sweet Science of Constraints, Joseph Lewandowski. The author jdlewandowski@fulbrightweb.org is with the Dept. of Philosophy, University of Central Missouri, Warrensburg, MO 64093.

[802] Elster, J. *Ulysses and the Sirens.* Cambridge, UK: Cambridge University Press, 1984, p.113.

[803] Of course, in stopping their ears Ulysses also forces his crew to be bound. The forced binding of others is an explicit theme of Elster's analysis of politics and constitution making (4) but beyond the narrow scope of the present inquiry.

[804] Elster, J. *Ulysses Unbound: Studies in Rationality, Precommitment, and Constraints.* Cambridge, UK: Cambridge University Press, 2000, p.176.

[805] Elster, J. *Ulysses Unbound: Studies in Rationality, Precommitment, and Constraints.* Cambridge, UK: Cambridge University Press, 2000, p.281.

[806] Elster, J. *Ulysses Unbound: Studies in Rationality, Precommitment, and Constraints.* Cambridge, UK: Cambridge University Press, 2000, p.200.

[807] Elster, J. *Ulysses Unbound: Studies in Rationality, Precommitment, and Constraints.* Cambridge, UK: Cambridge University Press, 2000, p.190.

[808] In fact, Elster, develops a more elaborate typology of constraints than I, for reasons of space and immediate relevance, shall consider here.

[809] This distinction between constitutive and regulative rules is, in my view, most fully elaborated in (14). It also has directly relevant antecedents in Suits's work (16) in the philosophy of sport. I shall consider Suits's work briefly at the close of this section.

[810] Elster, J. *Ulysses Unbound: Studies in Rationality, Precommitment, and Constraints.* Cambridge, UK: Cambridge University Press, 2000, p.177.

[811] Elster, J. *Ulysses Unbound: Studies in Rationality, Precommitment, and Constraints*. Cambridge, UK: Cambridge University Press, 2000, p.201.

[812] As Keating (6) unfairly suggests in the following passage:
In essence, play has for its direct and immediate end joy, pleasure, and delight which is dominated by a spirit of moderation and generosity. Athletics, on the other hand, is essentially competitive activity, which has for its end victory in the context and which is characterised by a spirit of dedication, sacrifice, and intensity. Keating, J. *Competition and Playful Activities*. Washington, DC: University Press of America, 1978, pp.43-44.

[813] Of course the goals of utility maximisation and "pleasure maximisation" can coincide in individual "player-athletes". As Feezell (5) points out in his objections to Keating, surely some athletes manage to fuse the pleasurable spirit of play and the winning spirit of competition. Their "purpose is to win the contest and to experience the playful and aesthetic delights of the experience". Feezell, R. *Sport, Play and Ethical Reflection*. Urbana: University of Illinois Press, 2001, p.89. but even here, such a complex, often contradictory set of purposes is not without deep tensions, as anyone who has played a game of pickup basketball against an elite former college ball player soon discovers.

[814] Clearly, there are obvious exceptions to this generalisation. Snowboarding, for example, is a case where a new athletic genre—a new set of constitutive constraints—was created by athletes. But such an exception, which is itself parasitic on existing constraints in the sport of skiing, does not undermine the central thesis being advanced here. New constraints are simply the introduction of a different *set* of constitutive soft constraints.

[815] Various artists also have "coaches" of a sort. Classical singers often have voice coaches, for example. To the extent that voice coaches are routinely used explicitly *as* constrainers in the ways that one finds coaches used in most sports, singing may also be a rich domain for the application of constraint theory.

[816] Suits, B. "The Elements of Sport." In *Philosophic Inquiry in Sport*, 2nd ed. W.J. Morgan and K.V. Meier (Eds.). Champaign, IL: Human Kinetics, 1995, 8-15, p.11.

[817] Boxers very rarely refuse to spar but some historically have refused to fight.

[818] Boxing and sparring are conceptually different because of the absence of competition in sparring.

[819] Simon R. "Violence in Sports". In *Ethics in Sport*, W.J. Morgan, K.V. Meier, and A. Schneider, (Eds.). Champaign, Il.: Human Kinetics, 2001, 345-356, p.355.

[820] No ordinary Joe, Century 2007.

[821] Mill J.S. on Liberty—Indianapolis IN: Hackett 1998.

[822] Such is the argument advanced in (2).

[823] There are more detailed arguments on this issue in other philosophical literature.

[824] Sugden J. Boxing and Society 1966. Manchester University Press, p.183.

[825] Ibid Why I am so wise—Ecce Homo.

[826] Outcome of Philosophy ibid p.177 ibid p.225.

[827] (ibid p.171).

[828] Russell ibid p.231.

[829] (Russell ibid p.232).

[830] Russell ibid p.232.

[831] Erwin Schrodinger (What is Life? With Mind and Matter and Autobiographical Sketches, Cambridge University Press, Seventeenth Printing 2007 Originally Published 1967, p.95).

[832] Provide reference.

[833] (ibid Schrodinger p.97).

[834] (Schrodinger p.99)

[835] Sigmund Freud quoted by Lionel Trilling "Authenticity and the Modern Consciousness in Commentary," New York, September, 1971, Vol.52, p.39.

[836] Arthur Koestler. The Art of Creation, p.178.

[837] See also "The quick and the dead: when reaction beats intention" Proceedings of the Royal Society Publishing Prg Feb 3 2020".

[838] (Libet *et al* 1979). "Subjective Referral of the Timing for a Conscious Sensory Experience" Brain 102 pp.193-224.

[839] A chart of the approximately millisecond values of some relevant durations.
saying "one, Mississippi"1000msec
a 90 mph fastball travels the 90 feet to home place680msec
unmyelinated fibre, fingertip to brain500msec
speaking a syllable200msec
starting and stopping a stopwatch175msec
a frame of motion picture film42msec
a frame of television33msec
fast (myelinated) fibre, fingertip to brain20msec
the basic cycle time of a neuron10msec
the basic cycle time of a personal computer.0001msec

[840] ibid Penrose 573.

[841] [bid Penrose 577.

[842] (Dennett, 1969 Content and Consciousness: London Routledge & Kegan Paul p.116ff.

[843] Dennett, 1984a *Elbow Room: The Varieties of Free Will Worth Wanting.* Cambridge, MA: MIT Press/A Bradford Book; 1991b "Producing Future by Telling Stories" in K.M. Ford and Z. Pylyshyn, eds, *Robot's Dilemma Revisited: The Frame Problem in Artificial Intelligence.* Ablex Series in Theoretical Issues in Cognitive Science. Norwood, N.J: Ablex).

[844] (Calvin, 1983, 1986) The Throwing Madonna; Essays on the Brain. New York. McGraw-Hill.

[845] (Libet et al., 1979, p.222) "The Timing of a Subjective Experience," *Behavioural and Brain Sciences.* 12, pp.183-185.

[846] (Popper K.R. and Eccles, J.C. 1977. *The Self and Its Brain.* Berlin: Springer-Verlag, p.361).

[847] (1985a, p.534). "Can Machines Think"—M. Shaftoed How We Know: New York; Harper & Row pp.121-145.

[848] (1982, p.241). "How to study Human Consciousness Empirically, or Nothing Comes to Mind" Syntheme 59 pp.159-180.

[849] Dennett [1987, p.785]. *The International Stance.* Cambridge, MA: MIT Press/A Bradford Book. "The Logical Geography of Computational Approaches: A View from the East Pole" in M. Harnish and M. Brand, eds, *Problems in the Representation of Knowledge.* Tucson: University of Arizona Press.

[850] Dennett (1981, 182 ("Reflections" on "Software" in Hofstadter and Dennett, 1981. "Wondering Where the Yellow Went" (commentary on W. Sellar's Carus Lectures), *Monist,* **64**, pp.102-108.

[851] Libet et al. 1979, p.222. "Subjective Referral of the Timing for a Conscious Sensory Experience," *Brain,* **102**, pp.193-224.

[852] Dennett (1985a, p.529). "Can Machines Think:" in M. Shafto, ed., *How We Know.* New York: Harper & Row, pp.121-145.

[853] Agency and Free Will (from Gallagher ibid).

[854] Libet 1999. Do we have free will? *Journal of Consciousness Studies,* 6(8-9): 47-57.

[855] Dennett (1985, "Can Machines Think?" in M. Shafto, ed., *How We Know.* New York: Harper & Row, pp.121-145, "Music of the Hemispheres," a review of M. Gazzaniga, *The Social Brain,* in New York Times Book Review, November 17, 1985, p.53; 2003).

[856] Hacking, I. 1995. The looping effects of human kinds. In D. Sperber, D. Premack, and A.J. Premack (eds.), *Casual cognition: A Multidisciplinary Debate* (pp.351-83). New York: Oxford University Press.

[857] Zhu, J. 2003. Reclaiming volition: An alternative interpretation of Libet's experiments. *Journal of Consciousness Studies*, 10(11):61-77.

[858] Dennett 2003, *Freedom Evolves*. New York: Viking.

[859] ibid 242.

[860] Ibid 242 n.3.

[861] Philosophy in the Flesh. The Embodied Mind and its Challenge to Western Thought. George Lakoff and Mark Johnson 1999, Basic Books p.3.

[862] Ibid Lakoff and Johnson p.4.

[863] Ibid Lakoff and Johnson p.4.

[864] Ibid Lakoff and Johnson p.6.

[865] Ibid Lakoff and Johnson p.8.

[866] Ibid Lakoff and Johnson p.10.

[867] Ibid Lakoff and Johnson p.11.

[868] Ibid Lakoff and Johnson p.13.

[869] Ibid Lakoff and Johnson p.13.

[870] Ibid Lakoff and Johnson p.15.

[871] Ibid Lakoff and Johnson p.17.

[872] Ibid Lakoff and Johnson p.17.

[873] Ibid Lakoff and Johnson p.18.

[874] Ibid Lakoff and Johnson p.18.

[875] Ibid Lakoff and Johnson p.18.

[876] (ibid p.20).

[877] Ibid Lakoff and Johnson p.24.

[878] Ibid Lakoff and Johnson p.24.

[879] Ibid Lakoff and Johnson p.30.

[880] Ibid Lakoff and Johnson p.31.

[881] Ibid Lakoff and Johnson p.37.

[882] Narayanan 1997 Embodiment in Language Understanding. Sensory—Motor Representations for Metaphoric Reasoning About Event Description.

[883] Ibid Lakoff and Johnson p.38.

[884] Ibid Lakoff and Johnson p.43.

[885] Visual Perception and Action in Sport, A.M. Williams, K. Davids, and J.G. Williams, Taylor & Francis 1999.

[886] Glencross and Cibich 1977; 'A decision analysis of games skills'. *Australian Journal of Sports Medicine* 9: 72-5. Bahill and LaRitz T. (1984) 'Why can't batters keep their eyes on the ball?," *American Scientist* 72: 249-53.

[887] Regan, D.M. (1986) 'The eye in ball games: Hitting and catching', *Proceedings of Conference in Vision and Sport*, Haarlem: De Vriesborch.

[888] Acquiring Ball Skill A Psychological Interpretation by H.T.A. Whiting, London & Bell & Sons Ltd 1969 p.1.

[889] Swanton, E.W. (1967) *Obituary of S.F. Barnes*. London: *Daily Telegraph*, Dec. 27th 1967.

[890] Belisle, J.L. (1963). Accuracy, reliability and refractoriness in a coincidence anticipation task. *Res. Quart*, 34, 3 [40].

[891] Mowrer, O.H. (1941) Preparatory set (expectancy)—further evidence of its central locus. *F. Exp. Psych.*, 28, 116-133 [41].

[892] Lange, L. (1888) Neue experimente uber den Vorgang dereinfachen reaction auf Sinnesreizen. *Philos. Stud.,* 4, 479-510 [41].

[893] Hick, W.E. & Bates, J.A.V. (1948) *The human operator of control mechanisms*. Ministry of Supply permanent records of research and development, No.17 [41].

[894] Knapp, B. *et al.* (1961) Simple reaction time of selected top-class sportsmen and research students. *Res Quart*, 32, 3, 409-412 [41].

[895] Slater-Hammel, A.T. & Stumpner, R.L. (1950) Batting reaction time. *Res Quart.*, 21, 353-357 [41].

[896] Miller, R.G. & Shay, C.T. (1964) Relativity of reaction time to the speed of a softball. *Res Quart.,* 35, 433-437 [41, 42].

[897] Poulton, E.C. (1965) Skill in fast ball games. *Biology and Human Affairs*, 31, 1, 1-5, [11, 38, 41].

[898] Eastwood, Entwhistle, Gill & Stephenson (1968). Relation of reaction time and movement time within a physical education concept. Advance Diploma in Physical Education Thesis, University of Leeds, p.43.

[899] Hay, J.G. (1988 'Approach strategies in the long jump', *International Journal of Sport Biomechanics* 4: 114-29.

[900] Bootsma R.J. and Weiringen, P.C.W. (1990) 'Timing an attacking forehand drive in table tennis', *Journal of Experimental Psychology: Human Perception and Performance* 16: 21-9.

[901] From Motor Learning and Human Performance by Robert N. Singer Second Edition MacMillan Publishing to New York 1975.

[902] Bruce, Green and Georgeson 1996 *Visual Perception: Physiology, Psychology and Ecology*. 3rd ed, London: Lawrence Erlbaum.

[903] Lee, D.N. and Lishman, R. 1975 'Visual proprioceptive control of stance', *Journal of Human Movement Studies* 1: 87-95; Forsberg H. and Nashner, L. (1982) 'Ontogenetic development of postural control in man: Adaptation to altered support and visual conditions during stance', *Journal of Neuroscience* 2: 545-52.

[904] Bootsma, R.J. (1988) 'The timing of rapid interceptive actions: Perception-action coupling in the control and acquisition of skill', PhD thesis, Amsterdam: Free University Press, Fitch, H. and Turvey M.T. (1978) 'On the control of activity: Some remarks from an ecological point of view', in D. Landers and R. Christina (eds.) *Psychology of Motor Behaviour and Sport*, Champaign II: Human Kinetics; Lee 1980s; Turvey, M.T. (1990) 'Coordination', *American Psychologist* 45: 938-53.

[905] Lee, D.N. (1978) 'The functions of vision', in H. Pick and E. Saltzman (eds) *Modes of Perceiving and Processing Information*, Hillsdale NJ. Lawrence Erlbaum; Cutting 1986.

[906] e.g. see Neisser, U.N. (1967) *Cognitive Psychology*, New York: Appleton Press; Marteniuk, R.G. (1976) *Information Processing in Motor Skills*, New York: Holt, Rinehart and Winston; Neumann, O. and Prinz, W. (1990) *Relationships Between Perception and Action: Current Approaches,* Berlin: Sprinter-Verlag; Williams, J.G. *et al* (1992) 'Effects of instruction and practice on ball catching skill: single-subject study of an eight-year-old', *Perceptual and Motor Skills* 76: 392-4.

[907] Silva, J.M. and Hardy, C.J. (1984) 'Precompetitive affect and athletic performance', in W.F. Straub and J.M. Williams (eds) *Cognitive Sport Psychology*, Ithaca NY: Sports Science Associates.

[908] e.g. see Edelman, G (1992) *Bright Air, Brilliant Fire: On the Matter of Mind*, New York, Penguin; van Gelder, 'The dynamical hypothesis in cognitive science', *Brain and Behavioural Sciences.*

[909] See Steier, D.M. and Mitchell, T.M. (eds) (1996) *Mind Matters: A Tribute to Allen Newell*, London: Lawrence Erlbaum.

[910] Newell, A. (1980) 'Physical symbol systems', *Cognitive Science* 4: 135-83; 136.

[911] (ibid: 136).

[912] e.g. Kelso, J.A.S. (1992) 'Theoretical concepts and strategies for understanding perceptual-motor skill: From informational capacity in closed systems to self-organisation in open, non-equilibrium systems', *Journal of Experimental Psychology (General)* 121: 260-1.

[913] Carello, C., Turvey, M.T., Kugler, P.N. and Shaw, R.E. (1984) 'Inadequacies of a computer metaphor', in M. Gazzaniga (ed.,) *Handbook of Cognitive Neuroscience*, New York: Plenum Press.

[914] See Williams *et al.* 1992.

[915] Helmholtz, H. von (1925) *Treatise on Psychological Optics* (English language version translated and edited by J.P. Southall), Rochester NY: Optical Society of America.

[916] e.g. Marteniuk, R.G. (1976) *Information Processing in Motor Skills*, New York: Holt, Rinehart and Winston.

[917] Bruce, V., Green, P.R. and Georgeson, M.A. (1996) *Visual Perception: Physiology, Psychology and Ecology*, 3rd ed, London: Lawrence Erlbaum p.233.

[918] Craik, K.J.W. (1948) 'The theory of human control systems: II. Man as element in a control system', *British Journal of Psychology* 38: 142-8.

[919] Stroud, J. 1955 'The find structure of psychological time', in H. Quastler (ed.) *Information Theory and Psychology*, New York: Free Press, Shallice 1974.

[920] Welford, A.T. (1952) 'The "psychological refractory period" and the timing of high-speed performance: A review and a theory', *British Journal of Psychology* 43: 2-19.

[921] Poulton, E.C. (1957) 'On prediction in skilled movements', *Psychological Bulletin* 54: 467-78; 91965) 'Skill in fast ball games', *Biological Affairs* 31: 1-5.

[922] See Bruce, V., Green, P.R. and Georgeson, M.A. 91996) *Visual Perception: Physiology, Psychology and Ecology*, 3rd ed, London: Lawrence Erlbaum, Gordon, I. (1989) *Theories of Visual Perception*, Chichester: Wiley; Savelsbergh 1990.

[923] Tresilian, J.R. (1991) 'Empirical and theoretical issues in the perception of time to contact', *Journal of Experimental Psychology: Human Perception and Performance* 17: 865-76.

[924] (1992) p.3.

[925] e.g. Whiting H.T.A. *et al* (1970) 'An operational analysis of a continuous ball throwing and catching task', *Ergonomics* 13: 445-54; Whiting H.T.A. and Sharp, R.H., (1974) 'Visual occlusion factors in a discrete ball catching task', *Journal of Motor Behaviour* 6(1): 11-16; Sharp R.H. and Whiting, H.T.A. (1975) 'Information processing and eye movement behaviour in ball catching skill, *journal of Human Movement Studies* 1: 124-31.

[926] e.g. Tyldesley D. and Whiting H.T.A. (1975) 'Operational Timing', *Journal of Human Movement Studies* 1: 172-7; Franks, L.M., Weicker, D. and Robertson, D.G. (1985) 'The kinematics, movement phasing and timing of a skilled action

in response to varying conditions of uncertainty', *Human Movement Science* 4: 91-105.

[927] e.g. Abernethy B. 1981, 1987a 'Anticipation in sport: A review', *Physical Education Review* 10: 5-16; 1987b 'Selective attention to fast ball sports: II Expert-novice differences', *Australian Journal of Science and Medicine in Sport* 19(4): 3-6; Abernethy, B and Russell, D.G. (1984) 'Advance cue utilisation by skilled cricket batsman, *Australian Journal of Science and Medicine in Sport* 16(2): 2-10.

[928] Whiting, H.T.A., Alderson, G.J.K. and Sanderson, F.H. (1973) 'Critical time intervals for viewing and individual differences in performance of a ball-catching task', *International Journal of Sport Psychology* 4: 155-6.

[929] Hubbard, A.W. and Seng, S.N. (1954) 'Visual movements of batters', *Research Quarterly of American Association of Health and Physical Education* 25: 42-57.

[930] Bahill, A. and LaRitz, T. (1984) 'Why can't batters keep their eyes on the ball?', *American Scientist* 72: 249-53.

[931] Tyldesley, D. and Whiting, H.T.A. (1975) 'Operational Timing', *Journal of Human Movement Studies* 1: 172-7.

[932] (1973) p.173.

[933] (1975 ibid p.176).

[934] (ibid p.174).

[935] Woods, D.L. (1990): 178. 'The physiological basis of selective attention: Implications of event-related potential studies', in J.W. Rohrbaugh, R. Parasuraman and R. Johnson (eds) *Event-Related Brain Potentials*, New York: Oxford University Press.

[936] Woods, D.L. (1990): 178. 'The physiological basis of selective attention: Implications of event-related potential studies', in J.W. Rohrbaugh, R. Parasuraman and R. Johnson (eds) *Event-Related Brain Potentials*, New York: Oxford University Press.

[937] Schneider, W., Dumais, S.T. and Shiffrin, R.M. (1984) 'Automatic and control processing and attention', in R. Parasuraman and R. Davies (eds) *Varieties of Attention*, Orlando FFL: Academic Press.

[938] Boutcher, S.H. (1992) 'Attention and athletic performance: An integrated approach', in T.S. Horn (ed) *Advances in sport psychology*, Champaign II: Human Kinetics.

[939] Boucher, S.H. (1992) 'Attention and athletic performance: An integrated approach', in T.S. Horn (ed) *Advances in sport psychology*, Champaign II: Human Kinetics; Nouguier, V., Stein, J.F. and Bonnel, A.M. (1991) 'Information

processing in sport and orienting of attention', *International Journal of Sport Psychology* 22: 307-27.

[940] Keele, S.W. and Hawkins, H. (1982) 'Exploration of individual differences relevant to high skill level', *Journal of Motor Behaviour* 14: 3-23.

[941] For a review, see Gould, D.R. and Krane, V. (1992) 'The arousal-athletic performance relationship: Current status and future directions', in T.S. Horn (ed) *Advances in Sport Psychology,* Champaign II: Human Kinetics.

[942] Hughes, C. (19940 *The Winning Formula*, London: William Collins.

[943] Williams, A.M. and Burwitz, K. (1993) 'Advance cue utilisation in soccer', in T. Reilly, J. Clarys and A. Stibbe (eds) *Science and Football*, vol. II, London: E & FN Spon.

[944] Williams, A.M. Davids, K., Burwitz, L. and Williams, J.G (1994) 'Visual search strategies of experienced and inexperienced soccer players, *Research Quarterly for Sport and Exercise* 65(2): 127-35.

[945] See Rumelhart, D.E. and McClelland, J.L. (eds) (1986) *Parallel Distributed Processing*, Cambridge MA: MIT Press; Rumelhart, D.E. (1989) 'The architecture of the mind: A connectionist approach' in M.I. Posner (ed) *Foundations of Cognitive Science*, Cambridge MA: MIT Press.

[946] Anderson, J.R. (1990) *Cognitive Psychology and Its Implications*, 3rd ed. New York: W.H. Freeman.

[947] Rumelhart, D.E. (1989) 'The architecture of the mind: A connectionist approach', in M.I. Posner (ed) *Foundations of Cognitive Science*, Cambridge MA: MIT Press.

[948] Glencross, D.J. (1978) 'Control and capacity in the study of skill', in D.J. Glencross (ed) *Psychology and Sport*, Sydney: McGraw-Hill.

[949] Baumeister, R.F. (1984) 'Choking under pressure: Self-consciousness and paradoxical effects of incentives on skilful performance', *Journal of Personality and Social Psychology* 46: 610-20; Gallway, T. and Kreigal, R. (1977) *Inner Skiing*, New York: Random House.

[950] e.g. Abernethy, B. (1993a) 'Attention', in R.N. Singer, M. Murphey and L.K. Tennant (eds) *Handbook of Research on Sport Psychology*, New York: Macmillan.

[951] Fitts, P.M. and Posner, M.I. (1967) *Human Performance*, Belmont CA: Books/Cale.

[952] e.g. Shiffrin, R.M. and Schneider, W. (1977) 'Controlled and automatic human information processing: II Perceptual learning, automatic attending, and a general theory', *Psychological Review*, 84: 127-90; Schneider, W., Dumais, S.T. and Shiffrin, R.M. (1984) 'Automatic and control processing and attention',

in R. Parasuraman and R. Davies (eds) *Varieties of Attention,* Orlando FL: Academic Press.

[953] Logan, G.D. (1988) 'Towards an instance theory of atomization', *Psychological Review* 95: 492-527.

[954] Page 238 Roger Bannister The Four Minute Mile (New York: Dodd, Mead 1955).

[955] "Wraparound; High: What's Up There" Harpers Vol.247 (Oct 1973) (pp.3-10).

[956] David Hemery Another Hurdle (New York: Taplinger 1976).

[957] (Ravizza, Kenneth "A Subjective Study of the Athlete's Greatest Moments in Sport," paper presented at Mouvement, Actes du 7ᵉ symposium en apprentisseage psycho-moteur et psychologie du sport, Octobre 1974, Quebec, Canada p.402).

[958] Quoted in Sheinberg, Lawrence, "Finding the Zone," New York Times Sunday Magazine, April 9 1989, p.35.

[959] (see p.198 Margharita Laski, Ecstasy (Bloomington University of Indiana Press 1961).

[960] Collins, Larry, and Downhapiere, Or I'll Dress You in the Morning (New York: Simon and Schuster, 1968).

[961] (Murphy, Michael and Joan Brodie: I Experience a Kind of Clarity, Intellectual Digest Vol.3, No.5 (Jan 1973) (pp.19-22)

[962] Arnold Palmer Go for Broke p.371 (New York, Simon and Schuster, 1973).

[963] (ibid The Psychic Side of Sports at page 27).

[964] The Joy of Sports (New York, Basic Books, 1976) at page 164.

[965] Zen The Art of Archery (New York: Pantheon 1953).

[966] (Johannson, Ingemar "New Challenger Scouts a Fight," Life Vol.46, no.19 (May 11ᵗʰ 1959) page 40).

[967] (Hunter, Catfish as told to George Vass "The Game I'll Never Forget," Baseball Digest Vol.32, no.6 (June 1973) pp.35-37).

[968] (ibid pages 39-40).

[969] (Neumann, Randy "Randy Neumann," Sport Vol.58 (July 1974) pp.85-89).

[970] (Neal, Patsy, Sport and Identity (Philadelphia: Dorrance, 1972).

[971] (Libby, Bill Purnelli (New York; Duton 1969).

[972] (Murphy, Michael and John Brodie "I Experience a Kind of Clarity," Intellectual Digest, Vol.3, no.5 (Jan 1973) pp.19-22).

[973] (Harriw, Dorothy V. "Sports Science: The Happy Addict" Women Sports Vol.5, no.1 (Jan 1978), p. 53).

[974] The World of Rugby (1979)—John Reason and Carwyn James.

[975] Fields of Praise.

[976] Ibid 603.

[977] From Frank Keating Another Time, Another Planet—from The Picador Book of Sport Writing 1996, p.334-336.

[978] The World of Rugby (1979).

[979] From Gerald Davies (The Times) Friday February 18th 2008.

[980] Sport/New York: Bartholomew House 1960 at pp.5.22.

[981] Netto, Aranjo and Claudio Melloe Souza "King of the Booters" Reader's Digest, Vol.43 (Oct 1964) pp.203-209.

[982] ("Scorecard: The Sporting Look," Sports Illustrated, Vol.31 (Nov 24th 1969, p.14).

[983] (Furlong, William Barry "The Fun in Fun" Psychology Today, Vol.10, no.1 (June 1976, p.35)).

[984] (Houts, Jo Ann, "Feeling and Perception in the Sport Experience," Journal of Health, Physical Education and Recreation, vol. III, no.8 (Oct 1972), pp.71-72).

[985] (Manso, Peter Vroom! Conversation with Grand Prix Champions (New York: Funk and Wagnalls 1969).

[986] (Moore, Kenny, "A Night for Stars, Both Born and Reborn" Sports Illustrated Vol.46 (May 23, 1977 pp.38-49).

[987] (Murphy, Michael and John Brodie, "I Experience a Kind of Clarity" Intellectual Digest Vol.3, no.5 (Jan 1973) pp.19-22).

[988] (Saal, Herbert "The Great Leap," Newsweek Vol.84, no.7 (Aug 12, 1974) p.84).

[989] (Floyd, Keith "Of Time and the Mind," Fields Within Fields Within Fields no.10 (Win 1973-1974) pp.47-57).

[990] (ibid p.213-214).

[991] (ibid page 52).

[992] Arnold Beisser, The Madness of Sports, Appleton Century Crofts 1967, p.166.

[993] (ibid p.167).

[994] (Quoted in the Sunday Times Nov 4 1973).

[995] (ibid).

[996] From Tackling Life Striving for Perfection, Headline Publishing Group 2008 Johnny Wilkinson with Steve Black at pages 230-231.

[997] Also at ibid p.173.

[998] (Johnny Wilkinson ibid p.21).

[999] Inside the Mind of the Grand Prix Driver by Christopher Hilton (2001) p.20.

[1000] p.162.

[1001] 2008 Ingram International.

[1002] Ernest Ross in The Psychobiology of Mind-Body Healing, p.53.

[1003] Wm Wenger and Richard Poe—The Einstein factor.

[1004] Randolph Feezell Sport, Play and Ethical Reflection Chicago, University of Illinois Press, 2004 (2006 Paperback Edition) p. xiii
See Christopher Lasch—The Culture of Narcissism (New York; Warner Books 1979).

[1005] Paul Weiss, *Sport: A Philosophic Inquiry* (Carbondale: Southern Illinois University Press 1969) 217.

[1006] See, for example, the following useful collections: Ellen Gerber and William Morgan, eds., *Sport and the Body: A Philosophic Symposium* 2d ed. (Philadelphia: Lea and Febiger, 1979); Robert G. Osterhoudt ed. *The Philosophy of Sport: A Collection of Original Essays* (Springfield III: Charles C Thomas 1973); William Morgan and Klaus Meier eds., *Philosophic Inquiry in Sport* [Champaign III: Human Kinetics 1988, 1995); David Vanderwerken and Spencer Wertz, eds. *Sport Inside Out: Readings in Literature and Philosophy* (Fort Worth: Texas Christian University Press, 1985); William Morgan Klaus Meier and Angela Schneider eds. *Ethics in Sport* (Champaign III: Human Kinetics 2001); and M. Andrew Holowchak ed., *Philosophy of Sport; Critical Readings, Crucial Issues (*Upper Saddle River, N.J. Prentice Hall 2002).

[1007] Michael Novak *The Joy of Sports* (New York: Basic Books, 1976) xiv

[1008] Weiss *Sport* 4

[1009] *Ibid*

[1010] *Ibid*10

[1011] Ibid Feezell

[1012] Ibid.3

[1013] Ibid 12

[1014] Ibid 13

[1015] Ibid Feezell

[1016] Ibid.14

[1017] Ibid Feezell p9

[1018] Keith Algozin "Man and Sport" *Philosophy Today* 20 no.3 (fall 1976): 190

[1019] Paul Kuntz calls this theory of sport "pleasure in play" in "Paul Weiss". 173

[1020] Ibid Feezell p.11

[1021] Weiss *Sport* 24

[1022] *Ibid* "Play, Sport and Game" chapter 9

[1023] Huizinga *Homo Ludens* 3

[1024] *Ibid* 13

[1025] *Ibid* 28

[1026] Caillois *Man, Play and Games* 3

[1027] Weiss, *Sport* 134

[1028] Ibid Feezell p.13

[1029] See Kenneth Schmitz, "Sport and Play: Suspension of the Ordinary," in *Sport and The Body,* ed. Gerber and Morgan, 22-29. The essay also appears in *Philosophic Inquiry in Sport,* ed. Morgan and Meier 29-38

[1030] Ibid Feezell p15

[1031] George Sheehan, *Running and Being: The Total Experience (New York:* Warner Books 1978)

[1032] *Ibid*

[1033] Ibid 18

[1034] Ibid 17

[1035] *Ibid 19*

[1036] Ibid Feezell p.23

[1037] *Ibid 37*

[1038] Ibid Feezell

[1039] *Ibid* 91

[1040] *Ibid* 92

[1041] Historical Review of Pennsylvania.

[1042] ibid Feezell, p.48.

[1043] Albert Camus, *The Myth of Sisyphus,* trans. Justin O'Brien (New York: vintage Books, 1955), 89.

[1044] Richard Taylor, *Good and Evil (New* York: Macmillan, 1970), 257-58.

[1045] Caillos, *Man, Play and Games,* 5-6.

[1046] Robert Osterhoudt, "The Term 'Sport': Some Thoughts on a Proper Name," in *Sport and the Body,* ed. Gerber and Morgan, 5.

[1047] ibid Feezell, p.50.

[1048] Taylor, *Good and Evil,* 260.

[1049] Taylor also discusses this possibility in *Good and Evil,* 259-60.

[1050] ibid Feezell, p.51.

[1051] Ibid Feezell, p.51

[1052] Camus, *Myth of Sisyphus,* 21.

[1053] Thomas Nagel, 'The Absurd', in *Mortal Questions* (New York: Cambridge University Press, 1979), 13.

[1054] Ibid

[1055] *Ibid,* 14.

[1056] Ibid Feezell, p52

[1057] Ibid, 15

308 ibid Feezell, p.53.

[1059] Ibid 19-20

[1060] ibid Feezell, p.55.

[1061] ibid Feezell, p.55.

[1062] ibid Feezell, p.55.

313 Nagel, *'The Absurd'*, 20.

314 Quoted ibid Feezell, p.58.

315 Thomas Nagel, *Mortal Questions* (Cambridge, UK: Cambridge University Press, 1979). I will refer to this book as *MQ* and cite all specific references to it in the body of the text.

[1066] Thomas Nagel, *The View from Nowhere* (New York: Oxford University Press, 1986). I will refer to this book as *VN* and cite all specific references to it in the body of the text.

[1067] See this essay in *Mortal Questions* 196-213

[1068] Ibid Feezell. p.59

[1069] Ibid Feezell p.60

[1070] Ibid Feezell, p.63

[1071] See chapter 7, "Freedom," in Nagel, *The View From Nowhere* and "Moral Luck" in *Mortal Questions.*

[1072] See Nagel's widely discussed "What is it Like to be a Bat" in *Mortal Questions.*

[1073] Ibid Feezell, p.64

[1074] Ibid Feezell, p.65

[1075] See Nagal, 'The Absurd'.

[1076] Ibid Feezell p.66

[1077] The judgments about sports that we might make from a relatively "objective" perspective—remember that the distinction is a matter of degree—are not unitary. I do not believe that sports are reduced to absurdity based on all objective views of it. Later I argue that the development of moral character in sport represents a form of "objective re-engagement" after reflective detachment has seemed to undermine the objective significance of sports participation. Some kind of objective viewpoint can obviously recognise valuable aspects of sports participation, including psychological, physical and social benefits, as well as the intrinsic value of non-instrumental activities. Nevertheless, at some point objectivity generates a perspective from which judgments about the relative triviality of sports participation are occasioned, especially when certain aspects are emphasised to the exclusion of others. Likewise, subjectivity in sports is a complex phenomenon. It includes intersubjective elements insofar as sports are social practices that involve shared goods, as well as more subjective elements like the joys of playful exuberance, personal well-being, and a sense of achievement.

[1078] Ibid Feezell

[1079] Ibid, 11

[1080] Huizinga, *Homo Ludens.* See also Schmitz, "Sport and Play."

[1081] Ibid Feezell p.70

[1082] Ibid Feezell, pp.70-71

[1083] Both Nagel and Joel Feinberg offer useful comments on irony. Nagel recommends an ironic view of life in 'The Absurd', while Feinberg comments favourably on this suggestion in "Absurd Self-Fulfilment" in *Freedom and Fulfilment*: *Philosophical Essays* (Princeton: Princeton University Press, 1994). The quote comes from Feinberg's essay.

[1084] Ibid Feezell, p.74

[1085] The reference is to the appearance of the article in Philosophy and the Human Condition, ed. Tom Beauchamp and Joel Feinberg (Englewood Cliffs, N.J.: Prentice-Hall, 1984) 601.2

[1086] Ibid Feezell, p.75

[1087] Ibid Feezell, p.76

[1088] See Clifford and Feezell, *Coaching for Character* chapter 3

[1089] See George Will's discussion in the conclusion to *Men at Work: The Craft of Baseball (*New York: Macmillan, 1990)

[1090] Ibid Feezell, p.77

[1091] Ibid Feezell, p.77

[1092] A Short Treatise on The Great Virtues—William Heinemann, London 2022, André Comte-Sponville.

[1093] Ibid Comte-Sponville p.45.

[1094] André Comte-Sponville ibid p.48.

[1095] André Comte-Sponville ibid p.48.

[1096] Ibid p.49.

[1097] Ibid Comte-Sponville, ibid, p.51.

[1098] Ibid Comte-Sponville, ibid, p.51.

[1099] Jankelevitch, Les Vertus et Clamour in Champs-Flammaria in edition 1986.

[1100] Aristotle's Magna Moralia.

[1101] Ibid Les Virtus et Clamour, Vol. 1, p.110.

[1102] Spinoza, The Ethics III, p.58-60, p.186-87.

[1103] Aristotle Niromachean Ethics III.

[1104] Francois Rabelaid, Gargantua Chpt.4 in the Complete Works of Francois Rabelaid, University of California Press, 1991, p.100).

[1105] E.H. Erikson—The Roots of Virtue—George Allen & Unwin by Julia Huxley, 1961, pp.147-165.

[1106] New Scientist, 2012, p.03—Jonathan Haidt.

[1107] Ibid E.H. Erikson p.149.

[1108] Ibid Erikson p.150.

[1109] Ibid Erikson, p.151.

[1110] French T. The Integration of Behaviour, Chicago, University of Chicago Press, 1952.

[1111] Abraham Cowley.

[1112] Richard Crashaw.

[1113] Ibid, The Roots of Virtue, p.157.

[1114] Ibid The Roots of Virtue, p.157.

[1115] William Barrett "Irrational Man".

[1116] Ibid The Roots of Virtue, p.157.

[1117] Beyond Good and Evil, Nietzsche, 1886.

[1118] Animal Automation.

[1119] Institiones Oratoria 18.14.

[1120] Meditations ix 16.

[1121] "Thoughts on Various Subjects"—Swift.

[1122] Viii 20.

[1123] Edward Young, 1684-1765, Night Thoughts Line 795.

[1124] See Gregory E. Pence, "Recent work on Virtues" *American Philosophical Quarterly* 21 no.4 (Oct.1984): 281-97 for an informative overview. A more recent collection, *Virtue Ethics, ed.* Roger Crisp and Michael Slote (New York: Oxford University Press, 1997), contains many of the influential pieces on virtue ethics.

[1125] MacIntyre, *After Virtue,* 196 Bloomsbury Academic (1981)

[1126] *Ibid* 8. See chapter 2

[1127] Ibid Feezell p.126

[1128] Ibid 117

[1129] Ibid 181

[1130] *Ibid* 187

[1131] *Ibid 187,* 190. It is interesting to note that MacIntyre's notion of a practice is helpful in distinguishing a game from a sport.

[1132] MacIntyre *After Virtue* 190

[1133] *Ibid.* The point is wonderfully exemplified in an anecdote described by David Halberstam in *Summer of '49* (New York; William Morrow 1989), 175. Ted Williams' passionate devotion to the art of hitting caused him to give tips to opposing players. The Boston owner, Tom Yawkey, asked him to stop helping the competition. Williams is quoted as having responded to his owner, "Come on…the more hitters we have in this game, the better it is for the game. Listen, when you're coming towards the park and you hear a tremendous cheer, that isn't because someone has thrown a strike. That's because someone has hit the ball." William was interested in a shared internal good; Yawkey's interest was in external goods.

[1134] *Ibid* 191

[1135] Ibid Feezell, p.128

[1136] Pascal Pensees, no.534.

[1137] *Ibid* 84

[1138] *Ibid* 87

[1139] *Ibid* 89

[1140] Ibid Feezell p.131

[1141] MacIntyre, *After Virtue,* 194

[1142] Ibid Feezell, p. 132

[1143] Ibid Feezell p.132

[1144] *Ibid* 191

[1145] Christopher Lasch, *The Culture of Narcissism* (New York: Warner Books, 1979), 194–97.

[1146] *Ibid,* 191

[1147] *Ibid 195* Lasch describes sport as "splendid futility."

[1148] Ibid Feezell p.133

[1149] Ibid Feezell, p.133

[1150] Ibid 614

[1151] *Ibid*

[1152] *Ibid*

[1153] *Ibid 615*

[1154] *Ibid*

[1155] "What is your dangerous idea," ed. John Brockman—Simon Schuster Ltd, 2006.

[1156] Ibid Feezell p.141

[1157] In *Quandaries and Virtues* (Lawrence: University of Kansas Press, 1986), Edmund Pincoffs distinguishes between instrumental virtues and non-instrumental virtues. Non-instrumental virtues include aesthetic virtues (both noble, e.g. dignity and charming, e.g. wittiness), meliorating virtues (including mediating virtues, e.g. tolerance, temperamental virtues e.g. cheerfulness, and formal virtues, e.g. politeness, and moral virtue. See chapter 5. In relation to the question of having character, see his interesting discussion "On Becoming the right Sort" chapter 9

[1158] Ibid Feezell p.141

[1159] Pulvis et Umbra.

[1160] (John Reason and Carwyn James-The World of Rugby 1979).

[1161] Ibid Feezell pp.143-144

[1162] Ibid Feezell

[1163] Ibid Feezell, p145

[1164] Ibid Feezell, p.145

[1165] See Alfie Kohn's critique of competition in *No Contest: The Case Against Competition* (Boston: Houghton Mifflin, 1986).

[1166] Ibid Feezell p.150

[1167] See Joel Kupperman's discussion of the relation between descriptive claims about the human condition and appropriate or apt attitudes in *Classic Asian Philosophy* (New York: Oxford University Press, 2001),149-59. He argues that the logical relation between facts and ethical conclusions is neither deductive nor inductive, yet there *is* some sort of logical relation involved. One page 158 he says that "certain kinds of guidance in life seem a natural continuation of the picture of the world that is provided. Should this be viewed as logic or as coherent storytelling? Any reader can explore her or his own view on this."

[1168] Ibid Feezell pp.150-151

[1169] William Lad Sessions—Sportsmanship, Honour. Journal of the Philosophy of Sport 2004, pp.47-59.

[1170] Pope, The Rape of the Lock cantoi Line 1.

[1171] Ibid Feezell, p.151

[1172] The Times, page 62, September 14th 2022 "Great rivalries tell us about who we are".

[1173] Ibid Times, Mathew Syed.

[1174] Times obituary, Jan 11th, 2023.

[1175] Franklin Foer—Harper Collins 2004.

[1176] Ibid Feezell

[1177] Ibid Feezell, p.153

[1178] Ibid Feezell, p.154

[1179] Will, *Men at Work* 329

[1180] Jackson *Sacred Hoops,* 162

[1181] *Ibid* 121

[1182] See *Journal of the Philosophy of Sport*, 2007, 34, 39-51 © 2007 International Association for the Philosophy of Sport. Michael Monahan.

[1183] Nietzsche, F. *Thus Spoke Zarathustra*. Trans. W. Kaufmann. New York: Random House, 1995, p.12.

[1184] It should be noted that Nietzsche's corpus, and *Thus Spoke Zarathustra*, in particular, are difficult works to interpret, and there is no settled common interpretation among scholars of Nietzsche's work.

[1185] The focus on *Thus Spoke Zarathustra* is motivated, in part, by the fact that this text is particularly focused on these more "positive" aspects of Nietzsche's ethical thought and by the desire to offer the novice a more coherent picture of a single Nietzschean text.

[1186] Nietzsche, F. *Thus Spoke Zarathustra*. Trans. W. Kaufmann. New York: Random House, 1995, p.12.

[1187] Nietzsche, F. *Thus Spoke Zarathustra*. Trans. W. Kaufmann. New York: Random House, 1995, p.115.

[1188] Nietzsche, F. *Thus Spoke Zarathustra*. Trans. W. Kaufmann. New York: Random House, 1995, pp.47-48.

[1189] Nietzsche, F. *Thus Spoke Zarathustra*. Trans. W. Kaufmann. New York: Random House, 1995, p.56.

[1190] Nietzsche, F. *Thus Spoke Zarathustra*. Trans. W. Kaufmann. New York: Random House, 1995, p.211; p.214.

[1191] Nietzsche, F. *Thus Spoke Zarathustra*. Trans. W. Kaufmann. New York: Random House, 1995, p.198.

[1192] Nietzsche, F. *Thus Spoke Zarathustra*. Trans. W. Kaufmann. New York: Random House, 1995, p.200.

[1193] Nietzsche, F. *Thus Spoke Zarathustra*. Trans. W. Kaufmann. New York: Random House, 1995, p.203.

[1194] Nietzsche, F. *Thus Spoke Zarathustra*. Trans. W. Kaufmann. New York: Random House, 1995, p.205-206.

[1195] Schutte, O. *Beyond Nihilism: Nietzsche Without Masks*. Chicago: University of Chicago Press, 1984, p.6.

[1196] Nietzsche, F. *Thus Spoke Zarathustra*. Trans. W. Kaufmann. New York: Random House, 1995, p.15.

[1197] Of course, there are "practical" effects of these examples: flexibility, strength, balance, mental focus, stamina, and so on. The point is not that there are no practical benefits at all, but rather that these practices transcend the purely practical in very important ways. There are much more direct and efficient ways to gain these benefits, just as there are more direct and efficient ways to protect oneself.

[1198] There is a distinction between martial arts and other practical systems of fighting and sport.

[1199] Nietzsche, F. *Thus Spoke Zarathustra*. Trans. W. Kaufmann. New York: Random House, 1995, pp.107-110 and p.195.

[1200] Nietzsche, f. *The Guy Science*. Trans. J. Nauckhoff. New York: Cambridge University Press, 2001, pp.245-248.

[1201] Nietzsche, F. *Thus Spoke Zarathustra*. Trans. W. Kaufmann. New York: Random House, 1995, pp.107-108 and pp.191-195.

[1202] This is often why people who claim to have "taught themselves" martial arts from books or videos are held in such low regard by practicing martial artists. Not out of some sense of elitism (though this may be so in some cases), but precisely because the practicing martial artists is aware of, even if unable to articulate, precisely this epistemic point.

[1203] Leiter, B *Nietzsche on Morality*, New York: Routledge, 2002, pp.12-21.

[1204] Schacht, R. *Nietzsche*. New York: Routledge, 1983, pp.52-117.

[1205] Nietzsche, F. *Twilight of the Idols*. Trans. R.J. Hollingdale. New York: Penguin Putnam Inc., 1968, pp.45-49.

[1206] Clark, M. *Nietzsche on Truth and Philosophy*, Cambridge, UK: Cambridge University Press, 1990.

[1207] Leiter, B. *Nietzsche on Morality*. New York: Routledge, 2002.

[1208] Poellner, P. *Nietzsche and Metaphysics*. Oxford, UK: Oxford University Press, 1995.

[1209] Schacht, r. *Nietzsche*. New York: Routledge, 1983.

[1210] There is a distinction between the martial arts and practice and training.

[1211] Nietzsche, F. Thus Spoke Zarathustra. Trans. W. Kaufmann. New York: Random House. 1995, p.239.

[1212] Awaken The Giant Within, Simon & Schuster Ltd, 1992, Pocket Books 2001, pp.232-233).

[1213] See *Journal of the Philosophy of Sport*, 2007, 34, 52-67 © 2007 International Association for the Philosophy of Sport. Ann E. Cudd.

[1214] Karl Pribram "Metaphors to Models: the used Analogy in Neuropsychology". Metaphors in the History of Psychology edited by David Leary, p.79.

[1215] Mandelbaum, M. *The Meaning of Sports: Why Americans Watch Baseball, Football, and Basketball, and What They See When They Do*. New York: Public Affairs, 2004.

[1216] Greider, W. *The Soul of Capitalism: Opening Paths to a Moral Economy.* New York: Simon and Schuster, 2003.

[1217] Lakoff, G., and M. Johnson. *Metaphors We Live By.* Chicago: University of Chicago, 1980, p.5.

[1218] Educational uses of metaphors, p.622.

[1219] Lakoff, G., and M. Johnson. *Metaphors We Live By.* Chicago: University of Chicago, 1980, p.22.

[1220] Lakoff, G., and M. Johnson. *Metaphors We Live By.* Chicago: University of Chicago, 1980, ch.5.

[1221] Roberts, T.J. 'It's Just Not Cricket!' Rorty and Unfamiliar Movements: History of Metaphors in a Sporting Practice." *Journal of the Philosophy of Sport*, XXIV, 1997, 67-78, p.70.

[1222] Sypnowich, C. "Law and Ideology." *Stanford Encyclopaedia of Philosophy.* Accessed November 2006. Available at http://plato.stanford.edu/entries/law-ideology/#1.

[1223] Marx, K. *Capital*, vol.1, S. Moore and E. Aveling (trans.). New York: International Publishers, 1967, pp.71-83.

[1224] Suits, B. "The Elements of Sport." In *Philosophic Inquiry in Sport,* W. Morgan and K. Meier (Eds.). Champaign, II: Human Kinetics, 1988, 39-48.

[1225] Boxill, J. "Introduction: The Moral Significance of Sport." In *Sports Ethics,* J. Boxill (Ed.). Malden, MA: Blackwell, 2003, 1-12.

[1226] Palmatier, R.A., and H.L. Ray. *Sports Talk: A Dictionary of Sports Metaphors.* New York: Greenwood Press. 1989.

[1227] Hardaway, F. "Foul Play: Sports Metaphors as Public Doublespeak." In *Sports Ethics*, J. Boxill (Ed.). Malden, MA: Blackwell, 2003: 56-60.

[1228] Suits, B. "The Elements of Sport." In *Philosophic Inquiry in Sport*, W. Morgan and K. Meier (Eds.). Champaign, II: Human Kinetics, 1988, 39-48, pp.46-47.

[1229] MEGA. Kohn, A. *No Contest: The Case Against Competition.* Rev. ed. Boston: Houghton Mifflin, 1992.

[1230] Kohn, A. *No Contest: The Case Against Competition.* Rev. ed. Boston: Houghton Mifflin, 1992, p.9.

[1231] Tännsjö, T. "Is It Fascistoid to Admire Sports Heroes?" In *Values in Sport*, T. Tännsjö and c. Tamburrini (Eds.). London: E&FN Spon, 2000, 9-23.

[1232] Nozick, R. *Anarchy, State, and Utopia.* New York: Basic Books, 1977.

[1233] Boxill, J. "Introduction: The Moral Significance of Sport." In *Sports Ethics,* J. Boxill (Ed.). Malden, MA: Blackwell, 2003, 1-12.

[1234] Butcher, R., and Schneider, A. "Fair Play as Respect for the Game." In *Sports Ethics*, J. Boxill (Ed.). Malden, MA: Blackwell, 2003, 153-171, p.163.

[1235] Hyland, D.A. "Opponents, Contestants, and Competitors: The Dialectic of Sport." *Journal of the Philosophy of Sport*, XI, 1985, 63-70.

[1236] Hyland, D.A. "Opponents, Contestants, and Competitors: The Dialectic of Sport." *Journal of the Philosophy of Sport*, XI, 1985, 63-70, p.66.

[1237] Simon R. *Fair Play: The Ethics of Sport*. Boulder, CO: Westview Press, 2004.

[1238] Lewis, C.T., and C. Short. *A Latin Dictionary*. 9 October 2005. Accessed online through the Perseus Project at www.perseus.tufts.edu.

[1239] Hyland, D.A. "Opponents, Contestants, and Competitors: The Dialectic of Sport." *Journal of the Philosophy of Sport*, XI, 1985, 63-70, p. 64.

[1240] Jensen, M.C., and W.H. Meckling. "Theory of the Firm: Managerial Behaviour, Agency Costs and Ownership Structure." *The Journal of Financial Economics*, 3, 1976, 305-360.

[1241] "Sledging" is an Australian sports term, primarily used in discussions of cricket. It is loosely analogous to the American use of trash-talking, though the latter is not necessarily limited to sports. For the purposes of this discussion, I take sledging generally to be a comment or series of comments, generally negative, taunting, or aggressive in tone, made by one or more competitors in the hope that said comment will affect the mental state of another competitor.

[1242] Ouch…You just Dropped the Ashes, Chuck Summers. For those who do not know, The Ashes are the ashes of English cricket. They are the prize given to the winner of the recurrent series of test matches between England and Australia. The author >scummes@pgrad.unimelh.edu.au> is with the Centre for Applied Philosophy and Public Ethics, University of Melbourne, Melbourne, Victoria 3000, Australia. *Journal of the Philosophy of Sport*, 2007, 34, 68-76 © 2007 International Association for the Philosophy of Sport.

[1243] See R.S. Kretchmar. Kretchmar, R.S. "In Defence of Winning." *In Sports Ethics: An Anthology*, Jan Boxill (Ed.). Oxford, UK: Blackwell Publishers, 2003, 130-135.

[1244] I am using a commonsense distinction between physical and mental here.

[1245] Consider any number of teams with a tendency for running on third and long.

[1246] Consider the example of the Carolina Panthers team of 2003—an aging quarterback, a weak secondary, a weak linebacker corps, and so on. Yet, they made it to the Superbowl because they adjusted their strategy to what they had, a very strong defensive line and a good running back and took advantage of their strengths rather than attempting to emulate a successful team of a different model, which is what they attempted when Seifert was in charge.

[1247] Simon, R.L. *Fair Play: The Ethics of Sport*. 2nd ed. Boulder, CO: Westview Press, 2004, p.29.

[1248] By instance of competition I mean the basic self-contained amount of play for the sport in question (i.e., one soccer match, a tennis match, a baseball game, a cricket test) as opposed to a series of competitions after which one side is declared the winner by virtue of winning more of the instances of competition (i.e., the World Series, The Ashes, any season of professional sports, etc.).

[1249] Simon, R.L. *Fair Play: The Ethics of Sport*. 2nd ed. Boulder, CO: Westview Press, 2004, pp.17-39.

[1250] Simon, R.L. *Fair Play: The Ethics of Sport*. 2nd ed. Boulder, CO: Westview Press, 2004, p.27.

[1251] I do not want to suggest that professional sports leagues should be the model for all sports, but I do think they have the potential, whether or not ever actually achieved, to be the model for pure win-at-all-costs competition and even competition as the mutual pursuit of excellence.

[1252] Contrary to the rules means gross violations of constitutive rules (i.e., Maradona's "Hand of God" or using a corked bat in baseball), although it is certain open to debate whether such things as intentional fouls should be considered smart tactics or unacceptable breaches of the rules.

[1253] By ignoring the economic reasons behind such leagues I am leaving out a significant factor in the explanation of how they actually operate and leaving considerable room for discussion of how they should operate.

[1254] At least at the level of competition that I am limiting this discussion to.

[1255] This last example is a bit controversial since it seems to go against the spirit of the rules of football as an ideally noncontact sport.

[1256] Such behaviour is not uncommon.

[1257] As long as it is within the rules or, possibly, the spirit of the game.

[1258] The examples of the "Broad Street Bullies" era Philadelphia Flyers and 1980 U.S. Olympic hockey team have been suggested by one reviewer in raising this objection.

[1259] Strange tactical formations are a prime example here. If a competitor is confronted with a formation never encountered before, her awareness of what is happening and what is likely to happen next is going to be severely limited.

[1260] Though not exclusively, since there is considerable room for overlap and crossover between innovative or unknown strategy and simply using the same strategy solely for the purpose of confusing or demoralising an opponent.

[1261] See Howe. Howe, L.A. "Gamesmanship". *Journal of the Philosophy of Sport*. XXXI, 2004, 212-225.

[1262] Though Howe might object to this example, since taunting seems to be included in the class of objectionable forms of "strong" gamesmanship. Howe, L.A. "Gamesmanship". *Journal of the Philosophy of Sport*. XXXI, 2004, 212-225, p.221.

[1263] Howe, L.A. "Gamesmanship". *Journal of the Philosophy of Sport*. XXXI, 2004, 212-225, p.220.

[1264] This is not an exhaustive list of the myriad understandings of sportsmanship.

[1265] Loland, S., and M. McNamee. "Fair Play and the Ethos of Sports: An Eclectic Philosophical Framework." *Journal of the Philosophy of Sport*, XXVIII, 2000, 63-80, p.76.

[1266] We may well want competitors to show the mental agility and toughness to be able to skilfully sledge and responds to sledging in certain areas. I certainly enjoy watching two sides compete with everything they have, including verbally.

[1267] Sessions, W.L. "Sportsmanship as Honor". *Journal of the Philosophy of Sport*, XXXI, 2004, 47-59, p.52.

[1268] Baseball, cricket, basketball, and football are all fairly obvious examples of professional sports in which sledging is a widespread activity. Though in the case of cricket, the ICC is attempting to limit and possibly eliminate such occurrences. See ITT(2).

[1269] Maxims.

[1270] Michael Atherton, The Times, February 18th 2023.

[1271] Ibid Why I am so Wise—Ecce Homo.

[1272] New York, Random House 2007.

[1273] Deserving to Be Lucky: Reflections on the Role of Luck and Desert in Sports, Robert Simon, *Journal of the Philosophy of Sport*, 2007, 34, 13-25 20 © 2007 International Association for the Philosophy of Sport.

[1274] Dixon, Nicholas, "On Winning and Athletic Superiority." *Journal of the Philosophy of Sport*, XXVI, 1999, 10-26, p.17.

[1275] Simon, Robert L. *Fair Play*. Boulder, CO: Westview Press, 2004, p.83.

[1276] Loland, Sigmund. *Fair Play in Sport: A Moral Norm System*. New York: Routledge, 2002, p.88.

[1277] In the Fraleight Lecture, Gunnar Breivik suggested that luck does and perhaps should play varying degrees in different sports. For further consideration of this issue, see Loland's discussion, Loland, Sigmund. *Fair Play in Sport: A Moral norm System*. New York: Routledge, 2002, particularly pp.87-92.

[1278] Loland, Sigmund. *Fair Play in Sport: A Moral norm System*. New York: Routledge, 2002, esp. pp.87–92.

[1279] Loland thinks that the influence of luck does not always spoil a sports context and can sometimes enhance it, and hence the influence of luck should not be eliminated but "optimised" Loland, Sigmund. *Fair Play in Sport: A Moral norm System*. New York: Routledge, 2002, p.91.

[1280] Loland, Sigmund. *Fair Play in Sport: A Moral norm System*. New York: Routledge, 2002, p.88.

[1281] Sher, George, "Effort, Ability, and Personal Desert" *Philosophy and Public Affairs*, 8(4), 1979, 361-376, pp.362-364.

[1282] Rawls, John, *A Theory of Justice*. Cambridge, MA: Harvard University Press, 1971, p.104.

[1283] Carr, David, "Where's the Merit if the Best Man Wins?" *Journal of the Philosophy of Sport*, XXVI, 1999, 1-9, p.6.

[1284] Sher, George, "Effort, Ability, and Personal Desert" *Philosophy and Public Affairs*, 8(4), 1979, 361-376, pp.370-373.

[1285] Chambliss, Dan, *Champions: The Making of Olympic Swimmers*. New York: Harper Collins, 1988.

[1286] Chambliss may assign too low a role to natural ability (after all, the athletes he studied may have been preselected at least partly on the basis of such ability), but even if that is right, he still can be correct in arguing that other factors also play a significant role in affecting athletic success.

[1287] Such a view is suggested, for example, by William Prosser when he writes that "the real basis of negligence is…behaviour which *should* be recognised as involving *unreasonable* danger to others" (italics my own). William Prosser, "Negligence" from his *The Law of Torts*, 4th ed. (West Publishing Company, 1971), reprinted in John Arthur and William Shaw, eds., *Readings in the Philosophy of Law* (NJ: Pearson, Prentice-Hall, 2006) p.409, and by Jules L. Coleman's and Jeffrie G. Murphy's discussions of assigning the cost of automobile accidents, particularly in their *Philosophy of Law* (Boulder, CO: Westview Press, 1990), pp.149-161.

[1288] Rawls, John. *A Theory of Justice*, Cambridge, MA: Harvard University Press, 1971.

[1289] Card, Claudia. "Individual Entitlements in Justice as Fairness" In Victoria Davion and Clark Wolf (Eds.) *The Idea of a Political Liberalism*. Lanham, MD: Rowman and Littlefield, 2000. 176-189, p.188.

[1290] Publications Ltd, London, 1948.

[1291] Fundamentals Ten Keys to Reality by Frank Wilczek Allen Lane, 2021.

[1292] Simon Ings, The Times Books, p.14, January 2, 2021.

[1293] Ibid Fundamentals 2021.